SCORE EASY

Innovative
MATHEMATICS
Class-X

R.P. Gupta

TGT (Mathematics)
M.Sc., B.Ed.
Diploma in I.B. (U.S.A.)
Life Member of Indian Mathematical Society.

Pharos Books Pvt. Ltd.

ISBN: 978-93-67006-70-2

©**Publishers**

Publisher: Pharos Books (P) Ltd.
Plot No.-63, 1st Floor, Main Mother Dairy Road
Pandav Nagar, East Delhi-110092
Phone: 011-40395855
WhatsApp: +91 9319228272
E-mail: sales@pharosbooks.in
Website: www.pharosbooks.in
Edition: 2024

Score Easy Innovative Mathematics Class-X
Author: R.P. Gupta

CONTENTS

Preface

Innovative Mathematics by R.P. GUPTA SCORE EASY is a series of workbook modules for Mathematics, carefully curated for prospective secondary school students and teachers to provide them with a platform where they can actively learn the structures, concepts, and foundations of secondary school Mathematics.

The series comes in a unique format, which makes it distinct from other books in the market. Each book, in the series, is dedicated to a single chapter of the curriculum. They comprise a detailed explanation of the topic at hand with thorough research, relevant examples, and thoughtful exercises. The books are formulated for students enabling them to learn in small collaborative group settings, where students read and learn together. This, assists in developing inquisitiveness in students as they question and help each other in solving problems. The books are crafted in a manner that the instructor does not have to necessarily lecture, but act more as a facilitator available to clarify doubts of the students.

This series intends to empower students to learn on their own and to minimize the extra cost they bear for supplementary tuition classes. The content of the books is such that it can be understood and solved by the student in a self-paced manner at his convenience, this also suites to the students who are highly motivated and already have a strong background in Mathematics.

Each module is self-contained and provides space to practise questions. The practising space in the books will enable students to organize their notes in a chapter-wise manner, which will prove to be more productive and beneficial.

This series includes all types of questions with solutions, thoughtfully categorized into various difficulty levels. This series provides a holistic understanding of all mathematical concepts.

Good Luck for your journey to learn Mathematics with SCORE EASY SERIES- INNOVATIVE MATHEMATICS by R.P. GUPTA.

Please call or write to us if you have any queries, suggestions or corrections.

R.P. GUPTA (TGT, Mathematics)
Mob No.: +91 9625914319 (WhatsApp only)
E-Mail: innovativemathematics@gmail.com.

MATHEMATICS (IX-X)
(CODE NO. 041)
Session 2023-24

The Syllabus in the subject of Mathematics has undergone changes from time to time in accordance with growth of the subject and emerging needs of the society. The present revised syllabus has been designed in accordance with National Curriculum Framework 2005 and as per guidelines given in the Focus Group on Teaching of Mathematics which is to meet the emerging needs of all categories of students. For motivating the teacher to relate the topics to real life problems and other subject areas, greater emphasis has been laid on applications of various concepts.

The curriculum at Secondary stage primarily aims at enhancing the capacity of students to employ Mathematics in solving day-to-day life problems and studying the subject as a separate discipline. It is expected that students should acquire the ability to solve problems using algebraic methods and apply the knowledge of simple trigonometry to solve problems of height and distances. Carrying out experiments with numbers and forms of geometry, framing hypothesis and verifying these with further observations form inherent part of Mathematics learning at this stage. The proposed curriculum includes the study of number system, algebra, geometry, trigonometry, mensuration, statistics, graphs and coordinate geometry, etc.

The teaching of Mathematics should be imparted through activities which may involve the use of concrete materials, models, patterns, charts, pictures, posters, games, puzzles and experiments.

Objectives

The broad objectives of teaching of Mathematics at secondary stage are to help the learners to:

- consolidate the Mathematical knowledge and skills acquired at the upper primary stage;
- acquire knowledge and understanding, particularly by way of motivation and visualization, of basic concepts, terms, principles and symbols and underlying processes and skills;
- develop mastery of basic algebraic skills;
- develop drawing skills;
- feel the flow of reason while proving a result or solving a problem;
- apply the knowledge and skills acquired to solve problems and wherever possible, by more than one method;
- to develop ability to think, analyze and articulate logically;
- to develop awareness of the need for national integration, protection of environment, observance of small family norms, removal of social barriers, elimination of gender biases;
- to develop necessary skills to work with modern technological devices and mathematical software's.
- to develop interest in mathematics as a problem-solving tool in various fields for its beautiful structures and patterns, etc.
- to develop reverence and respect towards great Mathematicians for their contributions to the field of Mathematics;
- to develop interest in the subject by participating in related competitions;
- to acquaint students with different aspects of Mathematics used in daily life;
- to develop an interest in students to study Mathematics as a discipline.

COURSE STRUCTURE CLASS –X

Units	Unit Name	Marks
I	NUMBER SYSTEMS	06
II	ALGEBRA	20
III	COORDINATE GEOMETRY	06
IV	GEOMETRY	15
V	TRIGONOMETRY	12
VI	MENSURATION	10
VII	STATISTICS & PROBABILTY	11
	Total	**80**

UNIT I: NUMBER SYSTEMS

1. REAL NUMBERS   (15) Periods

Fundamental Theorem of Arithmetic - statements after reviewing work done earlier and after illustrating and motivating through examples, Proofs of irrationality of $\sqrt{2}, \sqrt{3} \sqrt{5}$

UNIT II: ALGEBRA

1. POLYNOMIALS   (8) Periods

Zeros of a polynomial. Relationship between zeros and coefficients of quadratic polynomials.Recall of algebraic expressions and identities. Verification of identities:

2. PAIR OF LINEAR EQUATIONS IN TWO VARIABLES   (15) Periods

Pair of linear equations in two variables and graphical method of their solution, consistency/inconsistency.

Algebraic conditions for number of solutions. Solution of a pair of linear equations in two variables algebraically - by substitution, by elimination. Simple situational problems.

3. QUADRATIC EQUATIONS   (15) Periods

Standard form of a quadratic equation $ax^2 + bx + c = 0$, $(a \neq 0)$. Solutions of quadratic equations (only real roots) by factorization, and by using quadratic formula. Relationship between discriminant and nature of roots.

Situational problems based on quadratic equations related to day to day activities to be incorporated.

4. ARITHMETIC PROGRESSIONS   (10) Periods

Motivation for studying Arithmetic Progression Derivation of the nth term and sum of the first n terms of A.P. and their application in solving daily life problems.

UNIT III: COORDINATE GEOMETRY

COORDINATE GEOMETRY   (15) Periods

Review: Concepts of coordinate geometry, graphs of linear equations. Distance formula. Section formula (internal division).

UNIT IV: GEOMETRY

1. TRIANGLES   (15) Periods

Definitions, examples, counter examples of similar triangles.

1. (Prove) If a line is drawn parallel to one side of a triangle to intersect the other two sides in distinct points, the other two sides are divided in the same ratio.

2. (Motivate) If a line divides two sides of a triangle in the same ratio, the line is parallel to the third side.

3. (Motivate) If in two triangles, the corresponding angles are equal, their corresponding sides are proportional and the triangles are similar.

4. (Motivate) If the corresponding sides of two triangles are proportional, their corresponding angles are equal and the two triangles are similar.

5. (Motivate) If one angle of a triangle is equal to one angle of another triangle and the sides including these angles are proportional, the two triangles are similar.

2. CIRCLES   (10) Periods

Tangent to a circle at, point of contact

1. (Prove) The tangent at any point of a circle is perpendicular to the radius through the point of contact.

2. (Prove) The lengths of tangents drawn from an external point to a circle are equal.

UNIT V: TRIGONOMETRY

1. INTRODUCTION TO TRIGONOMETRY   (10) Periods

Trigonometric ratios of an acute angle of a right-angled triangle. Proof of their existence (well defined); motivate the ratios whichever are defined at 0o and 90o. Values of the trigonometric ratios of 300, 450 and 600. Relationships between the ratios.

2. TRIGONOMETRIC IDENTITIES   (15) Periods

Proof and applications of the identity $\sin^2 A + \cos^2 A = 1$. Only simple identities to be given.

3. HEIGHTS AND DISTANCES: Angle of elevation, Angle of Depression.   (10) Periods

Simple problems on heights and distances. Problems should not involve more than two right triangles. Angles of elevation / depression should be only 30°, 45°, and 60°.

UNIT VI: MENSURATION

1. AREAS RELATED TO CIRCLES   (12) Periods

Area of sectors and segments of a circle. Problems based on areas and perimeter / circumference of the above said plane figures. (In calculating area of segment of a circle, problems should be restricted to central angle of 60°, 90° and 120° only.

2. SURFACE AREAS AND VOLUMES (12) Periods

Surface areas and volumes of combinations of any two of the following: cubes, cuboids, spheres, hemispheres and right circular cylinders/cones.

UNIT VII: STATISTICS AND PROBABILITY

1. STATISTICS (18) Periods

Mean, median and mode of grouped data (bimodal situation to be avoided).

2. PROBABILITY (10) Periods

Classical definition of probability. Simple problems on finding the probability of an event.

MATHEMATICS-Standard
QUESTION PAPER DESIGN
CLASS – X (2023-24)

Time: 3 Hrs.   **Max. Marks: 80**

S. No.	Typology of Questions	Total Marks	% Weightage (approx.)
1	**Remembering:** Exhibit memory of previously learned material by recalling facts, terms, basic concepts, and answers. **Understanding:** Demonstrate understanding of facts and ideas by organizing, comparing, translating, interpreting, giving descriptions, and stating main ideas	43	54
2	**Applying:** Solve problems to new situations by applying acquired knowledge, facts, techniques and rules in a different way.	19	24
3	**Analysing:** Examine and break information into parts by identifying motives or causes. Make inferences and find evidence to support generalizations **Evaluating:** Present and defend opinions by making judgments about information, validity of ideas, or quality of work based on a set of criteria. **Creating:** Compile information together in a different way by combining elements in a new pattern or proposing alternative solutions	18	22
	Total	80	100

INTERNAL ASSESSMENT	20 MARKS
Pen Paper Test and Multiple Assessment (5 + 5)	10 Marks
Portfolio	05 Marks
Lab Practical (Lab activities to be done from the prescribed books)	05 Marks

MATHEMATICS-Basic
QUESTION PAPER DESIGN
CLASS – X (2023-24)

Time: 3 Hrs.   **Max. Marks: 80**

S. No.	Typology of Questions	Total Marks	% Weightage (approx.)
1	**Remembering:** Exhibit memory of previously learned material by recalling facts, terms, basic concepts, and answers. **Understanding:** Demonstrate understanding of facts and ideas by organizing, comparing, translating, interpreting, giving descriptions, and stating main ideas	43	54
2	**Applying:** Solve problems to new situations by applying acquired knowledge, facts, techniques and rules in a different way.	19	24
3	**Analysing:** Examine and break information into parts by identifying motives or causes. Make inferences and find evidence to support generalizations **Evaluating:** Present and defend opinions by making judgments about information, validity of ideas, or quality of work based on a set of criteria. **Creating:** Compile information together in a different way by combining elements in a new pattern or proposing alternative solutions	18	22
	Total	80	100

INTERNAL ASSESSMENT	20 MARKS
Pen Paper Test and Multiple Assessment (5 + 5)	10 Marks
Portfolio	05 Marks
Lab Practical (Lab activities to be done from the prescribed books)	05 Marks

PRESCRIBED BOOKS:

1. Mathematics - Textbook for class IX - NCERT Publication
2. Mathematics - Textbook for class X - NCERT Publication
3. Guidelines for Mathematics Laboratory in Schools, Class IX - CBSE Publication
4. Guidelines for Mathematics Laboratory in Schools, Class X - CBSE Publication
5. Laboratory Manual - Mathematics, secondary stage - NCERT Publication
6. Mathematics exemplar problems for Class IX, NCERT publication.
7. Mathematics exemplar problems for Class X, NCERT publication.

CHAPTER-1

REAL NUMBERS

LEARNING OBJECTIVES

1. To revisit number systems from naturals to real numbers.
2. To learn Euclid's Division Lemma.
3. To learn Fundamental theorem of Arithmetic.
4. To apply Euclid's Division Lemma/Algorithm for finding HCF of two numbers.
5. To find HCF and LCM using Prime Factorization Method.
6. To prove Irrational numbers.
7. Decimal expansion of real numbers.
8. Condition of terminating decimal expansion.

REAL NUMBERS

Real numbers a name in Mathematics which is continued since Class Vth but till now are you clear about it? Let's try to understand by this circle graph representation.

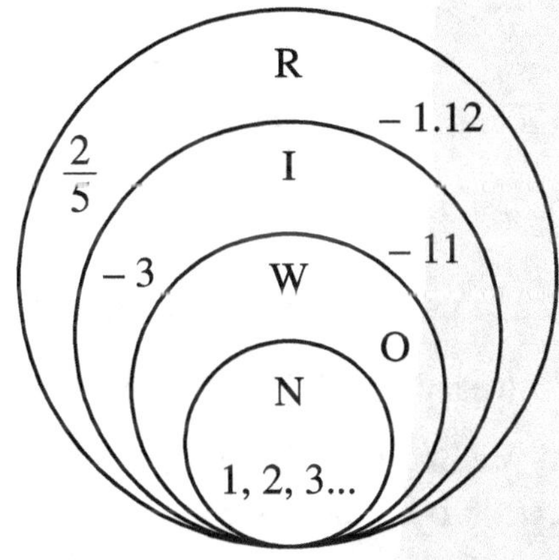

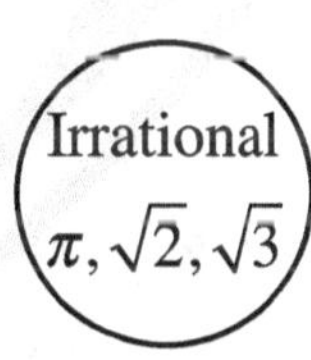

N → Natural Numbers
W → Whole Numbers
I → Integers
R → Rational Numbers

EUCLID'S DIVISION LEMMA

It is basically the proven statement of the division system.
According to Euclid's Division Lemma:-
For the given positive integers a and b, there exist unique integers q and r satisfying $a = bq + r$, $0 \leq r < b$;
where, q and r can also be zero;
where, 'a' is dividend, 'b' is divisor, 'q' is quotient and 'r' is remainder;
∴ Dividend = (Divisor × Quotient) + Remainder.
Let's understand like this

$$\text{Divisor}\overline{)\ \overset{\textstyle\text{Quotient}}{\text{Dividend}}\ (}$$

$$\underline{\hphantom{xxxxxxxxxxxxx}}$$
Remainder

OR

$$g(x)\overline{)\ p(x)\ (q(x)}$$

$$\underline{\hphantom{xxxxxx}}$$
r(x)

OR

$$\boxed{p(x) = g(x) \quad q(x) + r(x)}$$

Now, I am going to explain you regarding algorithm and lemma because, you will see and will use these words oftenly.

Lemma: A Lemma is a proven statement which is used to prove the other statement.

Algorithm: An algorithm gives us some definite steps to solve a particular type of problem in a well-defined manner.

Euclid's Division Algorithm: This concept is based on Euclid's Division Lemma. This is the method to calculate the HCF of given two numbers a and b where a > b by these steps.

Step 1– Apply Euclid's Division Lemma to a and b so that a = bq + r,
$0 \leq r < b.$

Step 2– Continue these steps till r = 0. At this stage divisor will be your HCF.

THE FUNDAMENTAL THEOREM OF ARITHMETIC

When we factorize each composite number as a product of some
prime numbers and, of course, this is unique as order of prime factors does
not matter.

If we have two numbers (positive) m and n, the property of their HCF and LCM will be

$$\boxed{\text{L.C.M.} \quad \text{H.C.F.} = m \quad n}$$

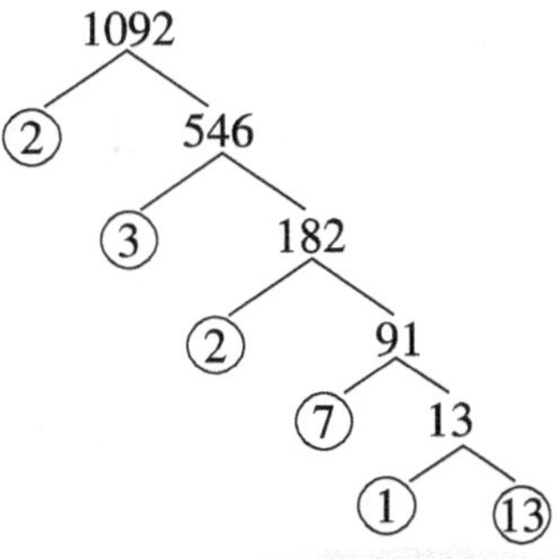

Prime factors are:

1092:- $2 \times 2 \times 3 \times 7 \times 13$

Uniqueness of Fundamental theorem of Arithmetic:-

By the uniqueness of fundamental theorem of arithmetic, we can show that the given number ends with zero or not.

Uniqueness states about the prime factors in the form of $2^m \times 5^n$.

As example: $(7)^n$

Here, $7 = 1 \times 7$

$\therefore \quad 7^n = (1 \times 7)^n$

So, the prime factors of 7 are 1 and 7 only so by uniqueness of fundamental theorem of arithmetic it cannot end with zero.

- **How to find LCM and HCF by Prime Factorization method.**

Example: Suppose two numbers whose prime factors are of the form.

$$b^2 a^3, \ b^2 ac \text{ and } c^2 ab$$

Here, $b^2 a^3 = \boxed{b} \times b \times a \times \boxed{a} \times a$

$b^2 ac = \boxed{b} \times b \times \boxed{a} \times c$

$c^2 ab = c \times c \times \boxed{a} \times \boxed{b}$

HCF = Lowest power of common number only = $\boxed{ab}$

LCM = Highest power of common number multiplied by rest numbers

$= a^3 b^2 c^2$

As we know, relationship between LCM and HCF with their two number.

Let a, b are two numbers and LCM and HCF are as same.

$$\boxed{\text{LCM} \times \text{HCF} = \text{Product of two numbers}}$$

OR

$$\boxed{\begin{array}{l} \text{LCM} \times \text{HCF} = \text{I} \times \text{II} \\[2mm] \text{HCF} = \dfrac{\text{I} \times \text{II}}{\text{LCM}} \\[2mm] \text{LCM} = \dfrac{\text{I} \times \text{II}}{\text{HCF}} \end{array}}$$

Application of LCM and HCF

Listen I am giving you some clue for word problems, if you see these words in word problems then do as directed.

LCM	**HCF**
• Minimum	• Maximum
• Lowest	• Greatest
• Smallest	• Largest
• Shortest	• Stacking of materials
• Traffic light	• Number of rooms
• Circular Path	
• Bell Questions	

IRRATIONALITY OF NUMBERS

Irrationality of Numbers are proved by contradiction method.

Step 1– Let the given number as $\dfrac{a}{b}$ where a, b are co-prime b ≠ 0.

Step 2– By squaring both sides, prove that a or b anyone is divisible by its denominator.

Step 3– Now prove another alphabet by same method that is divisible by denominator.

Step 4– Write the contradiction.

- **Revisiting rational number and their decimal expansion.**

Let a number in the form $\dfrac{p}{q}$ where q is in the form of 2^m or 5^n or $2^m \times 5^n$, then the given fraction is terminating otherwise non-terminating.

R.P. GUPTA DESK:–

- While deciding whether the given rational number is terminating or not, in this you must check given term should be lowest.

- Rational and Irrational between two number.

(a) and   (b) 0.12 and 0.13   (c) 2 and 3

1.414   1.732

R = 1.5, 1.6, 1.51, 1.52, ...   R = 2.1, 2.2, 2.3, 2.4, ...

I = 1.5010010001...   I = 2.010010001...

1.5020020002...   = 2.020010001...

                                                    Innovative Mathematics X-1

WARM UP – Level-I

1. Learn statement of Euclid's Division Lemma and write down _______________.

2. Complete the missing steps.

   Use Euclid's division algorithm to find the HCF of 324 and 12084.

   Since, $12084 > 324$, by Euclid's Division Lemma:

   $$\left[\begin{array}{l} a = bq + r \\ 0 \leq r < b \end{array}\right]$$

   $$12084 = 324 \times 37 + 96$$

   Since remainder is not zero so,

   $324 = 96 \times 3 + 36$

   $96 = $ ______________ $\times$ ______________ $+$ ______________

   $36 = $ ______________ $\times$ ______________ $+$ ______________

   $24 = $ ______________ $\times$ ______________ $+$ ______________

   $\text{HCF} = $ ______________ $\times$ ______________ $+$ ______________

3. The fundamental theorem of arithmetic states that every composite number can be expressed as ______________ and this factorization is ______________ apart from ______________.

4. Compute the following factor tree:

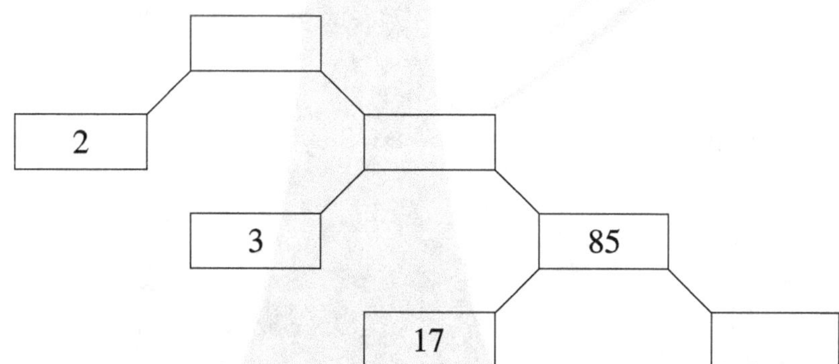

5. The decimal representation of $\dfrac{23}{2^2 5^2}$ will (i) terminate (ii) terminate after 3 decimal places.

   Reason

   (i) ____

   ____

   ____

   ____

   (ii) ____

   ____

6. Application of fundamental theorem of arithmetic:

(i) $7 \times 11 \times 13 + 13$

$= 13 \, (\underline{\hspace{4cm}} + \underline{\hspace{4cm}})$

$= 132 \times 2 \times 3 = \underline{\hspace{5cm}}$

Is this factorization unique? (Y/N)

So, the given expression is a $\underline{\hspace{4cm}}$ number.

(ii) $5 \times 7 \times 13 + 5$

$= \underline{\hspace{5cm}}$

$= \underline{\hspace{5cm}}$

Is this factorization unique? (Y/N)

So, the given expression is a $\underline{\hspace{4cm}}$ number.

(iii) $8 \times 7 \times 6 \times 5 \times 4 \times 3 \times 2 \times 1 + 6 \times 4 \times 3$

$= \underline{\hspace{5cm}}$ (take $6 \times 4 \times 3$ common)

$= \underline{\hspace{5cm}}$ (factorise 561)

$= \underline{\hspace{5cm}}$ (observe the product)

Is this factorization unique? (Y/N)

So, the given expression is a $\underline{\hspace{4cm}}$ number.

7. Proof of $\sqrt{2}$ is irrational.

Step 1. Let $\sqrt{2}$ be a rational number.

Let $\sqrt{2} = \dfrac{a}{b}$, a and b are integers $b \neq 0$ a and b are co-prime.

* Applying definition of rational number

(Focus on definition of co-prime numbers they do not have common factor other than 1).

Step 2. Squaring both sides $2 = \dfrac{a^2}{b^2}$

$\Rightarrow \quad a^2 = 2.b^2$ ...(1)

$\therefore \quad a^2$ is divisible by 2

$\therefore \quad a$ is divisible by 2 ...(2)

Observe, a^2 has factor 2

When a^2 is even then a is also even

Think: 4 is even

$\therefore$ 2 is even

Step 3. a is divisible by 2 we can write

$a = 2m, \ m \in N$ ...(3)

Step 4. Put (3) in (1),

$4m^2 = 2b^2$

$b^2 = 2m^2$

$\Rightarrow \quad b^2$ is divisible by 2

b is divisible by 2 ...(4)

Step 5. From (2) and (4), a and b have a common factor 2.

This is a contradiction.

$\therefore \quad$ Our assumption is false $\therefore \ \sqrt{2}$ is irrational.

8. Develop the proof of $\sqrt{5}$ is irrational.

(i) ____

____

____

____

Step 1.

Step 2.

Step 3.

Step 4.

Step 5.

9. Remember: Let $x = \dfrac{p}{q}$ be a rational number, such that the prime factorisation of q is of the form $2^n\ 5^m$, where n, m are non-negative integers. Then x has a decimal expansion which terminates.

Consider: $x = \dfrac{3}{8} = \dfrac{3}{2^3} = \dfrac{3 \times 5^3}{2^3 \times 5^3} = \dfrac{375}{10^3}$

Since the denominator can be expressed in the form of $2^n 5^m$, where n, m are non-negative integers, therefore the decimal expansion of x terminates.

Check whether the following rational numbers will have a terminating decimal expansion or a non-terminating repeating decimal expansion.

(i) $\dfrac{64}{455} =$ (Hint: First simplify then find factors of denominator)

$\dfrac{64}{\boxed{}} =$ Can you express denominator in the form $2^n 5^m$?

Now try:

(ii) $\dfrac{29}{343} =$ (iii) $\dfrac{35}{50} =$

(iv) $\dfrac{77}{210} =$ (v) $\dfrac{6}{15} =$

Consider: $x = \dfrac{3}{80} = \dfrac{3}{2^4 \times 5} = 0.0375$

Observe the decimal expansion of x terminates after 4 decimals and the highest value of n or m is also 4 in the denominator.

Consider: $x = \dfrac{14588}{625} = \dfrac{14588}{5^4} = 23.3408$

The decimal expansion terminates after 4 decimals.

10. Proof of $\sqrt{3}$ is irrational.

Step 1. Let $\sqrt{3}$ be rational.

Let $= \sqrt{3} = \dfrac{a}{b}$, a and b are ________________.

$b \neq 0$

a and b are ________________.

Step 2. Squaring both sides, $3 = \dfrac{a^2}{b^2}$

$a^2 = 3b^2$ ...(1)

$\Rightarrow \quad a^2$ is divisible by ________________.

a is divisible by ________________ ...(2)

Step 3. $a = 3m,\ m \in N$ ...(3)

Step 4. Put (3) in (1), $(\underline{\quad\quad})^2 = 3b^2$

$\Rightarrow \quad b^2$ is divisible by ________________.

$\therefore \quad$ b is divisible by ________________ ...(4)

Step 5. From (2) and (4), we get,

a and b have a common factor ________________.

11.

Rational Number $(x = \dfrac{a}{b}$, $b \neq 0$, a and b are integers, a and b are co-prime)	Decimal expansion will terminate (Put ✓ or ✗) If it terminates then after how many decimal places it will terminate?	Decimal expansion will not terminate (Put ✓ or ✗)
1. $\dfrac{13}{3125}$		
2. $\dfrac{29}{343}$		
3. $\dfrac{177}{210}$		
4. $\dfrac{23}{2^3 5^2}$		
5. $\dfrac{49}{2^7 5^2}$		
6. $\dfrac{7}{80}$		
7. $\dfrac{13}{125}$		
8. $\dfrac{120}{400}$		

Crossword Puzzle Sheet

Across

4. Fundamental theorem of ______________ states that every composite number can be uniquely expressed as a product of primes, apart from the order of factors.

7. The ____________ factorization of composite numbers is unique.

10. ____________ numbers have either terminating or non-terminating repeating decimal expansion.

Down

1. ____________ is a sequence of well-defined steps to solve any problem.

2. Numbers having non-terminating, non-repeating decimal expansion are known as ____________.

3. A proven statement used as a stepping stone towards the proof of another statement is known as ____________.

5. Decimal expansion of $\dfrac{3}{35}$ is ____________.

6. The ____________ expansion of rational numbers is terminating if the denominator has 2 and 5 as its only factors.

8. ____________ division algorithm is used to find the HCF of two positive numbers.

9. For any two numbers, HCF × LCM = ____________ of numbers.

Rating Scale	Student knows		Yes	No
	1.	Euclid's division lemma		
	2.	Fundamental theorem of arithmetic		
	3.	Decimal expansion of rational numbers		
	4.	Decimal expansion of irrational numbers		
	5.	Relation between LCM, HCF and two nos.		

WARM UP – Level-II

Q 1. **Explain why 132334563725 is a composite number.**

Ans. The given number ends with 5. Hence it is a multiple of 5. So we can say that it has minimum 1, 5 and itself are factor, so it is a composite number.

Q 2. **a and b are two positive integers such that the least prime factor of a is 3 and the least prime factor of b is 5. Then calculate the least prime factor of (a + b).**

Ans. Least prime factor of a and b are 3 and 5.

$$\therefore \quad (a + b) \Rightarrow 8$$

$$8 = 2 \times 2 \times 2$$

$\therefore$ Least prime factor $(a + b)$, i.e. 8 is 2.

Q 3. **What is the HCF of smallest composite and the smallest prime number?**

Ans. Smallest prime number = 2

Smallest composite number = 4

$$2 = 2^1 \; ; \; 4 = 2^2$$

$\therefore$ HCF $= 2^1 = 2$

Q 4. **Calculate the HCF of $3^3 \times 5$ and $3^2 \times 5^2$.**

Ans. $\left(\dfrac{3^3}{3^2}\right) \times \left(\dfrac{5^1}{5^2}\right)$

$$\text{HCF} = 3^2 \times 5^1 \Rightarrow 45$$
$$\text{LCM} = 3^3 \times 5^2 \Rightarrow 225$$

Q 5. **If HCF of (a, b) = 12 and Product of (a, b) = 1800, find their LCM.**

Ans. $\text{LCM} = \dfrac{\text{Pr oduct}}{\text{HCF}}$ 　　　　　　　　　　　　　　　　　[ref. my Notes]

Q 6. **Find HCF of 135 and 225 by using Euclid's algorithm.**

Ans. According to Euclid's division lemma,

$$a = bq + r, 0 \leq r < b$$

$$a = bq + r$$

$$225 = 135 \times 1 + 90$$

$$135 = 90 \times 1 + 45$$

$$90 = 45 \times 2 + 0$$

Hence, HCF = 45.

Q 7. **Explain whether $3 \times 12 \times 101 + 4$ is a prime or composite number.**

Ans. $3 \times 12 \times 101 + 4$

$3 \times 3 \times 4 \times 101 + 4$

$4[3 \times 3 \times 101 + 1]$

4[909 + 1]

4 × 910 × 1

It has more than two factors, so it is a composite number.

Q 8. **Can two numbers have 12 as their HCF and 121 as their LCM?**

Ans. No. Because, as we know, LCM of two numbers should be exactly divisible by their HCF.

Q 9. **Check whether 7^n can end with digit 0 for any natural number n.**

Ans. $(7)^n \Rightarrow (1 \times 7)^n$

It is not in the form of $2^m \times 5^n$

$\therefore$ By uniqueness of F.T.A., it cannot end with zero (0).

Q 10. **Show that any positive odd integer is of the form 6q + 1, 6q + 3, 6q + 5, where q is some integer.**

Ans. Let 'a' be any positive integer and b = 6.

By using E.D.L., a = bq + r, $0 \leq r < b$

Possible remainder = 0, 1, 2, 3, 4, 5

$\therefore$ a is written as:

6q + 0

$\Rightarrow$ 2(3q)

It is a multiple of 2, so it is even.

6q + 1

2(3q) + 1

It is not divisible by 2.

6q + 2

2(3q + 1)

It is divisible by 2.

6q + 3

2(3q + 1) + 1

It is not divisible by 2.

6q + 4

2(3q + 2)

It is divisible by 2.

6q + 5

2(3q + 2) + 1

It is not divisible by 2.

As we know, the numbers which are not divisible by 2 are odd numbers.

So, 6q + 1, 6q + 3, 6q + 5 are odd positive integers, where q is any integer.

Q 11. **Show that the square of any positive integers is of the form 4m or 4m + 1, where m is any integer.**

Ans. Let 'a' be any positive integer, b = 2

$\therefore$ By using E.D.L., a = bq + r, $0 \leq r < b$.

Possible r = 0.1

$\therefore$ a is written as:

$2q + 0 \Rightarrow 2q$

$2q + 1$

According to question squaring them,

Case I:

$(2q)^2 = 4q^2$

$\Rightarrow 4[q^2] \qquad \Rightarrow 4m$ [where m = q^2]

Case II:

$[2q + 1]^2$

$(2q)^2 + 4q + 1 = 4q^2 + 4q + 1$

$4[q^2 + q] + 1$

$4m + 1 [m = q^2 + q]$

$\therefore$ Every positive square number is written of the form 4m, 4m + 1.

Q 12. **Show that $n^2 - n$ is divisible by 2 for every positive integer n.**

Ans. Let 'a' be any positive integer, b = 2

$\therefore$ By using E.D.L., a = bq + r, $0 \leq r < b$.

r = 0, 1

$\therefore$ a = 2q, 2q + 1

Case I:

Let n = 2q

$\therefore$ $n^2 - n$

$= (2q)^2 - 2q = 4q^2 - 2q = 2[2q^2 - q]$

It is divisible by 2.

Case II:

Let n = 2q + 1

Thus, $n^2 - n$

$(2q + 1)^2 - (2q + 1)$

$4q^2 + 4q + 1 - 2q - 1$

$4q^2 + 2q \Rightarrow 2[2q^2 + 1]$

Therefore, every positive integer of the form $n^2 - n$ is divisible by 2 for every positive integer n.

Q 13. **Find HCF of 81 and 237 and express it as a linear combination of 81 and 237.**

Ans. By using E.D.L., first we will find HCF.

$237 = 81 \times 2 + 75$

$81 = 75 \times 1 + 6$

$75 = 6 \times 12 + 3$

$6 = 3 \times 2 + 0$

$\therefore \quad$ HCF $= 3$

Now for its linear combination always remember leave last line, then from 2nd last line find the value of HCF and then go up one-by-one.

$3 = 75 - 6 \times 12$

$3 = (81 - 6) - 6 \times 12$

$= 81 - 2 \times 6 \times 12$

$= 81 - 2(81 - 75) \times 12$

$= 81 - 2 \times 81 + 75 \times 12$

$= 81 \times -1 + 75 \times 12$

$= 81 \times -1 + (237 - 81 \times 2) \times 12$

$= 81 \times -1 + 237 \times 12 - 81 \times 24$

$= 81 \times -1 - 81 \times 24 + 237 \times 12$

$= 81 \times -25 + 237 \times 12$

$= 81x + 237y$

$\therefore \quad x = -25, y = 12$

Remember due to not uniqueness, the value of x and y can vary.

Q 14. **Show that there is no positive integer n for which $\sqrt{n-1} + \sqrt{n-1}$ is rational.**

Ans. Let us assume that $\sqrt{n-1} + \sqrt{n-1}$ is a rational number.

$\therefore \quad \sqrt{n-1} + \sqrt{n-1}$ $\qquad$ [where p, q are co-prime, $q \neq 0$] $\qquad$ ...(i)

or, $\dfrac{q}{p} = \dfrac{1}{\sqrt{n-1}+\sqrt{n+1}} = \dfrac{\sqrt{n-1}-\sqrt{n+1}}{\left(\sqrt{n-1}+\sqrt{n+1}\right)\left(\sqrt{n-1}-\sqrt{n+1}\right)}$

$= \dfrac{\sqrt{n-1}-\sqrt{n+1}}{(n-1)+(n+1)}$

or, $\dfrac{q}{p} \bcancel{=} \dfrac{\sqrt{n-1}-\sqrt{n+1}}{-2}$ (using By Cross Multiplication)

$\dfrac{2q}{p} = \left(\sqrt{n+1}\right) - \left(\sqrt{n-1}\right)$ $\qquad$ ...(ii)

Now adding and subtracting (ii) and (i),

$2\sqrt{n+1} = \dfrac{p}{q} + \dfrac{2q}{p}$

$= \dfrac{p^2 + 2q^2}{pq}$ $\qquad$ ...(iii)

$$2\sqrt{n-1} = \frac{p^2 - 2q^2}{pq} \qquad \qquad ...(iv)$$

From (iii) and (iv), we observe that $\sqrt{n+1}$ and $\sqrt{n-1}$ are rational because p and q both are rational.

Q 15. **Find the least number 11 the LCM of all numbers between 1 and 10.**

Ans. $1 = 1^1$

$2 = 2^1$

$3 = 3^1$

$4 = 2^2$

$5 = 5^1$

$6 = 2^1 \times 3^1$

$7 = 7^1$

$8 = 2^3$

$9 = 3^2$

$10 = 2^1 \times 5^1$

$LCM = 2 \times 2 \times 2 \times 3 \times 3 \times 5 \times 7 = 2520.$

FROM R.P. GUPTA DESK:

Decimal form of Real Numbers

RATIONAL NUMBERS

A number that can be expressed in the form of p/q where 'p' and 'q' are co-prime integers and q = 0, is called a rational number. Their decimal representation is either terminating or non-terminating and re-curring (repeating).

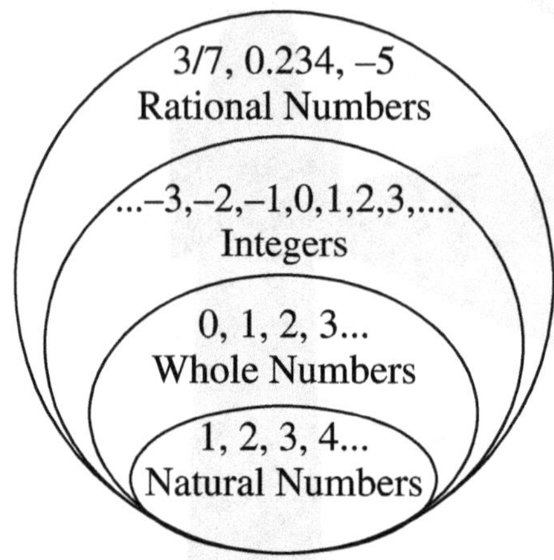

IRRATIONAL NUMBERS

A number that cannot be expressed in the form p/q where 'p' and 'q' are co-prime integers, is called an irrational number. Their decimal representation is non-terminating and non-recurring (non-repeating).

(e.g. $\sqrt{3}, \sqrt{2} + \sqrt{5}$, p, 0.102003102...)

FUNDAMENTAL THEOREM OF ARITHMETIC

Every composite number can be written as a product of primes in one and only one way, apart from the order in which the primes are written.

e.g. $286 = 2 \times 11 \times 13$

HCF USING PRIME FACTORIZATION

HCF is the product of the smallest power of each common prime factor of the given numbers,

e.g. HCF (240 and 360)

$240 = 2^4 \times 3 \times 5$

$360 = 2^3 \times 3^2 \times 5$

$HCF = 2^3 \times 3 \times 5 = 120$

LCM USING PRIME FACTORIZATION

LCM is the product of the greatest power of each prime factor of the given numbers.

e.g. LCM (240 and 360)

$240 = 2^4 \times 3 \times 5$

$360 = 2^3 \times 3^2 \times 5$

$LCM = 2^4 \times 3^2 \times 5 = 720$

Relationship Between HCF and LCM of Two Numbers

HCF is always a factor of the LCM of two numbers.

If 'a' and 'b' are two numbers, then

HCF (a, b) $\times$ LCM (a, b) = Product of 'a' and 'b'

e.g. For two numbers 24 and 36

Prime factorisation of $24 = 2^3 \times 3$ and $36 = 2^2 \times 3^2$

HCF $(24, 36) = 2^2 \times 3 = 12$

LCM $(24, 36) = 2^3 \times 3^2 = 72$

HCF $\times$ LCM $= 12 \times 72 = \mathbf{864}$

Product of 'a' and 'b' $= 24 \times 36 = \mathbf{864}$

Thus HCF (a, b) $\times$ LCM (a, b) = Product of 'a' and 'b'

VERY SHORT ANSWER TYPE QUESTIONS

Q 1. A number N when divided by 16 gives the remainder 5. ______ is the remainder when the same number is divided by 8.

Sol.

Q 2. HCF of $3^3 \times 5^4$ and $3^4 \times 5^2$ is __________ .

Sol.

Q 3. If $a = xy^2$ and $b = x^3y^5$ where x and y are prime numbers then LCM of (a, b) is ______ .

Sol.

Q 4. In the given factor tree find x and y

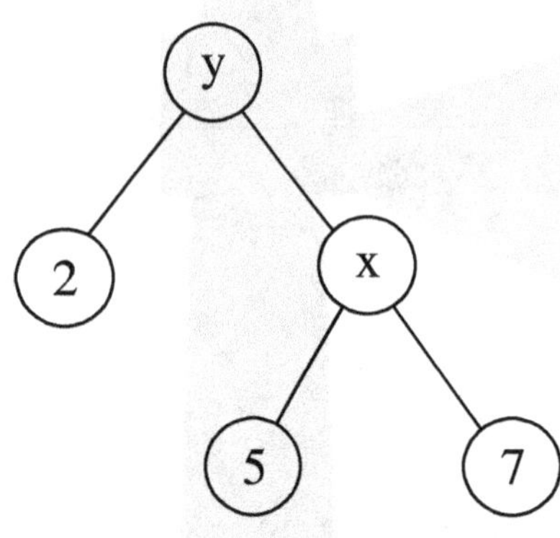

Sol.

Q 5. If n is a natural number, then $25^{2n} - 9^{2n}$ is always divisible by:

(a) 16   (b) 34   (c) both 16 or 34   (d) None of these

Sol.

Q 6. Given HCF (2520, 6600) = 120 and LCM (2520, 6600) is 252k, then the value of 'k' is

(a) 165   (b) 550   (c) 990   (d) 1650

Sol.

Q 7. The product of HCF and LCM of the smallest prime number and the smallest composite number is

(a) 2   (b) 4   (c) 6   (d) 8

Sol.

Q 8. If the LCM of two numbers is 3600, then which of the following cannot be their HCF?

(a) 600   (b) 500   (c) 400   (d) 150

Sol.

Q 9. $p^n = (a \times 5)^n$ For p^n to end with the digit zero a = ______ for natural number n.

(a) any natural number   (b) even number

(c) odd number   (d) none of these

Sol.

Innovative Mathematics X-1

Q 10. **HCF is always**

(a) multiple of LCM   (b) Factor of LCM

(c) divisible by LCM   (d) (a) and (c) both

Sol.

Q 11. **All decimal numbers are**

(a) rational numbers   (b) irrational numbers

(c) real numbers   (d) integers

Sol.

Q 12. **Which of these numbers always end with the digits 6.**

(a) 4^n (b) 2^n

(c) 6^n (d) 8^n

Sol.

Q 13. **Write the prime factor of $2 \times 7 \times 11 \times 13 \times 17 + 21$**

Sol.

Q 14. **Write the form in which every odd integer can be written taking t as variable.**

Sol.

Q 15. **Find the least number which is divisible by all numbers from 1 to 10 (both inclusive).**

Sol.

Q 16. **The numbers 525 and 3000 are divisible by 3, 5, 15, 25 and 75. What is the HCF of 525 and 3000?**

Sol.

Q 17. **What is x : y in the factor-tree?**

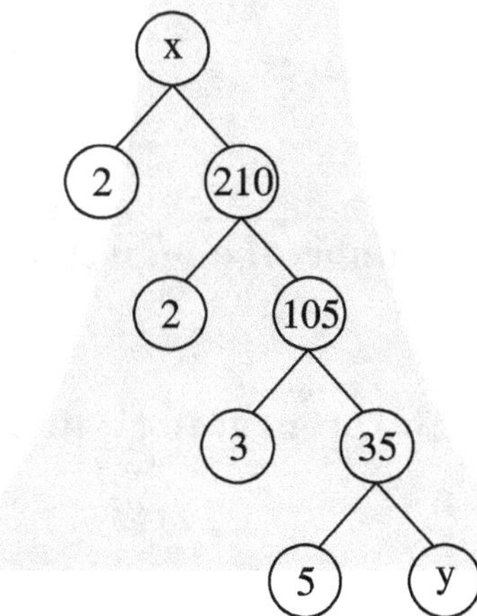

Sol.

SHORT ANSWER TYPE QUESTIONS-I

Q 18. **Show that 12^n cannot end with the digit 0 or 5 for any natural number n.**   (NCERT Exampler)

Sol.

Q 19. What is the smallest number by which $\sqrt{5} - \sqrt{3}$ is to be multiplied to make it a rational number? Also find the number so obtained?

Sol.

Q 20. Find one rational number and one irrational number between $\sqrt{2}$ and $\sqrt{5}$.

Sol.

Q 21. If HCF of 144 and 180 is expressed in the form 13m – 3, find the value of m. (CBSE 2014)

Sol.

Q 22. Find the value of: $(-1)^n + (-1)^{2n} + (-1)^{2n} + 1 + (-1)^{4n+2}$, where n is any positive odd integer.

(CBSE 2016)

Sol.

Q 23. Two tankers contain 850 litres and 680 litres of petrol respectively. Find the maximum capacity of a container which can measure the petrol of either tanker in exact number of times.

(CBSE 2016)

Sol.

SHORT ANSWER TYPE QUESTIONS-II

Q 24. Express 2658 as a product of its prime factors.

Sol.

Q 25. If $7560 = 2^3 \times 3^p \times q \times 7$, find p and q.

Sol.

Q 26. Prove that $\sqrt{3} + \sqrt{5}$ is irrational number.

Sol.

Q 27. Prove that $5 - \dfrac{3}{7}\sqrt{3}$ is an irrational number.

Sol.

Q 28. Prove that $\dfrac{1}{2 - \sqrt{5}}$ is an irrational number if it is given that $\sqrt{5}$ is irrational.

Sol.

Q 29. Find HCF and LCM of 56 and 112 by prime factorization method.

Sol.

Q 30. Explain why:

(i) $7 \times 11 \times 13 \times 15 + 15$ is a composite number

(ii) $11 \times 13 \times 17 + 17$ is a composite number.

(iii) $1 \times 2 \times 3 \times 5 \times 7 + 3 \times 7$ is a composite number.

Sol.

Q 31. On a morning walk, three persons steps off together and their steps measure 40 cm, 42 cm, and 45 cm respectively. What is the minimum distance each should walk, so that each can cover the same distance in complete steps? (NCERT Exemplar)

Sol.

Q 32. During a scale, colour pencils were being sold in the pack of 24 each and crayons in the pack of 32 each. If you want full packs of both and the same number of pencils and crayons, how many packets of each would you need to buy? (CBSE : 2017)

Sol.

Q 33. Find the largest number that divides 31 and 99 leaving remainder 5 and 8 respectively.

Sol.

Q 34. The HCF of 65 and 117 is expressible in the form $65\,m - 117$. Find the value of m. Also find the LCM of 65 and 117 using prime factorisation method. (NCERT Exemplar)

Sol.

Q 35. Find the largest number that divides 1251, 98733 and 15328 leaving remainder 1, 2 and 3 respectively.

Sol.

Q 36. Find the HCF of 180, 252 and 324.

Sol.

Q 37. Find the greatest number of six digits exactly divisible by 18, 24 and 36.

Sol.

Q 38. Three bells ring at intervals of 9, 12, 15 minutes respectively. If they start ringing together at a time, after how much time will they next ring together?

Sol.

Q 39. The length, breadth and height of a room are 8 m 25 cm, 6 m 75 cm and 4 m 50 cm respectively. Find the length of the longest rod that can measure the three dimensions of the room exactly.

Sol.

Q 40. Find HCF and LCM of 404 and 96 and verify that HCF × LCM = Product of two given number.

(CBSE 2018)

Sol.

LONG ANSWER TYPE QUESTIONS

Q 41. Find the HCF of 56, 96, 324 by prime factorization.

Sol.

Q 42. What will be the least possible number of the planks, if three pieces of timber 42 m. 49 m, and 63 m long have to be divided into planks of the same length?

Sol.

Q 43. Amit, Sunita and Sumit start preparing cards for all the persons in an old age home. In order to complete one card, they take 10, 16 and 20 minutes respectively. If they all started together, after what time will they begin preparing a new card together?

Sol.

Q 44. Aakriti decided to distribute milk in an orphanage on her birthday. The supplier brought two milk containers which contain 398 l and 436 l of milk. The milk is to be transferred to another containers so that 7 *l* and 11 *l* of milk is left in both the containers respectively. What will be the maximum capacity of the drum?

Sol.

Q 45. Find the smallest number, which when increased by 17, is exactly divisible by both 520 and 468.

Sol.

Q 46. A street shopkeeper prepares 396 Gulab jamuns and 342 ras-gullas. He packs them, in combination. Each container consists of either gulab jamuns or ras-gulla but have equal number of pieces.

Find the number of pieces he should put in each box so that number of boxes are least. How many boxes will be packed in all. (CBSE 2016)

Sol.

Q 47. Find the number nearest to 110000 but greater than 1 lakh, which is exactly divisible by 8, 15, 21.

Sol.

Q 48. In a seminar, the no. of participants in Hindi, English and Mathematics are 60, 84 and 108 respectively. Find the minimum number of rooms required if in each room the same number of participants are to be seated and all of the them being of the the same subject. (HOTS)

Sol.

Q 49. State fundamental theorem of Arithmetic. Is it possible that HCF and LCM of two numbers be 24 and 540 respectively. Justify your answer.

Sol.

Q 50. Find the smallest number which when increased by 20 is exactly divisible by 90 and 144. Is LCM, a multiple of 144?

Sol.

Q 51. If the HCF of 1032 and 408 is expressible in the form $1032\,p - 408 \times 5$, find p.

Sol.

Q 52. The LCM of two numbers is 14 times their HCF. The sum of LCM and HCF is 600. If one of the number is 280. Find the other number.

Sol.

1. 5

2. $3^3 \times 5^2$

3. $x^3 \times y^5$

4. $x = 35, y = 70$

5. (iii) $25^{2n} - 9^{2n}$ is of the form $a^{2n} - b^{2n}$ which is divisible by both $a - b$ and $a + b$ so, by both $25 + 9 = 34$ and $25 - 9 = 16$.

6. (b) 550

7. (d) 8

8. (b) 500

9. (b) even number

10. (b) Factor of LCM

11. (c) real numbers

12. (c) 6^n

13. 7

14. $2t + 1$ or $2t - 1$

15. 2520

16. 75

17. $60 : 1$

18. As 12 has factors 2, 2, 3. It doesnot has 5 as its factor so 12^n will never end with 0 or 5.

19. $\sqrt{5} + \sqrt{3},\ 2$

21. HCF of 180 and 144 is 36.

$$13m - 3 = 36$$
$$13\,m = 39$$
$$m = 3$$

22. Given that n is a positve odd integer

$\Rightarrow$ $2n$ and $4n + 2$ are even positive integers and n and $2n + 1$ are odd positive integers.

$\therefore$ $(-1)^n = -1,\ (-1)^{2n} = +1,\ (-1)^{2n+1} = -1,\ (-1)^{2n+2} = +1$

$\therefore$ $(-1)^n + (-1)^{2n} + (-1)^{2n+1} + (-1)^{4n+2} = -1 + 1 - 1 + 1 = 0$

23. HCF of 850 and 680 is $2 \times 5 \times 17 = 170$ litres.

24. $2658 = 2 \times 3 \times 443$

25. $p = 3$ and $q = 5$

26. Prove that $\sqrt{3}$ and $\sqrt{5}$ is irrational number separately.

27. 5 is rational no. and $\dfrac{3}{7}\sqrt{3}$ is an irrational number. Difference of a rational number and irrational number is an irrational number.

29. HCF : 56, LCM : 112

30. (1) $15 \times (7 \times 11 \times 13 + 1)$ as it has more than two factors so it is composite no. Similarly for part (ii) and (iii)

31. LCM of 40, 42, 45 = 2520

Minimum distance each should walk 2520 cm.

32. LCM of 24 and 32 is 96

96 crayons or $\dfrac{96}{32}$ = 3 packs of crayons

96 crayons or $\dfrac{96}{32}$ = 4 packs of pencils.

33. Given number = 31 and 99

$$31 - 5 = 26 \quad \text{and} \quad 99 - 8 = 91$$

Prime factors of $\quad 26 = 2 \times 13$

$$91 = 7 \times 13$$

HCF of $(26, 91) = 13$.

$\therefore$ 13 is the largest number which divides 31 and 99 leaving remainder and 8 respectively.

34. HCF $(117, 52) = 13$.

Given that $65\,m - 117 = 13 \Rightarrow 65\,m = 130 \Rightarrow m = 2$.

LCM $(65, 117) = 13 \times 3^2 \times 5 = 585$

35. $1251 - 1 = 1250$, $9377 - 2 = 9375$, $15628 - 3 = 15625$

HCF or $(15625, 9375) = 3125$

HCF of $(3125, 1250) = 625$

$\Rightarrow$ HCF of $(1250, 9375, 15625) = 625$

36. HCF $(324, 252, 180) = 36$

37. LCM of $(18, 24, 36) = 72$.

Greatest six digit number = 999999

$$72\,\overline{)\,999999\,}\,(\,13888$$

$-\,72$	
279	
$-\,216$	
639	Require six digit number
$-\,576$	999999
639	$-\,63$
$-\,576$	999936
639	
$-\,576$	
63	

38. LCM of $(9, 12, 15) = 180$ minutes.

39. HCF of 8 m 25 cm, 6 m 75 cm and 4 m 50 cm = 75 cm

40. HCF $(404, 96) = 4$

LCM $(404, 96) = 9696$

HCF $\times$ LCM $= 38,784$

Also, $404 \times 96 = 38,784$

42. HCF of 42 m, 49 m and 63 m = 7 m

Number of planks = 42/7 + 63/7 = 6 + 7 + 9 = 22

43. LCM of 10, 16 and 20 minutes = 80 minutes

44. 17

45. 4663

LCM of (468, 520) = 4680

$\therefore$ Required no. = 4680 – 17 = 4663

46. HCF (396, 342) = 18

No. of boxes = $\dfrac{396 + 342}{18} = \dfrac{738}{18} = 41$

47. 109200

48. HCF of 60, 84 and 108 is $2^2 \times 3 = 12$ = No. of participants in each row.

No. of rooms required = $\dfrac{\text{Total number of participants}}{12}$

$$= \dfrac{60 + 84 + 108}{12} = 21$$

49. $\qquad$ HCF = 24, $\quad$ LCM = 540

$\dfrac{\text{LCM}}{\text{HCF}} = \dfrac{540}{24} = 22.5$, not an integer.

Hence two numbers cannot have HCF and LCM as 24 and 540 respectively.

50. [The LCM of (90, 144) – 20] = Required No.

$\Rightarrow$ Required No. = 700

51. p = 2

52. HCF = 40, LCM = 560

$\therefore$ Other No. = 80

SECTION-A

Q 1. Check whether $17 \times 19 \times 21 \times 23 + 7$ is a composite number. (1)

Sol.

Q 2. What is the LCM of the smallest prime number and the smallest composite number? (1)

Sol.

Q 3. HCF of x^4y^5 and x^8y^3. (1)

Sol.

Q 4. LCM of 14 and 122. (1)

Sol.

SECTION-B

Q 5. Show that 9^n can never ends with unit digit zero. (2)

Sol.

Q 6. Find the pairs of the natural numbers whose least common multiple is 78 and the greatest divisor is 13. (2)

Sol.

Q 7. Find prime factors of 7650 using factor tree. (2)

Sol.

SECTION-C

Q 8. Prove that $\dfrac{1}{3 - 2\sqrt{5}}$ is an irrational number. (3)

Sol.

Q 9. Find the HCF of 36, 96 and 120 by prime factorization. (3)

Sol.

SECTION-D

Q 10. Once a sports goods retailer organized a campaign "Run to remember" to spread awareness about benefits of walking. In that Sohan and Baani participated. There was a circular path around a sports field. Sohan took 12 minutes to drive one round of the field, while Baani took 18 minutes for the same. Suppose they started at the same point and at the same time and went in the same direction. After how many minutes have they met again at the starting point? (4)

Sol.

CASE-STUDY

CASE STUDY 1.

To enhance the reading skills of grade X students, the school nominates you and two of your friends to set up a class library. There are two sections- section A and section B of grade X. There are 32 students in section A and 36 students in section B.

Q. 1. **What is the minimum number of books you will acquire for the class library, so that they can be distributed equally among students of Section A or Section B?**

(a) 144   (b) 128   (c) 288   (d) 272

Q. 2. **If the product of two positive integers is equal to the product of their HCF and LCM is true then, the HCF (32, 36) is**

(a) 2   (b) 4   (c) 6   (d) 8

Q. 3. **36 can be expressed as a product of its primes as**

(a) $2^2 \times 3^2$ (b) $2^1 \times 3^3$ (c) $2^3 \times 3^1$ (d) $2^0 \times 3^0$

Q. 4. **$7 \times 11 \times 13 \times 15 + 15$ is a**

(a) Prime number   (b) Composite number

(c) Neither prime nor composite   (d) None of the above

Q. 5. **If p and q are positive integers such that p = a and q = b, where a , b are prime numbers, then the LCM (p, q) is**

(a) ab   (b) a^2b^2 (c) a^3b^2 (d) a^3b^3

ANSWERS

1. (c) 288   **2.** (b) 4   **3.** (a) $2^2 \times 3^2$

4. (c) composite number   **5.** (b) a^2b^2

CASE STUDY 2.

A seminar is being conducted by an Educational Organisation, where the participants will be educators of different subjects. The number of participants in Hindi, English and Mathematics are 60, 84 and 108 respectively.

Q. 1. **In each room the same number of participants are to be seated and all of them being in the same subject, hence maximum number participants that can accommodated in each room are**

(a) 14   (b) 12   (c) 16   (d) 18

Q. 2. **What is the minimum number of rooms required during the event?**

(a) 11   (b) 31   (c) 41   (d) 21

Q. 3. **The LCM of 60, 84 and 108 is**

(a) 3780   (b) 3680   (c) 4780   (d) 4680

Q. 4. **The product of HCF and LCM of 60,84 and 108 is**

(a) 55360   (b) 35360   (c) 45500   (d) 45360

Q. 5. **108 can be expressed as a product of its primes as**

(a) $2^3 \times 3^2$ (b) $2^3 \times 3^3$ (c) $2^2 \times 3^3$ (d) $2^2 \times 3^3$

ANSWERS

1. (b) 12   **2.** (d) 21   **3.** (a) 3780

4. (d) 45360   **5.** (d) $2^2 \times 3^3$

CASE STUDY 3.

A Mathematics Exhibition is being conducted in your School and one of your friends is making a model of a factor tree. He has some difficulty and asks for your help in completing a quiz for the audience.

Observe the following factor tree and answer the following:

Innovative Mathematics X-1

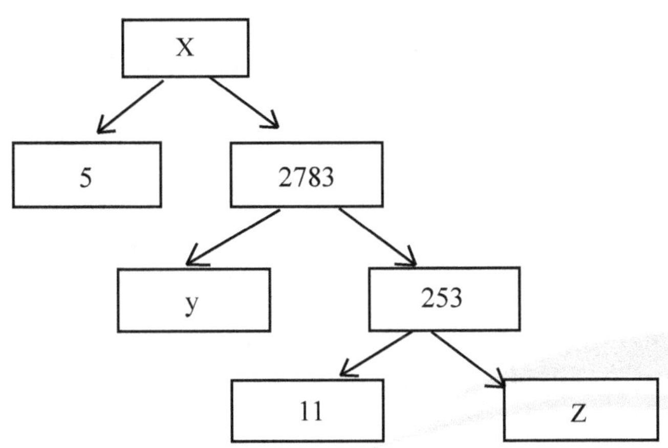

Q. 1. **What will be the value of x?**

(a) 15005   (b) 13915   (c) 56920   (d) 17429

Q. 2. **What will be the value of y?**

(a) 23   (b) 22   (c) 11   (d) 19

Q. 3. **What will be the value of z?**

(a) 22   (b) 23   (c) 17   (d) 19

Q. 4. **According to Fundamental Theorem of Arithmetic 13915 is a**

(a) Composite number   (b) Prime number

(c) Neither prime nor composite   (d) Even number

Q. 5. **The prime factorisation of 13915 is**

(a) $5 \times 11^3 \times 13^2$ (b) $5 \times 11^3 \times 23^2$ (c) $5 \times 11^2 \times 23$ (d) $5 \times 11^2 \times 13^2$

ANSWERS

1. (b) 13915   **2.** (c) 11   **3.** (b) 23

4. (a) composite number   **5.** (c) $5 \times 11^2 \times 23$

CHAPTER-2

POLYNOMIALS

LEARNING OBJECTIVES

- Recall and review polynomial, degree, coefficients, constants, variables, zeroes, terms and their factors.
- To learn geometrical meaning of a polynomial.
- Review and recall splitting the middle term of a quadration polynomial.
- To learn the verify relation, ship between zeroes and coefficients of a polynomial.
- To learn to find remaining zeroes of a biquadratic polynomials if two of its zeroes are given.

REVISION NOTES

Polynomial: If x is a variable, n is a natural number and a_0, a_1, a_2, a_3, ..., a_n are real numbers. Then $P(x) = a_n x_n + a^{n-1} x^{n-1} + ... + a_0$ is caused a polynomial.

According to degree of a polynomial it is divided into find forms:

Note: (Degree: Degree is known as the highest power of a given polynomial)

Degree	Form	Type of Polynomial
0	a	Constant
1	$ax + b$	Linear
2	$ax^2 + bx + c$	Q.uadrant
3	$ax^3 + bx^2 + cx + d$	Cybic
4	$ax^4 + bx^3 + cx^2 + dx + e$	Biquadrant

Zero of the Polynomial: If the value of $P(x)$ at $x = k$ is 0 that is $P(k) = 0$ then $y = k$ will be the zero of that polynomial $P(x)$.

Geometrical Meaning of the Zeroes of a Polynomial: Zeroes of the polynomials are the X co-ordinates of the point where the graph of that polynomial intersects the x-axis.

Graph of Q.uadratic Polynomial: When a > 0 Now, we will study about quadratic and cubic polynomials in this chapter:

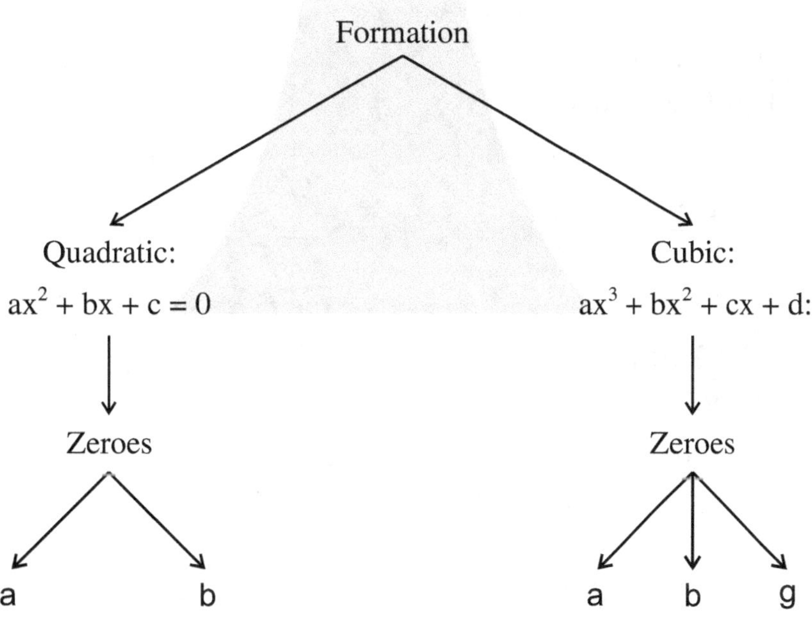

- Relationship between zeroes and coefficients of a polynomials

Case I:

Polynomial	Form	Zeroes
Linear	$ax + b$	1
	$a \neq 0$	

Relationships B/w zeroes and coefficients:

$$K = \frac{-b}{a} = \frac{\text{Constant term}}{\text{Coefficient of x}}$$

Case II:

Polynomial: Q.uadratic

Form: $ax^2 + bx + c, \ a \neq 0$

Zeroes: 2 (Two)

Relationship between zeroes and coefficients:

Sum of zeroes

$$\alpha + \beta = \frac{-b}{a} = \frac{\text{Coefficient of x}}{\text{Coefficient of } x^2}$$

Production of zeroes:

$$\alpha \times \beta = \frac{c}{a} = \frac{\text{Constant term}}{\text{Coefficient of } x^2}$$

Case III:

Polynomial: Cubic

Form: $ax^3 + bx^2 + cx + d, \ a \neq 0$

Zeroes: 3 (Three)

Relationship between zeroes and their coefficients:

Sum of zeroes:

$$\alpha + \beta + \gamma = \frac{-b}{a} = \frac{-[\text{Coefficient of } x^2]}{[\text{Coefficient of } x^3]}$$

Product of Zeroes:

$$\alpha\beta\gamma = \frac{-d}{a} = \frac{\text{Constant term}}{\text{Coefficient of } x^3}$$

Sum of product:

$$\alpha\beta + \beta\gamma + \gamma\alpha = \frac{c}{a} = \frac{\text{Coefficient of x}}{\text{Coefficient of x}}$$

$\Rightarrow$ Formation of Q.uadratic Polynomial

$$K[x^2 - (\alpha + \beta) \cdot x + (\alpha \cdot \beta)]$$

OR

$$K[x^2 - (\text{Sum of zeroes}) \cdot x + \text{Product of zeroes}]$$

Where,

K = Any non zero real number and α and β are the zeroes of the polynomial.

$\Rightarrow$ Formation of Cubic Polynomial:

$$K[x^3 - (\text{Sum of zeroes}) \, x^2 + (\text{sum of product}) \, x - (\text{Product})]$$

OR

$$K[x^3 - (\alpha + \beta + \gamma)x^2 + (\alpha\beta + \beta\gamma + \gamma\alpha)x - (\alpha\beta\gamma)]$$

Where, α, β, γ are the zeroes of the polynomial.

Division Algorithem for Polynomial: It states that given any polynomials p(x) and g(x) there exist polynomial q(x) and r(x) such that

$p(x) = g(x) \times q(x) + r(x)$

Where,

$r(x) = 0$

Like:

$$g(x) \, \overline{\smash{)}\, p(x)\, } \, (\, q(x)$$
$$\overline{}$$
$$r(x)$$

OR

As we studied in earliar classes

Dividend = Divisor $\times$ Q.uotient + Remainder

$\Rightarrow$ How to find more zeroes if two zeroes are given:

Step I: Let two given zeroes as a and b then make a quadratic polynomial using formula and let it g(x)

Step II: Now divide the given polynomial by g(x) using long division method.

Step III: Do factorize of the Q.uotient you will get the all required zeroes.

WARM UP – Level-I

Q. 1. Use the given constants and variables and make 3 polynomials.

Constants	Variables
$1, -7, \sqrt{5}$	x, y
$15, 5\sqrt{3}$	z
13	t, p

Q. 2. Complete it:

$P(x) = 5x^7 - 6x^5 + 7x - 6$	Coefficient of x5 = degree of P(x) = constant term = No. of terms:
$T(x) = 12x^2 + 5x - 3$	Coefficient of x = degree = constent term = Name of polynomial

$\Rightarrow$ Lets talk about splitting the middle term.

To factorize $x^2 + bx + c$ into factors of form $(x + p)$ and $(x + q)$ we find the numbers p and q such that $p + q = b$ and $p \times q = 0$

So, $x^2 + bx + c$

$= x^2 + (p + q)\, x + (p \times q)$

$= x^2 + px + qx + pq$

$= x(x + p) + q(x + p)$

$= (x + p)\,(x + q)$

$\Rightarrow$ Fill in the Blank Boxes:

(a) $x^2 + 7x + 12$

$= x^2 + \boxed{}\, x + \boxed{}\, x + 12$

$= x\,(x + \boxed{}) + (x + \boxed{})$

$= (x + \boxed{})\,(x + \boxed{})$

(b) $6x^2 + 19x + 10$

Product of factors $= 60$

Sum of factors $= 19$

$= 6x^2 + \boxed{}\, x + \boxed{}\, x + 10$

$= 3x\,(2x + \boxed{}) + \boxed{}\,(2x + \boxed{})$

$= (3x + \boxed{})\,(2x + \boxed{})$

(c) $6x^2 + 13x - 8$

Product of factor = $\square$

Sum of factors = $\square$

$= 6x^2 + \square\, x + \square\, x - 8$

$= 2x\,(3x + \square\,) - 1\,(3x + \square\,)$

$= (2x - 1)\,(3x + \square\,)$

Q. 3. Find the mistake:

(i) $2x^2 - 9 - 3x$

$= 2x^2 - 3x - 9$

$= 2x^2 - 6x - 3x - 9$

$= 2x\,(x + 3) - 3(x + 3)$

$= (x + 3)\,(2x - 3)$

$= 2x^2 - 3x - 9$

$= 2x^2 - 6x - 3x - 9$

$= 2x\,(x - 3) - 3(x - 3)$

$= (2x - 3)\,(x - 3)$

(ii) $2x^2 - 9 - 3x$

WARM UP – Level-II

Q. 1. If (-1) is a zero of the polynomial $f(x) = x^2 - 7x - 8$, then calculate the other zero.

Ans. $F(x) = x^2 - 7x - 8$

Let another zero is y.

$$\therefore \quad -1 + y = \left(\frac{-7}{1}\right)$$

$$-1 + y = 7$$

$$y = 8$$

Q. 2. If the zeroes of the polynomial $x^2 + 4x + 2a$ are a and $\dfrac{2}{a}$ then find the value of a.

Ans. Product of zeroes

$$\alpha \times \beta = \frac{c}{a}$$

$$\cancel{a} \times \frac{2}{\cancel{a}} = \frac{2a}{1} \Rightarrow a = 1$$

Q. 3. Find the Q.uadratic polynomial whose zeroes and $\dfrac{5}{3}$ and $\dfrac{1}{2}$.

Ans. $$\alpha + \beta = \frac{5}{3} + \frac{1}{2}$$

$$= \frac{10 + 3}{6}$$

$$= \frac{13}{6}$$

$$a \times b = \frac{5}{3} \times \frac{1}{2} = \frac{5}{6}$$

Q.uadratic Polynomial:

$$K[x^2 - (a + b)x + (ab)]$$

$$K[x^2 - \left(\frac{13}{6}\right)x + \frac{5}{6}]$$

$$K\left[\frac{6x^2 - 13x + 5}{6}\right]$$

$$P(x) = 6x^2 - 13x + 5$$

where $k = 6$

Q. 4. If α and β are the zeroes of a quadratic polynomial $2x^2 - 5x + 3 = 0$ then evaluate:

(i) $\alpha^2 + \beta^2$ (ii) $\dfrac{1}{\alpha} + \dfrac{1}{\beta}$ (iii) $\alpha^2\beta + \beta^2\alpha$

(iv) $\dfrac{\alpha}{\beta} + \dfrac{\beta}{\alpha}$ **(v)** $\alpha^3 + \beta^3$ **(vi)** $\alpha^4 + \beta^4.$

Ans. $P(x) = 2x^2 - 5x + 3 = 0$

$$\alpha + \beta = \frac{-(-5)}{2} \Rightarrow \frac{5}{2}$$

$$\alpha \times \beta = \frac{3}{2} \Rightarrow \frac{3}{2}$$

(i) $\alpha^2 + \beta^2$

$= (\alpha + \beta)^2 - 2 \times \beta$

$= \left(\dfrac{5}{2}\right)^2 - 2 \times \dfrac{3}{2}$

$= \dfrac{25}{4} - 3 \Rightarrow \dfrac{25 - 12}{4} \Rightarrow \dfrac{13}{4}$

(ii) $\dfrac{1}{\alpha} + \dfrac{1}{\beta}$

$= \dfrac{\beta + \alpha}{\alpha\beta}$

$= \dfrac{5}{2} \times \dfrac{2}{3}$

$= \dfrac{5}{3}$

(iii) $\alpha^2 b + \beta^2 \alpha$

$\alpha\beta\,(\alpha + \beta)$

$= \dfrac{3}{2}\left(\dfrac{5}{2}\right)$

$= \dfrac{15}{4}$

(iv) $\dfrac{\alpha}{\beta} + \dfrac{\beta}{\alpha}$

$= \dfrac{\alpha^2 + \beta^2}{\alpha\beta}$

$= \dfrac{\dfrac{13}{4}}{\dfrac{3}{2}}$

$$= \frac{13}{6}$$

(v) $\alpha^3 + \beta^3$

$(\alpha + \beta)^3 - 3\,\alpha\beta\,(\alpha + \beta)$

$$= \left(\frac{5}{2}\right)^3 - 3 \times \frac{3}{2}\left(\frac{5}{2}\right)$$

$$= \frac{125}{8} - \frac{45}{4}$$

$$= \frac{125 - 90}{8}$$

$$= \frac{35}{8}$$

(vi) $\alpha^4 + \beta^4 = (\alpha^2 + \beta^2)^2 - 2\alpha^2\beta^2$

$$= \left[\frac{13}{4}\right]^2 - 2\left(\frac{3}{2}\right)^2$$

$$= \frac{169}{16} - \frac{18}{4}$$

$$= \frac{169 - 72}{16}$$

$$= \frac{97}{16}$$

Q. 5. If one zero of the polynomial $3x^2 - 8x - (2k + 1)$ is seven times the other find zeroes and value of k.

Ans. $P(x) = 3x^2 - 8x - (2k + 1)$

Let one zero is α and other is 7α.

Sum of zeroes

$$\alpha + 7\alpha = \frac{-(-8)}{3}$$

$$8\alpha = \frac{8}{3}$$

$$\alpha = \frac{1}{3}$$

$$\therefore \quad 7\alpha = \frac{7}{3}$$

Now, product of zeroes.

$$\alpha \times \beta = \frac{c}{a}$$

$$\alpha \times 7\alpha = \frac{-(2k+1)}{3}$$

$$7\alpha^2 = \frac{-2k-1}{3}$$

$$7\left(\frac{1}{3}\right)^2 = \frac{-2k-1}{3}$$

$$\frac{7}{9} = \frac{-2k-1}{3}$$

$$\therefore \quad k = -5/3$$

Q. 6. **If sum of the squares of the zeroes of the quadratic polynomial $p(x) = x^2 - 8x + k$ is 34, find K.**

Ans. Let α and β are the zeroes of the polynomial $p(x) = x^2 - 8x + k$, then

$$\alpha + \beta = \frac{-b}{a}$$

$$= \frac{-(-8)}{1}$$

$$= 8$$

$$\alpha \times \beta = \frac{c}{a}$$

$$= k$$

ATQ.

$$\alpha^2 + \beta^2 = 34$$
$$\Rightarrow \quad 8(\alpha + \beta)^2 - 2\alpha\beta = 34$$
$$\Rightarrow \quad (8)^2 - 2(k) = 34$$
$$-2k = 34 - 64$$
$$-2k = -30$$
$$k = 15$$

$$
\begin{array}{r}
2x^2 + 3x - 2 \\
x^2 - 4x + 1 \overline{)\ 2x^4 - 5x^3 - 12x^2 + 11x - 2} \\
2x^4 - 8x^3 + 2x^2 \\
\underline{(-)\ (+)\ \ (-)} \\
3x^3 - 14x^2 + 11x - 2 \\
3x^3 - 12x^2 + 3x \\
\underline{(-)\ \ +\ \ \ \ -} \\
-2x^2 + 8x - 2 \\
-2x^2 + 8x - 2 \\
\underline{+\ \ (+)\ \ (+)} \\
0
\end{array}
$$

$q(x) = 2x^2 + 3x - 2$

after doing M.T.S.

$2x2 + 3x - 2$

$2x2 + 4x - x - 2 = 0$

$2x(x + 2) - 1(x + 2)$

$(2x - 1)(x + 2)$

$x = \dfrac{1}{2}, -2$

Q. 7. **If α, β and γ are zeroes of the polynomial $6x^3 + 3x^2 - 5x + 1$ then find $\alpha^{-1} + \beta^{-1} + \gamma^{-1}$.**

Ans. Since, α, β, γ are the zeroes of the given polynomial.

$$\therefore \quad \alpha + \beta + \gamma = \frac{-b}{a}$$

$$= \frac{-3}{6} = \frac{-1}{2}$$

$$\alpha\beta + \beta\gamma + \gamma\alpha = \frac{c}{a}$$

$$= \frac{-5}{6}$$

$$\alpha\beta\gamma = \frac{-d}{a}$$

$$= \frac{-1}{6}$$

Now, $\alpha^{-1} + \beta^{-1} + \gamma^{-1}$

$$= \frac{1}{\alpha} + \frac{1}{\beta} + \frac{1}{\gamma}$$

$$= \frac{\beta\gamma + \gamma\alpha + \alpha\beta}{\alpha\beta\gamma} \frac{-5}{6} \times \frac{-6}{1}$$

Q. 8. **If the zeroes of polynomial $x^3 - 3x^2 + x + 1$ are $a - b$, a and $a + b$. Find the values of a and b.**

Ans. Let α, β, γ are the zeroes of the given polynomial.

Let $\quad \alpha = a - b$

$\quad\quad \beta = a$

$\quad\quad \gamma = a + b.$

Then, $\quad \alpha + \beta + \gamma = \dfrac{-b}{a}$

$$a - \cancel{b} + a + a + \cancel{b} = \frac{-(-3)}{1}$$

$$3a = 3$$

$a = 1$

$$\alpha \times \beta \times \gamma = \frac{-d}{a}$$

$$\alpha \times \beta \times \gamma = \frac{-d}{a}$$

$(a - b)(a + b)(a) = -1$

$(a^2 - b^2)\, a = -1$

$a^3 - ab^2 = -1$

$(1)^3 - b^2 = -1$

$Tb^2 = +2$

$b = \pm\sqrt{2}$

Hence, $a = 1$, $b = \pm\sqrt{2}$

$$
\require{enclose}
\begin{array}{r}
x^2 + 2x + 1 \\
3x^2 - 5 \,\overline{\smash{\big)}\, 3x^4 + 6x^3 - 2x^2 - 10x - 5} \\
\end{array}
$$

$$
\begin{array}{l}
\ x^2 + 2x + 1 \\
3x^2 - 5\ \overline{)\ 3x^4 + 6x^3 - 2x^2 - 10x - 5} \\
\ 3x^4 - 5x^2 \\
\ \underline{\ - +} \\
\ \ 6x^3 + 3x^2 - 10x - 5 \\
\ \ 6x^3 - 10x \\
\ \ \underline{\ - +} \\
\ \ \ \ 3x^2 - 5 \\
\ \ \ \ 3x^2 - 5 \\
\ \ \ \ \underline{\ - +} \\
\ \ \ \ \ \ \ 0
\end{array}
$$

Hence, now perform M.T.S. of q(x)

$x^2 + 2x + 1$

$x^2 + x + x + 1$

$x(x + 1) + 1(x + 1)$

$(x + 1)(x + 1)$

$x = -1, -1$

$\therefore$ All possible zeroes are

$-\sqrt{5/3},\ -\sqrt{5/3},\ -1,\ -1$

Q. 9. **If α and β are two zeroes of the polynomial $25x^2 - 15x + 2$ find a quadratic polynomial whose zeroes are $\dfrac{1}{2\alpha}$ and $\dfrac{1}{2\beta}$.**

Ans. If α and β are two zeroes then

$$\alpha + \beta = \frac{-b}{a} = -\left(\frac{-15}{25}\right) = \frac{3}{5}$$

$$\alpha\beta = \frac{c}{a} = \frac{2}{25}$$

Now $\quad \alpha_1 = \dfrac{1}{2\alpha}$

$$\beta_1 = \frac{1}{2\beta}$$

So $\quad \alpha_1 + \beta_1 = \dfrac{1}{2\alpha} + \dfrac{1}{2\beta}$

$$= \frac{\beta + \alpha}{2 \times \beta}$$

$$= \frac{\dfrac{3}{5}}{} \times \frac{25}{4} = \frac{15}{4}$$

$$= x^2 - \frac{15}{4}x + \frac{25}{8}$$

$$= \frac{1}{8}(8x^2 - 30x + 25)$$

Hence, a quadratic polynomial whose zeroes are $\dfrac{1}{2\alpha}$ and $\dfrac{1}{2\beta}$

$\therefore \quad P(x) = 8x^2 - 30x + 25$

Q. 10. **Find a cubic polynomial with the sum, sum of the product of its zeroes taken two at a time and the product of zeroes as 2, – 7, – 14 respectively.**

Ans. Formula of cubic polynomial is $k[x^3 - (\alpha + \beta + \gamma)\,x^2 + (\alpha\beta + \beta\gamma + \gamma\alpha)\,x - (\alpha\beta\gamma)]$

Q. 11. **What must be subtracted or added to the polynomial $p(x) = x^4 + 2x^3 - 2x^2 + x - 1$ so that the resulting polynomial is divisible by $x^2 + 2x - 3$.**

Ans. Here, $p(x) = x^4 + 2x^3 - 2x^2 + x - 1$

$g(x) = x^2 + 2x - 3$

By using long division,

$$
\begin{array}{r}
x^2 + 1 \\
x^2 - 2x - 3 \enclose{longdiv}{x^4 + 2x^3 - 2x^2 + x - 1} \\
\underline{x^4 (+) 2x^3 (-) 3x^2 } \\
- - + \\
x^2 + x - 1 \\
\underline{x^2 + 2x - 3 } \\
- + \\
-x + 2
\end{array}
$$

$\therefore -x + 2$ should be subtracted and $-(-x + 2)$ should be added.

Q. 12. **If the polynomial $x^4 - 6x^3 + 16x^2 - 25x + 10$ is divided by another polynomial $x^2 - 2x + k$, the remainder comes out to be $x + a$, find k and a.**

Ans. By performing long division find remainder first

$$
\begin{array}{r}
x^2 - 4x + (8 - k) \\
x^2 - 2x + k \overline{\smash{)}\, x^4 - 6x^3 + 16x^2 - 25x + 10} \\
x^4 - 2x^3 + kx^2 \\
(+) \quad (-) \\
\hline
-4x^3 + (16 - k)\,x^2 - 25x + 10 \\
-4x^3 + 8x^2 - 4kx \\
(+) \quad (-) \quad (+) \\
\hline
(8 - k)\,x^2 + (4k - 25)\,x + 10 \\
(8 - k)\,x^2 + (16 - 2k)\,x + (8k - k) \\
\hline
(2k - 9)\,x + 10 - 8k + k^2
\end{array}
$$

Now, A/C to question remainder is $x + a$.

where its given that,

$\alpha + \beta + \gamma = 2$

$\alpha\beta + \beta\gamma + \gamma\alpha = -7$

$\alpha\beta\gamma = -14$

$\therefore$ required polynomial

$k[x^3 - (2)\,x^2 + (-7)\,x - (-14)]$

$k[x^3 - 2x^2 - 7x + 14]$

$\therefore \quad p(x) = x^3 - 2x^2 - 7x + 14$

where $k = 1$.

Q. 13. **Obtain all zeroes of the polynomial $3x^4 + 6x^3 - 2x^2 - 10x - 5$ if its two zeroes are:**

$$\sqrt{5/3} \text{ and } -\sqrt{5/3}.$$

Ans. Here,

$\alpha = \sqrt{5/3}$,

$\beta = -\sqrt{5/3}$

$\therefore \quad \alpha + \beta = 0, \quad \alpha \times \beta = \dfrac{-5}{3}$

Now, Q.uadratic polynomial is

$k[x^2 - (a + b)\,x + ab]$

$k\left[x^2 - (0)x + \left(\dfrac{-5}{3}\right)\right]$

$k\left[x^2 - \dfrac{5}{3}\right]$

$$k\left[\frac{3x^2-5}{3}\right]$$

$P(x) = 3x^2 - 5$, where $k = 3$

Now,

$P(x) = 3x^4 + 6x^3 - 2x^2 - 10x - 5$

$g(x) = 3x^2 - 5$.

After performing long division.

$$\alpha_1 \times \beta_1 = \frac{1}{2\alpha} \times \frac{1}{2\beta}$$

$$= \frac{1}{4\alpha\beta}$$

$$= \frac{1}{4 \times \dfrac{2}{25}}$$

$$= \frac{1 \times 25}{8} = \frac{25}{8}$$

Now Q.uadratic polynomial

$= k[x^2 - (\alpha + \beta)\, x + \alpha\beta]$

$$= k\left[x^2 - \left(\frac{1}{2\alpha} + \frac{1}{2\beta}\right)x + \frac{1}{2\alpha} \times \frac{1}{2\beta}\right]$$

$$= k\left[x^2 - \left(\frac{15}{4}\right)x + \frac{25}{8}\right]$$

Q. 14. Find all the zeroes of the polynomial $2x^4 - 5x^3 - 12x^2 + 11x - 2$ if its two zeroes are $2 \pm \sqrt{3}$.

Ans. Here two zeroes

$\alpha = 2 + \sqrt{3}$

$\alpha = 2 - \sqrt{3}$

$\alpha + \beta = 4$

and $\quad \alpha\beta = 1$

$\therefore$ Q.uadratic polynomial

$k[x^2 - (\alpha + \beta)\, x + \alpha\beta]$

$k[x^2 - (4)\, x + (1)]$

$k[x^2 - 4x + 1]$

$g(x) = x^2 - 4x + 1$

where $k = 1$.

Now perming long division method,

So, $(2k - 9)x + 10 - 8k + k^2 = x + a$

$\therefore \quad (2k - 9)x = x$

$2k - 9 = \dfrac{x}{x}$

$2k - 9 = 1$

$2k = 10$

$k = 5$

$\Rightarrow \quad 10 - 8k + k^2 = a$

$10 - 8(5) + (5)2 = a$

$10 - 40 + 25 = a$

$35 - 40 = a$

$a = -5$

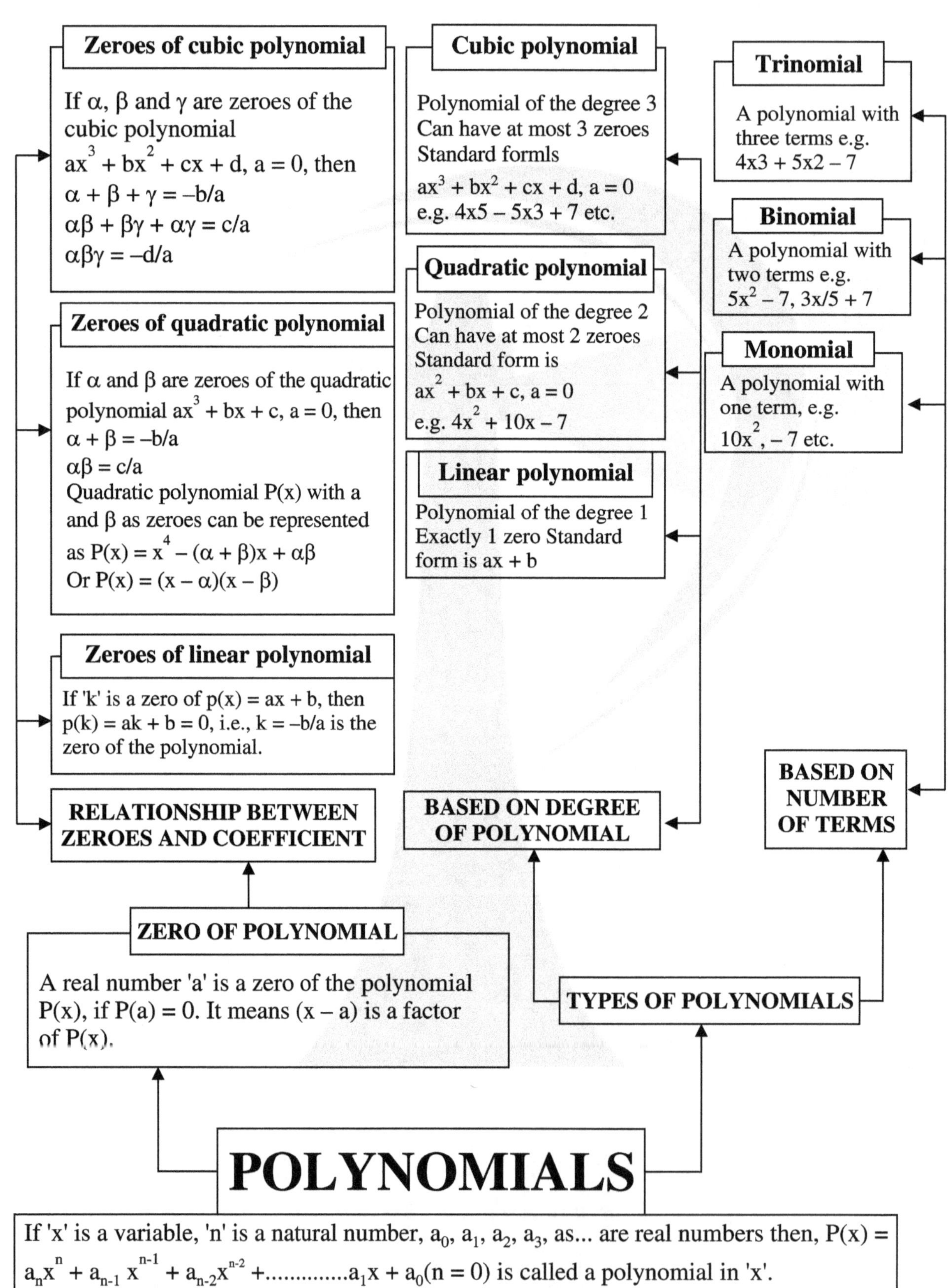

Zeroes of cubic polynomial
If α, β and γ are zeroes of the cubic polynomial
$ax^3 + bx^2 + cx + d, a = 0$, then
$α + β + γ = –b/a$
$αβ + βγ + αγ = c/a$
$αβγ = –d/a$

Cubic polynomial
Polynomial of the degree 3
Can have at most 3 zeroes
Standard formls
$ax^3 + bx^2 + cx + d, a = 0$
e.g. 4x5 – 5x3 + 7 etc.

Trinomial
A polynomial with three terms e.g.
4x3 + 5x2 – 7

Binomial
A polynomial with two terms e.g.
$5x^2 – 7, 3x/5 + 7$

Zeroes of quadratic polynomial
If α and β are zeroes of the quadratic polynomial $ax^3 + bx + c, a = 0$, then
$α + β = –b/a$
$αβ = c/a$
Quadratic polynomial P(x) with a and β as zeroes can be represented
as $P(x) = x^4 – (α + β)x + αβ$
Or $P(x) = (x – α)(x – β)$

Quadratic polynomial
Polynomial of the degree 2
Can have at most 2 zeroes
Standard form is
$ax^2 + bx + c, a = 0$
e.g. $4x^2 + 10x – 7$

Monomial
A polynomial with one term, e.g.
$10x^2, – 7$ etc.

Linear polynomial
Polynomial of the degree 1
Exactly 1 zero Standard form is ax + b

Zeroes of linear polynomial
If 'k' is a zero of p(x) = ax + b, then
p(k) = ak + b = 0, i.e., k = –b/a is the zero of the polynomial.

BASED ON NUMBER OF TERMS

RELATIONSHIP BETWEEN ZEROES AND COEFFICIENT

BASED ON DEGREE OF POLYNOMIAL

ZERO OF POLYNOMIAL
A real number 'a' is a zero of the polynomial P(x), if P(a) = 0. It means (x – a) is a factor of P(x).

TYPES OF POLYNOMIALS

POLYNOMIALS
If 'x' is a variable, 'n' is a natural number, a_0, a_1, a_2, a_3, as... are real numbers then, P(x) = $a_n x^n + a_{n-1} x^{n-1} + a_{n-2} x^{n-2} + \ldots\ldots\ldots a_1 x + a_0 (n = 0)$ is called a polynomial in 'x'.

GRAPH OF A POLYNOMIAL

Geometrical Representation of a Linear Polynomial

Graph of a linear polynomial $P(x) = ax + b$, $a = 0$ is a straight line cutting x-axis exactly at one point.

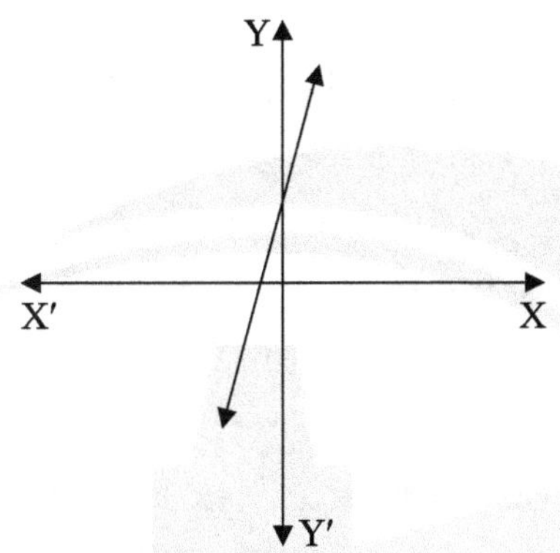

Geometrical Representation of a Q.uadratic Polynomial

Graph of a quadratic polynomial $P(x) = ax^2 + bx + c$, $a = 0$, is a parabola open upwards, if $a > 0$.
e.g. $5x^2 + 4x + 1$

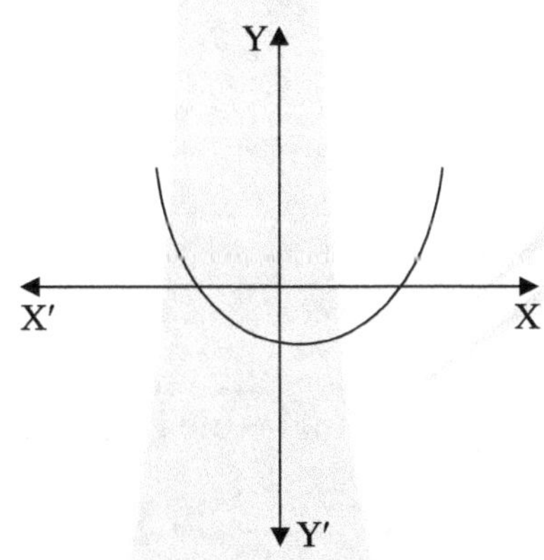

Graph of a quadratic polynomial $P(x) = -ax^2 + bx + c$, $a = 0$, is a parabola open downwards, if $a < 0$.
e.g. $-x^2 + 7x + 1$

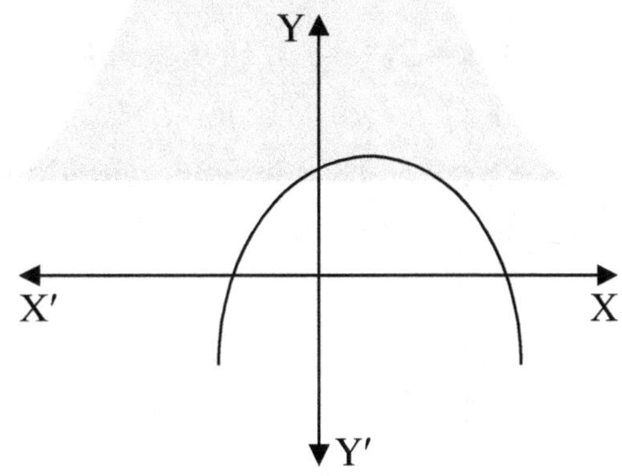

In general a polynomial $P(x)$ of degree 'n' crosses the x-axis at most 'n' points.

Q. 1. If one zero of the polynomial $P(x) = 5x^2 + 13x + k$ is reciprocal of the other, then value of k is

(a) 0   (b) 5   (c) $\dfrac{1}{6}$ (d) 6

Sol.

Q. 2. If α and β are the zeroes of the polynomial $p(x) = x^2 - p(x + 1) - c$ such that $(a + 1)(b + 1) = 0$, the c = _______ .

Sol.

Q. 3. If one zero of the quadratic polynomial $x^2 + 3x + k$ is 2, then the value of k is

(a) 10   (b) -10 (c) 5   (d) -5

Sol.

Q. 4. If the zeroes of the quadratic polynomial $x^2 + (a + 1)x + b$ are 2 and -3, then

(a) $a = -7, b = -1$ (b) $a = 5, \ b = -1$

(c) $a = 2, b = -6$ (d) $a = 0, b = -6$

Sol.

Q. 5. What should be added to the polynomial $x^2 - 5x + 4$, so that 3 is the zero of the resulting polynomial:

(a) 1   (b) 2   (c) 4   (d) 5

Sol.

Q. 6. If α and β are the zeroes of the polynomial

$$f(x) = x^2 + x + 1, \quad \text{then} \quad \frac{1}{\alpha} + \frac{1}{\beta} =$$

Sol.

Q. 7. If a quadratic polynomial $f(x)$ is not factorizable into linear factors, then it has no real zero.

(True/False)

Sol.

Q. 8. If a quadratic polynomial $f(x)$ is a square of a linear polynomial, then its two zeroes are coincident.

(True/False)

Sol.

Q. 9. If $p(x) = x^3 - 2x^2 - x + 2 = (x + 1)(x - 2)(x - d)$ then what is the value of d?

Sol.

Q. 10. The quadratic polynomial $ax^2 + bx + c$, $a \neq 0$ is represented by the graph then a is

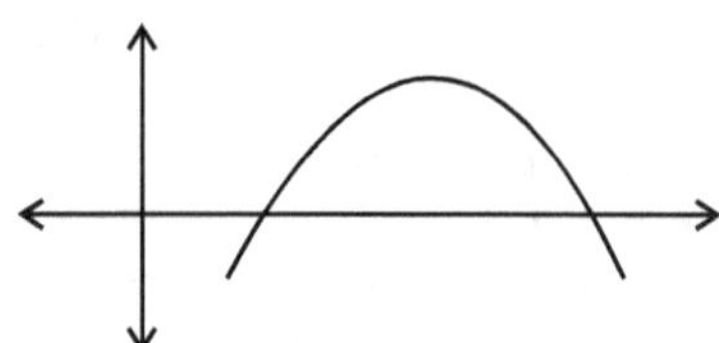

(a) Natural no.   (b) Whole no.   (c) Negative Integer   (d) Irrational no.

Innovative Mathematics X-2

Sol.

Q. 11. What will be the number of zeroes of a linear polynomial p(x) if its graph (i) passes through the origin. (ii) doesn't intersect or touch x-axis at any point?

Sol.

Q. 12. Find the quadratic polynomial whose zeroes are

$$\left(5+2\sqrt{3}\right) \text{ and } \left(5-2\sqrt{3}\right)$$

Sol.

Q. 13. If one zero of $p(x) = 4x^2 - (8k^2 - 40k)x - 9$ is negative of the other, find values of k.

Sol.

Q. 14. What number should be subtracted to the polynomial $x^2 - 5x + 4$, so that 3 is a zero of polynomial so obtained.

Sol.

Q. 15. How many (i) maximum (ii) minimum number of zeroes can a quadratic polynomial have?

Sol.

Q. 16. What will be the number of real zeroes of the polynomial $x^2 + 1$?

Sol.

Q. 17. If a and b are zeroes of polynomial $6x2 - 7x - 3$, then form a quadratic polynomial where zeroes are 2α and 2β (CBSE)

Sol.

Q. 18. If α and $\dfrac{1}{\alpha}$ are zeroes of $4x^2 - 17x + k - 4$, find the value of k.

Sol.

Q. 19. What will be the number of zeroes of the polynomials whose graphs are parallel to (i) y-axis (ii) x-axis?

Sol.

Q. 20. What will be the number of zeroes of the polynomials whose graphs are either touching or intersecting the axis only at the points:

(i)   (–3, 0), (0, 2) & (3, 0) (ii) (0, 4), (0, 0) and (0, –4)

Sol.

SHORT ANSWER TYPE (I) Q.UESTIONS

Q. 21. For what value of k, $x^2 - 4x + k$ touches x-axis.

Sol.

Q. 22. If the product of zeroes of $ax^2 - 6x - 6$ is 4, find the value of a. Hence find the sum of its zeroes.

Sol.

Q. 23. If zeroes of $x^2 - kx + 6$ are in the ratio 3 : 2, find k.

Sol.

Q. 24. If one zero of the quadratic polynomial $(k^2 + k)x^2 + 68x + 6k$ is reciprocal of the other, find k.

Sol.

Q. 25. If α and β are the zeroes of the polynomial $x^2 - 5x + m$ such that $\alpha - \beta = 1$, find m. **(CBSE)**

Sol.

Q. 26. If the sum of squares of zeroes of the polynomial $x^2 - 8x + k$ is 40, find the value of k.

Sol.

Q. 27. If α and β are zeroes of the polynomial $t^2 - t - 4$, form a quadratic polynomial whose zeroes are $\dfrac{1}{\alpha}$ and $\dfrac{1}{\beta}$.

Sol.

Q. 28. What should be added to the polynomial $x^3 - 3x^2 + 6x - 15$, so that it is completely divisible by $x - 3$? **(CBSE 2016)**

Sol.

Q. 29. If m and n are the zeroes of the polynomial $3x^2 + 11x - 4$, find the value of $\dfrac{m}{n} + \dfrac{n}{m}$.

(CBSE, 2012)

Sol.

Q. 30. Find a quadratic polynomial whose zeroes are $\dfrac{3+\sqrt{5}}{5}$ and $\dfrac{3-\sqrt{5}}{5}$. **(CBSE, 2013)**

Sol.

SHORT ANSWER TYPE (II) Q.UESTIONS

Q. 31. If $(k + y)$ is a factor of each of the polynomials $y^2 + 2y - 15$ and $y^3 + a$, find the values of k and a.

Sol.

Q. 32. Obtain zeroes of $4\sqrt{3}\, x^2 + 5x - 2\sqrt{3}$ and verify relation between its zeroes and coefficients.

Sol.

Q. 33. Form a quadratic polynomial, whose one zero is 8 and the product of zeroes is –56.

Sol.

Q. 34. –5 is one of the zeroes of $2x^2 + px - 15$, zeroes of $p(x^2 + x) + k$ are equal to each other. Find the value of k.

Sol.

Q. 35. Find the value of k such that $3x^2 + 2kx + x - k - 5$ has the sum of zeroes as half of their product.

Sol.

Q. 36. If zeroes of the polynomial $ax^2 + bx - c$, $a \neq 0$ are additive inverse of each other then what is the value of b?

Sol.

Q. 37. If α and β are zeroes of $x^2 - x - 2$, find a polynomial whose zeroes are $(2\alpha + 1)$ and $(2\beta + 1)$

Sol.

Q. 38. If α, β are zeroes of the quadratic polynomial $2x^2 + 5 + k$, then find the value of 'k' such that $(\alpha + \beta)^2 - \alpha\beta = 24$.

Sol.

Q. 39. If one zero of the polynomial $2x^2 - 3x + p$ is 3, find the other zero and the value of 'p'.

Sol.

Q. 40. Find a quadratic polynomial, whose zeroes are in the ratio 2 : 3 and their sum is 15.

Sol.

LONG ANSWER TYPE Q.UESTIONS

Q. 41. If $(x + a)$ is a factor of two quadratic polynomials $x^2 + px + q$ and $x^2 + mx + n$, then prove that $a = (n - q)/(m - p)$

Sol.

Q. 42. If one zero of the quadratic polynomial $4x^2 - 8kx + 8x - 9$ is the negative of the other, then find the zeroes of $kx^2 + 3kx^2 + 2$

Sol.

Q. 43. If α, β are zeroes of the quadratic polynomial $x^2 - 5x - 3$, then form a polynomial whose zeroes are $(2\alpha + 3\beta)$ and $(3\alpha + 2\beta)$.

Sol.

Q. 44. If one zero of the polynomial $(k + 1)\,x^2 - 5x + 5$ is multiplicative inverse of the other, then find the zeroes of $kx^2 - 3kx + 9$.

Sol.

Q. 45. If the product of the zeroes of the quadratic polynomial $kx^2 + 11x + 42$ is 7, then find the zeroes of the polynomial $(k - 4)x^2 + (k + 1)x + 5$.

Sol.

Q. 46. If α and β are zeroes of the polynomial $x^2 + 4x + 3$, find the polynomial whose zeroes are $1 + \dfrac{\beta}{\alpha}$ and $1 + \dfrac{\alpha}{\beta}$. **(CBSE)**

Sol.

Q. 47. Form a quadratic polynomial one of whose zero is $2 + \sqrt{5}$ and sum of the zeroes is 4.

Sol.

Q. 48. Form a polynomial whose zeroes are the reciprocal of the zeroes of $p(x) = ax^2 + bx + c, a \neq 0$.

Sol.

Q. 49. If $(x + 2)$ is a factor of $x^2 + px + 2q$ and $p + q = 4$ then what are the values of p and q?

Sol.

Q. 50. What should be subtracted from $x^3 - 3x^2 + 6x - 15$, so that it is completely divisible by $(x - 3)$?

Sol.

Q. 51. If sum of the zeroes of $5x^2 + (p + q + r)x + pqr$ is zero, then find $p^3 + q^3 + r^3$.

Sol.

Q. 52. If the zeroes of $x^2 + px + q$ are double in value to the zeroes of $2x^2 - 5x - 3$ find p and q.

Sol.

1. (b) 5
2. 1
3. (b) –10
4. (d) a = 0, b = –6
5. (b) 2
6. – 1
7. True
8. True
9. 1
10. (c) Negative Integer
11. (i) 1 (ii) 0
12. $x^2 - 10x + 13$
13. k = 0, 5
14. (– 2)
15. (i) 2 (ii) 0
16. 0
17. $[3x^2 - 7x - 6]\,k$
18. k = 8
19. (i) 1 (ii) 0
20. (i) 2 (ii) 1
21. 4
22. $a = -\dfrac{3}{2}$, sum of zeroes $= -4$
23. – 5, 5
24. 5
25. 6
26. 12
27. $4t^2 + t - 1$
28. On dividing $x^3 - 3x^2 + 6x - 15$ by $x - 3$, remainder is $+ 3$, hence $- 3$ must be added to $x^3 - 3x^2 + 6x - 15$.

29. $\dfrac{m}{n} + \dfrac{n}{m} = \dfrac{m^2 + n^2}{mn} = \dfrac{(m + n)^2 - 2mn}{mn} = \dfrac{\left(-\dfrac{11}{3}\right)^2 - 2\left(-\dfrac{4}{3}\right)}{-\dfrac{4}{3}} = -\dfrac{145}{12}$

30. $\alpha + \beta = \dfrac{6}{5}, \quad \alpha\beta = \dfrac{4}{25},$

    $25x^2 - 30x + 4$

31. k = –3, 5 and a = –27, 125

32. $-\dfrac{2}{\sqrt{3}}, \dfrac{\sqrt{3}}{4}$

33. $\alpha\beta = -56$ and $\beta = -7$

    so, $\alpha = 8$, Now $\alpha + \beta = 1$

    Required polynomial is $x^2 - x - 56$

34. $\dfrac{7}{4}$
35. 1
36. b = 0
37. $x^2 - 4x - 5$
38. $(\alpha + \beta) = -5/2$ and $\alpha\beta = k/2$

    Substituting the above values in $(a + b)2 - ab = 24$. Solve to get 'k' $= \dfrac{-71}{2}$

39. 3 is a zero, so $2(3)^2 - 3 \times 3 + p = 0$

$p = 9$, Now $\alpha\beta = \dfrac{c}{a}$, solve to get the other zero $\dfrac{-3}{2}$.

40. $\alpha : \beta = 2{:}3$. So $\alpha = 2\beta/3$

Using $(\alpha + \beta) = 15$, solve to get get α and β as 9 and 6 respecitvely.

Required polynomial is $x^2 - 15x + 54$

41. Since $(x + 2)$ is a factor of $x^2 + px + q$

$(-a)^2 - ap + q = 0$

$(-a)^2 = ap - q$ ...(1)

Similarly from $x^2 + mx + n$

$(a)^2 = am - n$ ...(2)

Comparing equation (1) and (2)

$a = (n - q)/(m - p)$

42. $f(x) = 4x^2 + (8 - 8k)x - 9$

$(\alpha + \beta) = -(8 - 8k) / 4$

$k = 1$

Substitute $k = 1$ in $kx^2 + 3kx^2 + 2$ and solve for $x = -2$ and -1

43. For given polynomial, $(\alpha + \beta) = 5$, $\alpha\beta = -3$

For Required polynomial

$$\text{Sum of zeroes} = (2\alpha + 3\beta) + (3\alpha + 2\beta)$$
$$= 5(a + p)$$
$$= 25$$
$$\text{Product of zeroes} = (2\alpha + 3\beta)(3\alpha + 2\beta)$$
$$= 6\alpha^2 + 6\beta^2 + 13\alpha\beta = 6(\alpha^2 + \beta^2) + 13\alpha\beta$$
$$= 6\,[(\alpha + \beta)^2 - 2\alpha\beta] + 13\alpha\beta$$
$$= 147$$

Required polynomial is $x^2 - 25x + 147$

44. $f(x) = (k + 1)x^2 - 5x + 5$

$(ab) = 1$

$5/(k + 1) = 1$

$k = 4$

Substituting $k = 4$ in $kx^2 - 3kx + 9$ solve to get zeroes $x = 3/2$ and $3/2$

45. $f(x) = kx^2 + 11x + 42$

$(\alpha\beta) = 7$

$k = 7$

Substituting $k = 6$ in $(k - 4)x^2 + (k + 1)x + 5$, solve to get zeroes $x = -1$ and $x = -5/2$

$x = -5/2$

46. $x^2 - \dfrac{16}{3}x + \dfrac{16}{3}$ or $\dfrac{1}{3}(3x^2 - 16x + 16)$

47. $\alpha + \beta = 4$

$(2 + \sqrt{5}) + \beta = 4$

$\beta = 2 - \sqrt{5}$

$\alpha\beta = -1 \qquad \therefore \quad$ Polynomial $= k[x^2 - 4x - 1]$

48. $k\left[x^2 + \dfrac{b}{c}x + \dfrac{a}{c}\right]$ **49.** $p = 3, q = 1$

50. 3 **51.** Product of the zeroes $= 3\,pqr$

52. $p = -5$ and $q = -6$

PRACTICE TEST

SECTION-A

Q. 1. If α and β are zeroes of a quadratic polynomial p(x), then factorize p(x). (1)

Sol.

Q. 2. If α and β are zeroes of $x^2 - x - 1$, find the value of $\dfrac{1}{\alpha} + \dfrac{1}{\beta}$. (1)

Sol.

Q. 3. If one of the zeroes of quadratic polynomial $(K - 1)x^2 + kx + 1$ is -3 then the value of K is, (1)

    (a) $\dfrac{4}{3}$ (b) $-\dfrac{4}{3}$ (c) $\dfrac{2}{3}$ (d) $-\dfrac{2}{3}$

Sol.

Q. 4. A quadratic polynomial, whose zeroes are -3 and 4, is (1)

    (a) $x^2 - x + 12$ (b) $x^2 + x + 12$ (c) $\dfrac{x^2}{2} - \dfrac{x}{2} - 6$ (d) $2x^2 + 2x - 24$

Sol.

SECTION-B

Q. 5. If α and β are zeroes of $x^2 - (k + 6)x + 2(2k - 1)$, find the value of k if $\alpha + \beta = \dfrac{1}{2}\alpha\beta$. (2)

Sol.

Q. 6. Find a quadratic polynomial one of whose zeroes is $(3 + \sqrt{2})$ and the sum of its zeroes is 6.

Sol.

Q. 7. If zeroes of the polynomial $x^2 + 4x + 2a$ are α and $\dfrac{2}{\alpha}$ then find the value of a. (2)

Sol.

SECTION-C

Q. 8. If α and β are zeroes of the polynomial $p(s) = 3s^2 - 6s + 4$, then find the value of $\alpha/\beta + \beta/\alpha + 2(1/\alpha + 1/\beta) + 3\alpha\beta$ (3)

Sol.

Q. 9. If truth and lie are zeroes of the polynomial $px^2 + qx + r$, $(p \neq 0)$ and zeroes are reciprocal to each other, Find the relation between p and r. (3)

Sol.

SECTION-D

Q. 10. Find the zeroes of the polynomial $\sqrt{3}x^2 + 10x + 7\sqrt{3}$. Also verify the relationship between the zeroes and their coefficients.

Sol.

CASE-STUDY

CASE STUDY 1.

The below picture are few natural examples of parabolic shape which is represented by a quadratic polynomial. A parabolic arch is an arch in the shape of a parabola. In structures, their curve represents an efficient method of load, and so can be found in bridges and in architecture in a variety of forms.

Q.. 1. In the standard form of quadratic polynomial, $ax^2 + bx + c$, a, b and c are

(a) All are real numbers.

(b) All are rational numbers.

(c) 'a' is a non zero real number and b and c are any real numbers.

(d) All are integers.

Q.. 2. If the roots of the quadratic polynomial are equal, where the discriminant $D = b^2 - 4ac$, then

(a) $D > 0$ (b) $D < 0$ (c) $D \geq 0$ (d) $D = 0$

Q.. 3. If α and $\dfrac{1}{\alpha}$ are the zeroes of the qudratic polynomial $2x^2 - x + 8k$, then k is

(a) 4   (b) $\dfrac{1}{4}$ (c) $\dfrac{-1}{4}$ (d) 2

Q.. 4. The graph of $x^2 + 1 = 0$

(a) Intersects x-axis at two distinct points.

(b) Touches x-axis at a point.

(c) Neither touches nor intersects x-axis.

(d) Either touches or intersects x- axis.

Innovative Mathematics X-2

Q.. 5. If the sum of the roots is –p and product of the roots is $-\dfrac{1}{p}$, then the quadratic polynomial is

    (a) $k\left(-px^2 + \dfrac{x}{p} + 1\right)$ (b) $k\left(px^2 - \dfrac{x}{p} - 1\right)$

    (c) $k\left(x^2 + px - \dfrac{1}{p}\right)$ (d) $k\left(x^2 - px + \dfrac{1}{p}\right)$

ANSWERS

1. (c) 'a' is a non zero real number and b and c are any real numbers

2. (d) $D = 0$

3. (b) $\dfrac{1}{4}$

4. (c) Neither touches nor intersects x-axis.

5. (c) $k\left(x^2 + px - \dfrac{1}{p}\right)$

CASE STUDY 2.

An asana is a body posture, originally and still a general term for a sitting meditation pose, and later extended in hatha yoga and modern yoga as exercise, to any type of pose or position, adding reclining, standing, inverted, twisting, and balancing poses. In the figure, one can observe that poses can be related to representation of quadratic polynomial.

Q.. 1. The shape of the poses shown is

    (a) Spiral   (b) Ellipse   (c) Linear   (d) Parabola

Q.. 2. The graph of parabola opens downwards, if _______

    (a) $a \geq 0$ (b) $a = 0$ (c) $a < 0$ (d) $a > 0$

Q.. 3. In the graph, how many zeroes are there for the polynomial?

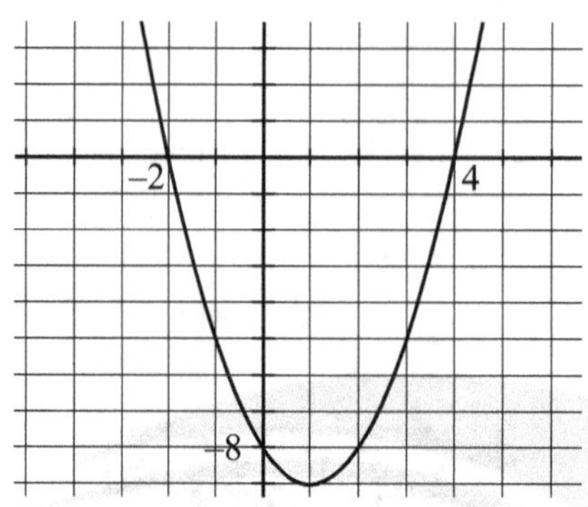

(a) 0   (b) 1   (c) 2   (d) 3

Q.. 4. The two zeroes in the above shown graph are

(a) 2, 4   (b) –2, 4   (c) –8, 4   (d) 2, –8

Q.. 5. The zeroes of the quadratic polynomial $4\sqrt{3}x^2 + 5x - 2\sqrt{3}$ are

(a) $\dfrac{2}{\sqrt{3}}, \dfrac{\sqrt{3}}{4}$ (b) $-\dfrac{2}{\sqrt{3}}, \dfrac{\sqrt{3}}{4}$ (c) $\dfrac{2}{\sqrt{3}}, -\dfrac{\sqrt{3}}{4}$ (d) $-\dfrac{2}{\sqrt{3}}, -\dfrac{\sqrt{3}}{4}$

ANSWERS

1. Parabola   **2.** (c) $a < 0$ **3.** (c) 2

4. (b) –2, 4   **5.** (b) $-\dfrac{2}{\sqrt{3}}, \dfrac{\sqrt{3}}{4}$

CASE STUDY 3.

Basketball and soccer are played with a spherical ball. Even though an athlete dribbles the ball in both sports, a basketball player uses his hands and a soccer player uses his feet. Usually, soccer is played outdoors on a large field and basketball is played indoor on a court made out of wood. The projectile (path traced) of soccer ball and basketball are in the form of parabola representing quadratic polynomial.

Q.. 1. The shape of the path traced shown is

    (a) Spiral   (b) Ellipse   (c) Linear   (d) Parabola

Q.. 2. The graph of parabola opens upwards, if _______

    (a) $a = 0$ (b) $a < 0$ (c) $a > 0$ (d) $a \geq 0$

Q.. 3. Observe the following graph and answer

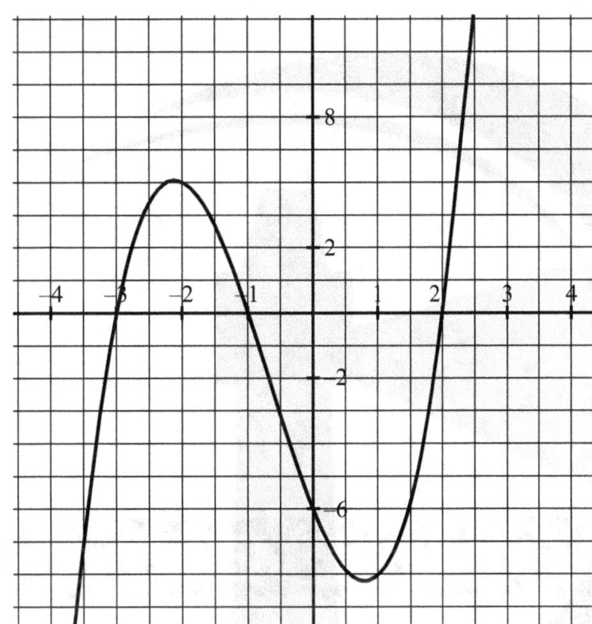

In the above graph, how many zeroes are there for the polynomial?

    (a) 0   (b) 1   (c) 2   (d) 3

Q.. 4. The three zeroes in the above shown graph are

    (a) $2, 3, -1$ (b) $-2, 3, 1$ (c) $-3, -1, 2$ (d) $-2, -3, -1$

Q.. 5. What will be the expression of the polynomial?

    (a) $x^3 + 2x^2 - 5x - 6$ (b) $x^3 + 2x^2 - 5x + 6$

    (c) $x^3 + 2x^2 + 5x - 6$ (d) $x^3 + 2x^2 + 5x + 6$

ANSWERS

1. (d) parabola   **2.** (c) $a > 0$ **3.** (d) 3

4. (c) $-3, -1, 2$ **5.** (a) $x^3 + 2x^2 - 5x - 6$

CHAPTER-3

PAIR OF LINEAR EQUATIONS IN TO VARIABLE

LEARNING OBJECTIVES

The standard form of Linear equation is (in two variable)

$$ax + by + c = 0$$

we also studied in Class IXth.

But in Class X^{th} we will study the comparision of two linear equation as.

$$a_1x + b_1y + c_1 = 0$$
$$a_2x + b_2y + c_2 = 0$$

Where a_1, a_2, b_1, b_2 and c_1, c_2 are real nos.

Here, we will study 3 criterias of comparision in two linear equational.

(i) $\dfrac{a_1}{a_2} = \dfrac{b_1}{b_2}$ $\Rightarrow$
- $\rightarrow$ Unique solution
- $\rightarrow$ Intersecting lines on the graph.
- $\rightarrow$ Consistent.

(ii) $\dfrac{a_1}{a_2} = \dfrac{b_1}{b_2} = \dfrac{c_1}{c_2}$ $\Rightarrow$
- $\rightarrow$ Infinite solutions
- $\rightarrow$ Co-incident lines
- $\rightarrow$ Dependent consisters

(iii) $\dfrac{a_1}{a_2} = \dfrac{b_1}{h_2} \neq \dfrac{c_1}{c_2}$ $\Rightarrow$
- $\rightarrow$ No solution
- $\rightarrow$ Parallel lines.
- $\rightarrow$ In consistent.

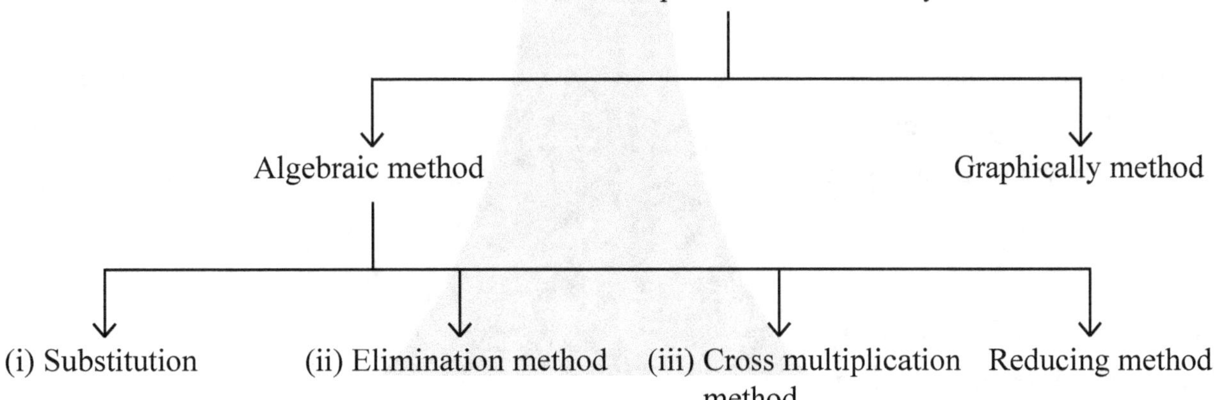

GRAPHICALLY METHOD

$\Rightarrow$ We put the value of one variable by our selves and find the another one and it will be the solutions of given equation.

$\Rightarrow$ To plot on the graph we will find at least 3 solutions.

Example:

$$x + 3y = 6 \qquad (1)$$
$$2x - 3y = 12 \qquad (2)$$

From 1

x	0	3	6
y	2	1	0

$$x = 6 - 3y$$

From 2

x	0	3	6
y	-4	-2	0

$$2x = 12 + 3y$$

$$x = \frac{12 + 3y}{2}$$

Let a the system be:

$$a_1 x + b_1 y + c_1 = 0$$
$$a_2 x + b_2 y + c_2 = 0$$

Case I: if $\dfrac{a_1}{a_2} \neq \dfrac{b_1}{b_2}$

(i)   The system has a unique solution.

(ii)   On the graph intersecting lines.

(iii)   System is known as conistent independent.

Case II: if $\dfrac{a_1}{a_2} = \dfrac{b_1}{b_2} \neq \dfrac{c_1}{c_2}$

(i)   System has no solution.

(ii)   On the graph lines are parallel.

(iii)   System is known as in consistent.

Case III: if $\dfrac{a_1}{a_2} = \dfrac{b_1}{b_2} = \dfrac{c_1}{c_2}$

(i)   The system has infinite no. of solutions.

(ii)   On the graph lines are coincident.

(iii)   System in known as consistent dependent.

Q. 1. **Solve the following systems of equations by eliminating 'y' (by substitution):**

(i) $3x - y = 3$
$7x + 2y = 20$

(ii) $7x + 11y - 3 = 0$
$8x + y - 15 = 0$

(iii) $2x + y - 17 = 0$
$17x - 11y - 8 = 0$

Ans. (i) We have;

$$3x - y = 3 \qquad \ldots(1)$$
$$7x + 2y = 20 \qquad \ldots(2)$$

From equation (1), we get ;

$$3x - y = 3 \qquad \Rightarrow \ y = 3x - 3$$

Substituting the value of 'y' in equation (2), we get:

$\Rightarrow \quad 7x + 2 \times (3x - 3) = 20$

$\Rightarrow \quad 7x + 6x - 6 = 20$

$\Rightarrow \quad 13x = 26 \qquad \Rightarrow \ x = 2$

Now, substituting $x = 2$ in equation (1), we get;

$$3 \times 2 - y = 3$$

$\Rightarrow \ y = 3$

Hence, $x = 2, \qquad y = 3$.

(ii) We have;

$$7x + 11y - 3 = 0 \qquad \ldots(1)$$
$$8x + y - 15 = 0 \qquad \ldots(2)$$

From equation (1), we get;

$$7x + 11y = 3$$

$$\Rightarrow \quad y = \frac{3 - 7x}{11}$$

Substituting the value of 'y' in equation (2), we get;

$$\Rightarrow \quad 8x + \frac{3 - 7x}{11} = 15$$

$\Rightarrow \quad 88x + 3 - 7x = 165$

$\Rightarrow \quad 81x = 162$

$\Rightarrow \quad x = 2$

Now, substituting, $x = 2$ in the equation (2), we get ;

$$8 \times 2 + y = 15$$

$\Rightarrow \ y = -1$

Hence, $x = 2, \qquad y = -1$.

(iii) We have,

$$2x + y = 17 \qquad \qquad \text{...(1)}$$
$$17x - 11y = 8 \qquad \qquad \text{...(2)}$$

From equation (1), we get;

$$2x + y = 17 \qquad \Rightarrow \quad y = 17 - 2x$$

Substituting the value of 'y' in equation (2), we get ;

$$17x - 11 (17 - 2x) = 8$$
$$\Rightarrow \quad 17x - 187 + 22x = 8$$
$$\Rightarrow \quad 39x = 195$$
$$\Rightarrow \quad x = 5$$

Now, substituting the value of 'x' in equation (1), we get ;

$$2 \times 5 + y = 17$$
$$\Rightarrow \quad y = 7$$

Hence, $x = 5, \qquad y = 7$.

Q. 2. **Solve the following systems of equations,**

(i) $\dfrac{15}{u} + \dfrac{2}{v} = 17 \qquad \dfrac{1}{u} + \dfrac{1}{v} = \dfrac{36}{5}$ (ii) $\dfrac{11}{v} - \dfrac{7}{u} = 1 \qquad \dfrac{9}{v} - \dfrac{4}{u} = 6$

Ans. (i) The given system of equation is ;

$$\dfrac{15}{u} + \dfrac{2}{v} = 17 \qquad \qquad \text{...(1)}$$

$$\dfrac{1}{u} + \dfrac{1}{v} = \dfrac{36}{5} \qquad \qquad \text{...(2)}$$

Considering $1/u = p$, $1/v = y$, the above system of linear equations can be written as:

$$15x + 2y = 17 \qquad \qquad \text{...(3)}$$

$$x + y = \dfrac{36}{5} \qquad \qquad \text{...(4)}$$

Multiplying (4) by 15 and (iii) by 1, we get ;

$$15x + 2y = 17 \qquad \qquad \text{...(5)}$$

$$15x + 15y = \dfrac{36}{5} \times 15 = 108 \qquad \qquad \text{...(6)}$$

Subtracting (6) form (5), we get;

$$-13y = -91 \qquad \Rightarrow \quad y = 7$$

Substituting $y = 7$ in (4), we get ;

$$x + 7 = \dfrac{36}{5} \qquad \Rightarrow \quad x = \dfrac{36}{5} - 7 = \dfrac{1}{5}$$

But, $y = \dfrac{1}{v} = 7 \qquad \Rightarrow \quad v = \dfrac{1}{7}$

and, $x = \dfrac{1}{u} = \dfrac{1}{5}$ $\Rightarrow u = 5$

Hence, the required solution of the given system is u = 5, v = 1/7.

(ii) The given system of equation is ;

$$\dfrac{11}{v} - \dfrac{7}{u} = 1 ; \qquad \dfrac{9}{v} - \dfrac{4}{u} = 6$$

Taking 1/v = x and 1/u = y, the above system of equations can be written as;

$$11x - 7y = 1 \qquad \qquad \text{...(1)}$$
$$9x - 4y = 6 \qquad \qquad \text{...(2)}$$

Multiplying (1) by 4 and (2) by 7, we get,

$$44x - 28y = 4 \qquad \qquad \text{...(3)}$$
$$63x - 28y = 42 \qquad \qquad \text{...(4)}$$

Subtracting (4) from (3) we get,

$$-19x = -38 \Rightarrow x = 2$$

Substituting the above value of x in (2), we get;

$$9 \times 2 - 4y = 6 \qquad \Rightarrow -4y = -12$$
$$\Rightarrow y = 3$$

But, $x = \dfrac{1}{v} = 2 \qquad \Rightarrow v = \dfrac{1}{2}$

and, $y = \dfrac{1}{u} = 3$

$\Rightarrow u = \dfrac{1}{3}$

Hence, the required solution of the given system of the equation is ;

$$v = \dfrac{1}{2} , \qquad u = \dfrac{1}{3}$$

Q. 3. **Solve the following system of equations by the method of elimination (substitution).**

$(a + b) x + (a - b) y = a^2 + b^2$
$(a - b) x + (a + b) y = a^2 + b^2$

Ans. The given system of equations is

$(a + b) x + (a - b) y = a^2 + b^2$ $\qquad \text{...(1)}$
$(a - b) x + (a + b) y = a^2 + b^2$ $\qquad \text{...(2)}$

From (2), we get $(a + b) y = a^2 + b^2 - (a - b) x$

$$\Rightarrow \quad y = \dfrac{a^2 + b^2}{a + b} - \dfrac{a - b}{a + b} x \qquad \qquad \text{...(3)}$$

Substituting $y = \dfrac{a^2 + b^2}{a + b} - \dfrac{a - b}{a + b} x$ in (1), we get

$$(a + b)\, x + (a - b)\left[\dfrac{a^2 + b^2}{a + b} - \dfrac{a - b}{a + b}\, x\right] = a^2 + b^2$$

$$\Rightarrow \quad (a + b)\, x + \dfrac{(a - b)(a^2 + b^2)}{a + b} - \dfrac{(a - b)^2}{(a + b)}\, x$$

Q. 4. **Solve the following system of linear equation by using the method of elimination by equating the coefficients:**

$$3x + 4y = 25 \; ; \qquad 5x - 6y = -9$$

Ans. The given system of equations is

$$3x + 4y = 25 \qquad\qquad \text{...(1)}$$
$$5x - 6y = -9 \qquad\qquad \text{...(2)}$$

Let us eliminate y. The coefficients of y are 4 and – 6. The LCM of 4 and 6 is 12.

So, we make the coefficients of y as 12 and – 12.

Multiplying equation (1) by 3 and equation (2) by 2, we get

$$9x + 12y = 75 \qquad\qquad \text{...(3)}$$
$$10x - 12y = -18 \qquad\qquad \text{...(4)}$$

Adding equation (3) and equation (4), we get

$$19x = 57 \quad \Rightarrow x = 3.$$

Putting x = 3 in (1), we get,

$$3 \times 3 + 4y = 25$$

$$\Rightarrow \quad 4y = 25 - 9 = 16 \Rightarrow y = 4$$

Hence, the solution is x = 3, y = 4.

Verfication : Both the equations are satisified by x = 3 and y = 4, which shows that the solutions is correct.

Q. 5. **Solve the following system of equations:**

$$15x + 4y = 61; \ 4x + 15y = 72$$

Ans. The given system of equation is

$$15x + 4y = 61 \qquad\qquad \text{...(1)}$$
$$4x + 15y = 72 \qquad\qquad \text{...(2)}$$

Let us eliminate y. The coefficients of y are 4 and 15. The L.C.M. of 4 and 15 is 60. So, we make the coefficients of y as 60. Multiplying (1) by 15 and (2) by 4, we get

$$225x + 60y = 915 \qquad\qquad \text{...(3)}$$
$$16x + 60y = 288 \qquad\qquad \text{...(4)}$$

Substracting (4) from (3), we get

$$209x = 627 \qquad \Rightarrow x = \dfrac{627}{209} = 3$$

Putting x = 3 in (1), we get

$$15 \times 3 + 4y = 61 \Rightarrow 45 + 47 = 61$$

$\Rightarrow \quad 4y = 61 - 45 = 16 \quad \Rightarrow \quad y = \dfrac{16}{4} = 4$

Hence, the solution is $x = 3$, $y = 4$.

Verification : On putting $x = 3$ and $y = 4$ in the given equations, they are satisified. Hence, the solution is correct.

Q. 6. **Solve the following system of linear equations by using the method of elimination by equating the coefficients.**

$$\sqrt{3}\, x - \sqrt{2}\, y = \sqrt{3} \; ; \; \sqrt{5}\, x + \sqrt{3}\, y = \sqrt{2}$$

Ans. The given equations are

$$\sqrt{3}\, x - \sqrt{2}\, y = \sqrt{3} \qquad \qquad \text{...(1)}$$

$$\sqrt{5}\, x + \sqrt{3}\, y = \sqrt{2} \qquad \qquad \text{...(2)}$$

Let us eliminate y. To make the coefficients of equal, we multiply the equation (1) by $\sqrt{3}$ and equation (2) by $\sqrt{2}$ to get

$$3x - \sqrt{6}\, y = 3 \qquad \qquad \text{...(3)}$$

$$\sqrt{10}\, x + \sqrt{6}\, y = 2 \qquad \qquad \text{...(4)}$$

Adding equation (3) and equation (4), we get

$$3x + \sqrt{10}\, x = 5 \;\Rightarrow\; (3 + \sqrt{10})\, x = 5$$

$$\Rightarrow \quad x = \frac{5}{3 + \sqrt{10}} = \left(\frac{5}{\sqrt{10} + 3}\right) \times \left(\frac{\sqrt{10} - 3}{\sqrt{10} - 3}\right)$$

$$= \frac{5(\sqrt{10} - 3)}{10 - 9} = 5(-3)$$

Putting $x = 5(\sqrt{10} - 3)$ in (1) we get

$$\sqrt{3} \times 5(\sqrt{10} - 3) - \sqrt{5}\, y = \sqrt{3}$$

$$\Rightarrow \quad 5\sqrt{30} - 15\sqrt{3} - \sqrt{2}\, y = \sqrt{3}$$

$$\Rightarrow \quad \sqrt{2}\, y = 5\sqrt{30} - 15\sqrt{3} - \sqrt{3}$$

$$\Rightarrow \quad \sqrt{2}\, y = 5\sqrt{30} - 16\sqrt{3}$$

$$\Rightarrow \quad y = \frac{5\sqrt{30}}{\sqrt{2}} - \frac{16\sqrt{3}}{\sqrt{2}} = 5\sqrt{15} - 8\sqrt{6}$$

Hence, the solution is $x = 5(\sqrt{10} - 3)$ and $y = 5\sqrt{15} - 8\sqrt{6}$.

Verification : After verifying, we find the solution is correct.

Q. 7. **Solve for x and y :**

$$\frac{ax}{b} - \frac{by}{a} = a + b \; ; \; ax - by = 2ab$$

Ans. The given system of equation is

$$\frac{ax}{b} - \frac{by}{a} = a + b \qquad \qquad \text{...(1)}$$

$$ax - by = 2ab \qquad \qquad \text{...(2)}$$

Dividing (2) by a, we get

$$x - \frac{by}{a} = 2b \qquad \qquad \text{...(3)}$$

On subtracting (3) from (1), we get

$$\frac{ax}{b} - x = a - b \;\Rightarrow\; x\left(\frac{a}{b} - 1\right) = a - b$$

$$\Rightarrow\; x = \frac{(a-b)b}{a-b} = b \;\Rightarrow\; x = b$$

On substituting the value of x in (3), we get

$$b - \frac{by}{a} = 2b \qquad \Rightarrow \quad b\left(1 - \frac{y}{a}\right) = 2b$$

$$\Rightarrow\; 1 - \frac{y}{a} = 2 \qquad \Rightarrow \quad \frac{y}{a} = 1 - 2$$

$$\Rightarrow\; \frac{y}{a} = -1 \qquad \Rightarrow \quad y = -a$$

Hence, the solution of the equations is

$$x = b, \qquad y = -a$$

Q. 8. **Solve the following system of linear equations:**

$$2(ax - by) + (a + 4b) = 0$$

$$2(bx + ay) + (b + 4a) = 0$$

Ans.

$$2ax - 2by + a + 4b = 0 \qquad \qquad \text{...(1)}$$

$$2bx + 2ay + b - 4a = 0 \qquad \qquad \text{...(2)}$$

Multiplying (1) by b and (2) by a and subtracting, we get

$$2(b^2 + a^2)\, y = 4\, (a^2 + b^2) \Rightarrow y = 2$$

Multiplying (1) by a and (2) by b and adding, we get

$$2(a^2 + b^2)\, x + a^2 + b^2 = 0$$

$$\Rightarrow\; 2(a^2 + b^2)\, x = -(a^2 + b^2) \Rightarrow\; x = -\frac{1}{2}$$

Hence x = –1/2, and y = 2

Q. 9. Solve $(a - b) x + (a + b) y = a^2 - 2ab - b^2$

$\qquad (a + b) (x + y) = a^2 + b^2$

Ans. The given system of equation is

$(a - b) x + (a + b) y = a^2 - 2ab - b^2$ $\hfill ...(1)$

$(a + b) (x + y) = a^2 + b^2$ $\hfill ...(2)$

$\Rightarrow \quad (a + b) x + (a + b) y = a^2 + b^2$ $\hfill ...(3)$

Subtracting equation (3) from equation (1), we get

$(a - b) x - (a + b) x = (a^2 - 2ab - b^2) - (a^2 + b^2)$

$\Rightarrow \quad -2bx = - 2ab - 2b^2$

$\Rightarrow \quad x = \dfrac{-2ab}{-2b} - \dfrac{2b^2}{-2b} = a + b$

Putting the value of x in (1), we get

$\qquad (a - b) (a + b) + (a + b) y = a^2 - 2ab - b^2$

$\Rightarrow \quad (a + b) y = a^2 - 2ab - b^2 - (a^2 - b^2)$

$\Rightarrow \quad (a + b) y = - 2ab$

$\Rightarrow \quad y = \dfrac{-2ab}{a + b}$

$\Rightarrow \quad$ Hence, the solution is $x = a + b$,

$\qquad y = \dfrac{-2ab}{a + b}$

Q. 10. Solve the following system of equations

$$\frac{1}{2x} - \frac{1}{y} = - 1; \frac{1}{x} + \frac{1}{2y} = 8$$

Ans. We have ;

$$\frac{1}{2x} - \frac{1}{y} = - 1 \hfill ...(1)$$

$$\frac{1}{x} + \frac{1}{2y} = 8 \hfill ...(2)$$

Let us consider $1/x = u$ and $1/y = v$.

Putting $1/x = u$ and $1/y = v$ in the above equations, we get;

$$\frac{u}{2} - v = - 1 \hfill ...(3)$$

$$u + \frac{v}{2} = 8 \hfill ...(4)$$

Let us eliminate v from the system of equations. So, multiplying equation (3) with 1/2 and (4) with 1, we get ;

$$\frac{u}{4} - \frac{v}{2} = -\frac{1}{2} \qquad \text{...(5)}$$

$$u + \frac{v}{2} = 8 \qquad \text{...(6)}$$

Adding equation (5) and (6), we get ;

$$\frac{u}{4} + u = \frac{-1}{2} + 8$$

$$\Rightarrow \quad \frac{5u}{4} = \frac{15}{2}$$

$$\Rightarrow \quad u = \frac{15}{2} \times \frac{4}{5}$$

$$\Rightarrow \quad u = 6$$

We know,

$$\frac{1}{x} = u \Rightarrow \frac{1}{x} = 6$$

$$\Rightarrow \quad x = -2$$

$$\Rightarrow \quad \frac{-y}{-4} = \frac{1}{4}$$

$$\Rightarrow \quad y = 1$$

Hence, the solution is $x = -2$, $y = 1$

Q. 1. **Solve the following system of equations by the method of cross-multiplication.**

$$11x + 15y = -23;\ 7x - 2y = 20$$

Ans. The given system of equations is

$$11x + 15y + 23 = 0$$

$$7x - 2y - 20 = 0$$

Now, by cross-multiplication method, we have

$$\frac{x}{\begin{matrix}15 & 23\\ -2 & -20\end{matrix}} = \frac{-y}{\begin{matrix}11 & 23\\ 7 & -20\end{matrix}} = \frac{1}{\begin{matrix}11 & 15\\ 7 & -2\end{matrix}}$$

$$\Rightarrow \frac{x}{15\times(-20)-(-2)\times23} = \frac{-y}{11\times(-20)-7\times23}$$

$$= \frac{1}{11\times(-2)-7\times15}$$

$$\Rightarrow \frac{x}{-300+46} = \frac{-y}{-220-161} = \frac{1}{-22-105}$$

$$\Rightarrow \frac{x}{-254} = \frac{-y}{-381} = \frac{1}{-127}$$

$$\Rightarrow \frac{x}{-254} = \frac{1}{-127} \Rightarrow x = 2$$

and $\dfrac{-y}{-381} = \dfrac{1}{-127} \Rightarrow y = -3$

Hence, $x = 2$, $y = -3$ is the required solution.

Q. 2. **Solve the following system of equations by cross-multiplication method.**

$$ax + by = a - b;\ bx - ay = a + b$$

Ans. Rewriting the given system of equations, we get

$$ax + by - (a - b) = 0$$

$$bx - ay - (a + b) = 0$$

By cross-multiplication method, we have

$$\frac{x}{\begin{matrix}b & -(a-b)\\ -a & -(a+b)\end{matrix}} = \frac{-y}{\begin{matrix}a & -(a-b)\\ b & -(a+b)\end{matrix}} = \frac{1}{\begin{matrix}a & b\\ b & -a\end{matrix}}$$

$$\Rightarrow \quad \frac{x}{b \times \{-(a+b)\} - (-a) \times \{-(a-b)\}}$$

$$= \frac{-y}{-a(a+b) + b(a-b)} = \frac{1}{-a^2 - b^2}$$

$$\Rightarrow \quad \frac{x}{-ab - b^2 - a^2 + ab} = \frac{-y}{-a^2 - ab + ab - b^2} = \frac{1}{-(a^2 + b^2)}$$

$$\Rightarrow \quad \frac{x}{-(a^2 + b^2)} = \frac{-y}{-(a^2 + b^2)} = \frac{1}{-(a^2 + b^2)}$$

$$\Rightarrow \quad \frac{x}{-(a^2 + b^2)} = \frac{1}{-(a^2 + b^2)} \Rightarrow x = 1$$

and $\dfrac{-y}{-(a^2 + b^2)} = \dfrac{1}{-(a^2 + b^2)}$

$$\Rightarrow \quad y = -1$$

Hence, the solution is $x = 1$, $y = -1$.

Q. 3. **Solve the following system of equations by cross-multiplication method.**

$$x + y = a - b; \ ax - by = a^2 + b^2$$

Ans. The given system of equations can be rewritten as:

$$x + y - (a - b) = 0$$

$$ax - by - (a^2 + b^2) = 0$$

By cross-multiplication method, we have

$$\frac{x}{\begin{matrix} 1 & -(a-b) \\ -b & -(a^2+b^2) \end{matrix}} = \frac{-y}{\begin{matrix} 1 & -(a-b) \\ a & -(a^2+b^2) \end{matrix}} = \frac{1}{\begin{matrix} 1 & 1 \\ a & -b \end{matrix}}$$

$$\Rightarrow \quad \frac{x}{-(a^2+b^2) - (-b) \times \{-a-b\}} = \frac{-y}{-(a^2+b^2) - a \times \{-(a-b)\}} = \frac{1}{-b-a}$$

$$\Rightarrow \quad \frac{x}{-(a^2+b^2) - b(a-b)} = \frac{-y}{-(a^2+b^2) + a(a-b)} = \frac{1}{-(b+a)}$$

$$\Rightarrow \quad \frac{x}{-a^2 - b^2 - ab + b^2} = \frac{-y}{-a^2 - b^2 + a^2 - ab} = \frac{1}{-(a+b)}$$

$$\Rightarrow \quad \frac{x}{-a(a+b)} = \frac{-y}{-b(a+b)} = \frac{1}{-(a+b)}$$

$$\Rightarrow \quad \frac{x}{-a(a+b)} = \frac{1}{-(a+b)} \Rightarrow x = a$$

and $\dfrac{-y}{-b(a+b)} = \dfrac{1}{-(a+b)} \Rightarrow y = -b$

Hence, the solution is $x = a$, $y = -b$.

Q. 4. **Show that the following system of equations has unique solution**

$$2x - 3y = 6;\ x + y = 1.$$

Ans. The given system of equation can be written as

$$2x - 3y - 6 = 0$$
$$x + y - 1 = 0$$

The given equations are of the form

$$a_1 x + b_1 y + c_1 = 0$$
$$a_2 x + b_2 y + c_2 = 0$$

Where, $a_1 = 2$, $b_1 = -3$, $c_1 = -6$

and $a_2 = 1$, $b_2 = 1$, $c_2 = -1$

$$\dfrac{a_1}{a_2} = \dfrac{2}{1} = 2,\quad \dfrac{b_1}{b_2} = \dfrac{-3}{1} = 3$$

$$\dfrac{c_1}{c_2} = \dfrac{-6}{-1} = 6$$

Clearly, $\dfrac{a_1}{a_2} \neq \dfrac{b_1}{b_2}$

So, the given system of equations has a unique solution, i.e., it is consistent.

Q. 15. **Show that the following system of equations has unique solution:**

$$x - 2y = 2 \ ;\ 4x - 2y = 5$$

Ans. The given system of equations can be written as

$$x - 2y - 2 = 0$$
$$4x - 2y - 5 = 0$$

The given equations are of the form

$$a_1 x + b_1 y + c_1 = 0$$
$$a_2 x + b_2 y + c_2 = 0$$

Where, $a_1 = 1$, $b_1 = -2$, $c_1 = -2$

and $a_2 = 4$, $b_2 = -2$, $c_1 = -5$

$$\dfrac{a_1}{a_2} = \dfrac{1}{4},\ \dfrac{b_1}{b_2} = \dfrac{-2}{-2} = 1,\ \dfrac{c_1}{c_2} = \dfrac{-2}{-5} = \dfrac{2}{5}$$

Clearly, $\dfrac{a_1}{a_2} \neq \dfrac{b_1}{b_2}$

So, the given system of equations has a unique solution i.e. It is consistent.

Q. 16. For what value of k the following system of equations has a unique solution:

$$x - ky = 2 \; ; \; 3x + 2y = -5$$

Ans. The given system of equation can be written as

$$x - ky - 2 = 0$$
$$3x + 2y + 5 = 0$$

The given system of equations is of the form

$$a_1x + b_1y + c_1 = 0$$
$$a_2x + b_2y + c_2 = 0$$

where, $a_1 = 1$, $b_1 = -k$, $c_1 = -2$

and $a_2 = 3$, $b_2 = 2$, $c_2 = 5$

Clearly, for unique solution $\dfrac{a_1}{a_2} \neq \dfrac{b_1}{b_2}$

$$\Rightarrow \quad \frac{1}{3} \neq \frac{-k}{2} \qquad \Rightarrow \quad k \neq \frac{-2}{3}$$

Q. 17. Show that the following system has infinitely many solutions.

$$x = 3y + 3 \; ; \; 9y = 3x - 9$$

Ans. The given system of equations can be written as

$$x - 3y - 3 = 0$$
$$3x - 9y - 9 = 0$$

The given equations are of the form

$$a_1x + b_1y + c_1 = 0$$
$$a_2x + b_2y + c_2 = 0$$

Where, $a_1 = 1$, $b_1 = -3$, $c_1 = -3$

and $a_2 = 3$, $b_2 = -9$, $c_2 = -9$

$$\frac{a_1}{a_2} = \frac{1}{3}, \frac{b_1}{b_2} = \frac{-3}{-9} = \frac{1}{3}, \frac{c_1}{c_2} = \frac{-3}{-9} = \frac{1}{3} = f$$

Clearly, $\dfrac{a_1}{a_2} = \dfrac{b_1}{b_2} = \dfrac{c_1}{c_2}$ so the given system of equations has infinitely many solutions.

Q. 18. Show that the following system has infinitely many solutions:

$$2y = 4x - 6 \; ; \; 2x = y + 3$$

Ans. The given system of equations can be written as

$$4x - 2y - 6 = 0$$
$$2x - y - 3 = 0$$

The given equations are of the form

$$a_1x + b_1y + c_1 = 0$$
$$a_2x + b_2y + c_2 = 0$$

Where, $a_1 = 4$, $b_1 = -2$, $c_1 = -6$

and $\quad a_2 = 2$, $b_2 = -1$, $c_2 = -3$

$$\frac{a_1}{a_2} = \frac{4}{2} = 2, \quad \frac{b_1}{b_2} = \frac{-2}{-1} = 2, \quad \frac{c_1}{c_2} = \frac{-6}{-3} = 2$$

Clearly, $\dfrac{a_1}{a_2} = \dfrac{b_1}{b_2} = \dfrac{c_1}{c_2}$, so the given system of equations has infinitely many solutions.

Q. 19. **Find the value of k for which the following system of equations has infinitely many solutions.**

$(k - 1) x + 3y = 7;\ (k + 1) x + 6y = (5k - 1)$

Ans. The given system of equations can be written as

$$(k - 1)x + 3y - 7 = 0$$
$$(k + 1) x + 6y - (5k - 1) = 0$$

Here $a_1 = (k - 1)$, $b_1 = 3$, $c_1 = -7$

and $a_2 = (k + 1)$, $b_2 = 6$, $c_2 = -(5k - 1)$

For the system of equations to have infinite number of solutions.

$$\frac{a_1}{a_2} = \frac{b_1}{b_2} = \frac{c_1}{c_2}$$

$$\Rightarrow \quad \frac{k-1}{k+1} = \frac{3}{6} = \frac{-7}{-(5k-1)}$$

$$\Rightarrow \quad \frac{k-1}{k+1} = \frac{1}{2} = \frac{7}{5k-1}$$

Taking I and II

$$\frac{k-1}{k+1} = \frac{1}{2}$$

$\Rightarrow \quad 2k - 2 = k + 1 \Rightarrow k = 3$

Taking II and III

$$\frac{1}{2} = \frac{7}{5k-1} \quad \Rightarrow \quad 5k - 1 = 14$$

$\Rightarrow \quad 5k = 15 \Rightarrow k = 3$

Hence, $k = 3$.

Q. 20. **For what values of a and b, the following system of equations have an infinite number of solutions:**

$2x + 3y = 7;\ (a - 7) x + (a + b) y = 3a + b - 2$

Ans. The given system of linear equations can be written as

$$2x + 3y - 7 = 0$$

$(a - b) x + (a + b) y - (3a + b - 2) = 0$

The above system of equations is of the form

$$a_1 x + b_1 y + c_1 = 0$$
$$a_2 x + b_2 y + c_2 = 0,$$

where $a_1 = 2$, $b_1 = 3$, $c_1 = -7$

$a_2 = (a - b)$, $b_2 = (a + b)$, $c_2 = -(3a + b - 2)$

For the given system of equations to have an infinite number of solutions

$$\frac{a_1}{a_2} = \frac{b_1}{b_2} = \frac{c_1}{c_2}$$

Here, $\dfrac{a_1}{a_2} = \dfrac{2}{a - b}$, $\dfrac{b_1}{b_2} = \dfrac{3}{a + b}$ and

$$\frac{c_1}{c_2} = \frac{-7}{-(3a + b - 2)} = \frac{7}{3a + b - 2}$$

$\Rightarrow \quad \dfrac{2}{a - b} = \dfrac{3}{a + b} = \dfrac{7}{3a + b - 2}$

$\Rightarrow \quad \dfrac{2}{a - b} = \dfrac{3}{a + b}$ and $\dfrac{3}{a + b} = \dfrac{7}{3a + b - 2}$

$\Rightarrow \quad 2a + 2b = 3a - 3b$ and $9a + 3b - 6 = 7a + 7b$

$\Rightarrow \quad 2a - 3a = -3b - 2b$ and $9a - 7a = 7b - 3b + 6$

$\Rightarrow \quad -a = -5b$ and $2a = 4b + 6$

$\Rightarrow \quad a = 5b$ .... (3) and $a = 2b + 3$ ....(4)

Solving (3) and (4) we get

$5b = 2b + 3 \Rightarrow b = 1$

Substituting $b = 1$ in (3), we get $a = 5 \times 1 = 5$

Thus, $a = 5$ and $b = 1$

Hence, the given system of equations has infinite number of solutions when

$a = 5$, $b = 1$

VERY SHORT ANSWER TYPE Q.UESTIONS

Q. 1. If the lines given by $3x + 2ky = 2$ and $2x + 5y = 1$ are parallel, then the value of k is _______ .

Sol.

Q. 2. If $x = a$ and $y = b$ is the solution of th equation $x - y = 2$ and $x + y = 4$, then the values of a and b are respectively _______ .

Sol.

Q. 3. A pair of linear equations which has a unique solution $x = 2$ and $y = -3$ is

    (a) $x + y = 1$ and $2x - 3y = -5$ (b) $2x + 5y = -11$ and $2x - 3y = -22$

    (c) $2x + 5y = -11$ and $4x + 10y = -22$ (d) $x - 4y - 14 = 0$ and $5x - y - 13 = 0$

Sol.

Q. 4. The area of the triangle formed by the lines $x = 3$, $y = 4$ and $x = y$ is _____ .

Sol.

Q. 5. The value of k for which the system of equations $3x + 5y = 0$ and $kx + 10y = 0$ has a non-zero solutions is _____ .

Sol.

Q. 6. If a pair of linear equations in two variables is consistent, then the lines represented by two equations are:

    (a) Intersecting   (b) Parallel

    (c) always coincident   (d) intersecting or coincident

Sol.

Q. 7. For $2x + 3y = 4$, y can be written in terms of x as _____ .

Sol.

Q. 8. One of the common solution of $ax + by = c$ and y axis is

    (a) $\left(0, \dfrac{c}{b}\right)$ (b) $\left(0, \dfrac{b}{c}\right)$ (c) $\left(\dfrac{c}{b}, 0\right)$ (d) $\left(0, -\dfrac{c}{b}\right)$

Sol.

Q. 9. If $ax + by = c$ and $lx + my = n$ has unique solution then the relation between the coefficient will be:

    (a) $am \neq lb$ (b) $am = lb$ (c) $am = lm$ (d) $ab = lm$

Sol.

Q. 10. In $\triangle ABC$, $\angle C = 3\angle B$, $\angle C = 2(\angle A + \angle B)$ then, $\angle A$, $\angle B$, $\angle C$ are respectively.

    (a) $30°, 60°, 90°$ (b) $20°, 40°, 120°$ (c) $45°, 45°, 90°$ (d) $110°, 40°, 50°$

Sol.

Q. 11. If $x = 3m - 1$ and $y = 4$ is a solution of the equation $x + y = 6$, then find the value of m.

Sol.

Q. 12. What is the point of intersection of the line represented by $3x - 2y = 6$ and the y-axis?

Sol.

Q. 13. For what value of p, system of equations $2x + py = 8$ and $x + y = 6$ have no solution.

Sol.

Q. 14. A motor cyclist is moving along the line $x - y = 2$ and another motor cyclist is moving along the line $x - y = 4$ find out their moving direction.

Sol.

Q. 15. Find the value of k for which pair of linear equations $3x + 2y = -5$ and $x - ky = 2$ has a unique solution.

Sol.

Q. 16. Write the solution of $y = x$ and $y = -x$.

Sol.

Q. 17. If $2x + 5y = 4$, write another linear equation, so that lines represented by the pair are coincident.

Sol.

Q. 18. Check whether the graph of the pair of linear equations $x + 2y - 4 = 0$ and $2x. 4y - 12 = 0$ is intersecting lines or parallel lines.

Sol.

Q. 19. What is the value of p, for which the pair of linear equations $x + y = 3$ and $3x + py = 9$ is inconsistent.

Sol.

Q. 20. If we draw lines of $x = 2$ and $y = 3$ what kind of lines do we get?

Sol.

SHORT ANSWER TYPE (I) Q.UESTIONS

Q. 21. Form a pair of linear equations for: The sum of the numerator and denominator of the fraction is 3 less than twice the denominator. If the numerator and denominator both are decreased by 1, the numerator becomes half the denominator.

Sol.

Q. 22. For what value of p the pair of linear equations $(p + 2)x - (2p + 1)y = 3(2p - 1)$ and $2x - 3y = 7$ has a unique solution.

Sol.

Q. 23. ABCDE is a pentagon with BE || CD and BC || DE, BC is perpendicular to CD. If the perimeter of ABCDE is 21 cm, find x and y.

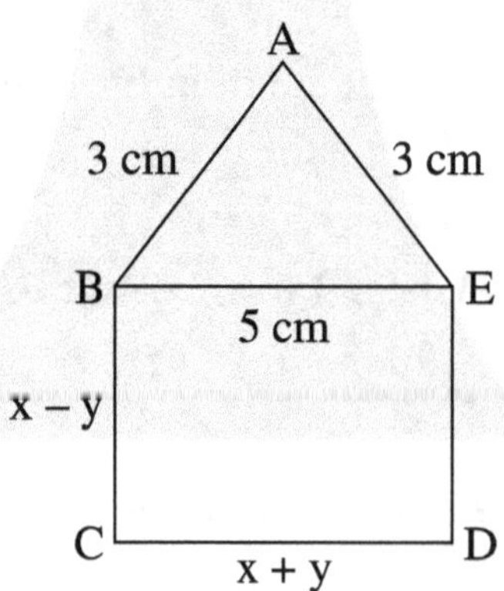

Sol.

Q. 24. Solve for x and y

$$x - \frac{y}{2} = 3 \text{ and } \frac{x}{2} - \frac{2y}{3} = \frac{2}{3}$$

Innovative Mathematics X-3

Sol.

Q. 25. **Solve for x and y**

$3x + 2y = 11$ and $2x + 3y = 4$

Also find p if $p = 8x + 5y$

Sol.

Q. 26. **Solve the pair of linear equations by substitution method $x - 7x + 42 = 0$ and $x - 3y - 6 = 0$.**

Sol.

Q. 27. **Ram is walking along the line joining (1, 4) and (0, 6).**

Sol.

Q. 28. **Rahim is walking along the line joining (3, 4) and (1, 0).**

Sol.

Q. 29. **Represent on graph and find the point where both of them cross each other. Given the linear equation $2x + 3y - 12 = 0$, write another linear equation in these variables, such that geometrical representation of the pair so formed is**

(i)   Parallel Lines   (ii)   Coincident Lines

Sol.

Q. 30. **The difference of two numbers is 66. If one number is four times the other, find the numbers.**

Sol.

Q. 31. **For what value of k, the following system of equations will be inconsistent**

$kx + 3y = k - 3$

$12x + ky = k$

Sol.

SHORT ANSWER TYPE (II) Q.UESTIONS

Q. 32. **Solve graphically the pair of linear equations $5x - y = 5$ and $3x - 2y = -4$. Also find the co-ordinates of the points where these lines intersect y-axis**   (CBSE)

Solve: $\dfrac{x}{a} + \dfrac{y}{b} = a + b$ $\dfrac{x}{a^2} + \dfrac{y}{b^2} = 2$

Sol.

Q. 33. **For what values of a and b the following pair of linear equations have infinite number of solutions?**   (CBSE)

$$2x + 3y = 7$$

$$a(x + y) - b(x - y) = 3a + b - 2$$

Sol.

Q. 34. **Find the value of k for no solutions**

$$(3k + 1)x + 3y - 2 = 0$$

$$(k^2 + 1)x + (k - 2)y - 5 = 0$$

Sol.

Q. 35. Solve the pair of linear equations

$$152x - 378y = -74 \qquad\qquad -378x + 152y = -604$$

Sol.

Q. 36. Pinky scored 40 marks in a test getting 3 marks for each right answer and losing 1 mark for each wrong answer. Had 4 marks been awarded for each correct answer and 2 marks were deducted for each wrong answer, then pinky again would have scored 40 marks. How many questions were there in the test?

Sol.

Q. 37. Father's age is three times the sum of ages of his two children. After 5 years his age will be twice the sum of ages of two children. Find the age of the father.

Sol.

Q. 38. On selling a T.V. at 5% gain and a fridge at 10% gain, a shopkeeper gain ₹ 2000. But if he sells the T.V. at 10% gain and fridge at 5% loss, he gains ₹ 1500 on the transaction. Find the actual price of the T.V. and the fridge.

Sol.

Q. 39. Sunita has some ₹ 50 and ₹ 100 notes amounting to a total of ₹ 15,500. If the total number of notes is 200, then find how many notes of ₹ 50 and ₹ 100 each, she has.

Sol.

LONG ANSWER TYPE Q.UESTIONS

Q. 40. Solve graphically the pair of linear equations $3x - 4y + 3 = 0$ and $3x + 4y - 21 = 0$

Find the co-ordinates of vertices of triangular region formed by these lines and x-axis. Also calculate the area of this angle.

Sol.

Q. 41. A and B are two points 150 km apart on a highway. Two cars start with different speeds from A and B at same time. If they move in same direction, they meet in 15 hours. If they move in opposite direction, they meet in one hour. Find their speeds

Sol.

Q. 42. The ratio of incomes of two persons A and B is 3 : 4 and the ratio of their expenditures is 5 : 7. If their savings are ₹ 15,000 annually find their annual incomes.

Sol.

Q. 43. Vijay had some bananas and he divided them into two lots A and B. He sold the first lot at the rate of ₹ 2 for 3 bananas and the second lot at the rate of ₹ 1 per banana and got a total of ₹ 400. If he had sold the first lot at the rate of ₹ 1 per banana and the second lot at the rate of ₹ 4 for 5 bananas, his total collection would have been ₹ 460. Find the total number of bananas he had. (HOTS, Exampler)

Sol.

Q. 44. A railway half ticket cost half the full fare but the reservation changes are the same on a half ticket as on a full ticket. One reserved first class ticket costs ₹ 2530. One reserved first class ticket and one reserved first class half ticket from stations A to B costs ₹ 3810. Find the full first class fare from stations A to B and also the reservation charges for a ticket. (Exemplar)

Sol.

Q. 45. Determine graphically, the vertices of the triangle formed by the lines $y = x$, $3y = x$ and $x + y = 8$. **(NCERT Exemplar)**

Sol.

Q. 46. Draw the graphs of the equations $x = 3$, $x = 5$ and $2x - y - 4 = 0$. Also find the area of the quadrilateral formed by the lines and the x-axis. **(NCERT Exemplar, HOTS)**

Sol.

Q. 47. Anirudh takes 3 hours more than Nishi to walk 30 km. But if Anirudh doubles his speed, he is ahead of Nishi by 1½ hours. Find their speed of walking.

Sol.

Q. 48. In a two digit number, the ten's place digit is 3 times the unit's place digit. When the number is decreased by 54, digits get reversed. Find the original number.

Sol.

Q. 49. A two-digit number is 3 more than 4 times the sum of the digits. If 18 is added to the number, digits reversed. Find the number.

Sol.

Q. 50. Find the values of a and b for infinite solutions

    (i) $2x - (a - 4)y = 2b + 1$ $4x - (a - 1)y = 5b - 1$

    (ii) $2x + 3y = 7$ $2ax + ay = 28 - by$

Sol.

1. $k = \dfrac{15}{4}$

2. $a = 3$ and $b = 1$

3. (c) $2x + 5y = -11$ and $4x + 10y = -22$

4. $\dfrac{1}{2}$ sq. unit

5. $k \ne 6$

6. (d) intersecting or coincident

7. $y = \dfrac{4 - 2x}{3}$

8. (a) $\left(0, \dfrac{c}{b}\right)$

9. (a) $am \ne lb$

10. (b) $20°, 40°, 120°$

11. $m = 1$

12. $(0, -3)$

13. $p = 2$

14. Move Parallel

15. $k \ne \dfrac{-2}{3}$

16. $(0, 0)$

17. $4x + 10y = 8$

18. Parallel lines

19. $P = 3$

20. intersecting lines

21. $x - y = -3, 2x - y = 1$

22. $P \ne 4$

23. $x = 5, y = 0$

24. $4, 2$

25. $x = 5, y = -2, p = 30$

26. $14, 12$

27. $(2, 2)$

28. (i) $4x + 6y + 10 = 0$

(ii) $4x + 6y - 24 = 0$

29. $88, 22$

30. $k = -6$

31. $(2, 5) (0, -5)$ and $(0, 2)$

32. $x = a^2, y = b^2$

33. $a = 5, b = 1$

34. $k = -1$

35. $2, 1$

36. 40 Q.uestions

37. 45 years

38. T.V. = ₹ 20,000, Fridge = ₹ 20,000

39. $90, 110$

40. Solution: $(3, 3)$

Vertices: $(-1, 0) (7, 0)$ and $(3, 3)$

Area = 12 sq.units

41. 80 km/hr ; 70 km/hr.

42. ₹ 90,000, ₹ 1,20,000

43. $300, 200$

44. $x = 2500, y = 30$

45. Vertices of triangle are: $(0, 0) (4, 4) (6, 2)$

46. Area = 8 sq.units.

47. $\dfrac{10}{3}$ km/hrs, 5 km/hrs.

48. 93

49. 35

50. (i) $7, 3$, (ii) $4, 8$

SECTION-A

Q. 1. For what value of k system of equations $x + 2y = 3$ and $5x + ky + 7 = 0$ has a unique solution. (1)

Sol.

Q. 2. Does the point (2, 3) lie on line represented by the graph of $3x - 2y = 5$. (1)

Sol.

Q. 3. The pair of equations $x = a$ and $y = b$ graphically representes lines which are: (1)

   (a) Parallel   (b) Intersecting at (b, a)

   (c) Coincident   (d) Intersecting at (a, b)

Sol.

Q. 4. For what value of K, the equations $3x - y + 8 = 0$ and $6x - Ky = -16$ represent coincident lines? (1)

   (a) $\dfrac{1}{2}$ (b) $-\dfrac{1}{2}$ (c) 2 (d) -2

Sol.

SECTION-B

Q. 5. For what value of a and b the pair of linear equations have infinite number of solutions (2)

$$2x - 3y = 7$$
$$ax + 3y = b$$

Sol.

Q. 6. Solve for x and y (2)

$$0.4x + 0.3y = 1.7$$
$$0.7x - 0.2y = 0.8$$

Sol.

Q. 7. If the system of equations $6x + 2y = 3$ and $kx + y = 2$ has a unique solution, find the value of k. (2)

Sol.

SECTION-C

Q. 8. Solve for x and y (3)

$$x + y = a + b$$
$$ax - by = a^2 - b^2$$

Sol.

Q. 9. Sum of the ages of a father and the son is 40 years. If fathers age is three times that of his son, then find their ages. (3)

Sol.

Q. 10. **Solve the following pair of equations graphically.** (4)

$3x + 5y = 12$ and $3x - 5y = -18$.

Also shade the region enclosed by these two lines and x-axis.

Sol.

CASE STUDY

CASE STUDY 1.

A test consists of 'True' or 'False' questions. One mark is awarded for every correct answer while ¼ mark is deducted for every wrong answer. A student knew answers to some of the questions. Rest of the questions he attempted by guessing. He answered 120 questions and got 90 marks.

Type of Q.uestion	Marks given for correct answer	Marks deducted for wrong answer
True/False	1	0.25

Q. 1. If answer to all questions he attempted by guessing were wrong, then how many questions did he answer correctly?

Q. 2. How many questions did he guess?

Q. 3. If answer to all questions he attempted by guessing were wrong and answered 80 correctly, then how many marks he got?

Q. 4. If answer to all questions he attempted by guessing were wrong, then how many questions answered correctly to score 95 marks?

ANSWERS

Let the no of questions whose answer is known to the student x and questions attempted by cheating be y

$x + y = 120$ $x - 1/4y = 90$

solving these two $x = 96$ and $y = 24$

1. He answered 96 questions correctly.

2. He attempted 24 questions by guessing.

3. Marks $= 80 - ¼$ of $40 = 70$

4. $x - ¼$ of $(120 - x) = 95$ $5x = 500,\ x = 100$

CASE STUDY 2.

Amit is planning to buy a house and the layout is given below. The design and the measurement has been made such that areas of two bedrooms and kitchen together is 95 sq.m.

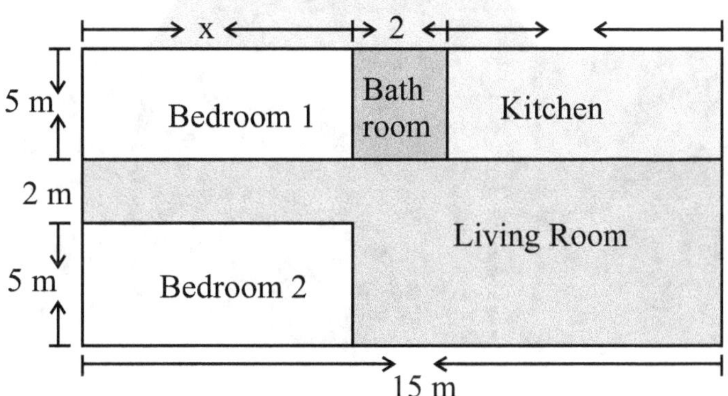

Based on the above information, answer the following questions:

Q. 1. Form the pair of linear equations in two variables from this situation.

Q. 2. Find the length of the outer boundary of the layout.

Q. 3. Find the area of each bedroom and kitchen in the layout.

Q. 4. Find the area of living room in the layout.

Q. 5. Find the cost of laying tiles in kitchen at the rate of Rs. 50 per sq.m

ANSWERS

1. Area of two bedrooms= $10 \times$ sq.m

   Area of kitchen = 5y sq.m

   $10x + 5y = 95$

   $2x + y = 19$

   Also, $x + 2 + y = 15$

   $x + y = 13$

2. Length of outer boundary = $12 + 15 + 12 + 15 = 54$ m

3. On solving two equation part (i)

   $x = 6$ m and $y = 7$ m

   area of bedroom = $5 \times 6 = 30$ m

   area of kitchen = $5 \times 7 = 35$ m

4. Area of living room = $(15 \times 7) - 30 = 105 - 30 = 75$ sq.m

5. Total cost of laying tiles in the kitchen = Rs $50 \times 35 =$ Rs 1750

CASE STUDY 3.

It is common that Governments revise travel fares from time to time based on various factors such as inflation (a general increase in prices and fall in the purchasing value of money) on different types of vehicles like auto, Rickshaws, taxis, Radio cab etc. The auto charges in a city comprise of a fixed charge together with the charge for the distance covered. Study the following situations.

Name of the city	Distance travelled (Km)	Amount paid (Rs.)
City A	10	75
	15	110
City B	8	91
	14	145

Situation 1: In city A, for a journey of 10 km, the charge paid is Rs 75 and for a journey of 15 km, the charge paid is Rs 110.

Situation 2: In a city B, for a journey of 8 km, the charge paid is Rs 91 and for a journey of 14km, the charge paid is Rs 145.

Refer situation 1

Q. 1. If the fixed charges of auto rickshaw be Rs x and the running charges be Rs y km/hr, the pair of linear equations representing the situation is

(a) $x + 10y = 110$, $x + 15y = 75$

(b) $x + 10y = 75$, $x + 15y = 110$

(c) $10x + y = 110$, $15x + y = 75$

(d) $10x + y = 75$, $15x + y = 110$

Q. 2. A person travels a distance of 50km. The amount he has to pay is

(a) Rs.155   (b) Rs.255   (c) Rs.355   (d) Rs.455

Refer situation 2

Q. 3. What will a person have to pay for travelling a distance of 30 km?

(a) Rs.185   (b) Rs.289   (c) Rs.275   (d) Rs.305

Q. 4. The graph of lines representing the conditions are: (situation 2)

(i)
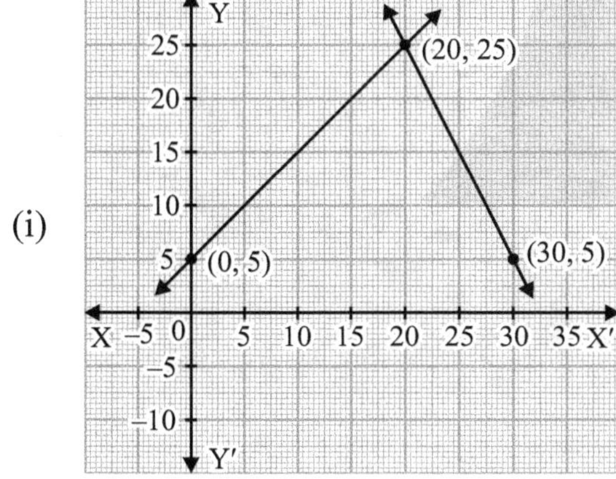

(ii)
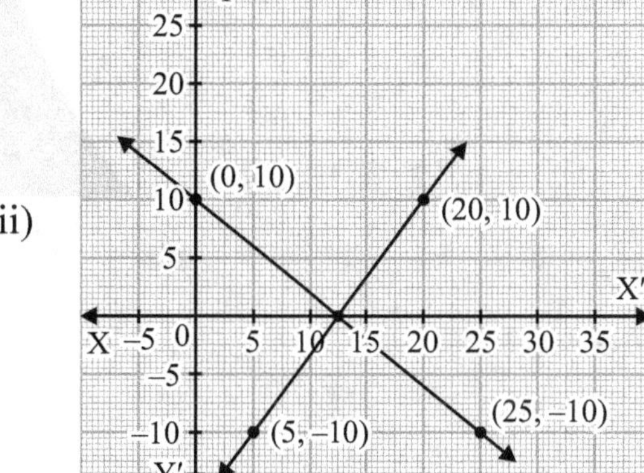

(iii) 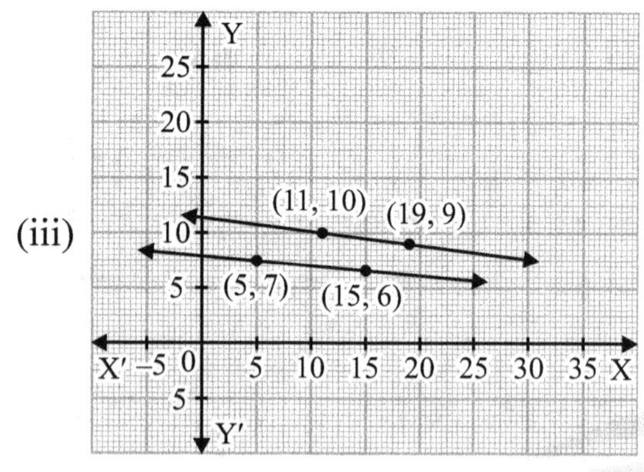

(iv) 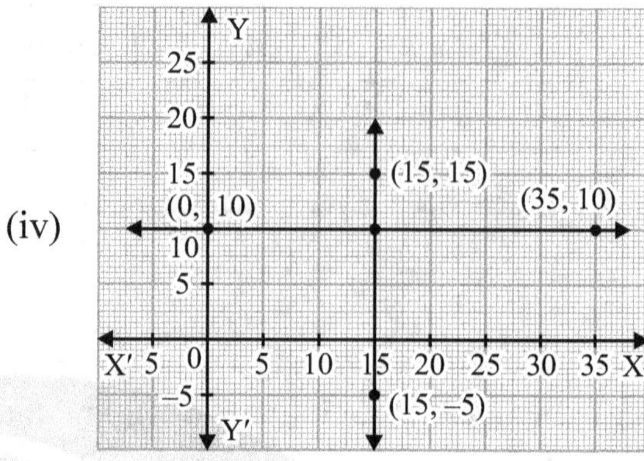

ANSWERS

1. B   **2.** C   **3.** B   **4.** (iii)

Innovative Mathematics X-3

CHAPTER-4

QUADRATIC EQUATIONS

LEARNING OBJECTIVES

As we know the format of $ax^2 + bx + c = 0$, Here degree of the polynomial is 2 so its possible zeroes are 2 (two) as example α, β or α, β also zeroes are known as roots of the polynomial.

METHODS

1. Factorization Method

2. Completing the square method

3. Q.uadratic formula.

1. FACTORICATION METHOD

EXAMPLE: $3x^2 + 2x - 1 = 0$

Here, the coefficient of x^2 is 3 and constant term is -1.

Step-1: We multiply the coefficient of x^2 and constant term and then we find the factors of product.

Like: $\qquad 3 \times -1 = -3$

Step-2: Now we will Add/subtract the factors and try to find out the middle term as coefficient of x.

Like: $\qquad +3 - 1$

Step-3: Write them as four terms and then regroup.

Like:
$$3x^2 + 3x - 1x - 1 = 0$$
$$3x(x + 1) - 1(x + 1) = 0$$
$$(x + 1)(3x - 1) = 0$$
$$x = -1, \quad x = \frac{1}{3}$$

Eg:
$$6x^2 - x - 2 = 0$$
$$6x^2 - 4x + 3x - 2 = 0$$
$$2x(3x - 2) + 1(3x - 2) = 0$$
$$(2x + 1)(3x - 2) = 0$$
$$x = \frac{-1}{2}, \quad x = \frac{2}{3}$$

Completing the Square Method:

$$ax^2 + bx + c = 0$$

Dividing the polynomial by coefficient of x^2 i.e. by a.

$$\frac{ax^2 + bx + c}{a} = \frac{0}{a}$$

$$x^2 + \frac{b}{a}x + \frac{c}{a} = 0$$

$$(x)^2 + 2 \cdot x \cdot \frac{b}{2a} + \left(\frac{b}{2a}\right)^2 - \left(\frac{b}{2a}\right)^2 + \frac{c}{a} = 0$$

Here what you should do is multiply and divide by 2, also add and subtract the square of $\frac{b}{2a}$ and then write $\frac{c}{a}$.
Now making combination as $(x + y)^2$ as.

$$x^2 + 2(x) \cdot \frac{b}{2a} + \left(\frac{b}{2a}\right)^2 - \frac{b^2}{4a^2} + \frac{c}{a} = 0$$

$$\left(x + \frac{b}{2a}\right)^2 = \frac{b^2}{4a^2} - \frac{c}{a}$$

$$\left(x + \frac{b}{2a}\right)^2 = \frac{b^2 - 4ac}{4a^2}$$

$$x + \frac{b}{2a} = \sqrt{\frac{b^2 - 4ac}{4a^2}}$$

$$x + \frac{b}{2a} = \pm\sqrt{\frac{b^2 - 4ac}{2a}}$$

$$x = \frac{-b}{2a} \pm \sqrt{\frac{b^2 - 4ac}{2a}}$$

$$\boxed{x = \frac{-b + \sqrt{b^2 - 4ac}}{2a}}$$

Eg: $3x^2 - 11x - 20 = 0$

Dividing all terms by coefficient of x^2 i.e. 3

$$\frac{3x^2 - 11x - 20}{3} = \frac{0}{3}$$

$$x^2 - \frac{11}{3}x - \frac{20}{3} = 0$$

$$(x)^2 - 2 \times x \times \frac{11}{6} + \left(\frac{11}{6}\right)^2 - \left(\frac{11}{6}\right)^2 - \frac{20}{3} = 0$$

$$\left(x - \frac{11}{6}\right)^2 = \frac{121}{36} + \frac{20}{3}$$

$$\left(x - \frac{11}{6}\right)^2 = \frac{121 + 240}{36}$$

$$\left(x - \frac{11}{6}\right)^2 = \frac{361}{36}$$

$$x - \frac{11}{6} = \sqrt{\frac{361}{36}}$$

$$x - \frac{11}{6} = \pm\frac{19}{36}$$

$$x = \frac{11}{6} \pm \frac{19}{36}$$

$$= \frac{11}{6} + \frac{19}{36}, \frac{11}{6} - \frac{19}{36}$$

$$x = \frac{66 + 19}{36}, \frac{66 - 19}{36}$$

$$x = \frac{85}{36}, \frac{47}{36}$$

Q. 2. $5x^2 + 3x - 1 = 0$

Dividing all terms by coefficient of x^2 i.e. 5

$$\frac{5x^2 + 3x - 1}{5} = \frac{0}{5}$$

$$x^2 + \frac{3}{5}x - \frac{1}{5} = 0$$

$$(x)^2 + 2 \bullet (x) \times \frac{3}{10} + \left(\frac{3}{10}\right)^2 - \left(\frac{3}{10}\right)^2 - \frac{1}{5} = 0$$

$$\left(x + \frac{3}{10}\right)^2 - \frac{9}{100} - \frac{1}{5} = 0$$

$$\left(x + \frac{3}{10}\right)^2 = \frac{9}{100} + \frac{1 \times 20}{5 \times 20}$$

$$= \frac{9}{100} + \frac{20}{100}$$

$$x + \frac{3}{10} = \sqrt{\frac{29}{100}}$$

$$= \pm\sqrt{\frac{29}{100}} = \pm\frac{29}{10}$$

$$x = \frac{-3}{10} \pm \frac{\sqrt{29}}{10}$$

$$x = \frac{-3 \pm \sqrt{29}}{10}$$

(iii) Q.uadratic formula:-

Earliear you all seen that by using completing the square method, the final step was its quadratic formula.

$$x = \frac{-b \pm \sqrt{b^2 - 4ac}}{2a}$$

or, $\quad x = \dfrac{-b \pm \sqrt{D}}{2a}$

Where D is known as discriminant

and $\quad D = \sqrt{b^2 - 4ac}$

Eg: $\quad x^2 - 7x + 12 = 0$

$\qquad a - 1, b - 7, c - 12$

$\therefore \quad D = b^2 - 4ac$

$\qquad = (-7)^2 - 4a \times 1 \times 12$

$\qquad = 49 - 48$

$\qquad = 1$

$\therefore \quad x = \dfrac{-b \pm \sqrt{D}}{2a}$

$\qquad = \dfrac{-(-7) \pm \sqrt{1}}{2 \times 1}$

$\qquad = \dfrac{7 \pm 1}{2}$

$\qquad = \dfrac{7+1}{2}, \dfrac{7-1}{2}$

$\qquad x = 4, 3$

Eg–2: $\quad 3x^2 - x - 10 = 0$

$\qquad a = 3, b = -1, c = -10$

$\qquad D = b^2 - 4ac$

$\qquad = (-1)^2 - 4 \times 3 \times (-10)$

$$= 1 + 120 = 121$$

So, $\quad x = \dfrac{-b \pm \sqrt{b^2 - 4ac}}{2a}$

$$= \dfrac{-(-1) \pm \sqrt{121}}{2 \times 3}$$

$$= \dfrac{1 \pm 11}{6}$$

$$= \dfrac{1 + 11}{6}, \dfrac{1 - 11}{6}$$

$$= \dfrac{12}{6}, \dfrac{-10}{6}$$

$$x = 2, \dfrac{-5}{3}$$

NATURE OF THE ROOTS OF A Q.UADRATIC EQ.UATIONS:

The nature of roots depends on the value of $b^2 - 4ac$ is called discriminant of the Q.uadratic equation $ax^2 + bx + c = 0$ and is generally, denoted by D.

$\therefore \quad D = b^2 - 4ac$

Case I: If $b^2 - 4ac$ or D > 0

In this case i.e. when D > 0 then it has two distant or distinct roots and real roots.

Case II: If $b^2 - 4ac$ or D = 0

In this case two equal and real roots are possible.

Case III: If $b^2 - 4ac$ or D < 0

In this case No real roots are possible.

BY R.P. GUPTA DESK:

1. For Q.uadratic Equation if it is in the form $ax^4 + bx^2 + c = 0$

   Then solve, it by substituting $x^2 = y$ and solve.

2. Equations involving on radical $\sqrt{a - x^2} = bx + c$

   Method: Square both sides to start solving and simplify it to get a quadratic equation:

WARM UP – LEVEL-I

Q. 1. **Show that:**

(i) $x = 3$ is a zero of quadratic polynomial $x^2 - 2x - 3$.

(ii) $x = -2$ is a zero of quadratic polynomial $3x^2 + 7x + 2$.

(iii) $x = 4$ is not a zero of quadratic polynomial $2x^2 - 7x - 5$.

Ans. (i) The value of $x^2 - 2x - 3$ at $x = 3$ is

$(3)^2 - 2 \times 3 - 3 = 9 - 6 - 3 = 0$

$\Rightarrow$ $x = 3$ is a zero of quadratic polynomial $x^2 - 2x - 3$.

(ii) The value of $3x^2 + 7x + 2$ at $x = -2$ is

$3(-2)^2 + 7(-2) + 2 = 12 - 14 + 2 = 0$

$\Rightarrow$ $x = -2$ is a zero of quadratic polynomial $3x^2 + 7x + 2$

(iii) The value of $2x^2 - 7x - 5$ at $x = 4$ is

$2(4)^2 - 7(4) - 5 = 32 - 28 - 5 = -1 \neq 0$

$\Rightarrow$ $x = 4$ is not a zero of quadratic polynomial $2x^2 - 7x - 5$.

Q. 2. **Find the value of m, if $x = 2$ is a zero of quadratic polynomial $3x^2 - mx + 4$.**

Ans. Since, $x = 2$ is a zero of $3x^2 - mx + 4$

$\Rightarrow$ $3(2)^2 - m \times 2 + 4 = 0$

$\Rightarrow$ $12 - 2m + 4 = 0$, i.e., $m = 8$.

Q. 3. **Which of the following are quadratic equations, give reason:**

(i) $x^2 - 8x + 6 = 0$

(ii) $3x^2 - 4 = 0$

(iii) $2x + \dfrac{5}{x} = x^2$

(iv) $x^2 + \dfrac{2}{x^2} = 3$

Ans. (i) Since, $x^2 - 8x + 6$ is a quadratic polynomial

$\Rightarrow$ $x^2 - 8x + 6 = 0$ is a quadratic equation.

(ii) $3x^2 - 4 = 0$ is a quadratic equation.

(iii) $2x + \dfrac{5}{x} = x^3$

$\Rightarrow$ $2x^2 + 5 = x^3$

$\Rightarrow$ $x^3 - 2x^2 - 5 = 0$; which is cubic and not a quadratic equation.

(iv) $x^2 + \dfrac{2}{x^2} = 3$

$\Rightarrow$ $x^4 + 2 = 2x^2$

$\Rightarrow$ $x^4 - 2x^2 + 2 = 0$; which is biquadratic and not a quadratic equation.

Q. 4. **Solve the following quadratic equations:**

(i) $7x^2 = 8 - 10x$ (ii) $3(x^2 - 4) = 5x$

(iii) $x(x + 1) + (x + 2)(x + 3) = 42$

Ans. (i) $7x^2 = 8 - 10x$

$\Rightarrow \quad 7x^2 + 10x - 8 = 0$

$\Rightarrow \quad 7x^2 + 14x - 4x - 8 = 0$

$\Rightarrow \quad 7x(x + 2) - 4(x + 2) = 0$

$\Rightarrow \quad (x + 2)(7x - 4) = 0$

$\Rightarrow \quad x + 2 = 0 \quad$ or $\quad 7x - 4 = 0$

$\Rightarrow \quad x = -2 \quad$ or $\quad x = \dfrac{4}{7}$

(ii) $3(x^2 - 4) = 5x$

$\Rightarrow \quad 3x^2 - 5x - 12 = 0$

$\Rightarrow \quad 3x^2 - 9x + 4x - 12 = 0$

$\Rightarrow \quad 3x(x - 3) + 4(x - 3) = 0$

$\Rightarrow \quad (x - 3)(3x + 4) = 0$

$\Rightarrow \quad x - 3 = 0 \quad$ or $\quad 3x + 4 = 0$

$\Rightarrow \quad x = 3 \quad$ or $\quad x = -\dfrac{4}{3}$

(iii) $x(x + 1) + (x + 2)(x + 3) = 42$

$\Rightarrow \quad x^2 + x + x^2 + 3x + 2x + 6 - 42 = 0$

$\Rightarrow \quad 2x^2 + 6x - 36 = 0$

$\Rightarrow \quad x^2 + 3x - 18 = 0$

$\Rightarrow \quad x^2 + 6x - 3x - 18 = 0$

$\Rightarrow \quad x(x + 6) - 3(x + 6) = 0$

$\Rightarrow \quad (x + 6)(x - 3) = 0$

$\Rightarrow \quad x = -6 \quad$ or $\quad x = 3$

Q. 5. **Solve for x : $12\,abx^2 - (9a^2 - 8b^2)\,x - 6ab = 0$**

Ans. Given equation is:

$12abx^2 - 9a^2x + 8b^2x - 6ab = 0$

$\Rightarrow \quad 3ax(4bx - 3a) + 2b(4bx - 3a) = 0$

$\Rightarrow \quad (4bx - 3a)(3ax + 2b) = 0$

$\Rightarrow \quad 4bx - 3a = 0 \quad$ or $\quad 3ax + 2b = 0$

$\Rightarrow \quad x = \dfrac{3a}{4b} \quad$ or $\quad x = -\dfrac{2b}{3a}$

Q. 6. **Find the roots of the following quadratic equations (if they exist) by the method of completing the square.**

(i) $2x^2 - 7x + 3 = 0$ (ii) $4x^2 + 4\sqrt{3}\,x + 3 = 0$ (iii) $2x^2 + x + 4 = 0$

Ans. (i) $2x^2 - 7x + 3 = 0 \quad \Rightarrow x^2 - \dfrac{7}{2}x + \dfrac{3}{2} = 0$ [Dividing each term by 2]

$\Rightarrow \quad x^2 - 2 \times x \times \dfrac{7}{4} + \dfrac{3}{2} = 0$

$$\Rightarrow \quad x^2 - 2 \times x \times \frac{7}{4} + \left(\frac{7}{4}\right)^2 - \left(\frac{7}{4}\right)^2 + \frac{3}{2} = 0$$

$$\Rightarrow \quad \left(x - \frac{7}{4}\right)^2 - \frac{49}{16} + \frac{3}{2} = 0$$

$$\Rightarrow \quad \left(x - \frac{7}{4}\right)^2 - \left(\frac{49 - 24}{16}\right) = 0$$

$$\Rightarrow \quad \left(x - \frac{7}{4}\right)^2 - \frac{25}{16} = 0$$

i.e., $\left(x - \dfrac{7}{4}\right)^2 - \dfrac{25}{16} \quad \Rightarrow x - \dfrac{7}{4} = \pm \dfrac{5}{4}$

i.e., $x - \dfrac{7}{4} = \dfrac{5}{4} = \qquad$ or $\quad x - \dfrac{7}{4} = -\dfrac{5}{4}$

$$\Rightarrow \quad x = \frac{7}{4} + \frac{5}{4} \qquad \text{or } x = \frac{7}{4} - \frac{5}{4}$$

$$\Rightarrow \quad x = 3 \qquad \text{or } x = \frac{1}{2}$$

(ii) $4x^2 + 4\sqrt{3}\, x + 3 = 0$

$$\Rightarrow \quad x^2 + \sqrt{3}\, x + \frac{3}{4} = 0$$

i.e., $x^2 + 2 \times x \times \dfrac{\sqrt{3}}{2} + \left(\dfrac{\sqrt{3}}{2}\right)^2 - \left(\dfrac{\sqrt{3}}{2}\right)^2 + \dfrac{3}{4} = 0$

$$\Rightarrow \quad \left(x + \frac{\sqrt{3}}{2}\right)^2 - \frac{3}{4} + \frac{3}{4} = 0$$

i.e., $\left(x + \dfrac{\sqrt{3}}{2}\right)^2 = 0$

$$\Rightarrow \quad x + \frac{\sqrt{3}}{2} = 0 \qquad \text{and} \qquad x = \frac{-\sqrt{3}}{2}$$

$\therefore$ Roots are: $\dfrac{-\sqrt{3}}{2}$ and $\dfrac{-\sqrt{3}}{2}$

(iii) $2x^2 + x + 4 = 0 \quad \Rightarrow x^2 + \dfrac{x}{2} + 2 = 0$

i.e., $x^2 + 2 \times x \times \dfrac{1}{4} + \left(\dfrac{1}{4}\right)^2 - \left(\dfrac{1}{4}\right)^2 + 2 = 0$

$\Rightarrow \left(x + \dfrac{1}{4}\right)^2 - \dfrac{1}{16} + 2 = 0$

$\Rightarrow \left(x + \dfrac{1}{4}\right)^2 + \dfrac{31}{16} = 0$

$$\left[-\dfrac{1}{16} + 2 = \dfrac{-1 + 32}{16} = \dfrac{31}{16}\right]$$

i.e., $\left(x + \dfrac{1}{4}\right)^2 = -\dfrac{31}{16}$

This is not possible as the square of a real number can not be negative.

Q. 7. **Solve the following quadratic equations by using quadratic formula:**

(i) $x^2 - 7x + 12 = 0$

(ii) $3x^2 - x - 10 = 0$

Ans.   (i) Comparing the given equation $x^2 - 7x + 12 = 0$ with standard quadratic equation $ax^2 + bx + c = 0$;

we get : $a = 1$, $b = -7$ and $c = 12$

$\therefore \quad b^2 - 4ac = (-7)^2 - 4 \times 1 \times 12$

$\qquad\qquad = 49 - 48 = 1$

and $\sqrt{b^2 - 4ac} = \sqrt{1} = 1$

$\because \quad x = \dfrac{-b \pm \sqrt{b^2 - 4ac}}{2a}$

$\Rightarrow \quad x = \dfrac{7 \pm 1}{2 \times 1} = \dfrac{7 + 1}{2}$ or $\dfrac{7 - 1}{2} = 4$ or 3

(ii) Comparing the given equation

$3x^2 - x - 10 = 0$ with equation

$ax^2 + bx + c = 0$; we get : $a = 3$, $b = -1$ and $c = -10$

$\therefore \quad b^2 - 4ac = (-1)^2 - 4 \times 3 \times -10 = 1 + 120 = 121$

and $\sqrt{b^2 - 4ac} = \sqrt{121} = 11$

Hence, $x = \dfrac{-b \pm \sqrt{b^2 - 4ac}}{2a}$

$\Rightarrow \quad x = \dfrac{1 \pm 11}{2} = \dfrac{1 + 11}{2}$ or $\dfrac{1 - 11}{2} = 6$ or -5

Q. 8. For a quadratic equation $ax^2 + bx + c = 0$, where $a \neq 0$, prove that:

$$x = \frac{-b \pm \sqrt{b^2 - 4ac}}{2a}$$

Ans. $ax^2 + bx + c = 0$

$\Rightarrow \quad 4a^2x^2 + 4abx + 4ac = 0$ [Multiplying by '4a']

$\Rightarrow \quad (2ax)^2 + 2 \times 2ax \times b + b^2 - b^2 + 4ac = 0$

$\Rightarrow \quad (2ax + b)^2 - b^2 + 4ac = 0$

$\Rightarrow \quad (2ax + b)^2 = b^2 - 4ac$

$\Rightarrow \quad 2ax + b = \pm \sqrt{b^2 - 4ac}$

$\Rightarrow \quad 2ax = -b \pm \sqrt{b^2 - 4ac}$

$\Rightarrow \quad x = \dfrac{-b + \sqrt{b^2 - 4ac}}{2a}$

Q. 9. Solve, by using quadratic formula, each of the following equations:

(i) $2x^2 + 5\sqrt{3}\,x + 6 = 0$

(ii) $3x^2 + 2\sqrt{5}\,x - 5 = 0$

Ans. (i) Comparing $2x^2 + 5\sqrt{3}\,x + 6 = 0$ with

$ax^2 + bx + c = 0$, we get:

$a = 2, b = 5\sqrt{3}$ and $c = 6$

$b^2 - 4ac = (5\sqrt{3})^2 - 4 \times 2 \times 6$

$\qquad\qquad = 25 \times 3 - 48 = 27$

$\sqrt{b^2 - 4ac} = \sqrt{27} = \sqrt{3 \times 3 \times 3} = 3\sqrt{3}$

$\therefore \quad x = \dfrac{-b \pm \sqrt{b^2 - 4ac}}{2a} = \dfrac{-5\sqrt{3} \pm 3\sqrt{3}}{2 \times 4}$

$\qquad = \dfrac{-5\sqrt{3} + 3\sqrt{3}}{4} \quad \text{or} \quad \dfrac{-5\sqrt{3} - 3\sqrt{3}}{4}$

$\qquad = \dfrac{-2\sqrt{3}}{4} \text{ or } \dfrac{-8\sqrt{3}}{4} = -\dfrac{\sqrt{3}}{2} \text{ or } -2\sqrt{3}$

(ii) Comparing $3x^2 + 2\sqrt{5}\,x - 5 = 4 \times 5 + 60 = 80$

$ax^2 + bx + c = 0$, we get

$b^2 - 4ac = (2\sqrt{5})^2 - 4 \times 3 \times -5 = 4 \times 5 + 60 = 80$

$\sqrt{b^2 - 4ac} = \sqrt{80} = \sqrt{16 \times 5} = 4\sqrt{3}$

$$\therefore \quad x = \frac{-b \pm \sqrt{b^2 - 4ac}}{2a} = \frac{-2\sqrt{5} \pm 4\sqrt{5}}{2 \times 3}$$

$$= \frac{-2\sqrt{5} + 4\sqrt{5}}{6} \quad \text{or} \quad \frac{-2\sqrt{5} - 4\sqrt{5}}{6}$$

$$= \frac{2\sqrt{5}}{6} \quad \text{or} \quad \frac{-6\sqrt{5}}{6} = \frac{\sqrt{5}}{3} \quad \text{or} \quad -\sqrt{5}$$

Q. 10. **Using the quadratic formula, solve the equation:**

$$a^2b^2x^2 - (4b^2 - 3a^4)x - 12a^2b^2 = 0$$

Ans. Comparing given equations with

$Ax^2 + Bx + C = 0$, we get:

$A = a^2b^2$, $B = -(4b^4 - 3a^4)$ and $C = -12\,a^2b^2$

$\therefore \quad B^2 - 4AC = (4b^4 - 3a^4)^2 - 4 \times a^2b^2 \times (-12a^2b^2)$

$= 16b^8 + 9a^8 - 24a^4b^4 + 48a^4b^4$

$= 16b^8 + 9a^8 + 24a^4b^4 = (4b^4 + 3a^4)^2$

$$\sqrt{B^2 - 4AC} = 4b^4 + 3a^4$$

$$\therefore \quad x = \frac{-B \pm \sqrt{B^2 + 4AC}}{2A}$$

$$= \frac{(4b^4 - 3a^4) \pm (4b^2 + 3a^4)}{2 \times a^2b^2}$$

$$= \frac{4b^4 - 3a^4 + 4b^4 + 3a^4}{2a^2b^2}$$

$$\text{or} \quad \frac{4b^4 - 3a^4 - 4b^4 - 3a^4}{2a^2b^2}$$

$$= \frac{8b^4}{2a^2b^2} \quad \text{or} \quad \frac{-6a^4}{2a^2b^2} = \frac{4b^2}{a^2} \quad \text{or} \quad \frac{-3a^2}{b^2}$$

Q. 11. **Without solving, examine the nature of roots of the equations:**

(i) $2x^2 + 2x + 3 = 0$

(ii) $2x^2 - 7x + 3 = 0$

(iii) $x^2 - 5x - 2 = 0$

(iv) $4x^2 - 4x + 1 = 0$

Ans. (i)   Comparing $2x^2 + 2x + 3 = 0$

with $ax^2 + bx + c = 0$; we get : $a = 2$, $b = 2$ and $c = 3$

$D = b^2 - 4ac = (2)^2 - 4 \times 2 \times 3 = 4 - 24$

$= -20$; which is negative.

$\therefore$ The roots of the given equation are imaginary.

(ii) Comarping $2x^2 - 7x + 3 = 0$

with $ax^2 + bx + c = 0$;

we get : $a = 2$, $b = -7$ and $c = 3$

$D = b^2 - 4ac = (-7)^2 - 4 \times 2 \times 3$

$= 49 - 24 = 25$, which is perfect square.

$\therefore$ The roots of the given equation are rational and unequal.

(iii) Comarping $x^2 - 5x - 2 = 0$

with $ax^2 + bx + c = 0$;

we get : $a = 1$, $b = -5$ and $c = -2$

$D = b^2 - 4ac = (-5)^2 - 4 \times 1 \times -2$

$= 25 + 8 = 33$; which is positive but not a perfect square.

$\therefore$ The roots of the given equation are irrational and unequal.

(iv) Comparing $4x^2 - 4x + 1 = 0$

with $ax^2 + bx + c = 0$;

we get : $a = 4$, $b = -4$, and $c = 1$

$D = b^2 - 4ac = (-4)^2 - 4 \times 4 \times 1$

$= 16 - 16 = 0$

$\therefore$ Roots are real and equal.

Q. 12. **For what value of m, are the roots of the equation $(3m + 1) x^2 + (11 + m) x + 9 = 0$ equal?**

Ans. Comparing the given equation

with $ax^2 + bx + c = 0$;

we get : $a = 3m + 1$, $b = 11 + m$ and $c = 9$

$\therefore$ Discriminant, $D = b^2 - 4ac$

$= (11 + m)^2 - 4(3m + 1) \times 9$

$= 121 + 22m + m^2 - 108\,m - 36$

$= m^2 - 86m + 85$

$= m^2 - 85m - m + 85$

$= m(m - 85) - 1\,(m - 85)$

$= (m - 85)\,(m - 1)$

Since the roots are equal, $D = 0$

$\Rightarrow$ $(m - 85)\,(m - 1) = 0$

$\Rightarrow$ $m - 85 = 0$ or $m - 1 = 0$

$\Rightarrow$ $m = 85$ or $m = 1$

Q. 13. **For each quadratic equation given below, find the sum of the roots and the product of the roots:**

(i) $x^2 + 3x - 6 = 0$ (ii) $2x^2 + 5\sqrt{3}\,x + 6 = 0$ (iii) $3x^2 + 2\sqrt{5}\,x - 5 = 0$

Ans. (i) Comparing $x^2 + 3x - 6 = 0$

with $ax^2 + bx + c = 0$,

we get:

$a = 1$, $b = 3$ and $c = -6$

$\therefore$ The sum of the roots $= -\dfrac{b}{a} = -\dfrac{3}{2}$

And, the product of the roots $= \dfrac{c}{a} = \dfrac{-6}{1} = -6$

(ii) Comparing $2x^2 + 5\sqrt{3}\,x + 6 = 0$ with $ax^2 + bx + c = 0$; we get

$a = 2, b = 5\sqrt{3}$ and $c = 6$

$\therefore$ The sum of the roots $= -\dfrac{b}{a} = -\dfrac{5\sqrt{3}}{2}$

And, the product of the roots $= \dfrac{c}{a} = \dfrac{6}{2} = 3$

(iii) Comparing $3x^2 + 2\sqrt{5}\,x - 5 = 0$

with $ax^2 + bx + c = 0$; we get :

$a = 3, b = 2\sqrt{5}$ and $c = -5$

$\therefore$ The sum of the roots $= -\dfrac{b}{a} = -\dfrac{2\sqrt{5}}{3}$ and, the product of the roots $= \dfrac{c}{a} = \dfrac{-5}{3}$

Q. 14. **Construct the quadratic equation whose roots are given below.**

(i) $3, -3$

(ii) $3 + \sqrt{3}, 3 - \sqrt{3}$

(iii) $\dfrac{2 + \sqrt{5}}{2}, \dfrac{2 - \sqrt{5}}{2}$

Ans. (i) Since, the sum of the roots

$= (3) + (-3) = 3 - 3 = 0$

and, the product of the roots

$= (3)(-3) = -9$

$\therefore$ The required quadratic equation is:

$x^2 - (\text{sum of roots})\, x + (\text{product of roots}) = 0$

$\Rightarrow x^2 - (0)\, x + (-9) = 0, \quad \text{i.e., } x^2 - 9 = 0$

(ii) Since, the sum of the roots

$= 3 + \sqrt{3} + 3 - \sqrt{3} = 6$

and, the product of the roots

$= (3 + \sqrt{3})(3 - \sqrt{3}) = 9 - 3 = 6$

$\therefore$ The required quadratic equation is:

$x^2 - (\text{sum of roots})\, x + (\text{product of roots}) = 0$

$\Rightarrow x^2 - 6x + 6 = 0$

(iii) Since, the sum of the roots

$$= \frac{2+\sqrt{5}}{2} + \frac{2-\sqrt{5}}{2} = \frac{2+\sqrt{5}+2-\sqrt{5}}{2} = \frac{4}{2}$$

and, the product of the roots

$$= \left(\frac{2+\sqrt{5}}{2}\right)\left(\frac{2-\sqrt{5}}{2}\right) = \frac{4-5}{4} - \frac{1}{4}$$

$\therefore$ The required quadratic equation is:

$x^2 - (\text{sum of roots}) \, x + (\text{product of roots}) = 0$

$$\Rightarrow \quad x^2 - 2x + \left(-\frac{1}{4}\right) = 0$$

$$\Rightarrow \quad x^2 - 2x - \frac{1}{4} = 0$$

i.e., $4x^2 - 8x - 1 = 0$

WARM UP – LEVEL-II

Q. 1. **If one of the roots of the quadratic equation $2x^2 + px + 4 = 0$ is 2, find the the value of p, also find the value of the other roots.**

Ans. As, 2 is one of the roots, $x = 2$ will satisfy the equation $2x^2 + px + 4 = 0$

$\Rightarrow \quad 2(2)^2 + p(2) + 4 = 0$

$\Rightarrow \quad 8 + 2p + 4 = 0$

i.e., $2p = -12$ and $p = -6$

Substituting $p = -6$ in the equation

$2x^2 + px + 4 = 0$; we get : $2x^2 - 6x + 4 = 0$

$\Rightarrow \quad x^2 - 3x + 2 = 0$ [Dividing each term by 2]

$\Rightarrow \quad x^2 - 2x - x + 2 = 0$

$\Rightarrow \quad x - 2 = 0$ or $x - 1 = 0$

$\Rightarrow \quad x = 2$ or $x = 1$

$\therefore$ The other (second) root is 1.

Q. 2. **The sum of two natural numbers is 8. If the sum of their reciprocals is $\dfrac{8}{15}$, find the two numbers.**

Ans. Let the numbers be x and $8 - x$.

$\therefore \quad \dfrac{1}{x} + \dfrac{1}{8-x} = \dfrac{8}{15} \Rightarrow \dfrac{8-x+x}{x(8-x)} = \dfrac{8}{15}$

$\Rightarrow \quad \dfrac{8}{8x - x^2} = \dfrac{8}{15}$ i.e., $120 = 64x - 8x^2$

$\Rightarrow \quad 8x^2 - 64x + 120 = 0$

$\Rightarrow \quad x^2 - 8x + 15 = 0$ [Dividing by 8]

$\Rightarrow \quad (x - 5)(x - 3) = 0$ [On factorizing]

$\Rightarrow \quad x = 5,$ or $x = 3$

When $x = 5$, the number are x and $8 - x = 5$ and 3, and when $x = 3$, the numbers are x and $8 - x = 3$ and 5.

$\therefore$ Required numbers are 5 and 3.

Q. 3. **Divide 16 into two parts such that twice the square of the larger part exceeds the square of the smaller part by 164.**

Ans. Let larger part be x, therefore the smallar of part $= 16 - x$.

Given : $2x^2 - (16 - x)^2 = 164$

$\rightarrow \quad 2x^2 - (256 + x^2 - 32x) - 164 = 0$

i.e., $2x^2 - 256 - x^2 + 32x - 164 = 0$

$\Rightarrow \quad x^2 + 32x - 420 = 0$

On factorizing, it gives : $(x + 42)(x - 10) = 0$

i.e., $x = -42$ or $x = 140$

$\therefore \quad x = 10$

Hence the larger part $= 10$ and the amllar part

$= 16 - x = 16 - 10 = 6$

          Innovative Mathematics X-4

Q. 4. **Two positive numbers are in the ratio 2 : 5. If difference between the squares of these numbers is 189 ; find the numbers.**

Ans. Let numbers be 2x and 5x

$$\therefore \quad (5x)^2 - (2x)^2 = 189$$

$$\Rightarrow \quad 25x^2 - 4x^2 = 189 \text{ and } 21x^2 = 189$$

$$\text{i.e., } x^2 = \frac{189}{21} = 9 \quad \Rightarrow \quad x = \pm 3$$

Since, the required numbers are positive,

$$\therefore \quad x = 3$$

And, required numbers = 2x and 5x = 2 × 3 and 5 × 3 = 6 and 15

Q. 5. **A two digit number is such that the product of the digits is 35. When 18 is added to this number the digits interchange their places. Determine the number.**

Ans. Let ten's digit of the numbers = x and its unit digit = y.

$$\therefore \quad \text{The two digit number is } 10x + y.$$

Given : x . y = 35 and 10x + y + 18 = 10y + x

$$\Rightarrow \quad y = \frac{35}{x} \text{ and } 9x + 18 = 9y$$

i.e., x + 2 = y

On substituting $y = \dfrac{35}{x}$ in x + 2 = y ; we get:

$$x + 2 = \frac{35}{x}$$

$$\Rightarrow \quad x^2 + 2x = 35$$

and $x^2 + 2x - 35 = 0$

On factorising, we get : (x + 7)(x − 5) = 0

i.e., x = − 7 or x = 5

Since, x is digit, therefore x = 5 and

$$y = \frac{35}{x} = 7$$

$$\therefore \quad \text{The required two digit number} = 10x + y$$

$$= 10 × 5 + 7 = 57$$

Q. 6. **The sides (in cm) of a right triangle are x − 1, x and x + 1. Find the sides of triangle.**

Ans. It is clear that the largest side x + 1 is hypotenuse of the right triangle.

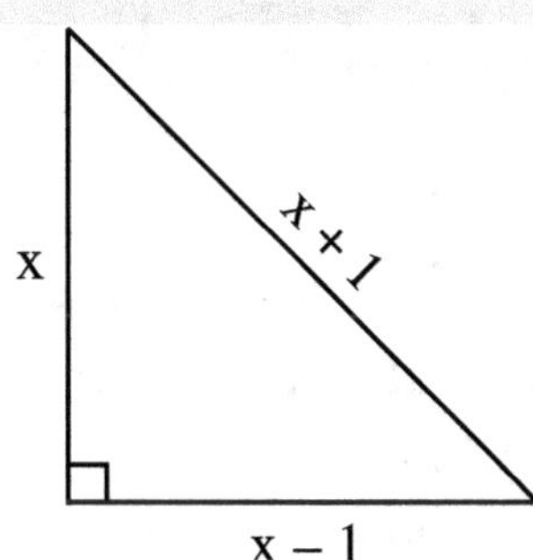

According to Pythagoras Theorem, we have:

$x^2 + (x - 1)^2 = (x + 1)^2$

$\Rightarrow \quad x^2 + x^2 - 2x + 1 = x^2 + 2x + 1$

This gives $x^2 - 4x = 0$

$\Rightarrow \quad x(x - 4) = 0$ i.e., $x = 0$ or $x = 4$

Since, with $x = 0$ the triangle is not possible; hence $x = 4$.

$\therefore \quad$ Sides, are $x - 1$, x and $x + 1 = 4 - 1$

i.e., 3 cm 4 cm and 5 cm

Q. 7. **If the perimeter of a rectangular plot is 68 m and its diagonal is 26 m. Find its area.**

Ans. Let the length of plot = x m

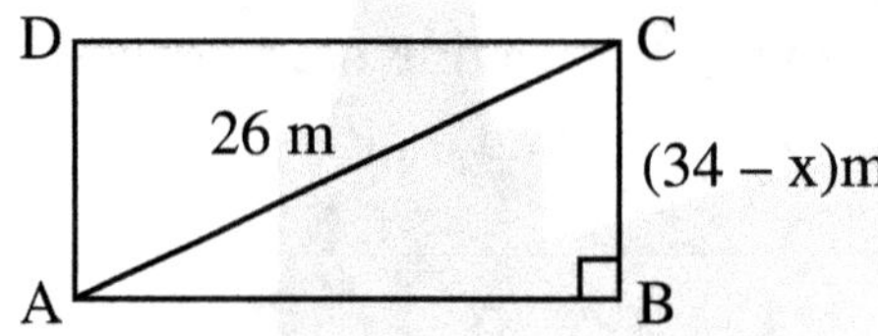

$\ominus \quad$ 2(length + breadth) = perimeter

$\Rightarrow \quad$ 2(x + breadth) = 68

$\Rightarrow \quad$ x + breadth = $\dfrac{68}{2}$ and breadth = (34 – x)m

Given its diagonal = 26 m and we know each angle of the rectangle = 90°.

$\therefore \quad x^2 + (34 - x)^2 = 262$ [Applying Pythagoras Theorem]

$\Rightarrow \quad x^2 + 1156 - 68x + x^2 - 676 = 0$

$\Rightarrow \quad 2x^2 - 68x + 480 = 0$

$\Rightarrow \quad x^2 - 34x + 240 = 0$

i.e., $x^2 - 34x + 240 = 0$

On factorising, we get : $(x - 24)(x - 10) = 0$

i.e., $x = 24$ or $x = 10$

$\qquad x = 24$

$\Rightarrow \quad$ length = 24 m and breadth

$\qquad = (34 - 24)$ m = 10 m

and, $x = 10$

$\Rightarrow \quad$ length = 10 m and breadth

$\qquad = (34 - 10)$ m = 24 m

$\therefore \quad$ Dimensions of the given rectangular plot are 24 m and 10 m.

Hence, its area = length × breadth

$= 24$ m × 10 m = 240 m²

Q. 8. **A train travels a distance of 300 km at a uniform speed. If the speed of the train is increased by 5 km an hour, the journey would have taken two hours less. Find the original speed of the train.**

Ans. Let the original speed of the train be x km/hr.

In 1st case, Distance = 300 km and speed = x km/hr.

$\Rightarrow$ Time taken = $\dfrac{\text{distance}}{\text{speed}} = \dfrac{300}{x}$ hrs.

In 2nd case, Distance = 300 km and speed = $(x + 5)$ km/hr.

$\therefore$ Time taken = $\dfrac{\text{distance}}{\text{speed}} = \dfrac{300}{x + 5}$ hrs.

Given : $\dfrac{300}{x} - \dfrac{300}{x + 5} = 2$

$\Rightarrow \dfrac{300(x + 5) - 300x}{x(x + 5)} = 2$

i.e., $\dfrac{300x + 1500 - 300x}{x^2 + 5x} = 2$

$\Rightarrow 2(x^2 + 5x) = 1500$

$\Rightarrow x^2 + 5x - 750 = 0$

On factorising, we get : $(x + 30)(x - 25) = 0$

i.e., $x = -30$ or $x = 25$

Neglecting $x = -30$; we get $x = 25$

i.e., $x = 25$ km per hour.

Q. 8. **A motor boat, whose speed is 15 km/hr in still water, goes 30 km downstream and comes back in a total of 4 hours 30 minutes. Determine the speed of the stream.**

Ans. Let the speed of the stream = x km/hr

$\Rightarrow$ The speed of the boat downstream

$= (15 + x)$ km/hr.

and the speed of the boat upstream

$= (15 - x)$ km/hr

Now, time taken to go 30 km downstream

$= \dfrac{30}{15 + x}$ hrs.

and, time take to come back 30 km upstream

$= \dfrac{30}{15 - x}$ hrs.

Given : the time taken for both the journeys = 4 hours 30 min. = $4\dfrac{1}{2}$ hrs = $\dfrac{9}{2}$ hrs.

$\therefore \dfrac{30}{15 + x} + \dfrac{30}{15 - x} = \dfrac{9}{2}$

$\Rightarrow \dfrac{30(15 - x) + 30(15 + x)}{(15 + x)(15 - x)} = \dfrac{9}{2}$

i.e., $\dfrac{450 - 30x + 450 + 30x}{225 - x^2} = \dfrac{9}{2}$

$\Rightarrow \quad 2 \times 900 = 9\,(225 - x^2)$

On dividing both the sides by 9, we get:

$\quad\quad 2 \times 100 = 225 - x^2$

i.e., $x^2 = 225 - 200 \Rightarrow x^2 = 25$ and $x = \pm\,5$

Rejecting the negative value of x,

we get : x = 5

i.e., the speed of the stream = 5 km/hr

Q. 9. **The hotel bill for a number of people for overnight stay in Rs. 4, 800. If there were 4 people more, the bill each person had to pay would have reduced by Rs. 200. Find the number of people staying overnight.**

Ans. Let the number of people staying overnight be x.

$\therefore$ For x people, the hotel bill = Rs 4,800

$\Rightarrow$ For 1 person, the hotel bill = Rs $\dfrac{4,800}{x}$

When 4 people were more :

For (x + 4) people, the hotel bill = Rs 4,800

$\Rightarrow$ For 1 person, the hotel bill = Rs $\dfrac{4,800}{x + 4}$

It is given that now the bill paid by each person is reduced by Rs 200.

$\therefore \quad \dfrac{4,800}{x} - \dfrac{4,800}{x + 4} = 200$

$\Rightarrow \quad \dfrac{4,800(x + 4) - 4,800x}{x(x + 4)} = 200$

i.e., $200(x^2 + 4x) = 4800x + 19200 - 4800x$

$\Rightarrow \quad x^2 + 4x = \dfrac{19200}{200} = 96$

i.e., $x^2 + 4x - 96 = 0$

On factorising, we get : (x + 12) (x − 8) = 0

i.e., x = − 12 \quad or \quad x = 8

$\ominus$ No. of people can not be negative

$\Rightarrow$ No. of people staying overnight = 8

Q. 10. **In an auditorium, the number of rows was equal to the number of seats in each row. If the number of roots is doubled and the number of seats in each row is reduced by5, then the total number of seats is increased by 375. How many rows were there?**

Ans. Let the number of rows be x

$\Rightarrow$ No. of seats in each rows = x

$\therefore$ The total number of seats in the auditorium

$= x \times x = x^2$

Now, the new no. of rows $= 2x$ and the new no. of seats in each row $= x - 5$

$\therefore$ The new no. of total seats in the auditorium $= 2x(x - 5)$

Given: $2x (x - 5) - x^2 = 375$

$\Rightarrow$ $2x^2 - 10x - x^2 = 375$ and $x^2 - 10x - 375 = 0$

On factorising, we get : $(x - 25) (x + 15) = 0$

i.e., $x = 25$ or $x = -15$

Neglecting $x = -15$, we get : no. of rows $= 25$

Q. 11. **Two years ago, a man's age was three times the square of his son's age. In three years time, his age will be four times his son's age. Find their present ages.**

Ans. Let persent age of son $= x$ years

Two years ago : The age of son was $(x - 2)$ years and so the age of the man was $3(x - 2)^2$

$\therefore$ Man's present age $= 3 (x - 2)^2 + 2$

$= 3(x^2 - 4x + 4) + 2 = 3x^2 - 12x + 14$

In 3 years time : The age of son will be $(x + 3)$ years and the age of man will be

$(3x^2 - 12x + 14) + 3 = 3x^2 - 12x + 17$ years

Given : $3x^2 - 12x + 17 = 4 (x + 3)$

$\Rightarrow$ $3x^2 - 12x + 17 = 4x + 12$ i.e., $3x^2 - 16x + 5 = 0$

On factorising, we get : $(x - 5) (3x - 1) = 0$

i.e., $x = 5$ or $x = \dfrac{1}{3}$

Since, $x = \dfrac{1}{3}$ is not possible ; $x = 5$

$\therefore$ The present age of amn $= 3x^2 - 12x + 14$

$= 3 \times 5^2 - 12 \times 5 + 14 = 29$ years.

And, the present age of son $= x = 5$ years

Q. 12. **Find the nature of the roots of the equation $(b + c) x^2 - (a + b + c) x + a = 0$, $(a,b,c \in Q.)$?**

Ans. The discriminant of the equation is $(a + b + c)^2 - 4(b + c) (a)$

$= a^2 + b^2 + c^2 + 2ab + 2bc + 2ca - 4(b + c)a$

$= a^2 + b^2 + c^2 + 2ab + 2bc + 2ca - 4ab - 4ac$

$= a^2 + b^2 + c^2 - 2\,ab + 2\,bc - 2\,ca$

$(a - b - c)^2 > 0$

So roots are rational and different.

Q. 13. **If the product of the roots of the quadratic equation $mx2 - 2x + (2m - 1) = 0$ is 3 then find the alue of m is -**

Ans. Product of the roots $c/a = 3 = \dfrac{2m - 1}{m}$

$\therefore$ $3m - 2m = -1 \Rightarrow m = -1$

Q. 14. If the equation $(k-2)x^2 - (k-4)x - 2 = 0$ has difference of roots as 3 then find the value of k.

Ans. $(\alpha - \beta) = \sqrt{(\alpha + \beta)^2 - 4\alpha\beta}$

Now $\alpha + \beta = \dfrac{(k-4)}{(k-2)}, \ \alpha\beta = \dfrac{-2}{k-2}$

$\therefore \quad (\alpha - \beta) = \sqrt{\left(\dfrac{k-4}{k-2}\right)^2 + \dfrac{8}{(k-2)}}$

$\qquad \qquad = \dfrac{\sqrt{k^2 + 16 - 8k + 8(k-2)}}{(k-2)}$

$\Rightarrow \quad 3 = \dfrac{\sqrt{k^2 + 16 - 8k + 8k - 16}}{(k-2)}$

$\Rightarrow \quad 3k - 6 = \pm k$

$\therefore \quad k = 3, 3/2$

Q. 15. If α, β are roots of the equation $ax^2 + bx + c = 0$ then find the value of $\dfrac{1}{(a\alpha + b)^2} + \dfrac{1}{(a\beta + b)^2}$.

Ans. Since α, β are the root of the $ax^2 + bx + c$ then $a\alpha^2 + b\alpha + c = 0$

$\Rightarrow \quad \alpha(a\alpha + b) + c = 0$

$\Rightarrow \quad (a\alpha + b) = -c / \alpha$ ...(1)

Similarly

$(a\beta + b) = -c / \beta$ ...(2)

$\therefore \quad \dfrac{1}{(a\alpha + b)^2} + \dfrac{1}{(a\beta + b)^2} = \dfrac{1}{(-c/\alpha)^2} + \dfrac{1}{(-c/\beta)^2}$

$\Rightarrow \quad \dfrac{\alpha^2}{c^2} + \dfrac{\beta^2}{c^2} = \dfrac{\alpha^2 + \beta^2}{c^2} = \dfrac{(\alpha + \beta)^2 - 2\alpha\beta}{c^2}$

$\qquad \qquad = \dfrac{b^2/a^2 - 2c/a}{c^2} = \dfrac{b^2 - 2ac}{a^2 c^2}$

Q. 16. Find the quadratic equation with rational coefficients whose one root is $2 \pm \sqrt{3}$.

Ans. The required equation is

$x^2 - \{(2 + \sqrt{3}) + (2 - \sqrt{3})\} x + (2 + \sqrt{3})(2 - \sqrt{3}) = 0$

or $\quad x^2 - 4x + 1 = 0$

Q. 17. If α, β are the root of a quadratic equation $x^2 - 3x + 5 = 0$ then find the equation whose roots are $(\alpha^2 - 3\alpha + 7)$ and $(\beta^2 - 3\beta + 7)$.

Ans. Since α, β are the roots of equation $x^2 - 3x + 5 = 0$

So $\quad \alpha^2 - 3\alpha + 5 = 0 \ \& \ \beta^2 - 3\beta + 5 = 0$

$\therefore \quad \alpha^2 - 3\alpha = -5 \ \& \ \beta^2 - 3\beta = -5$

putting in $(\alpha^2 - 3\alpha + 7)$ & $(\beta^2 - 3\beta + 7)$ ...(1)

$-5 + 7, -5 + 7$

2 and 2 are the roots the required equation is $x^2 - 4x + 4 = 0$.

Q. 18. **Find the nature of both roots of the equation (x–b) (x–c) + (x–c) (x–a) + (x–a) (x–b) = 0.**

Ans. The given equation can be written in the following form:

$3x^2 - 2(a + b + c)x + (ab + bc + ca) = 0$

Here discriminant

$= 4(a + b + c)^2 - 12(ab + bc + ca)$

$= 4[(a^2 + b^2 + c^2) - (ab + bc + ca)] > 0$

$$[a^2 + b^2 + c^2 > ab + bc + ca]$$

$\therefore$ Both roots are real.

Q. 19. **If one root of the equation $4x^2 + 2x - 1 = 0$ is α, then find the other root.**

Ans. Let α and β are roots of the given equation, then

$$\alpha + \beta = -\frac{1}{2} \quad \Rightarrow \beta = -\frac{1}{2} - \alpha$$

Now $4\alpha^2 + 2\alpha - 1 = 0$

$$4\alpha^2 = 1 - 2\alpha \qquad\qquad ...(1)$$

Now $4\alpha^3 = \alpha - 2\alpha^2$

$$= \quad \alpha - \frac{1}{2}(1 - 2\alpha) \qquad\qquad \text{[from (1)]}$$

$$\therefore \quad 4\alpha^3 - 3\alpha = -2\alpha - \frac{1}{2}(1 - 2\alpha)$$

$$= -\frac{1}{2} - \alpha = \beta$$

VERY SHORT ANSWER TYPE Q.UESTIONS

Q. 1. **Which of the following is not a Q.uadratic Equation?**

(a) $2(x-1)^2 = 4x^2 - 2x + 1$ (b) $3x - x^2 = x^2 + 6$

(c) $\left(\sqrt{3}x + \sqrt{2}\right)^2 = 2x^2 - 5x$ (d) $(x^2 + 2x)^2 = x^4 + 3 + 4x^2$

Sol.

Q. 2. **Which of the following equation has 2 as a root**

(a) $x^2 + 4 = 0$ (b) $x^2 - 4 = 0$

(c) $x^2 + 3x - 12 = 0$ (d) $3x^2 - 6x - 2 = 0$

Sol.

Q. 3. **If $\dfrac{1}{2}$ is a root of $x^2 + px - \dfrac{5}{4} = 0$ then value of p is**

(a) 2 (b) -2 (c) $\dfrac{1}{4}$ (d) $\dfrac{1}{2}$

Sol.

Q. 4. **Every Q.uadratic Equation can have at most**

(a) Three roots   (b) One root   (c) Two roots   (d) Any number of roots

Sol.

Q. 5. **Roots of Q.uadratic equation $x^2 - 7x = 0$ will be**

(a) 7 (b) $0, -7$ (c) $0, 5$ (d) $0, 7$

Sol.

Q. 6. **The value(s) of k for which the quadratic equation $2x^2 + kx + 2 = 0$ has equal roots, is**

(a) 4 (b) ± 4 (c) -4 (d) 0

Sol.

Q. 7. **Fill in the blanks:**

(a) If $px^2 + qx + r = 0$ has equal roots then value of r will be _______ .

(b) The quadratic equation $x^2 - 5x - 6 = 0$ if expressed as $(x + p)\,(x + q) = 0$ then value of p and q respectively are _______ and _______ .

(c) The value of k for which the roots of quadratic equations $x^2 + 4x + k = 0$ are real is _______ .

(d) If roots of $4x^2 - 2x + c = 0$ are reciprocal of each other then the value of c is _______ .

(e) If in a quadratic equation $ax^2 + bx + c = 0$, value of a is zero then it becomes a _______ equation.

Sol.

Q. 8. **Write whether the following statements are true or false. Justify your answer.**

(a) Every quadratic equation has atleast one real roots.

(b) If the coefficient of x^2 and the constant term of a quadratic equation have opposite signs, then the quadratic equation has real roots.

(c) 0.3 is a root of $x^2 - 0.9 = 0$.

(d) The graph of a quadratic polynomial is a straight line.

(e) The discriminant of $(x - 2)^2 = 0$ is positive.

Sol.

Q. 9. **Match the following:**

(i) Roots of $3x^2 - 27 = 0$ (a) 169/9

(ii) D of $2x^2 + \dfrac{5}{3}x - 2 = 0$ (d) 0

(iii) Sum of roots of $8x^2 + 2x - 3 = 0$ (c) $x^2 - (a + b)x + ab = 0$

(iv) A quadratic equation with roots a and b   (d) $3, -3$

(v) The product of roots of $x^2 + 8x = 0$ (e) $\dfrac{-1}{4}$

Sol.

SHORT ANSWER TYPE Q.UESTIONS-I

Q. 10. **If the Q.uadratic equation $px^2 - 2\sqrt{5}\,px + 15 = 0\ (p \neq 0)$ has two equal roots then find the value of p.**

Sol.

Q. 11. **Solve for x by factorisation**

(a) $8x^2 - 22x - 21 = 0$

(b) $3\sqrt{5}x^2 + 25x + 10\sqrt{5} = 0$

Sol.

(c) $3x^2 - 2\sqrt{6}x + 2 = 0$

Sol.

(d) $2x^2 + ax - a^2 = 0$

Sol.

(e) $\sqrt{3}x^2 + 10x + 7\sqrt{3} = 0$

Sol.

(f) $\sqrt{2}x^2 + 7x + 5\sqrt{2} = 0$

Sol.

(g) $(x - 1)^2 - 5(x - 1) - 6 = 0$

Sol.

Q. 12. **For what value of 'a' quadratic equation $3ax^2 - 6x + 1 = 0$ has no real roots?**   **(CBSE 2020)**

Sol.

Q. 13. **If -5 is a root of the quadratic equation $2x^2 + px - 15 = 0$ and the quadratic equation $p(x^2 + x) + k = 0$ has equal roots find the value of k.**   **(CBSE 2014, 2016)**

Sol.

Q. 14. If $x = \dfrac{2}{3}$ and $x = -3$ are roots of the quadratic equation $ax^2 + 7x + b = 0$. Find the value of a and b.

(CBSE 2016)

Sol.

Q. 15. Find value of p for which the product of roots of the quadratic equation $px^2 + 6x + 4p = 0$ is equal to the sum of the roots.

Sol.

Q. 16. The sides of two squares are x cm and (x + 4) cm. The sum of their areas is 656 cm² Find the sides of these two squares.

Sol.

Q. 17. Find K if the difference of roots of the quadratic equation $x^2 - 5x + (3k - 3) = 0$ is 11.

Sol.

SHORT ANSWER TYPE Q.UESTIONS-II

Q. 18. Find the positive value of k for which the quadratic equation $x^2 + kx + 64 = 0$ and the quadratic equation $x^2 - 8x + k = 0$ both will have real roots.

Sol.

Q. 19. Solve for x

(a) $\dfrac{1}{a+b+x} = \dfrac{1}{a} + \dfrac{1}{b} + \dfrac{1}{x}$ $a + b + x \neq 0,$ (CBSE 2005)

$a, b, x \neq 0$

(b) $\dfrac{1}{2a+b+2x} = \dfrac{1}{2a} + \dfrac{1}{b} + \dfrac{1}{2x}$ $2a + b + 2x \neq 0, \ a, b, x \neq 0$

Sol.

(c) $\dfrac{2x}{x-3} + \dfrac{1}{2x+3} + \dfrac{3x+9}{(x-3)(2x+3)} = 0,$ $x \neq 3, \dfrac{-3}{2}$

Sol.

(d) $\dfrac{1}{x-1} - \dfrac{1}{x+5} = \dfrac{6}{7}, x \neq 1, 5$ (CBSE 2010)

Sol.

(e) $4x^2 + 4bx - (a^2 - b^2) = 0$

Sol.

(f) $4x^2 - 2(a^2 + b^2)\, x + a^2 b^2 = 0$

Sol.

(g) $\dfrac{2}{x+1} + \dfrac{3}{2(x-2)} = \dfrac{23}{5x}, x \neq 0, -1, 2$

Sol.

Innovative Mathematics X-4

(h) $\left(\dfrac{2x}{x-5}\right) + \dfrac{10x}{(x-5)} - 24 = 0, x \neq 5$

Sol.

(i) $4x^2 - 4a^2x + a^4 - b^4 = 0$

Sol.

(j) $2a^2x^2 + b(6a^2 + 1)x + 3b^2 = 0$

Sol.

(k) $3\left(\dfrac{7x+1}{5x-3}\right) - 4\left(\dfrac{5x-3}{7x+1}\right) = 11, x \neq \dfrac{3}{5}, \dfrac{-1}{7}$

Sol.

(l) $\dfrac{1}{x+4} - \dfrac{1}{x-7} = \dfrac{11}{30}, x \neq -4, 7$ **(NCERT)**

Sol.

(m) $\dfrac{x-4}{x-5} + \dfrac{x-6}{x-7} = \dfrac{10}{3}, x \neq 5, 7$

Sol.

(n) $\dfrac{1}{x+1} + \dfrac{2}{x+2} = \dfrac{4}{x+4},$, $x \neq -1, -2, -4$

Sol.

(o) $\dfrac{1}{2x-3} + \dfrac{1}{x-5} = 1,$ $x \neq \dfrac{3}{2}, 5$

Sol.

(p) $x^2 + 5\sqrt{5}x - 70 = 0$

(q) $\dfrac{16}{x} - 1 = \dfrac{15}{x+1}, x \neq 0, -1$ **(CBSE 2014)**

Sol.

Q. 20.   Solve by using quadratic formula $abx^2 + (b^2 - ac)x - bc = 0$.   **(CBSE 2005)**

Sol.

Q. 21.   If the roots of the quadratic equation $(p + 1)x^2 - 6(p + 1)x + 3(p + 9) = 0$ are equal find p and then find the roots of this quadratic equation.

Sol.

Q. 22.   Find the nature of roots of the quadratic equation $3x^2 - 4\sqrt{3}x + 4 = 0$

If the roots are real, find them.   **(CBSE 2020)**

Sol.

Q. 23. Solve $9x^2 - 6a^2x + a^4 - b^4 = 0$ using quadratic formula. (CBSE 2020)

Sol.

LONG ANSWER TYPE Q.UESTIONS

Q. 24. A train travels at a certain average speed for a distance of 54 km and then travels a distance of 63 km at an average speed of 6 km/hr more than the first speed. If it takes 3 hours to complete the total journey, what is its first speed?

Sol.

Q. 25. A natural number, when increased by 12, equals 160 times its reciprocal. Find the number.

Sol.

Q. 26. A thief runs with a uniform speed of 100 m/minute. After one minute a policeman runs after the thief to catch him. He goes with a speed of 100 m/minute in the first minute and increases his speed by 10 m/minute every succeeding minute. After how many minutes the policemen will catch the thief?

Sol.

Q. 27. Two water taps together can fill a tank in 6 hours. The tap of larger diameter takes 9 hours less than the smaller one to fill the tank separately. Find the time in which each tap can separately fill the tank. (CBSE 2020)

Sol.

Q. 28. In the centre of a rectangular lawn of dimensions 50 m × 40 m, a rectangular pond has to be constructed, so that the area of the grass surrounding the pond would be 1184 m². Find the length and breadth of the pond.

Sol.

Q. 29. A farmer wishes to grow a 100 m² recangular garden. Since he has only 30 m barbed wire, he fences three sides of the rectangualr garden letting compound wall of this house act as the fourth side fence. Find the dimensions of his garden.

Sol.

Q. 30. A peacock is sitting on the top of a pillar, which is 9 m high. From a point 27 m away from the bottom of a pillar, a snake is coming to its hole at the base of the pillar. Seeing the snake the peacock pounces on it. If their speeds are equal at what distance from the hole is the snake caught?

Sol.

Q. 31. If the price of a book is reduced by ₹ 5, a person can buy 5 more books for ₹ 300. Find the original list price of the book.

Sol.

Q. 32. ₹ 6500 were divided equally among a certain number of persons. If there been 15 more persons, each would have got ₹ 30 less. Find the original number of persons.

Sol.

Q. 33. In a flight of 600 km, an aircraft was slowed down due to bad weather. Its average speed was reduced by 200 km/hr and the time of flight increased by 30 minutes. Find the duration of flight. (CBSE 2020, Outside Delhi)

Sol.

 Innovative Mathematics X-4

Q. 34. A fast train takes 3 hours less than a slow train for a journey of 600 km. If the speed of the slow train is 10 km/hr less than the fast train, find the speed of the two trains.   (CBSE 2020, Outside Delhi)

Sol.

Q. 35. The speed of a boat in still water is 15 km/hr. It can go 30 km upstream and return downstream to the original point in 4 hrs 30 minutes. Find the speed of the stream.

Sol.

Q. 36. Sum of areas of two squares is 400 cm^2. If the difference of their perimeter is 16 cm. Find the side of each square.

Sol.

Q. 37. The area of an isosceles triangle is 60 cm^2. The length of equal sides is 13 cm find length of its base.

Sol.

Q. 38. The denominator of a fraction is one more than twice the numerator. If the sum of the fraction and its reciprocai is $2\dfrac{16}{21}$. Find the fraction.

Sol.

Q. 39. A girl is twice as old as her sister. Four years hence, the product of their ages (in years) will be 160. Find their present ages.

Sol.

Q. 40. A two digit number is such thattheproductofits digits is 18. When 63 is subtracted from the number, the digits interchange their places. Find the number.   (CBSE 2006)

Sol.

Q. 41. Three consecutive positive integers are such that the sum of the square of the first and the product ofothertwo is 46, find the integers.   (CBSE 2010)

Sol.

Q. 42. A piece of cloth costs, ₹ 200. If the piece was 5 m longer and each metre of cloth costs, ₹ 2 less, then the cost of the piece would have remained unchanged. How long is the piece and what is the piece and what is the original rate per metre?

Sol.

Q. 43. A motor boat whose speed is 24 km/hr in still water takes 1 hour more to go 32 km upstream than to return downstream to the same spot. Find the speed of the stream   (CBSE 2016)

Sol.

Q. 44. If the roots of the quadratic equation $(b - c)x^2 + (c - a)x + (a - b) = 0$ are equal, prove $2b = a + c$.

Sol.

Q. 45. If the equation $(1 + m^2)n^2x^2 + 2mncx + (c^2 - a^2) = 0$ has equal roots, prove that $c^2 = a^2 (1 + m^2)$.

Sol.

Q. 46. A train covers a distance of 480 km at a uniform speed. If the speed had been 8 km/hr less, then it would have taken 3 hours more to cover the same distance. Find the original speed of the train.   (CBSE 2020)

Sol.

Q. 47. A rectangular park is to be designed whose breadth is 3 m less than its length. Its area is to be 4 square metres more than the area of a park that has already been made in the shape of an isosceles triangle with its base as the breadth of the rectangular park and of altitude 12 m. Find the length and breadth of the park. **(CBSE 2020)**

Sol.

1. (d)   **2.** (b)   **3.** (a)   **4.** (c)

5. (d)   **6.** (b)

7. (a) $r = g$ (b) $p = -6, q = 1$ (c) $k \leq 4$ (d) $c = 4$

     (e) Linear equation

8. (a) False   (b) True   (c) False   (d) False

     (e) True

9. (i) (d), (ii) (a), (iii) (e), (iv) (c), (v) (b)   **10.** $D = 0$

11. (a) $x = \dfrac{7}{2}, \dfrac{-3}{4}$ (b) $x = -\sqrt{5}, \dfrac{-2\sqrt{5}}{3}$ (c) $x = \sqrt{\dfrac{2}{3}}, \sqrt{\dfrac{2}{3}}$ (d) $x = \dfrac{a}{2}, -a$

     (e) $x = -\sqrt{3}, \dfrac{-7\sqrt{3}}{3}$ (f) $x = -\sqrt{2}, \dfrac{-5\sqrt{2}}{2}$ (g) $x = 0, 7$

12. $D < 0$ **13.** $k = \dfrac{7}{4}$ **14.** $a = 3, b = -6$ **15.** $P = \dfrac{-3}{2}$

16. $x = \dfrac{-4 \pm 36}{2}$ **17.** $k = -7$ **18.** $k = 16$

19. (a) $x = -a, -b$ (b) $x = -a, \dfrac{-b}{2}$ (c) $x = -1, x \neq \dfrac{-3}{2}$ (d) $x = 2, -6$

     (e) $x = \dfrac{-(a+b)}{2}, \dfrac{a-b}{2}$ (f) $x = \dfrac{b^2}{2}, \dfrac{a^2}{2}$ (g) $x = 4, x = \dfrac{-23}{11}$ (h) $x = 15, x = 4$

     (i) $x = \dfrac{a^2 \pm b^2}{2}$ (j) $x = \dfrac{-b}{2a^2}, -3b$ (k) $y = \dfrac{-1}{3}, 4$ (l) $x = 1, 2$

     (m) $x = 8, \dfrac{11}{2}$ (n) $x = 2 \pm 2\sqrt{3}$ (o) $x = \dfrac{-8 \pm 3\sqrt{2}}{2}$ (p) $x = 2\sqrt{5}, -7\sqrt{5}$

     (q) $x = \pm 4$ **20.** $x = \dfrac{-b}{a}, \dfrac{c}{b}$ **21.** $P = 3$ **22.** $\dfrac{2}{\sqrt{3}}, \dfrac{2}{\sqrt{3}}$

23. $x = \dfrac{a^2 \pm b^2}{3}$ **24.** $x = 36, x \neq -3$ **25.** $x = 8, x = (-20)$ Rejected

26. $x = 6, x \neq -3$ **27.** $x = 18, \therefore x < 0 \therefore x = 3$ is Rejected

28. $x = 37, 8$ But $x = 37$ Rejected   **29.** $y = 20$ m, 10 m.   **30.** $x = 12$ m

31. $x = 20, x = -15$ (Rejected)   **32.** $x = 50, x = -65$ (Rejected)

33. 1 hour.   **34.** 40 km/hrs, 50 km/hr.   **35.** 5 km/hr.

36. x = 12 m, y = 16 m   **37.** x = 12, 5   **38.** $\dfrac{3}{7}$

39. x = 6 years, x = –12 (Rejected)   **40.** Number is 92

41. Numbers: 4, 5, 6   **42.** x = 20, x = –25 (Rejected)   Rate = ₹ 10

43. x = 8 km/hr. x = –72 km/hr (Rejected)   **44.** a + c = 2b

45. $c^2 = a^2(1 + m^2)$ **46.** x = 40 km/hr. x = –32 km/hr. (Rejected)

47. Length = 7m, Breadth = 4m.

SECTION-A

Q. 1. The value of k is .... if $x = 3$ is one root of $x^2 - 2kx - 6 = 0$. (1)

Sol.

Q. 2. If the discriminant of $3x^2 + 2x + \alpha = 0$ is double the discriminant of $x^2 - 4x + 2 = 0$ then value of a is .... (1)

Sol.

Q. 3. If discriminant of $6x^2 - bx + 2 = 0$ is 1 then value of b is .... (1)

Sol.

Q. 4. $(x - 1)^3 = x^3 + 1$ is quadratic equation. (T/F) (1)

Sol.

SECTION-B

Q. 5. If roots of $x^2 + kx + 12 = 0$ are in the ratio 1 : 3 find k. (2)

Sol.

Q. 6. Solve for x : $21x^2 - 2x + \dfrac{1}{21} = 0$ (2)

Sol.

Q. 7. Find k if the quadratic equation has equal roots : $kx(x - 2) + 6 = 0$. (2)

Sol.

SECTION-C

Q. 8. Solve using quadratic formula (3)

$$4\sqrt{3}x^2 + 5x - 2\sqrt{3} = 0$$

Sol.

Q. 9. For what value of k, $(4 - k)x^2 + (2k + 4)x + (8k + 1) = 0$ is a perfect square. (3)

Sol.

SECTION-D

Q. 10. Two water taps together can fill a tank in $1\dfrac{7}{8}$ hours. The tap with longer diameter takes 2 hours less than the tap with smaller one to fill the tank separately. Find the time in which each tap can fill the tank separately. (CBSE 2018)

Sol.

CASE STUDY

CASE STUDY 1.

Raj and Ajay are very close friends. Both the families decide to go to Ranikhet by their own cars. Raj's car travels at a speed of x km/h while Ajay's car travels 5 km/h faster than Raj's car. Raj took 4 hours more than Ajay to complete the journey of 400 km.

Q. 1. **What will be the distance covered by Ajay's car in two hours?**

    (a) $2(x + 5)$km   (b) $(x - 5)$km   (c) $2(x + 10)$km   (d) $(2x + 5)$km

Q. 2. **Which of the following quadratic equation describe the speed of Raj's car?**

    (a) $x^2 - 5x - 500 = 0$ (b) $x^2 + 4x - 400 = 0$

    (c) $x^2 + 5x - 500 = 0$ (d) $x^2 - 4x + 400 = 0$

Q. 3. **What is the speed of Raj's car?**

    (a) 20 km/hour   (b) 15 km/hour   (c) 25 km/hour   (d) 10 km/hour

Q. 4. **How much time took Ajay to travel 400 km?**

    (a) 20 hour   (b) 40 hour   (c) 25 hour   (d) 16 hour

ANSWERS

1. (a) $2(x + 5)$km   **2.** (c) 25km/ hour   **3.** (a) 20km/ hour   **4.** (d) 16 hour

CASE STUDY 2:

The speed of a motor boat is 20 km/hr. For covering the distance of 15 km the boat took 1 hour more for upstream than downstream.

 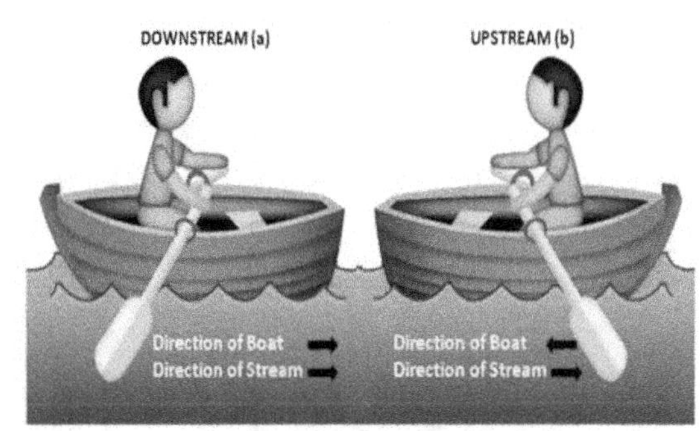

Q. 1. Let speed of the stream be x km/hr. then speed of the motorboat in upstream will be

(a) 20 km/hr   (b) $(20 + x)$ km/hr   (c) $(20 - x)$ km/hr   (d) 2 km/hr

Q. 2. What is the relation between speed ,distance and time?

(a) speed = (distance)/time   (b) distance = (speed)/time

(c) time = speed × distance   (d) speed = distance × time

Q. 3. Which is the correct quadratic equation for the speed of the current ?

(a) $x^2 + 30x - 200 = 0$ (b) $x^2 + 20x - 400 = 0$

(c) $x^2 + 30x - 400 = 0$ (d) $x^2 - 20x - 400 = 0$

Q. 4. What is the speed of current ?

(a) 20 km/hour   (b) 10 km/hour   (c) 15 km/hour   (d) 25 km/hour

Q. 5. How much time boat took in downstream?

(a) 90 minute   (b) 15 minute   (c) 30 minute   (d) 45 minute

ANSWERS

1. (c) $(20 - x)$km/hr
2. (b) distance = (speed)/time
3. (c) $x^2 + 30x - 400 = 0$
4. (b) 10 km/hour
5. (c) 45 minute

CHAPTER-5

ARITHIMATIC PROGRESSIONS

LEARNING OBJECTIVES

Arithmetic progression means patterns or sequence. In this chapter we will see special type of pattern or sequence called Arithmetic progression. (A.P.)

OBJECTIVES

After studying this lesson, you will be able to

* Identify arithmetic progression from a given list of numbers.
* Determine the general term of an arithmetic progression.
* Find the n terms, sum of n terms of an arithmetic progression.

General (n^{th}) term of an A.P.:

Let us consider an A.P. whose first term is 'a' and common difference is 'd'. Let us denote the terms of AP as t_1, t_2, t_3, ... t_n denotes the nth term of the AP. Since first term is a, second term is obtained by adding 'd' to 'a' i.e. a + d, the third term will be obtained by adding 'd' to a + d. So, third therm will be $(a + d) + d = a + 2d$ and so on:

With this

$$t_1 = a$$
$$t_2 = a + d$$
$$t_3 = a + 2d$$
$$t_4 = a + 3d$$
$$t_5 = a + 4d$$
$$t_n = a + (n - 1)\, d$$

where $d = a_2 - a_1$

$\therefore$ When you observe the pattern you will see that

$$t_3 = a + 2d$$

1 less

$$t_4 = a + 3d$$

1 less

$\therefore$

$$t_n = a + (n - 1)d$$

1 less

Q. 1. **Find the 15th and nth term of A.P.**

16, 11, b, 1, – 4, – 9, ...

Ans. Here,

$$a = 16$$
$$d = a_2 - a_1$$
$$= 11 - 16$$

$$= -5$$
$$t_n = a + (n-1)\,d$$
$$t_{15} = a + (15-1)d$$
$$= 16 + 14 \times (-5)$$
$$= 16 - 70$$
$$= -54$$
$$t_n = a + (n-1)d$$
$$= 16 + (n-1)(-5)$$
$$= 16 - 5n + 5$$
$$= 21 - 5n$$

Q. 2. **The first term of an A.P. is – 3 and 12th term is 41. Find c.d.**

Ans.
$$t_{12} = a + (12-1)\,d$$
$$41 = -3 + 11d$$
$$\frac{44}{11} = d$$
$$\therefore \qquad d = 4$$

SUM OF FIRST n TERMS OF AN A.P.:

Carl Friedrich Gauss, the great German mathematician was in elementry school, when his teacher asked the class to find the sum of first 100 natural numbers. While the rest of the class was struggling with the problem, gauss found the answer within no time. How lets see...

$$S = 1 + 2 + 3 + \dots + 99 + 100 \qquad \qquad \dots(1)$$

Also wniting these numbers inreverse order we get,

$$S = 100 + 99 + 98 + \dots + 2 + 1 \qquad \qquad \dots(2)$$

Adding 1 and 2 term by term we get

$$25 = 101 + 101 + 101 + 101 + \dots + 101 \ (100 \ \text{times})$$
$$= 100 \times 101$$
$$\cancel{2}5 = \frac{100 \times 101}{2}$$
$$= \underline{5050}$$

Similarly Lets Derive it:

Let A.P. are

$$a, a+d, a+2d, \dots a+(n-2)d, a+(n-1)d \dots$$
$$S_n = a + (a+d) + (a+2d) + \dots + [a+(n-2)d] + [a+(n-1)d] \qquad \dots(1)$$

Writing these term in reverse order we get,

$$S_n = [a+(n-1)d] + [a+(n-2)d] + \dots + (a+d) + a \qquad \dots(2)$$

Innovative Mathematics X-5

Adding 1 and 2

$$2S_n = [2a + (n-1)d] + [2a + (n-1)d] + ... + [2a + (n-1)d] + [2a + (n-1)d].... \text{ n times}$$

or $\quad 2S_n = n[2a + (n-1)d]$

$$S_n = \frac{n}{2}\big[2a + (n-1)d\big]$$

or $\quad S_n = \frac{n}{2}\big[a + \{a + (n-1)d\}\big]$

$$S_n = \frac{n}{2}\big[a + t_n\big]$$

Sometimes tn is also known as last term denoted by l

Thus, $\quad S_n = \frac{n}{2}\big[a + l\big]$

Q. 1. **Find the sum of the first 12 terms of the AP:**

$-151, -148, -145.....$

Ans. $a = -151,$

$d = -148 - (-151),$

$n = 12$

$= -148 + 151$

$= 3$

$$S_n = \frac{n}{2}\big[2a + (n-1)d\big]$$

$$= \frac{12}{2}\big[2 \times (-151) + (12-1)3\big]$$

$= 6[-302 + 33]$

$= 6 \times (-269)$

$= -1614$

Q. 2. **How many terms of the AP 2, 4, 6, 8, 10..... are needed to get sum 210.**

Ans. Here,

$a = 2,$

$d = 2,$

$S_n = 210$

Using formula,

$$S_n = \frac{n}{2}\big[2a + (n-1)d\big]$$

or, $\quad 210 = \frac{n}{2}\big[2 \times 2 + (n-1)2\big]$

or, $\quad 420 = n[2n + 2]$

or,	$420 = 2n^2 + 2n$

or,	$2n^2 + 2n - 420 = 0$

or,	$n^2 + n - 210 = 0$

or,	$n^2 + 15n - 14n - 210 = 0$

or,	$n(n + 15) - 14(n + 15) = 0$

or,	$(n - 14)(n + 15) = 0$

$n = 14, -15$

Since n cannot be negative so, n = 14.

WARM UP – LEVEL-I

Q. 1. Examine that the list of numbers 3, 10, 7, 4,... form an AP. If it form an AP, write the text two terms.

Ans. Here, we have $\quad a_2 - a_1 = 10 - 13 = -3,$

$$a_3 - a_2 = 7 - 10 = -3,$$

$$a_4 - a_3 = 4 - 7 = -3 \text{ and so on}$$

Since, difference of any two consecutive terms is same, so the given list of numbers form an AP.

Now, the next two terms are $4 + (-3) = 1$ and $1 + (-3) = -2$.

Q. 2. Examine that the list of numbers obtained from following situation, will be in the form of an AP. "Amount left with Sandeep (in ₹) out of the total amount of ₹ 12000 which he had in the beginning, when he spends ₹ 500 in the beginning of every month."

Ans. Given, in the beginning Sandeep had = ₹ 12000

Also, in the beginning of every month, he spend = ₹ 500

So, in the beginning of 2nd month, he had amount, $t_1 = ₹\ 12000$

In the beginning of 2nd month, he had amount,

$$t_2 = 12000 - 500 = ₹\ 11500$$

In the beginning of 3rd month, he had amount,

$$t_3 = 11500 - 500 = ₹\ 11000$$

In the beginning of 4th month, he had amount,

$$t_4 = 11000 - 500 = ₹\ 10500 \text{ and so on.}$$

Now, the list of amounts is 12000, 11500, 11000, 10500, ...

Here, $\quad t_2 - t_1 = t_3 - t_2 = t_4 - t_3 = -500$

i.e. $\quad t_{k+1} - t_k$ is the same everytime.

So, the above list of numbers forms an AP.

Q. 3. Find the common difference of the following AP's.

(i) $3, -2, -7, -12,...$ (ii) $11, 11, 11, 11, ...$

(iii) $5\dfrac{1}{2}, 9\dfrac{1}{2}, 13\dfrac{1}{2}, 17\dfrac{1}{2}, ...$ (iv) $\sqrt{3}, \sqrt{12}, \sqrt{27}, \sqrt{48}, ...$

Ans. (i) Given, AP is $3, -2, -7, -12,...$

Here, $a_1 = 3, a_2 = -2, a_3 = -7, a_4 = -12$ and so on.

$\therefore\quad$ Common difference $(d) = a_2 - a_1 = -2 - 3 = -5$

(ii) Given, AP is $11, 11, 11, 11, ...$

Here, $a_1 = 11, a_2 = 11, a_3 = 11, a_4 = 11$ and so on.

$\therefore\quad$ Common difference $(d) = a_2 - a_1 = 11 - 11 = 0$

(iii) Given, AP is $5\dfrac{1}{2}, 9\dfrac{1}{2}, 13\dfrac{1}{2}, 17\dfrac{1}{2},......$

Here, $a_1 = 5\dfrac{1}{2}, a_2 = 9\dfrac{1}{2}, a_3 = 13\dfrac{1}{2}, a_4 = 17\dfrac{1}{2}$ and so on.

$\therefore$ Common difference (d) $= a_2 - a_1 = 9\dfrac{1}{2} - 5\dfrac{1}{2}$

$$= \dfrac{19}{2} - \dfrac{11}{2} = \dfrac{8}{2} = 4$$

(iv) Given, AP is $\sqrt{3}, \sqrt{12}, \sqrt{27}, \sqrt{48}$ ...

Here, $a_1 = \sqrt{3}$, $a_2 = \sqrt{12} = 2\sqrt{3}$, $a_3 = \sqrt{27} = 3\sqrt{3}$

and $a_4 = \sqrt{48} = 4\sqrt{3}$ and so on.

$\therefore$ Common difference (d) $= a_2 - a_1$

$$= 2\sqrt{3} - \sqrt{3} = \sqrt{3}$$

Q. 4. **Write an AP having 4 as the first term and – 3 as the common difference.**

Ans. Given, first term (a) = 4 and common difference (d) = – 3

On putting the values of a and d in general form

a, a + d, a + 2d, a + 3d,..., we get

$$4, 4 - 3, 4 + 2(- 3), 4 + 3(- 3),...$$
$$4, 1, 4 - 6, 4 - 9, \text{ or }\quad 4, 1, - 2, - 5,...$$

which is the required AP.

Q. 5. **Find the 20th term of the AP:**

$$7, 3, - 1, - 5 \dots .$$

Ans. Given, AP is $7, 3, - 1, - 5,... .$

Here, a = 7 and d = 3 – 7 = – 4.

Since, nth term, $a_n = a + (n - d)d$

On putting n = 20, we get

$$a_{20} = a + (20 - 1)d = 7 + 19(- 4) \qquad\qquad [\because a = 7, d = - 4]$$
$$= 7 - 19 \times 4$$
$$= 7 - 76 = - 69$$

Hence, 20th term of given sequence is – 69.

Q. 6. **How many terms are there in the AP 3, 6, 9, 12, ..., 111?**

Ans. Given, AP is 3, 6, 9, 12, ..., 111.

Here, a = 3 and d = 6 – 3 = 3

Let there be n terms in the given AP.

Then, nth term = 111

$\Rightarrow$ $a + (n - 1)d = 111$ $\qquad\qquad [\because a_n = [a + (n - 1)d]]$

$\Rightarrow$ $3 + (n - 1) \times 3 = 111$

$\Rightarrow$ $3(1 + n - 1) = 111$

$\Rightarrow$ $n = \dfrac{111}{3}$

$\Rightarrow$ $n = 37$

Hence, the given AP contains 37 terms.

Q. 7. **Which term of the AP: 21, 18, 15,... is – 81? Also, is any term 0? Give reason.**

Ans. Given, AP is 21, 18, 15,... .

Here, $a = 21$ and $d = 18 - 21 = -3$

Let nth term of given AP be -81.

Then, $\qquad a_n = -81$

$\Rightarrow \qquad a + (n - 1)d = -81 \qquad\qquad [\because a_n = a + (n - 1)d]$

On putting the values of a and d, we get

$\qquad\qquad 21 + (n - 1)(-3) = -81 \Rightarrow 21 - 3n + 3 = -81$

$\Rightarrow \qquad 24 - 3n = -81$

$\Rightarrow \qquad -3n = -81 - 24 = -105$

$\Rightarrow \qquad n = \dfrac{-105}{-3} = 35$

Hence, 35th term of given AP is -81.

Now, we want to know that if there is any n for which $a_n = 0$.

If such an n is there, then we have

$\qquad 21 + (n - 1)(-3) = 0 \Rightarrow 3(n - 1) = 21$

$\Rightarrow \qquad n - 1 = 7 \Rightarrow n = 8$

Q. 8. **Check whether 200 is a term of the list of numbers 7, 11, 15, 19, ...**

Ans. Given, list of numbers 7, 11, 15, 19, ...

Here, $11 - 7 = 15 - 11 = 19 - 15... = 4$

So, it an AP with first term, $a = 7$ and common difference, $d = 4$

Let 200 be a term, say the nth term of this AP.

We know that, $a_n = a + (n - 1)d$

$\therefore \quad 200 = 7 + (n - 1)(4) \Rightarrow 193 = (n - 1)(4)$

$\Rightarrow \quad n - 1 = \dfrac{193}{4} \Rightarrow n = \dfrac{197}{4} = 49\dfrac{1}{4}$

But, the number of terms cannot be a fraction.

$\therefore \quad$ 200 is not a term of the given AP.

Q. 9. **Which term of the AP 3, 15, 27, 39,... will be 120 more than its 21st term?**

Ans. Given, AP is 3, 15, 27, 39, ...

Here, $a = 3$ and $d = (15 - 3) = 12$

$\therefore \quad$ 21st term is given by

$\qquad T_{21} = a + (21 - 1)d = a + 20d = 3 + 20 \times 12 = 243$

Required term $= (243 + 120) = 363$

Let it be nth term.

Then, $T_n = 363 \Rightarrow a + (n - 1)d = 363$

$\Rightarrow \quad 3 + (n - 1) \times 12 = 363 \Rightarrow 12n = 372 \Rightarrow n = 31$

Hence, 31st term is the required term.

Q. 10. **Determine the general term of an AP whose 7th term is – 1 and 16th term is 17.**

[Here, with the help of given two terms of AP, we will form two equations in a and d and then solve them to find the values of a and d. By sing these values of a and d, write general term of AP.]

Ans. Let a be the first term and d be the common difference of the AP, whose 7th term is – 1 and 16th term is 17.

Since $\qquad a_7 = -1 \quad$ and $\quad a_{16} = 17$

$\therefore$ We have

$$a + (7 - 1)d = -1 \quad \Rightarrow \quad a + 6d = -1 \qquad \qquad \text{...(i)}$$

and $\qquad a + (16 - 1)d = 17 \quad \Rightarrow \quad a + 15d = 17 \qquad \qquad \text{...(ii)}$

$$[\because \ a_n = a + (n - 1)d]$$

On subtracting Eq. (i) from Eq. (ii), we get

$$a + 15d - a - 6d = 17 + 1$$

$\Rightarrow \qquad \qquad 9d = 18 \quad \Rightarrow d = 2$

On substituting d = 2 in Eq. (i), we get

$$a + 6 \times 2 = -1$$

$\Rightarrow \qquad \qquad a + 12 = -1 \quad \Rightarrow \quad a = -13$

Hence, general term, $a_n = a + (n - 1)d$

$$= -13 + (n - 1)2 \qquad \qquad [\because \ a = -13 \text{ and } d = 2]$$

$$= -13 + 2n - 2 = 2n - 15$$

Q. 11. **How many numbers of two digits are visible by 7?**

Ans. Two-digits numbers are 10, 11, 12, 13, 14, 15,..., 97, 98, 99 in which only 14, 21, 28,..., 98 are divisible by 7.

Here, 21 – 14 = 28 – 21... = 7.

So, this list of numbers forms an AP, whose first term (a) = 14, common difference (d) = 7.

Let there are n terms in the above sequence, then $a_n = 98$

$\Rightarrow \qquad \qquad a + (n - 1)d = 98 \qquad \qquad [\because \ a_n = a + (n - 1)d]$

$\Rightarrow \qquad \qquad 14 + (n - 1)7 = 98 \quad \Rightarrow \quad 14 + 7n - 7 = 98$

$\Rightarrow \qquad \qquad 7n = 91 \quad \Rightarrow \quad n = \dfrac{91}{7} = 13$

Hence, 13 numbers of two digits are divisible by 7.

Q. 12. **A sum of ₹ 2000 is invested at 7% simple interest per year. Calculate the interest at the end of each year. Do these interest form an AP? If so, then find the interest at the end of 20th year making use of this fact.**

Ans. Given initial money P = ₹ 2000

Rate of interest, R = 7% per year; Time, T = 1, 2, 3, 4,...

We know that, simple interest is given by the following formula

$$SI = \frac{PRT}{100}$$

$\therefore \quad$ SI at the end of 1st year $= \dfrac{2000 \times 7 \times 1}{100} = ₹ \, 140$

SI at the end of 2nd year $= \dfrac{2000 \times 7 \times 2}{100} = ₹\ 280$

SI at the end of 3rd year $= \dfrac{2000 \times 7 \times 3}{100} = ₹\ 420$

Thus required list of numbers is 140, 280, 420,... .

Here, $280 - 140 = 420 - 280... = 140$

So, above list of numbers forms an AP, whose first term (a) = 140 and common difference (d) = 140.

Now, SI at the end of 20th year will be equal to 20th term of the above AP.

$\therefore \quad a_{20} = a + (20 - 1)d = 140 + 19 \times 140 = 140 + 2660 = 2800$

Hence, the interest at the end of 20th year will be ₹ 2800.

Q. 13. **Determine the 10th term from the end of the AP : 4, 9, 14, ..., 254.**

Ans. Given, AP is 4, 9, 14,....., 254.

Here, $\qquad l = $ last term $= 254$

$\qquad\qquad d = $ common difference $= 9 - 4 = 45$

$\therefore \quad$ 10th term from the end $= l - (10 - 1)d = l - 9d$

$\qquad\qquad\qquad = 254 - 9 \times 5 = 254 - 45 = 209$

Alternate Method

On reversing the given AP, new AP is 254, ..., 14, 9, 4.

Here, first term (a) = 254

and common difference (d) $= - 5$

Now, 10th term of new AP $= a_{10}$

$\qquad\qquad\qquad = 254 + (10 - 1)(- 5)$

$\qquad\qquad\qquad = 254 - 9 \times 5 = 209$

Hence, 10th term from the end of given AP is 209.

Q. 14. **Find the sum of the first 22 terms of the AP : 8, 3, – 2, ...**

Ans. Given, AP is 8, 3, – 2, ...

Here, first term, a = 8

Common difference d $= 3 - 8 = - 5$ and n = 22

$\therefore \quad$ Sum of first n terms, $S_n = \dfrac{n}{2}[2a + (n - 1)d]$

$\therefore \quad$ Sum of first 22 terms, $S_{22} = \dfrac{22}{2}[2 \times 8 + (22 - 1) \times (- 5)]$

$\qquad\qquad\qquad = 11\,[16 + 21 \times (- 5)] = 11\,[16 - 105]$

$\qquad\qquad\qquad = 11\,(- 89) = - 979$

Hence, sum of first 22 terms of an AP is – 979.

WARM UP – LEVEL-II

Q. 1. **Find the sum of first 20 terms of an AP in which a = 1 and 20th term = 58.**

Ans. Given, $a = 1$, $a_{20} = 58$ and $n = 58$.

Now, sum of first 20 terms,

$$S_{20} = \frac{20}{2}(1 + 58)$$

$$\left[\because S_n = \frac{n}{2}(a + a_n)\right]$$

$$= 10\,(59) = 590$$

Q. 2. **Find the sum of first 24 terms of an AP, whose nth term is given by $a_n = 3 + 2n$**

Ans. We have, $\qquad a_n = 3 + 2n$

$\therefore \qquad a_1 = 3 + 2 = 5$

$a_2 = 3 + 2 \times 2 = 7$

$a_3 = 3 + 2 \times 3 = 9$

$a_4 = 3 + 2 \times 4 = 11$

So, list of number becomes 5, 7, 9, 11, ...

Here, $7 - 5 = 9 - 7 = 11 - 9 = 2$ and so on.

So, it form an AP with common difference, $d = 2$

To find S_{24}, we have, $n = 24$, $a = 5$ and $d = 2$

$$\therefore \quad S_{24} = \frac{24}{2}[2 \times 5 + (24 - 1) \times 2]$$

$$\left[\because S_n = \frac{n}{2}[2a + (n-1)d]\right]$$

$$= 12(10 + 46) = 672$$

So, sum of first 24 terms of the list of numbers is 672.

Alternate Method

Given, nth term of an AP, $a_n = 3 + 2n$

Clearly, sum of first 24 terms, (S_{24})

$$= \frac{24}{2}(a + a_{24}) = 12(5 + 51)$$

$$[\because a_1 = 3 + 2 = 5 \text{ and } a_{24} = 3 + 2 \times 24 = 3 + 48 = 51]$$

$$= 12 \times 56 = 672$$

Q. 3. **If the sum of first four terms of an Ap is 40 and that of first 14 terms is 280. Find the sum of its first n terms.**

Ans. Given, $S_4 = 40$ and $S_{14} = 280$

Let a be the first term and d be the common difference of given AP.

Then, $\qquad S_4 = 40 \Rightarrow \dfrac{4}{2}[2a + 3d] = 40$

$$\left[\because S_n = \frac{n}{2}[2a + (n-1)d]\right]$$

$\Rightarrow \qquad 2[2a + 3d] = 40 \Rightarrow 2a + 3d = 20 \qquad \qquad \text{...(i)}$

and $\qquad S14 = 280$

$\Rightarrow \qquad \dfrac{14}{2}[2a + 13d] = 280 \Rightarrow 2a + 13d = 40 \qquad \qquad \qquad \text{...(ii)}$

On subtracting Eq. (i) from Eq. (ii), we get

On subtracting Eq. (i) from Eq. (ii), we get

$$10d = 20 \Rightarrow d = 2$$

On subtracting $d = 2$ in Eq. (i), we get

$$a = 7$$

Now, $\qquad S_n = \dfrac{n}{2}[2a + (n-1)d] = \dfrac{n}{2}[2(7) + (n-1)2]$

$$= \dfrac{n}{2}[14 + 2n - 2] = n[6 + n] = 6n + n^2$$

Hence, the sum of first n terms is $n^2 + 6n$.

Q. 4. **How many terms of the AP 20, $19\dfrac{1}{3}$, $18\dfrac{2}{3}$, ... must be taken, so that their sum is 300?**

Ans. Given, AP is 20, $19\dfrac{1}{3}, 18\dfrac{2}{3}, ...$.

Here, $\qquad a = 20$ and $d = 19\dfrac{1}{3} - 20 = \dfrac{58}{3} - 20$

$$= \dfrac{58 - 60}{3} = \dfrac{-2}{3}$$

Let n terms of given AP be required to get sum 300.

We know that, $S_n = \dfrac{n}{2}[2a + (n-1)\,d]$

$\Rightarrow \quad 300 = \dfrac{n}{2}\left[2(20) + (n-1)\left(\dfrac{-2}{3}\right)\right] \qquad \qquad [\because \ a = 20$ and $d = -2/3]$

$\Rightarrow \quad 600 = n\left[40 - \dfrac{2}{3}n + \dfrac{2}{3}\right] \Rightarrow 600 = \dfrac{1}{3}[120n - 2n^2 + 2n]$

$\Rightarrow \quad 600 \times 3 = 122n - 2n^2 \quad \Rightarrow \quad 1800 + 2n^2 - 122n = 0$

$\Rightarrow \quad 2[n^2 - 61n + 900] = 0 \quad \Rightarrow \quad n^2 - 61n + 900 = 0$

$\Rightarrow \quad n^2 - 36n - 25n + 900 = 0 \Rightarrow n(n - 36) - 25(n - 36) = 0$

$\Rightarrow \quad (n - 36)(n - 25) = 0 \Rightarrow n = 36$ or 25

Since, a is positive and d is negative, so both values of n are possible.

Hence, sum of 25 terms of given AP

= Sum of 36 terms of given AP = 300.

Q. 5. **If S_n, the sum of first n terms of an AP is given by $S_n = 3n^2 - 4n$, find the nth term.**

Ans. Given, $\qquad \qquad S_n = 3n^2 - 4n$

On replacing n by $(n - 1)$ in Eq. (i), we get

$$S_{n-1} = 3(n-1)^2 - 4(n-1)$$

nth term of the AP $a_n = S_n - S_{n-1}$

$$\therefore \quad a_n = (3n^2 - 4n) - [3(n-1)^2 - 4(n-1)]$$
$$\Rightarrow \quad a_n = 3[n^2 - (n-1)^2] - 4[n - (n-1)]$$
$$\Rightarrow \quad a_n = 3[n^2 - n^2 + 2n - 1] - 4[n - n + 1]$$
$$\Rightarrow \quad a_n = 3(2n-1) - 4$$
$$\Rightarrow \quad a_n = 6n - 3 - 4 \Rightarrow a_n = 6n - 7.$$

Thus, the nth term of the AP $= 6n - 7$.

Q. 6. **If the sum of first n terms of an AP is n^2, then find its 10th term.**

Ans. Given, $\qquad S_n = n^2$

We know that,

$$T_n = S_n - S_{n-1} = n^2 - (n-1)^2$$
$$= n^2 - n^2 + 2n - 1$$
$$\Rightarrow \quad T_n = 2n - 1$$

Now, 10th term $= T_{10} = 2 \times 10 - 1 = 20 - 1 = 19$

Q. 7. **Find the sum of all the two digit numbers which leave the remainder 2 when divided by 5.**

Ans. The sequence of two digit number which divided by 5 and leave the remainder 2 is

12, 17, 22, ..., 97 which is an A.P

Here, $a = 12$, $d = 17 - 12 = 5$ and $l = 97$

$$\therefore \qquad\qquad l = a + (n-1)\,d$$
$$\therefore \qquad\qquad 97 = 12 + (n-1)5$$
$$\Rightarrow \qquad\qquad 85 = (n-1)\,5$$
$$\Rightarrow \qquad\qquad (n-1) = 17 \Rightarrow n = 17 + 1 = 18$$

$\therefore$ required sum of two digit number which divided by 5 leave the remainder $= \dfrac{n}{2}(a+l)$

$$= \frac{18}{2}(12 + 97) = 9 \times 109 = 981$$

Q. 8. **A man repays a loan of ₹ 3250 by ₹ 20 in the first month and then increases the payment by ₹ 15 every month. How long will it take him to clear the loan?**

Ans. Given, total amount of loan $= ₹\,3250$

Amount paid in first month $= ₹\,20$

and amount increase every month $= ₹\,15$

Clearly, the amounts of repayment form an AP with first term, $(a) = 20$ and common difference, $(d) = 15$

Let the loan be cleared in n months.

Then, $\qquad\qquad S_n = 3250$

$$\Rightarrow \qquad\qquad \frac{n}{2}[2a + (n-1)\,d] = 3250$$

$$\Rightarrow \qquad\qquad \frac{n}{2}[2(20) + (n-1)15] = 3250$$

$\Rightarrow \qquad n[40 + 15n - 15] = 3250 \times 2$

$\Rightarrow \qquad 25n + 15n^2 = 6500$

$\Rightarrow \qquad 3n^2 + 5n - 1300 = 0 \qquad\qquad\qquad\qquad$ [dividing both sides by 5]

$\Rightarrow \qquad 3n^2 + 65n - 60n - 1300 = 0$

$\Rightarrow \qquad n(3n + 65) - 20(3n + 65) = 0$

$\Rightarrow \qquad (n - 20)(3n + 65) = 0$

$\Rightarrow \qquad n = 20 \text{ or } n = -\dfrac{65}{3}$

Since, n should be a positive integer, so neglect $n = -\dfrac{65}{3}$.

$\therefore \qquad n = 20$

Q. 9. Kanika was given her pocket money on Jan 1st, 2008. She puts ₹ 1 on day 1, ₹ 2 on day 2, ₹ 3 on day 3 and continued doing so till the end of the month, from this money into her piggy bank. She also spent ₹ 204 of her pocket money and found that at the end of the month she still had ₹ 100 with her. How much was her pocket money for the month?

Ans. Let her pocket money be ₹ x.

Now, she takes ₹ 1 on day 1, ₹ 2 on day 2, ₹ 3 on day 3 and so on till the end of the month, from this money.

Clearly, the amounts that she takes every day of the month, form an AP, in which number of terms is 31, first term (a) = 1 and common difference(d) = 2 – 1 = 1.

Now, sum of first 31 terms, $(S_{31}) = \dfrac{31}{2}[2 \times 1 + (31 - 1) \times 1]$

$$\left[\because \text{ sum of n terms}, (S_n) = \dfrac{n}{2}[2a + (n - 1)\, d]\right]$$

$$= \dfrac{31}{2}(2 + 30) = \dfrac{31 \times 32}{2}$$

$$= 31 \times 16 = 496$$

So, Kanika takes ₹ 496 till the end of the month from her pocket money.

Also, she spent ₹ 204 of her pocket money and found that at the end of the month, she still has ₹ 100 with her.

According to the above condition, we have

$$(x - 496) - 204 = 100$$

$\Rightarrow \qquad x - 700 = 100 \Rightarrow x = ₹\,800$

Hence, ₹ 800 was her pocket money for the month.

Q. 10. Find the arithmetic mean of 2 and 12.

Ans. Let the arithmetic mean of 2 and 12 be b.

Then, $\qquad b = \dfrac{2 + 12}{2} = \dfrac{14}{2}$

$\therefore \qquad b = 7$

Q. 11. Find four numbers in AP whose sum is 20 and the sum of whose squares is 120.

Ans. Let the numbers be a – 3d, a – d, a + d and a + 3d. Then, according to the given condition, we have

$$(a - 3d) + (a - d) + (a + d) + (a + 3d) = 20 \qquad \text{...(i)}$$

and $(a - 3d)^2 + (a - d)^2 + (a + d)^2 + (a + 3d)^2 = 120$ ...(ii)

From Eq. (i), we get
$$4a = 20 \Rightarrow a = 5$$
From Eq. (ii), we get
$$a^2 + 9d^2 - 6da + a^2 + d^2 - 2ad + a^2 + d^2 + 2ad + a^2 + 9d^2 + 6ad = 120$$
$$\Rightarrow \qquad 4a^2 + 20d^2 = 120$$
$$\Rightarrow \qquad a^2 + 5d^2 = 30$$
$$\Rightarrow \qquad 25 + 5d^2 = 30 \qquad\qquad [\because a = 5]$$
$$\Rightarrow \qquad 5d^2 = 5 \Rightarrow d2 = 1 \Rightarrow d = \pm 1$$

If $d = 1$, then the numbers are 2, 4, 6, 8 and if $d = -1$, then the numbers are 8, 6, 4, 2.

Hence, the numbers are 2, 4, 6, 8 or 8, 6, 4, 2.

Q. 12. **Which term of the AP : 121, 117, 113, ... is its first negative term?**

[Hint. Find n for which $a_n < 0$]

Ans. Given, AP is 121, 117, 113, ...

Here, first term, $a = 121$

and common difference, $d = 117 - 121 = -4$

Let, the nth term of this AP be the first negative term. Then, $a_n < 0$

$\Rightarrow \quad a + (n - 1) d < 0 \Rightarrow 121 + (n - 1) (-4) < 0$

$\Rightarrow \quad 125 - 4n < 0 \Rightarrow 125 < 4n \Rightarrow 4n > 125$

$$\Rightarrow \qquad n > \frac{125}{4} \quad \Rightarrow \quad n > 31\frac{1}{4}$$

So, least integral value of n is 32.

Hence, 32nd term of the given AP is the first negative term.

Q. 13. **The sum of the third and the seventh terms of an AP is 6 and their product is 8. Find the sum of first sixteen terms of the AP.**

Ans. Let the first term and the common difference of the given AP be a and d, respectively.

According to the question,

Third term + Seventh term = 6

$\Rightarrow \qquad [a + (3 - 1)d] + [a + (7 - 1)d] = 6 \qquad [\because a_n = a + (n - 1) \, d]$

$\Rightarrow \qquad (a + 2d) + (a + 6d) = 6$

$\Rightarrow \qquad a + 4d = 3 \qquad\qquad\qquad \text{...(i)}$

and Third term $\times$ Seventh term = 8

$\Rightarrow \qquad (a + 2d) (a + 6d) = 8$

$\Rightarrow \qquad \{(a + 4d) - 2d\} \{(a + 4d) + 2d\} = 8$

$\Rightarrow \qquad (3 - 2d) (3 + 2d) = 8 \qquad\qquad \text{[using Eq. (i)]}$

$\Rightarrow \qquad 9 - 4d^2 = 8 \qquad\qquad [\because (a - b) (a + b) = a^2 - b^2]$

$\Rightarrow \qquad 4d^2 = 9 - 8$

$$\Rightarrow \qquad d^2 = \frac{1}{4}$$

$$\Rightarrow \qquad d = \pm \frac{1}{2}$$

When, $d = \dfrac{1}{2}$, then from Eq. (i), we get

$$a + 4\left(\dfrac{1}{2}\right) = 3$$

$$\Rightarrow \qquad a + 2 = 3$$
$$\Rightarrow \qquad a = 3 - 2 \quad \Rightarrow \quad a = 1$$

Now, sum of first sixteen terms of the AP,

$$S_{16} = \dfrac{16}{2}\left[2a + (16 - 1)\,d\right] \qquad \left[\because S_n = \dfrac{n}{2}[2a + (n - 1)\,d]\right]$$

$$= 8[2a + 15d] = 8\left[2(1) + 15\left(\dfrac{1}{2}\right)\right] \qquad \left[\because a = 1, d = \dfrac{1}{2}\right]$$

$$= 8\left(2 + \dfrac{15}{2}\right) = 8\left(\dfrac{19}{2}\right) = 76$$

When, $d = \dfrac{1}{2}$, then from Eq. (i), we get

$$a + 4\left(-\dfrac{1}{2}\right) = 3 \quad \Rightarrow \quad a - 2 = 3 \quad \Rightarrow \quad a = 5$$

Now, sum of first sixteen terms of this AP,

$$S_n = \dfrac{16}{2}\left[2a + (16 - 1)\,d\right] \qquad \left[\because S_n = \dfrac{n}{2}\{2a + (n - 1)d\}\right]$$

$$= 8\,(2a + 15d)$$

$$= 8\left[2(5) + 15\left(-\dfrac{1}{2}\right)\right] = 8\left(10 - \dfrac{15}{2}\right) = 8\left(\dfrac{5}{2}\right) = 20$$

Hence, sum of first sixteen terms, $S_{16} = 20$ or 76.

Q. 14. **A ladder has rungs 25 cm apart (see figure). The rungs decrease uniformly in length from 45 cm at the bottom to 25 cm at the top. If the top and the bottom rungs are $2\dfrac{1}{2}$ m apart, then what is the length of the wood required for the rungs?**

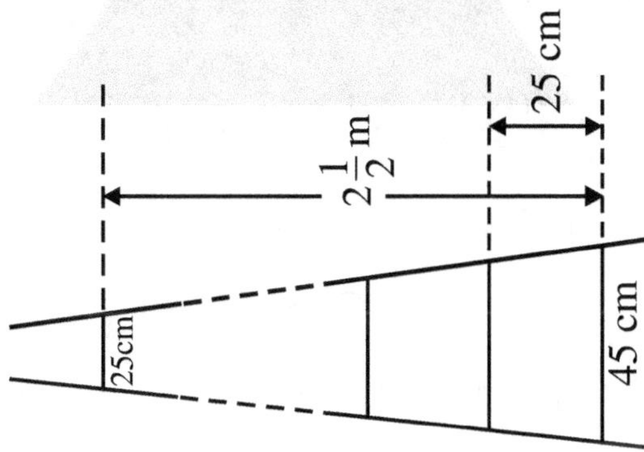

[**Hint.** Number of rungs = $\dfrac{\text{Distance between top and bottom rungs}}{\text{Distance betwenn two rungs}} + 1$]

Ans. According to the question,

Number of rungs

$$= \frac{\text{Distance between top and bottom rungs}}{\text{Distance between two rungs}} + 1$$

$$= \frac{2\frac{1}{2}\,\text{m}}{25\,\text{cm}} + 1 = \frac{\frac{5}{2} \times 100\,\text{cm}}{25\,\text{cm}} + 1 \qquad\qquad [\because 1\,\text{m} = 100\,\text{cm}]$$

$$= \frac{250\,\text{cm}}{25\,\text{cm}} + 1 = 10 + 1 = 11$$

Hence, there are 11 rungs.

The length of the wood required for the rungs

= Sum of length of 11 rungs.

$$= \frac{11}{2}(25 + 45) \qquad \left[\begin{array}{l} \because \text{ length of rungs forms an AP with first term} \\ (a) = 25 \text{ and last term}(1) = 45 \text{ and } S_n = \frac{n}{2}(a + 1) \end{array}\right]$$

$$= \frac{11}{2} \times 70 = 11 \times 35 = 385\,\text{cm}$$

Hence, the length of the wood required for the rungs is 385 cm.

Q. 15. **The houses of a row are numbered consecutively from 1 to 49. Show that there is a value of x such that the sum of the numbers of the houses preceding the house numbered x is equal to the sum of the numbers of the houses following it. Find the value of x. [Hint. $S_{x-1} = S_{49} - S_x$]**

Ans. The numbers on the houses of a row are 1, 2, 3, ..., 49. Clearly, this list of numbers forms an AP with

$$a = 1 \text{ and } d = 2 - 1 = 1$$

According to the question, we have

$$S_{x-1} = S_{49} - S_x \qquad\qquad\qquad ...(i)$$

$$\because \quad S_n = \frac{n}{2}[2a + (n-1)\,d]$$

$$\therefore \quad S_{x-1} = \frac{x-1}{2}[2 \times 1 + (x - 1 - 1) \times 1]$$

$$= \frac{x-1}{2}(2 + x - 2) = \frac{(x-1)\,x}{2} = \frac{x^2 - x}{2}$$

$$S_x = \frac{x}{2}[2 \times 1 + (x-1) \times 1] = \frac{x}{2}(x + 1) = \frac{x^2 + x}{2}$$

and $\quad S_{49} = \dfrac{49}{2}[2 \times 1 + (49 - 1) \times 1] = \dfrac{49}{2}[2 + 48]$

$$= \dfrac{49}{2} \times 50 = 49 \times 25 = 1225$$

Now, on substituting above values in Eq(i), we get

$$\dfrac{x^2 - x}{2} = 1225 - \dfrac{x^2 + x}{2}$$

$$\Rightarrow \qquad \dfrac{x^2 - x}{2} + \dfrac{x^2 + x}{2} = 1225$$

$$\Rightarrow \qquad \dfrac{x^2 - x + x^2 + x}{2} = 1225$$

$$\Rightarrow \qquad x_2 = 1225$$
$$\Rightarrow \qquad x = + 35$$

Since, x is a counting number.

$\therefore \quad$ Taking positive sign, we get x = 35.

Q. 16. **A small terrace at a football ground comprises of 15 steps each of which is 50 m long and built of solid concrete. Each step has a rise of $\dfrac{1}{4}$ m and a tread of $\dfrac{1}{2}$ m (see figure). Calculate the total volume of concrete required to build the terrace.**

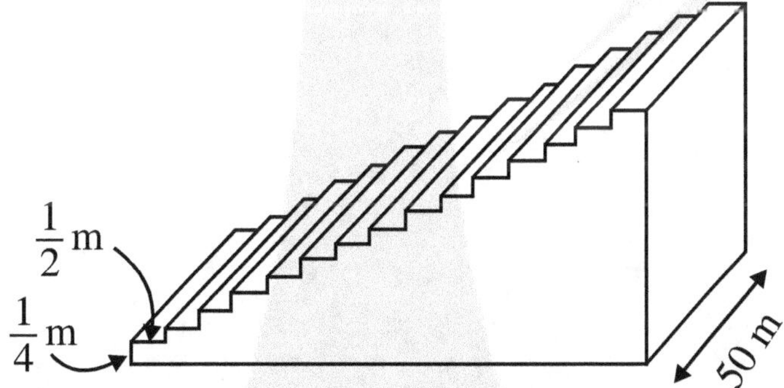

[Hint. Volume of concrete required to build the first step $= \dfrac{1}{4} \times \dfrac{1}{2} \times 50 \text{ m}^3$]

Ans. Clearly, volume of concrete required to build the I step, II step, III step,.. are respectively,

$$\dfrac{1}{4} \times \dfrac{1}{2} \times 50,$$

$$\left(2 \times \dfrac{1}{4}\right) \times \dfrac{1}{2} \times 50, \left(3 \times \dfrac{1}{4}\right) \times \dfrac{1}{2} \times 50, \dots ,$$

i.e. $\qquad \dfrac{50}{8}, 2 \times \dfrac{50}{8}, 3 \times \dfrac{50}{8}, \dots$

Now, total volume of concrete,

$$V = \frac{50}{8} + 2 \times \frac{50}{8} + 3 \times \frac{50}{8} + \ldots = \frac{50}{8}[1 + 2 + 3 + \ldots]$$

Note that the numbers in the bracket forms an AP with first term (a) = 1, common difference (d) = 2 – 1 = 1 and number of terms (n) = 15

$$\therefore \quad V = \frac{50}{8} \times \frac{15}{2}[2 \times 1 + (15 - 1) \times 1] \qquad \left[\because S_n = \frac{n}{2}[2a + (n - 1)d]\right]$$

$$= \frac{50}{8} \times \frac{15}{2} \times (2 + 14) = \frac{25 \times 15}{8} \times 16 = 750 \text{ m}^3$$

Hence, the total volume of concrete required to build the terrace is 750 m³.

VERY SHORT ANSWER TYPE Q.UESTIONS

Q. 1. Find 5th term of an A.P. whose n^{th} tenn is $3n - 5$

Sol.

Q. 2. Find the sum of first 10 even numbers.

Sol.

Q. 3. Write the nth term of odd numbers.

Sol.

Q. 4. Write the sum of first n natural numbers.

Sol.

Q. 5. Write the sum of first n even numbers.

Sol.

Q. 6. Find the nth term of the A.P. $- 10, - 15, - 20, - 25, \ldots\ldots$

Sol.

Q. 7. Find the common difference of A.P. $4\dfrac{1}{9}, 4\dfrac{2}{9}, 4\dfrac{1}{3}, \ldots\ldots$

Sol.

Q. 8. Write the common difference of an A.P. whose n^{th} term is $a_n = 3n + 7$

Sol.

Q. 9. What will be the value of $a_8 - a_4$ for the following A.P.

$4, 9, 14, \ldots\ldots, 254$

Sol.

Q. 10. What is value of a_{16} for the A.P. $- 10, - 12, - 14, - 16, \ldots\ldots$

Sol.

Q. 11. $3, k - 2, 5$ are in A.P. find k.

Sol.

Q. 12. For what value of p, the following terms are three consecutive terms of an A.P. $\dfrac{4}{5}, p, 2.$

Sol.

Q. 13. Detennine the 36th term of the A.P. whose first two tenns are -3 and 4 respectively.

Sol.

Q. 14. Multiple Choice Q.uestions:

    (a) 30th term of the A.P. $10, 7, 4 \ldots$ is

        (a) 97   (b) 77   (c) -77 (d) -87

    (b) 11th term of an A.P. $-3, \dfrac{1}{2}, \,, \ldots$ is

        (a) 28   (b) 22   (c) -38 (d) $-48\dfrac{1}{2}$

    (c) In an A.P. if $d = -4, n = 7, a_n = 4$, then a is

        (a) 6   (b) 7   (c) 120   (d) 28

    (d) The first three terms of an A.P. respectively are $3y - 1, 3y + 5$ and $5y + 1$ then y equals:

      (a) –3   (b) 4   (c) 5   (d) 2

(e)   The list of numbers $-10, -6, -2, 2, \ldots$ is

      (a) An A.P. with $d = -16$ (b) An A.P. with $d = 4$

      (c) An A.P. with $d = -4$ (d) Not an A.P.

(f)   The 11th term from the last term of an A.P. $10, 7, 4, \ldots, -62$ is   **(NCERT)**

      (a) 25   (b) –32   (c) 16   (d) 0

(g)   The famous mathematidan associated with finding the sum of the first 100 natural numbers is

      (a) Pythagoras   (b) Newton   (c) Gauss   (d) Euclid

(h)   What is the common difference of an A.P. in which $a_{18} - a_{14} = 32$?

      (a) 8   (b) – 8   (c) – 4   (d) 4

(i)   The nth term of the A.P. $\left(1+\sqrt{3}\right), \left(1+2\sqrt{3}\right), \left(1+3\sqrt{3}\right), \ldots$ is

      (a) $1 + n\sqrt{3}$ (b) $n + \sqrt{3}$ (c) $n\left(1+\sqrt{3}\right)$ (d) $n\sqrt{3}$

(j)   The common difference of the A.P. $\sqrt{2}, 2\sqrt{2}, 3\sqrt{2}, 4\sqrt{2} \ldots$ is

      (a) $\sqrt{2}$ (b) 1   (c) $2\sqrt{2}$ (d) $-\sqrt{2}$

(k)   The first term of an A.P. is p and the common difference is q, then its 10th term is

      (a) $a + 9p$ (b) $p - 9q$ (c) $p + 9q$ (d) $2p + 9q$

Sol.

Q. 15.   **Match the following:**

          Column A   **Column B**

(a) $a = -18$, $n = 10$, $d = 2$ then a_n of A.P.   (a) $\dfrac{a+c}{2}$

(b) a, b and c are in A.P. then their Arithmetic mean is   (b) 0

(c) If 2, 4, 6, are in A.P. then 4, 8, 12 will also be an   (c) – 41

(d) If $a_n = 9 - 5n$ of an A.P. then a_{10} will be   (d) 8

(e) If $d = -2$, $n = 5$ and $a_n = 0$ in A.P. then a is   (e) A.P.

Sol.

Q. 16.   **State True/False and justify**

(a) 301 isatennofanA.P. 5, 11, 17, 23 ...   **(NCERT)**

(b) Difference of mth and n^{th} term of an A.P. $= (m - n)\, d$.

(c) 2, 5, 9, 14, ... is an A.P.

(d) Sum of first 20 natural numbers is 410.

(e) n^{th} term of an A.P. 5, 10, 15, 20 .... n terms and n^{th} term of A.P. 15, 30, 45, 60, ... n terms are same.

Sol.

SHORT ANSWER TYPE Q.UESTIONS

Q. 17.   **Is 144 a term of the A.P. 3, 7, 11, ... ? Justify your answer.**

Sol.

Q. 18. Show that $(a - b)^2$, $(a^2 + b^2)$ and $(a + b^2)$ are in A.P.

Sol.

Q. 19. Which term of the A.P. 5, 15, 25, ... will be 130 more than its 31st term?

Sol.

Q. 20. The first term, common difference and last term of an A.P. are 12, 6 and 252 respectively, Find the sum of all terms of this A.P.

Sol.

Q. 21. Find the sum of first 15 multiples of 8.

Sol.

Q. 22. Is the sequence formed in the following situations an A.P.

    (i) Number of students left in the school auditorium from the· total strength of IQ.00 students when they leave the auditorium in batches of 25.

    (ii) The amount of money in the account every year when Rs. 100 are deposit annually to accumulate at compound interest at 4% per annum.

Sol.

Q. 23. Find the sum of even positive integers between I and 200.

Sol.

Q. 24. If $4m + 8$, $2m^2 + 3m + 6$, $3m^2 + 4m + 4$ are three consecutive terms of an A.P. find m.

Sol.

Q. 25. How many terms oftbe A.P. 22, 20, 18, ... should be taken so that their sum is zero.

Sol.

Q. 26. If 10 times of 10th term is equal to 20 times of 20th term of an A.P. Find its 30th term.

Sol.

Q. 27. Solve $1 + 4 + 7 + 10 + ... + x = 287$ (CBSE 2020)

Sol.

Q. 28. In an A.P., the sum of first n terms is $\dfrac{3n^2}{2} + \dfrac{5n}{2}$. Find its 25th terms.   (NCERT)

Sol.

Q. 29. Find how many two digit numbers are divisible by 6?   (CBSE 2011)

Sol.

Q. 30. If $\dfrac{1}{x + 2}, \dfrac{1}{x + 3}$ are $\dfrac{1}{x + 5}$ in A.P. find x.   (CBSE 2011)

Sol.

Q. 31. Find the middle term of an A.P. –6, –2, 2, ... 58.   (CBSE 2011)

Sol.

Q. 32. In an A.P. find S_n, where $a_n = 5n - 1$. Hence find the sum of the first 20 terms.   (CBSE 2011)

Sol.

Q. 33. Which term of A.P. 3, 7, 11, 15 .... is 79? Also find the sum $3 + 7 + 11 + ... + 79$.   (CBSE 2011C)

Sol.

Q. 34. Which term of the A.P. : 121, 117, 113 ... is the first negative terms? (NCERT)

Sol.

Q. 35. Find the 15th term from the last term of the A.P. 3, 8, 13, ... 253. (CBSE 2022)

Sol.

SHORT ANSWER TYPE Q.UESTIONS-II

Q. 36. Find the middle terms of the A.P. 7, 13, 19, ..., 241.

Sol.

Q. 37. Find the sum of integers between 10 and 500 which are divisible by 7.

Sol.

Q. 38. The sum of 5th and 9th terms of an A.P. is 72 and the sum of 7th and 12th term is 97. Find the A.P.

Sol.

Q. 39. If the mth term of an A.P. be $\dfrac{1}{n}$ and nth term be $\dfrac{1}{m}$, show that its (mn)th is 1.

Sol.

Q. 40. If the mth term of an A.P. is $\dfrac{1}{n}$ and the nth terms is $\dfrac{1}{m}$, show that the sum of mn term is $\dfrac{1}{2}$ (mn + 1).

Sol.

Q. 41. If the pth term A.P. is q and the qth term is p, prove that its nth term is (p + q – n).

Sol.

Q. 42. Find the number of natural numbers between 101 and 999 which are divisible by both 2 and 5.

Sol.

Q. 43. The sum of 5th and 9th terms of an A.P. is 30. If its 25th term is three times its 8th term, find the A.P.

Sol.

Q. 44. If m times the mth terms of an A.P. is equal to n times of nth term and m ≠ n, show that its (m + n)th term is zero. (CBSE 2014)

Sol.

Q. 45. Which term of the A.P. 3, 15, 27, 39 .... will be 120 more than its 21st term? (CBSE 2014)

Sol.

Q. 46. If S$_n$, the sum of first n terms of an A.P. is given by S$_n$ = 3n^2 – 4n, find the nth term. (CBSE 2018)

Sol.

Q. 47. The sum of first n terms of an A.P. is given by S$_n$ = 3n^2 – 2n. Find the A.P. (CBSE 2022)

Sol.

Q. 48. In an A.P., the first term is 12 and the common difference is 6. If the last term of the A.P. is 252, then find its middle term. (CBSE 2022)

Sol.

Q. 49. The 17th term of an A.P. is 5 more than twice its 8th term. If the 11th term of the A.P. is 43, then find the nth term of the A.P. (CBSE 2020)

Sol.

Q. 50. If the sum of the first 14 terms of an A.P. is 1050 and its fourth term is 40, find its 20th term.

(CBSE 2020)

Sol.

Q. 51. Find the number of terms in the series $20 + 19\dfrac{1}{3} + 18\dfrac{2}{3} + ...$ of which the sum is 300, explain the double answer.

(NCERT)

Sol.

Q. 52. The first term of an A.P. is 5, the last term is 45 and the sum is 400. Find the number of tenns and the common difference.

(NCERT)

Sol.

Q. 53. Find the sum of n terms of the series: $\left(4 - \dfrac{1}{n}\right) + \left(4 - \dfrac{2}{n}\right) + \left(4 - \dfrac{3}{n}\right) + ...$

(CBSE 2017)

Sol.

LONG ANSWER TYPE Q.UESTIONS

Q. 54. The sum of third and seventh terms of an A.P. is 6 and their product is 8. Find the sum of first 16 terms of the A.P.

Sol.

Q. 55. Determine the A.P. whose 4th term is 18 and the difference of 9th term from the 15th tennis 30.

Sol.

Q. 56. The sum of first 9 terms of an A.P. is 162. The ratio ofits 6th term to its 13th term is 1 : 2. Find the first and fifteenth terms of the A.P.

Sol.

Q. 57. The sum of the first 9 terms of an A.P. is 171 and the sum ofits first 24 terms is 996. Find the first term and common difference of the A.P.

(CBSE 2020)

Sol.

Q. 58. The sum of first 7 terms of an A.P. is 63 and the sum ofits next 7 tennis 161. Find the 28th term of this A.P.

Sol.

Q. 59. The sum of first 20 terms of an A.P. is one third of the sum of next 20 term. If first term is 1, find the sum of first 30 terms of this A.P.

Sol.

Q. 60. If the sum of the first four terms of an AP is 40 and the sum of the first fourteen terms of an AP is 280. Find the sum of first n terms of the A.P.

(CBSE 2018)

Sol.

Q. 61. Ramkali required ₹ 2500 after 12 weeks to send her daughter to school. She saved ₹ 100 in the first week and increased her weekly savings by ₹ 20 every week. Find wheather she will be able to send her daughter to school after 12 weeks.

(CBSE 2015)

Sol.

Q. 62. In an AP of 50 terms, the sum of first 10 terms is 210 and the sum of last 15 terms is 2565. Find the A.P.

(CBSE 2014)

Sol.

Q. 63. The sum of first n terms of an A.P. is $5n^2 + 3n$. If the mth term is 168, find the value of m. Also find the 20th term of the A.P. **(CBSE 2013)**

Sol.

Q. 64. If the sum of the first seven terms of an A.P. is 49 and the sum of its first 17 terms is 289. Find the sum of first n terms of an A.P. **(CBSE 2016)**

Sol.

Q. 65. If the 4th term of an A.P. is zero, prove that the 25th term of the A.P. is three times its 11th term. **(CBSE 2016)**

Sol.

Q. 66. In an A.P. if $S_5 + S_7 = 167$ and $S_{10} = 235$. Find the A.P., where S_n denotes the sum of its first n terms. **(CBSE 2015)**

Sol.

Q. 67. In an AP prove $S_{12} = 3(S_8 - S_4)$ where S_n represent the sum of first n terms of an A.P. **(CBSE 2015)**

Sol.

Q. 68. The sum of four consecutive numbers in A.P. is 32 and the ratio of the product of the first and last term to the product of two middle terms is 7 : 15. Find the numbers.

Sol.

Q. 69. Find the sum of first 16 terms of an Arithmetic Progression whose 4th and 9th terms are –15 and –30 respectively. **(CBSE 2020)**

Sol.

Q. 70. An A.P. consists of 37 terms. The sum of the three middle most terms is 225 and the sum of the last three terms is 429. Find the A.P.

Sol.

1. (d)   2. (b)   3. (a)   4. (c)

5. (d)   6. (b)

7. (a) $\left[r = \dfrac{q^2}{4p} \ (D = 0 \Rightarrow q^2 - 4pr = 0)\right]$ (b) $p = 6, q = 1$

   (c) $k \leq 4$ (d) $c = 4$ (e) Linear equation

8. (a) False   (b) True   (c) False

   (d) False   (e) True

9. (i) (d)   (ii) (a)   (iii) (e)

   (iv) (c)   (v) (b)

10. $D = 0$

11. (a) $x = \dfrac{7}{2}, \dfrac{-3}{4}$ (b) $-\sqrt{5}, \dfrac{-2\sqrt{5}}{3}$ (c) $\sqrt{\dfrac{2}{3}}, \sqrt{\dfrac{2}{3}}$ (d) $\dfrac{a}{2}, -a$

   (e) $-\sqrt{3}, \dfrac{-7\sqrt{3}}{3}$ (f) $-\sqrt{2}, \dfrac{-5\sqrt{2}}{2}$ (g) $x = 0, x = 7$

12. $D < 0, a > 3$ 13. $k = \dfrac{7}{4}$ 14. $a = 3, b = -6$

15. $p = \dfrac{-6}{4} = \dfrac{-3}{2}$ 16. $16, 20$ (cm)   17. $k = -7$ 18. $k = 16$

19. (a) $x = -a, -b$ (b) $x = a, \dfrac{-b}{2}$ (c) $x = -1, x \neq \dfrac{-3}{2}$ (d) $x = 2, -6$

   (e) $x = \dfrac{-(a+b)}{2}, \dfrac{a-b}{2}$ (f) $x = \dfrac{b^2}{2}, \dfrac{a^2}{2}$ (g) $x = 4, \dfrac{-23}{11}$ (h) $x = 15, 4$

   (i) $x = \dfrac{a^2 \pm b^2}{2}$ (j) $x = \dfrac{-b}{2a^2}, -3b$ (k) $y = \dfrac{-1}{3}, y$ (l) $x = 1, 2$

   (m) $x = 8, \dfrac{11}{2}$ (n) $x = 2 \pm 2\sqrt{3}$ (o) $\dfrac{-8 \pm 3\sqrt{2}}{2}$

   (p) $x = 2\sqrt{5}, -7\sqrt{5}$ (q) $x = \pm 4$

20. $x = \dfrac{-b}{a}, \dfrac{c}{b}$ 21. $P = 3$ 23. $\dfrac{a^2 \pm b^2}{3}$

24. $x = 36, x \neq -3$ 25. $x = 8, x = -20$ (Rejected)

26. $x = 6, x \neq -3$ 27. $x = 3, x = 18, (x = 3,$ Rejected$)$

28. Length $= 34$ m, breadth $= 24$ m

29. $y = 20$ m, 10 m   **30.** $x = 12$ m   **31.** ?

32. ?   **33.** ?   **34.** ?   **35.** ?

36. $x = 12$ m, $y = 16$ m

37. $x = 12, 5$ and $(-12, -5$ are Rejected)

38. $\dfrac{3}{7}$

39. $x = 6$ years, $x = -12$ (Rejected)

40. 92

41. $x = 4$, $x = \dfrac{-11}{2}$ (Rejected)   Numbers: 4, 5, 6.

42. $x = 20$, $x = -25$ (Rejected)

43. $x = 8$ km/hr., $x = -72$ km/hr (Rejected)

44. $a + c = 2b$

45. $c^2 = a^2 (1 + m^2)$

46. $x = 40$, $x = -32$ (Rejected)

47. Length = 7 m, breadth = 4 m.

PRACTICE-TEST

SECTION-A

Q. 1. Find the sum of first 10 natural numbers. (1)

Sol.

Q. 2. What is the common difference of an A.P. $8\dfrac{1}{8}, 8\dfrac{2}{8}, 8\dfrac{3}{8}, \ldots\ldots$ (1)

Sol.

Q. 3. If k, 2k – 1 and 2k + 1 are in A.P. them value of k is .... (1)

Sol.

Q. 4. The 10th term from the end of the AP 8, 10, 12, ..., 126 is .... (1)

Sol.

SECTION-B

Q. 5. How many 2 digit number are there in between 6 and 102 which are divisible by 6. (2)

Sol.

Q. 6. The sum of n terms of an A.P. is $n^2 + 3n$. Find its 20th term. (2)

Sol.

Q. 7. Find the sum (–5) + (–8) + (–11) + ... + (–230) (2)

Sol.

SECTION-C

Q. 8. Find the five terms of an A.P. whose sum is $12\dfrac{1}{2}$ and first and last term ratio is 2 : 3. (3)

Sol.

Q. 9. Find the middle term of an A.P. 20, 16, 12,........., – 176. (3)

Sol.

SECTION-D

Q. 10. The sum of three numbers in A.P. is 24 and their product is 440. Find the numbers. (4)

Sol.

CASE STUDY

CASE STUDY 1.

India is competitive manufacturing location due to the low cost of manpower and strong technical and engineering capabilities contributing to higher quality production runs. The production of TV sets in a factory increases uniformly by a fixed number every year. It produced 16000 sets in 6th year and 22600 in 9th year.

Based on the above information, answer the following questions:

Q. 1.   **Find the production during first year.**

Q. 2.   **Find the production during 8th year.**

Q. 3.   **Find the production during first 3 years.**

Q. 4.   **In which year, the production is Rs 29,200.**

Q. 5.   **Find the difference of the production during 7th year and 4th year.**

ANSWERS

1. Rs 5000
2. Production during 8th year is $(a + 7d) = 5000 + 2(2200) = 20400$
3. Production during first 3 year $= 5000 + 7200 + 9400 = 21600$
4. $N = 12$
5. Difference $= 18200 - 11600 = 6600$

CASE STUDY 2.

Your friend Veer wants to participate in a 200m race. He can currently run that distance in 51 seconds and with each day of practice it takes him 2 seconds less. He wants to do in 31 seconds.

Q. 1.   **Which of the following terms are in AP for the given situation**

(a) 51,53,55….   (b) 51, 49, 47….   (c) -51, -53, -55….   (d) 51, 55, 59…

Q. 2.   **What is the minimum number of days he needs to practice till his goal is achieved**

(a) 10   (b) 12   (c) 11   (d) 9

Q. 3.   **Which of the following term is not in the AP of the above given situation**

(a) 41   (b) 30   (c) 37   (d) 39

Q. 4.   **If n^{th} term of an AP is given by $a_n = 2n + 3$ then common difference of an AP is**

(a) 2   (b) 3   (c) 5   (d) 1

Q. 5.   **The value of x, for which 2x, x+ 10, 3x + 2 are three consecutive terms of an AP**

(a) 6   (b) –6   (c) 18   (d) –18

ANSWERS

1. (b)   **2.** (c)   **3.** (b)

4. (a)   **5.** (a)

CASE STUDY 3.

Your elder brother wants to buy a car and plans to take loan from a bank for his car. He repays his total loan of Rs 1,18,000 by paying every month starting with the first instalment of Rs 1000. If he increases the instalment by Rs 100 every month , answer the following:

Q. 1. The amount paid by him in 30th installment is

(a) 3900 (b) 3500 (c) 3700 (d) 3600

Q. 2. The amount paid by him in the 30 installments is

(a) 37000 (b) 73500 (c) 75300 (d) 75000

Q. 3. What amount does he still have to pay offer 30th installment?

(a) 45500 (b) 49000 (c) 44500 (d) 54000

Q. 4. If total installments are 40 then amount paid in the last installment?

(a) 4900 (b) 3900 (c) 5900 (d) 9400

Q. 5. The ratio of the 1st installment to the last installment is

(a) 1:49 (b) 10:49 (c) 10:39 (d) 39:10

ANSWERS

1. (a) 3900 **2.** (b) 73500 **3.** (c) 44500

4. (a) 4900 **5.** (b) 10 : 49

CHAPTER-6

TRIANGLES

LEARNING OBJECTIVES

CONCEPT OF SIMILARITY

Geometric figures having the same shape but different sizes arc known as similar figures. Two congruent figures are always similar but similar figures need not be congruent.

Illustration I: Any two line segments are always similar but they need not be congruent. They are congruent, if their lengths are equal.

Illustralion 2: Any two circles are similar but not necessarily congruent. They are congruent if their are equal.

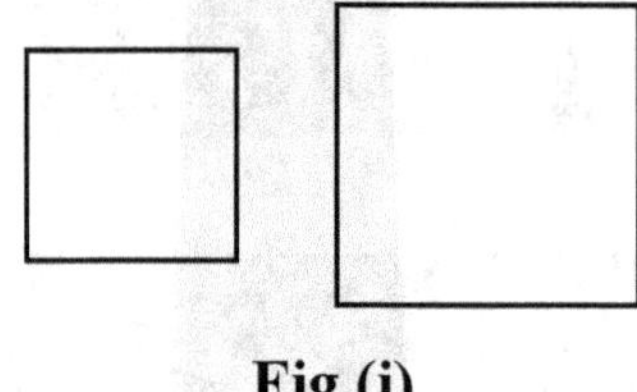

Illustration 3:

(i)   Any two square arc similar (sec fig. (i))

Fig.(i)

Fig.(ii)

(ii)   Any two equilateral triangles arc similar (see fig. (ii))

SIMILAR POLYGONS

Definition: Two polygons are said to be similar 10 each other, if

(i)   their corresponding angles are equal, and

(ii)   the lengths of 1heir corresponding sides arc proponional.

If two polygons ABCDE and PQ.RST are similar, then from the above definition it follows that :

Angle at A = Angle at P, Angle at B = Angle at Q., Angle at C = Angle at R. Angle at D = Angle at S, Angle at E = Angle at T

and, $\dfrac{AB}{PQ} = \dfrac{BC}{QR} = \dfrac{CD}{RS} = \dfrac{DE}{ST}$

Innovative Mathematics X-6

If two polygons ABCDE and PQ.RST, are similar, we write ABCED ~ PQ.RST.

Here, the symbol '-' stands for is similar to.

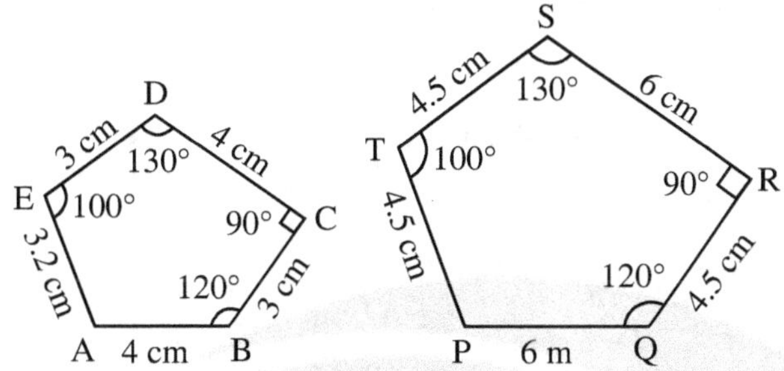

SIMILAR TRIANGLE AND THEIR PROPERTIES

Definition: Two triangles are said to be similar, if their

(i)   corresponding angles arc equal and,

(ii)   corresponding sides are proportional.

Two triangles ABC and DEF are similar, if

(i) $\angle A = \angle D$, $\angle B = \angle E$, $\angle C = \angle F$ and,

(ii) $\dfrac{AB}{DE} = \dfrac{BC}{EF} = \dfrac{AC}{DF}$

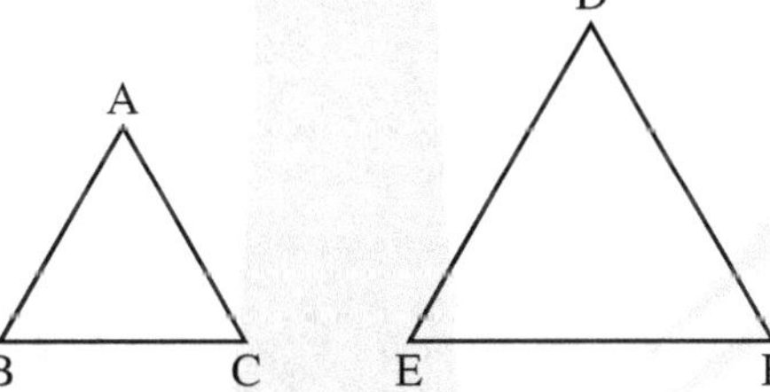

SOME BASIC RESULTS ON PROPORTIONALI1Y

Basic Proportionality Theorem or Thales Theorem

If a line is drawn parallel to one side of a triangle intersecting the other two sides, then it divides the two sides in the same ratio.

Given: A triangle ABC in which DE || BC, and intersects AB in D and AC in E.

To Prove: $\dfrac{AD}{DB} = \dfrac{AE}{EC}$

Construction: Join BE, CD and draw EF $\perp$ BA and DO $\perp$ CA.

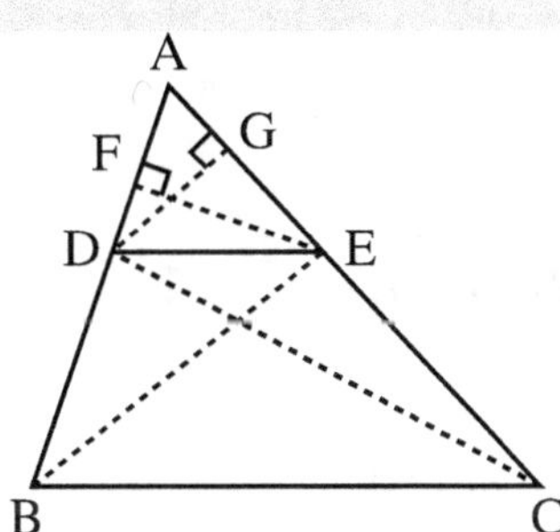

Proof: Since EF is perpendicular to AB. Therefore, EF is the height of triangles ADE and DBE.

Now, Area$(\Delta ADE) = \dfrac{1}{2}$ (base $\times$ height) $= \dfrac{1}{2}$ (AD.EF)

and, Area$(\Delta DBE) = \dfrac{1}{2}$ (base $\times$ height) $= \dfrac{1}{2}$ (DB.EF)

$\therefore \quad \dfrac{\text{Area}(\Delta ADE)}{\text{Area}(\Delta DBE)} = \dfrac{\dfrac{1}{2}(AD.EF)}{\dfrac{1}{2}(DB.BF)} = \dfrac{AD}{DB}$ ...(i)

Similarly, we have

$\dfrac{\text{Area}(\Delta ADE)}{\text{Area}(\Delta DBE)} = \dfrac{\dfrac{1}{2}(AE.DG)}{\dfrac{1}{2}(EC.DG)} = \dfrac{AE}{EC}$ ...(ii)

But, ΔDBE and ΔDEC are on the same base DE and between the same parallels DE and BC.

$\therefore \quad$ Area $(\Delta DBE) =$ Area (ΔDEC)

$\Rightarrow \quad \dfrac{1}{\text{Area}(\Delta DBE)} = \dfrac{1}{\text{Area}(\Delta DEC)}$

[Taking reciprocals of both sides]

$\Rightarrow \quad \dfrac{\text{Area}(\Delta ADE)}{\text{Area}(\Delta DBE)} = \dfrac{\text{Area}(\Delta ADE)}{\text{Area}(\Delta DEC)}$

[Multiplying both sides by Area (ΔADE)]

$\Rightarrow \quad \dfrac{AD}{DB} = \dfrac{AE}{EC}$ [Using (i) and (ii)]

Corollary: If in a ΔABC, a line DE $\parallel$ BC, intersects AB in D and AC in E, then:

(i) $\quad \dfrac{AB}{AD} = \dfrac{AC}{AE}$ $\qquad\qquad$ (ii) $\quad \dfrac{AB}{DB} = \dfrac{AC}{EC}$

Proof : (i) From the basic proportionality theorem, we have

$\dfrac{AD}{DB} = \dfrac{AE}{EC}$

$\Rightarrow \quad \dfrac{DB}{AD} = \dfrac{EC}{AE}$ [Taking reciprocals of both sides]

$\Rightarrow \quad 1 + \dfrac{DB}{AD} = 1 + \dfrac{EC}{AE}$ [Adding 1 on both sides]

$$\Rightarrow \quad \frac{AD+DB}{AD}=\frac{AE+EC}{AE} \Rightarrow \frac{AB}{AD}=\frac{AC}{AE}$$

(ii) From the basic proportionality theorem, we have

$$\frac{AD}{DB}=\frac{DE}{EC}$$

$$\Rightarrow \quad \frac{AD}{DB}+1=\frac{AE}{EC}+1 \qquad\qquad \text{[Adding 1 on both sides]}$$

$$\Rightarrow \quad \frac{AD+DB}{DB}=\frac{AE+EC}{EC} \Rightarrow \frac{AB}{DB}=\frac{AC}{EC}$$

So, if in a $\triangle ABC$, DE $\parallel$ BC, and intersect AB in D and AC in E, then we have

(i) $\dfrac{AD}{DB}=\dfrac{AE}{EC}$ $\qquad\qquad$ (ii) $\dfrac{DB}{AD}=\dfrac{EC}{AE}$

(iii) $\dfrac{AB}{AD}=\dfrac{AC}{AE}$ $\qquad\qquad$ (iv) $\dfrac{AD}{AB}=\dfrac{AE}{AC}$

(v) $\dfrac{AB}{DB}=\dfrac{AC}{EC}$ $\qquad\qquad$ (vi) $\dfrac{DB}{AB}=\dfrac{EC}{AC}$

Converse of Basic Proportionality Theorem

If a line divides any two sides of a triangle in the same ratio, then the line must be parallel to the third side.

Given : A $\triangle ABC$ and a line 1 intersecting AB in D and AC in E, such that $\dfrac{AB}{DB}=\dfrac{AE}{EC}$

To prove : $\lambda \parallel$ BC i.e. DE $\parallel$ BC

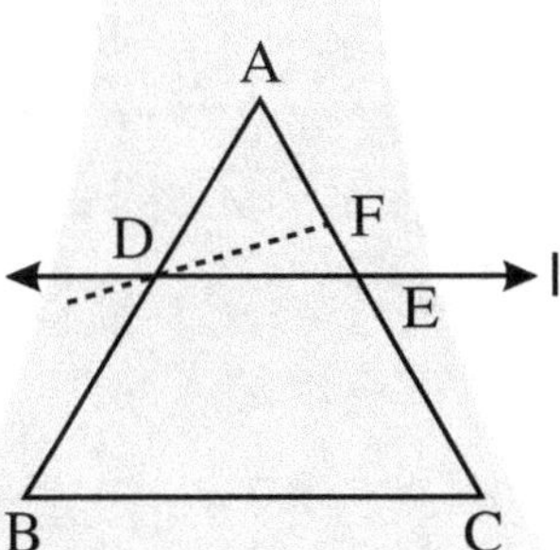

Proof : If possible, let DE be not parallel to BC. Then, there must be another line parallel to BC. Let DF $\parallel$ BC.

Since DF $\parallel$ BC. Therefore from Basic Proportionality Theorem, we get

$$\frac{AD}{DB}=\frac{AF}{FC} \qquad\qquad\qquad\qquad\qquad ...(i)$$

But, $\dfrac{AD}{DB}=\dfrac{AE}{EC}$ $\qquad$ (Given) $\qquad\qquad\qquad\qquad ...(ii)$

From (i) and (ii), we get

$$\frac{AF}{FC} = \frac{AE}{EC}$$

$$\Rightarrow \quad \frac{AF}{FC} + 1 = \frac{AE}{EC} + 1 \text{ [Adding 1 on both sides]}$$

$$\Rightarrow \quad \frac{AF + FC}{FC} = \frac{AE + EC}{EC}$$

$$\Rightarrow \quad \frac{AC}{FC} = \frac{AC}{EC} \Rightarrow FC = EC$$

This is possible only when F and E coincide i.e. DF is the line l itself. But, DF ∥ BC. Hence, l ∥ BC.

WARM – UP LEVEL-I

Q. 1. D and E are points on the sides AB and AC respectively of a ΔABC such that DE || BC. Find the value of x, when

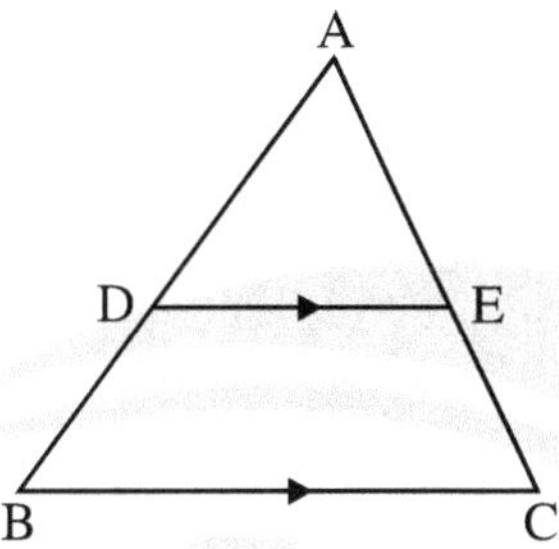

(i) AD = 4 cm, DB = (x – 4) cm, AE = 8 cm and EC = (3x – 19) cm

(ii) AD = (7x – 4) cm, AE = (5x – 2) cm, DB = (3x + 4) cm and EC = 3x cm.

Ans. (i) In ΔABC, DE || BC

$$\therefore \quad \frac{AD}{DB} = \frac{AE}{EC} \qquad \text{(By thales theorem)}$$

$$\frac{4}{x-4} = \frac{8}{3x-19}$$

$$4(3x - 19) = 8(x - 4)$$
$$12x - 76 = 8x - 32$$
$$4x = 44$$
$$x = 17$$

(ii) In ΔABC, DE || BC

$$\therefore \quad \frac{AD}{DB} = \frac{AE}{EC} \qquad \text{(By thales theorem)}$$

$$\frac{7x-4}{3x+4} = \frac{5x-2}{3x}$$

$$21x^2 - 12x = 15x^2 - 6x + 20x - 8$$
$$6x^2 - 26x + 8 = 0$$
$$3x^2 - 13x + 4 = 0$$
$$(x - 4)(3x - 1) = 0 \Rightarrow x = 4, 1/3$$

Q. 2. Let X be any point on the side BC of a triangle ABC. If XM, XN are drawn parallel to BA and CA meeting CA, BA in M, N respectively; MN meets BC produced in T, prove that $TX^2 = TB \times TC$.

Ans. In ΔTXM, we have

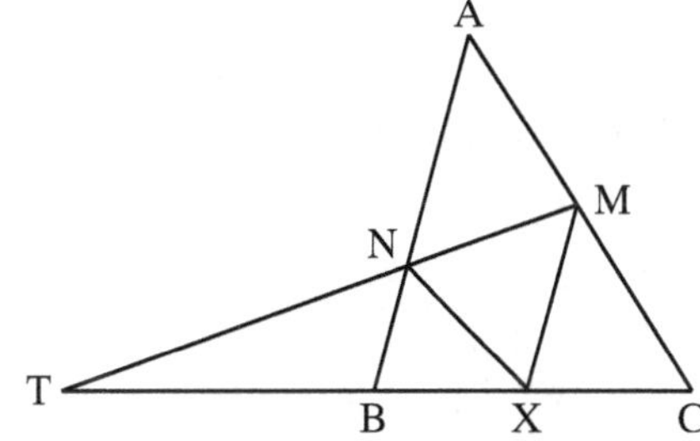

XM ‖ BN

$$\therefore \quad \frac{TB}{TX} = \frac{TM}{TN} \qquad \qquad \text{...(i)}$$

In $\triangle$TMC, we have

XN ‖ CM

$$\therefore \quad \frac{TX}{TC} = \frac{TN}{TM} \qquad \qquad \text{...(ii)}$$

From equations (i) and (ii), we get

$$\frac{TB}{TX} = \frac{TX}{TC}$$

$$\Rightarrow \quad TX^2 = TB \times TC$$

Q. 3. **In fig., EF ‖ AB ‖ DC. Prove that $\dfrac{AE}{ED} = \dfrac{BF}{FC}$.**

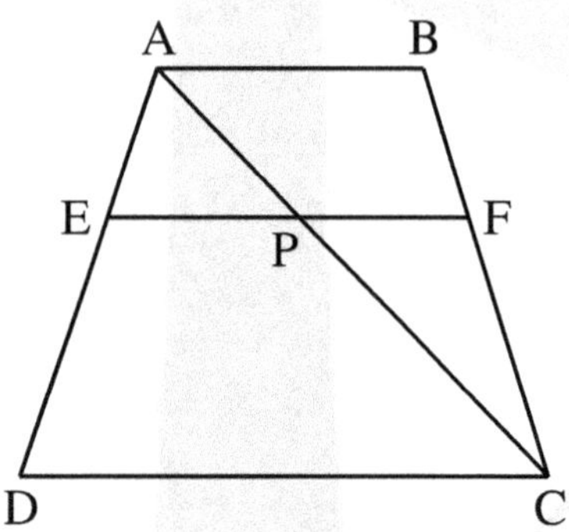

Ans. We have,

EF ‖ AB ‖ DC

$\Rightarrow$ EP ‖ DC

Thus, in $\triangle$ADC, we have

EP ‖ DC

Therefore, by basic proportionality theorem, we have

$$\frac{AE}{ED} = \frac{AP}{PC} \qquad \qquad \text{...(i)}$$

Again, EF ‖ AB ‖ DC

$\Rightarrow$ FP ‖ AB

Thus, in $\triangle$CAB, we have

FP ‖ BA

Therefore, by basic proportionality theorem, we have

$$\frac{BF}{FC} = \frac{AP}{PC} \qquad \qquad \text{...(ii)}$$

From (i) and (ii), we have

$$\frac{AE}{ED} = \frac{BF}{FC}$$

Q. 4. **In figure, $\angle A = \angle B$ and DE $\parallel$ BC. Prove that AD = BE**

Ans. $\angle A = \angle B$ (given)

$\Rightarrow$ BC = AC ...(i)

(Sides opposite to equal angles are equal)

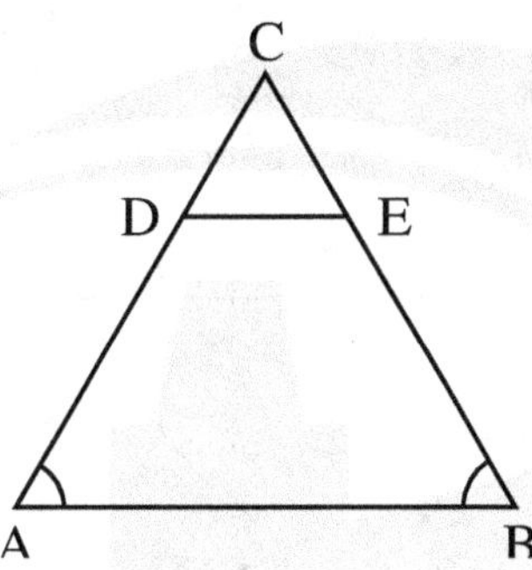

Now, DE $\parallel$ AB

$\Rightarrow \dfrac{CD}{DA} = \dfrac{CE}{EB}$ (By basic proportionality theorem)

$\Rightarrow \dfrac{CD}{DA} + 1 = \dfrac{CE}{EB} + 1$ (Adding 1 on both sides)

$\Rightarrow \dfrac{CD + DA}{DA} = \dfrac{CE + EB}{EB} \Rightarrow \dfrac{CA}{DA} = \dfrac{CE}{EB}$

$\Rightarrow \dfrac{AC}{AC} = \dfrac{BC}{BE} \Rightarrow \dfrac{AC}{AD} = \dfrac{AC}{BE}$

$\Rightarrow \dfrac{1}{AD} = \dfrac{1}{BE} \Rightarrow AD = BE$

Q. 5. **In fig, DE $\parallel$ BC. If AD = 4x – 3, DB = 3x – 1, AE = 8x – 7 and EC = 5x – 3, find the value of x.**

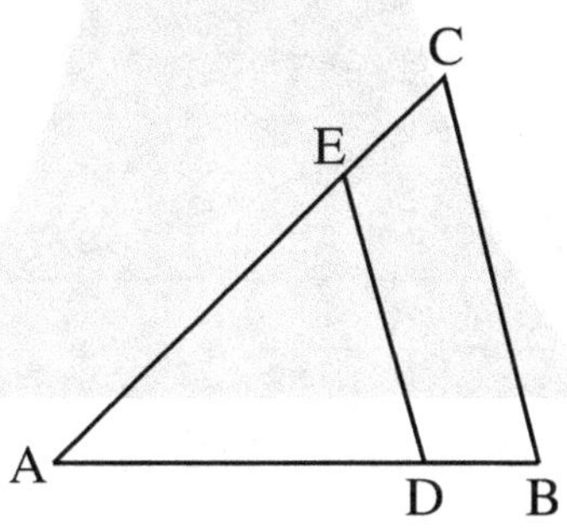

Ans. In $\triangle ABC$, we have

DE $\parallel$ BC

$\therefore \quad \dfrac{AD}{DB} = \dfrac{AE}{EC}$ [By Thale's Theorem]

$$\Rightarrow \quad \frac{4x-3}{3x-1} = \frac{8x-7}{5x-3}$$

$$\Rightarrow \quad (4x-3)(5x-3) = (3x-1)(8x-7)$$

$$x = 1$$

Q. 6. **Prove that the line segment joining the midpoints of the adjacent sides of a quadrilateral form a parallelogram.**

Ans. **Given :** A quadrilateral ABCD in which P, Q., R, S are the midpoints of AB, BC, CD and DA respectively.

To prove : PQ.RS is a parallelogram.

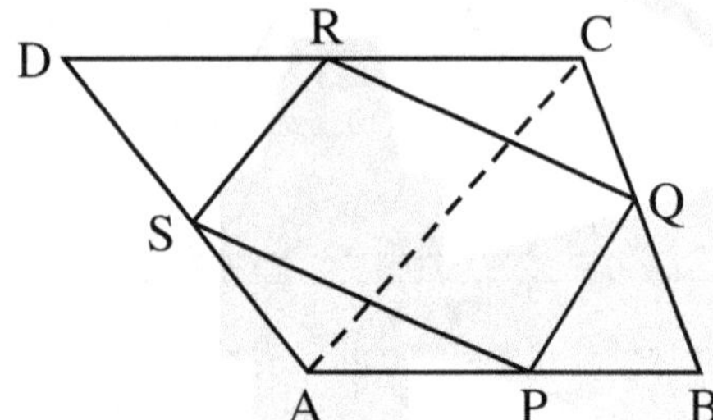

Construction : Join AC.

Proof : In $\triangle ABC$, P and Q. are the midpoints of AB and BC respectively.

$\therefore \quad$ PQ. $\parallel$ AC $\hspace{8cm}$ ...(i)

In $\triangle DAC$, S and R are the midpoints of AD and CD respectively.

$\therefore \quad$ SR $\parallel$ AC $\hspace{8cm}$ ...(ii)

From (i) and (ii), we get PQ. $\parallel$ SR.

Similarly, PS $\parallel$ Q.R.

Hence, PQ.RS is a parallelogram

Q. 7. **In fig. DE $\parallel$ BC and CD $\parallel$ EF. Prove that $AD^2 = AB \times AF$.**

Ans. In $\triangle ABC$, we have

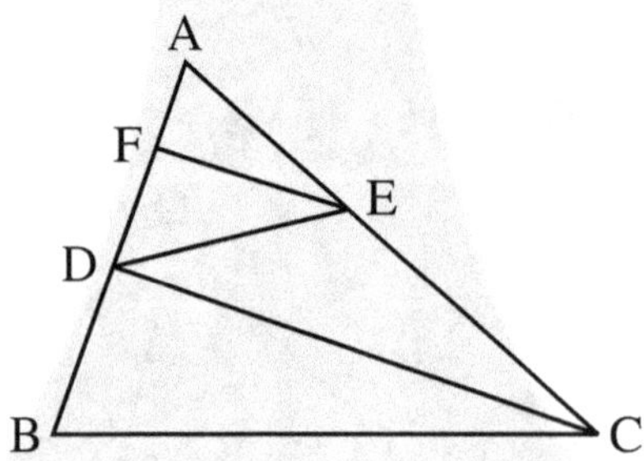

DE $\parallel$ BC

$$\Rightarrow \quad \frac{AB}{AD} = \frac{AC}{AE} \hspace{7cm} ...(i)$$

In $\triangle ADC$, we have

$\quad$ FE $\parallel$ DC

$$\Rightarrow \quad \frac{AD}{AF} = \frac{AC}{AE} \hspace{7cm} ...(ii)$$

From (i) and (ii), we get

$$\frac{AB}{AD} = \frac{AD}{AF} \implies AD^2 = AB \times AF$$

Q. 8. In the given figure PA, Q.B and RC each is perpendicular to AC such that PA = x, RC = y, Q.B = z, AB = a and BC = b. Prove that $\frac{1}{x} + \frac{1}{y} = \frac{1}{z}$.

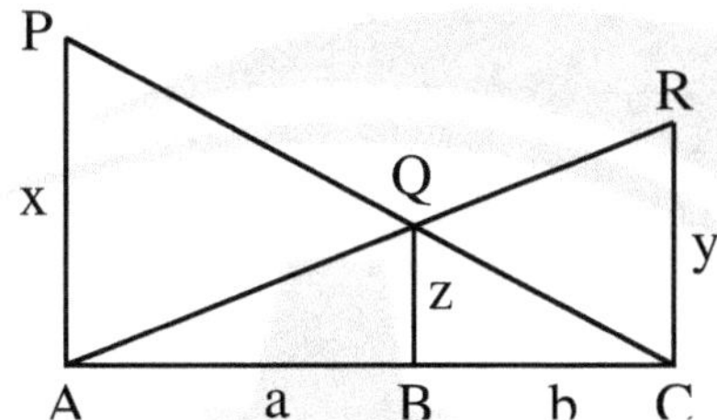

Ans. PA $\perp$ AC and Q.B $\perp$ AC $\implies$ Q.B $\parallel$ PA.

Thus, in ΔPAC, Q.B $\parallel$ PA. So, ΔQ.BC $\sim$ ΔPAC

$\therefore \quad \dfrac{QB}{PA} = \dfrac{BC}{AC} \implies \dfrac{z}{x} = \dfrac{b}{a+b}$ ...(i)

  [By the property of similar Δ]

In ΔRAC, Q.B $\parallel$ RC. So, ΔQ.BC $\sim$ ΔRAC

$\therefore \quad \dfrac{QB}{RC} = \dfrac{AB}{AC} \implies \dfrac{z}{y} = \dfrac{a}{a+b}$ ...(ii)

  [By the property of similar Δ]

From (i) and (ii), we get

$$\frac{z}{x} + \frac{z}{y} = \left(\frac{b}{a+b} + \frac{a}{a+b} \right) = 1$$

$$\implies \frac{z}{x} + \frac{z}{y} = 1 \implies \frac{1}{x} + \frac{1}{y} = \frac{1}{z}.$$

Hence, $\dfrac{1}{x} + \dfrac{1}{y} = \dfrac{1}{z}$.

Q. 9. In fig., LM $\parallel$ AB. If AL = x – 3, AC = 2x, BM = x – 2 and BC = 2x + 3, find the value of x.

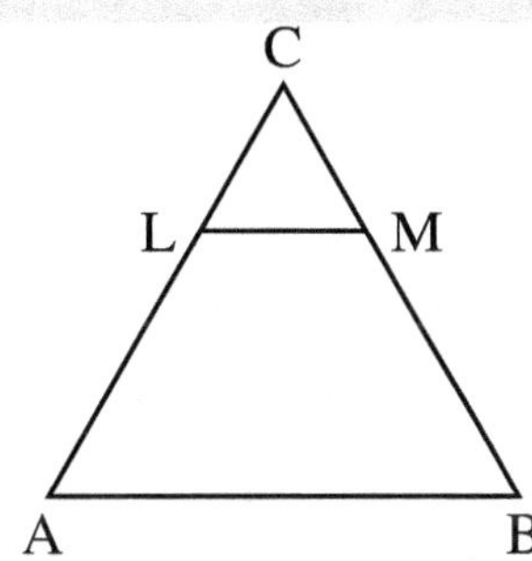

Ans. In $\triangle ABC$, we have

$$LM \parallel AB$$

$\therefore \quad \dfrac{AL}{LC} = \dfrac{MB}{MC}$ [By Thale's Theorem]

$\Rightarrow \quad \dfrac{AL}{AC - AL} = \dfrac{BM}{BC - BM}$

$\Rightarrow \quad \dfrac{x-3}{2x - (x-3)} = \dfrac{x-2}{(2x+3) - (x-2)}$

$\Rightarrow \quad \dfrac{x-3}{x+3} = \dfrac{x-2}{x+5}$

$\Rightarrow \quad (x-3)(x+5) = (x-2)(x+3)$

$\Rightarrow \quad x^2 + 2x - 15 = x^2 + x - 6$

$\Rightarrow \quad x = 9$

Q. 10. **In a given $\triangle ABC$, DE $\parallel$ BC and $\dfrac{AD}{DB} = \dfrac{3}{4}$. If AC = 14 cm, find AE.**

Ans. In $\triangle ABC$, we have

$$DE \parallel BC$$

$\therefore \quad \dfrac{AD}{DB} = \dfrac{AE}{EC}$ [By Thales Theorem]

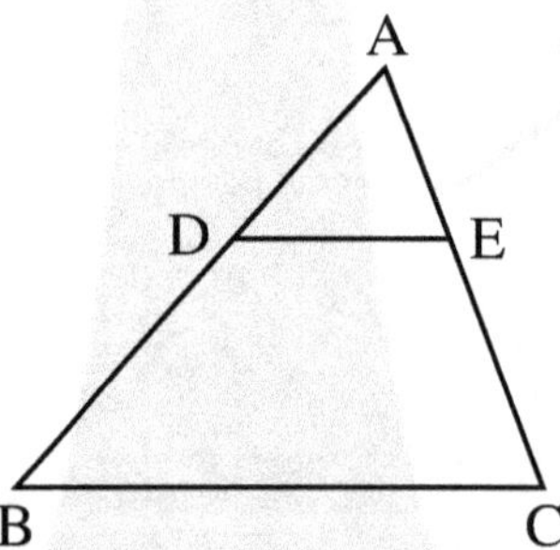

$\Rightarrow \quad \dfrac{AD}{DB} = \dfrac{AE}{AC - AE}$

$\Rightarrow \quad \dfrac{3}{4} = \dfrac{AE}{14 - AE}$ $[\Theta\ AC = 5.6]$

$\Rightarrow \quad 3(14 - AE) = 4AE$

$\Rightarrow \quad 42 - 3AE = 4AE$

$\Rightarrow \quad 42 = 7AE \Rightarrow AE = \dfrac{42}{7} = 6 \text{ cm}$

WARM – UP LEVEL-II

Q. 1. In figure, if $\angle A = \angle C$, then prove that $\triangle AOB \sim \triangle COD$.

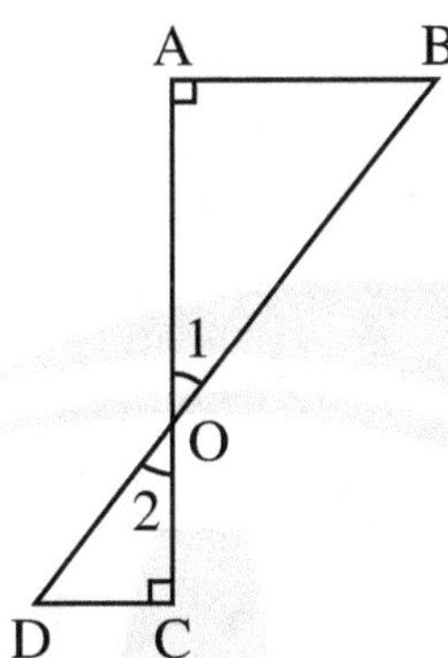

Ans. In triangles AOB and COD, we have

$$\angle A = \angle C \qquad \text{[Given]}$$

and, $\angle 1 = \angle 2$ [Vertically opposite angles]

Therefore, by AA-criterion of similarly, we have

$$\triangle AOB \sim \triangle COD$$

Q. 2. In figure, $\dfrac{AO}{OC} = \dfrac{BO}{OD} = \dfrac{1}{2}$ and AB = 5 cm. Find the value of DC.

Ans. In $\triangle AOB$ and $\triangle COD$, we have

$$\angle AOB = \angle COD \qquad\qquad\qquad\qquad\qquad\qquad \text{[Vertically opposite angles]}$$

$$\frac{AO}{OC} = \frac{OB}{OD} \qquad\qquad\qquad\qquad\qquad\qquad \text{[Given]}$$

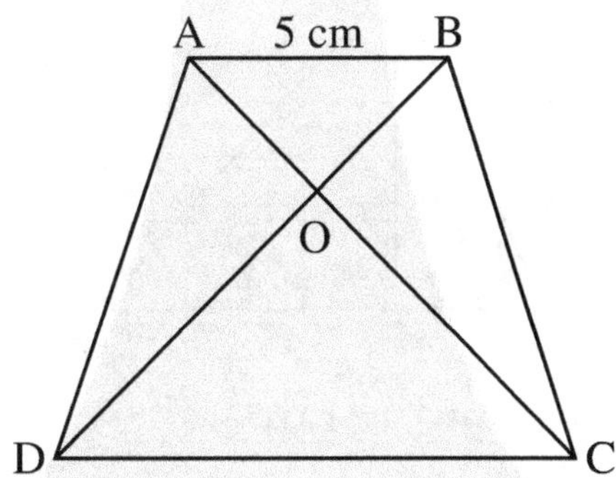

So, by SAS-criterion of similarity, we have

$$\triangle AOB \sim \triangle COD$$

$$\Rightarrow \quad \frac{AO}{OC} = \frac{BO}{OD} = \frac{AB}{DC}$$

$$\Rightarrow \quad \frac{1}{2} = \frac{5}{DC} \qquad\qquad\qquad\qquad\qquad\qquad [\because AB = 5 \text{ cm}]$$

$$\Rightarrow \quad DC = 10 \text{ cm}$$

Q. 3. In figure, considering triangles BEP and CPD, prove that BP × PD = EP × PC.

Ans. **Given :** A $\triangle$ABC is which BD $\perp$ AC and CE $\perp$ AB and BD and CE intersect at P.

To Prove : BP × PD = EP × PC

Proof : In $\triangle$EPB and $\triangle$DPC, we have

$\angle$PEB = $\angle$PDC [Each equal to 90°]

$\angle$EPB = $\angle$DPC [Vertically opposite angles]

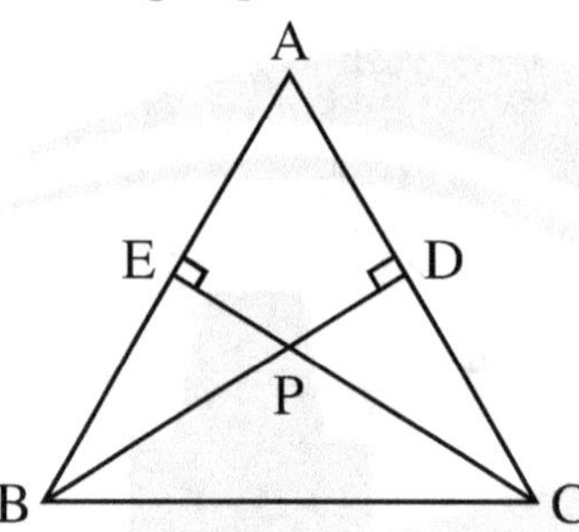

Thus, by AA-criterion of similarity, we have

$\triangle$EPB ~ $\triangle$DPC

$$\frac{EP}{DP} = \frac{PB}{PC}$$

$\Rightarrow$ BP × PD = EP × PC

Q. 4. **D is a point on the side BC of $\triangle$ABC such that $\angle$ADC = $\angle$BAC. Prove that $\dfrac{CA}{CD} = \dfrac{CB}{CA}$ or, CA² = CB × CD.**

Ans. In $\triangle$ABC and $\triangle$DAC, we have

$\angle$ADC = $\angle$BAC and $\angle$C = $\angle$C

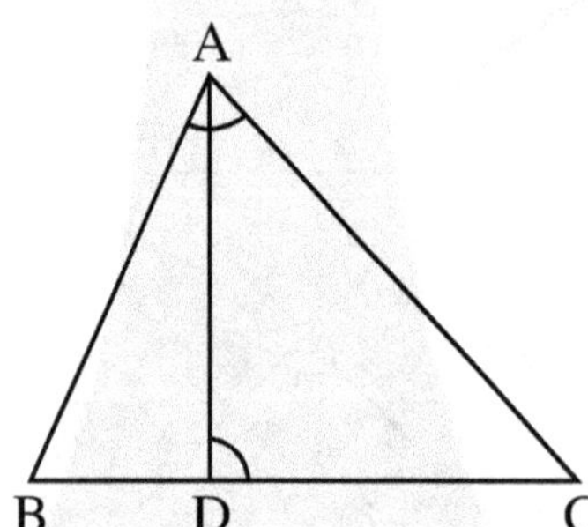

Therefore, by AA-criterion of similarity, we have

$\triangle$ABC ~ $\triangle$DAC

$\Rightarrow \quad \dfrac{AB}{DA} = \dfrac{BC}{AC} = \dfrac{AC}{DC}$

$\Rightarrow \quad \dfrac{CB}{CA} = \dfrac{CA}{CD}$

Q. 5. **P and Q. are points on sides AB and AC respectively of $\triangle$ABC. If AP = 3 cm, PB = 6 cm. AQ. = 5 cm and Q.C = 10 cm, show that BC = 3PQ.**

Ans. We have,

$$AB = AP + PB = (3 + 6)\text{ cm} = 9\text{ cm}$$

and, $AC = AQ. + Q.C = (5 + 10)\text{ cm} = 15\text{ cm}.$

$$\therefore \quad \frac{AP}{AB} = \frac{3}{9} = \frac{1}{3} \text{ and } \frac{AQ}{AC} = \frac{5}{15} = \frac{1}{3}$$

$$\Rightarrow \quad \frac{AP}{AB} = \frac{AQ}{AC}$$

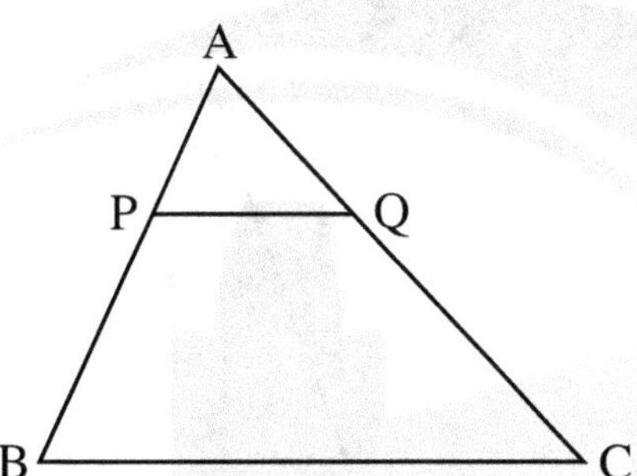

Thus, in triangles APQ. and ABC, we have

$$\frac{AP}{AB} = \frac{AQ}{AC} \text{ and } \angle A = \angle A \text{ [Common]}$$

Therefore, by SAS-criterion of similarity, we have

$$\triangle APQ. \sim \triangle ABC$$

$$\rightarrow \quad \frac{AP}{AB} = \frac{PQ}{BC} = \frac{AQ}{AC}$$

$$\Rightarrow \quad \frac{PQ}{BC} = \frac{AQ}{AC} \Rightarrow \frac{PQ}{BC} = \frac{5}{15}$$

$$\Rightarrow \quad \frac{PQ}{BC} = \frac{1}{3} \quad \Rightarrow \quad BC = 3PQ.$$

Q. 6. **In figure, $\angle A = \angle CED$, prove that $\triangle CAB \sim \triangle CED$. Also, find the value of x.**

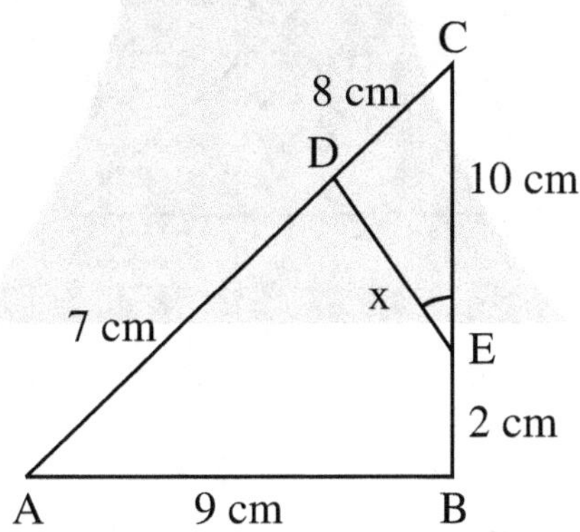

Ans. In $\triangle CAB$ and $\triangle CED$, we have

$$\angle A = \angle CED \text{ and } \angle C = \angle C \text{ [common]}$$

$$\therefore \quad \triangle CAB \sim \triangle CED$$

$$\Rightarrow \quad \frac{CA}{CE} = \frac{AB}{DE} = \frac{CB}{CD}$$

$$\Rightarrow \quad \frac{AB}{DE} = \frac{CB}{CD} \Rightarrow \frac{9}{x} = \frac{10+2}{8}$$

$$\Rightarrow \quad x = 6 \text{ cm}$$

Q. 7. In the figure, E is a point on side CB produced of an isosceles $\triangle$ABC with AB = AC. If AD $\perp$ BC and EF $\perp$ AC, prove that $\triangle$ABD ~ $\triangle$ECF.

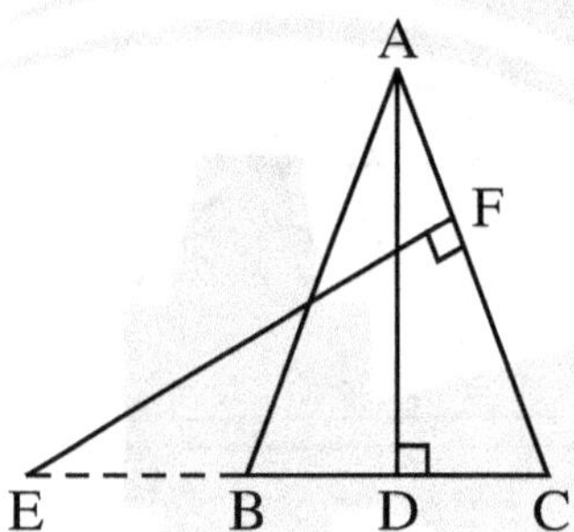

Ans. **Given :** A $\triangle$ABC in which AB = AC and AD $\perp$ BC. Side CB is produced to E and EF $\perp$ AC.

To prove : $\triangle$ABD ~ $\triangle$ECF.

Proof : we known that the angles opposite to equal sides of a triangle are equal.

$\therefore \quad \angle B = \angle C$ $\hfill$ [Θ AB = AC]

Now, in $\triangle$ABD and $\triangle$ECF, we have

$\quad \angle B = \angle C$ $\quad$ [proved above]

$\quad \angle ADB = \angle EFC = 90°$

$\therefore \quad \triangle ABD \sim \triangle ECF$ [By AA-similarity]

Q. 8. In figure, $\angle$BAC = 90° and segment AD $\perp$ BC. Prove that AD^2 = BD $\times$ DC.

Ans. In $\triangle$ABD and $\triangle$ACD, we have

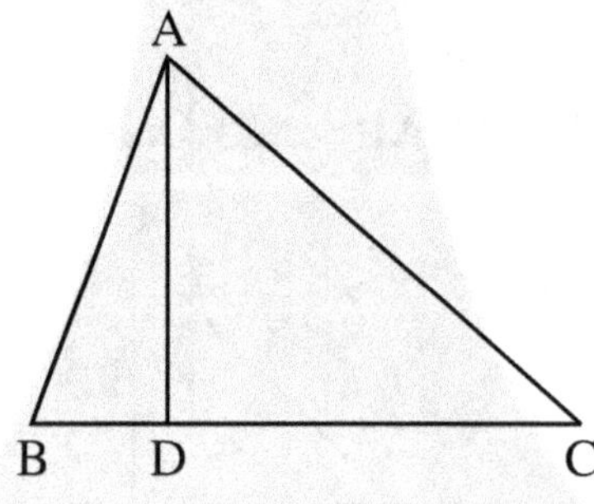

$\quad \angle ADB = \angle ADC$ $\hfill$ [Each equal to 90°]

and, $\angle DBA = \angle DAC$ $\hfill$ $\left[\begin{array}{l}\text{Each equal to complement of} \\ \angle BAD \text{ i.e., } 90° - \angle BAD\end{array}\right]$

Therefore, by AA-criterion of similarity, we have

$\triangle DBA \sim \triangle DAC$ $\hfill$ $\left[\begin{array}{l}\therefore \angle D \leftrightarrow \angle D, \angle DBA \leftrightarrow \angle DAC \\ \text{and } \angle BAD \leftrightarrow \angle BAD\end{array}\right]$

Innovative Mathematics X-6

$$\Rightarrow \quad \frac{DB}{DA} = \frac{DA}{DC} \qquad \left[\begin{array}{l}\text{In similar triangles corresponding}\\ \text{sides are proportional}\end{array}\right]$$

$$\Rightarrow \quad \frac{BD}{AD} = \frac{AD}{DC} \quad \Rightarrow \quad AD^2 = BD \times DC$$

Q. 9. In an isosceles $\triangle ABC$, the base AB is produced both ways in P and Q. such that $AP \times BQ. = AC^2$ and CE are the altitudes. Prove that $\triangle ACP \sim \triangle BCQ$.

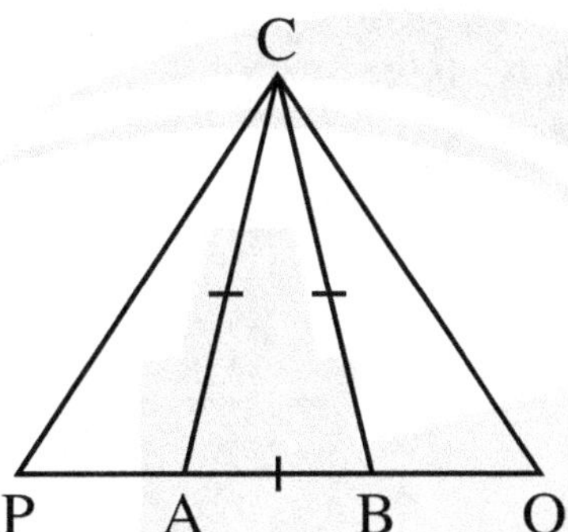

Ans. $CA = CB \Rightarrow \angle CAB = \angle CBA$

$\Rightarrow \quad 180° - \angle CAB = 180° - \angle CBA$

$\Rightarrow \quad \angle CAP = \angle CBQ.$

Now, $AP \times BQ. = AC^2$

$$\Rightarrow \quad \frac{AP}{AC} = \frac{AC}{BQ} \Rightarrow \frac{AP}{AC} = \frac{BC}{BQ} \qquad [\Theta\, AC = BC]$$

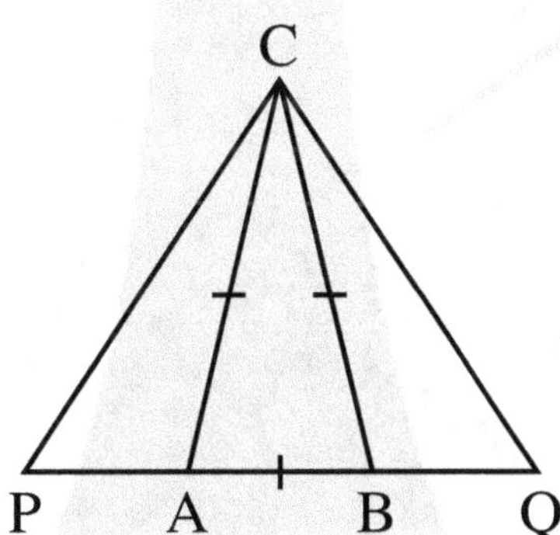

Thus, $\angle CAP = \angle CBQ.$ and $\dfrac{AP}{AC} = \dfrac{BC}{BQ}.$

$\therefore \quad \triangle ACP \sim \triangle BCQ.$

Q. 10. The diagonal BD of a parallelogram ABCD intersects the segment AE at the point F, where E is any point on the side BC. Prove that $DF \times EF = FB \times FA.$

Ans. In $\triangle AFD$ and $\triangle BFE$, we have

$\angle 1 = \angle 2$ [Vertically opposite angles]

$\angle 3 = \angle 4$ [Alternate angles]

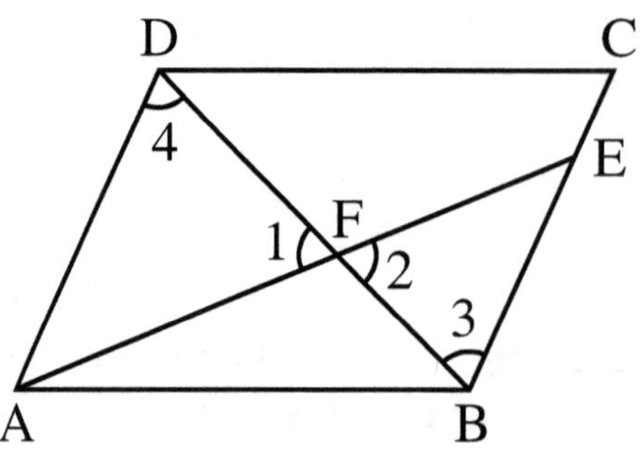

So, by AA-criterion of similarity, we have

$$\Delta FBE \sim \Delta FDA$$

$$\Rightarrow \quad \frac{FB}{FD} = \frac{FD}{FA} \quad \Rightarrow \quad \frac{FB}{DF} = \frac{EF}{FA}$$

$$\Rightarrow \quad DF \times EF = FB \times FA$$

VERY SHORT ANSWER TYPE Q.UESTIONS

Q. 1.   **Fill in the blanks:**

(i)   All equilateral triangles are ___________ .

(ii)   If $\triangle ABC \sim \triangle FED$, then $\underline{AB}$ = _____ .ED

(iii) Circles with equal radii are ___________ .

(iv) If a line is drawn parallel to one side of a triangle to intersect the other two sides in distinct points, the other two sides are divided in the ___________ ratio.

(v)   If two triangles are similar, their corresponding sides are _________ .   **(CBSE 2020)**

(vi) In $\triangle ABC$, $AB = 6\sqrt{3}$, $AC = 12$ cm and $BC = 6$ cm, then $\angle B =$ _______ .

Sol.

Q. 2.   **State True or False :**

(i)   All the similar figures are always congruent.

(ii)   The Basic Proportionality Theorem was given by Pythagoras.

(iii) The mid-point theorem can be proved by Basic Proprotionality Theorem.

(iv) Two squares are similar figures.

(v)   If all the sides of a triangle are proportional to the sides of other triangle, then the two triangles are congruent.

Sol.

Q. 3.   **Match the following:**

Column I		**Column II**
(a)	If corresponding angles are equal in two triangles, then the two triangles are similar.	(i) SAS similarity criterion
(b)	If sides of one triangle are proportional to the sides of the other triangle, then the two triangles are similar.	(ii) ASA similarity criterion
(c)	If one angle of a triangle is equal to one angle of the other triangle and the sides including these angles are proportional, then the two triangles are similar.	(iii) AAA similarity criterion
		(iv) SSS similarity criterion

Sol.

Q. 4.   **In the following figure, XY ‖ Q.R and** $\dfrac{PX}{XQ} = \dfrac{PY}{YR} = \dfrac{1}{2}$**, then**

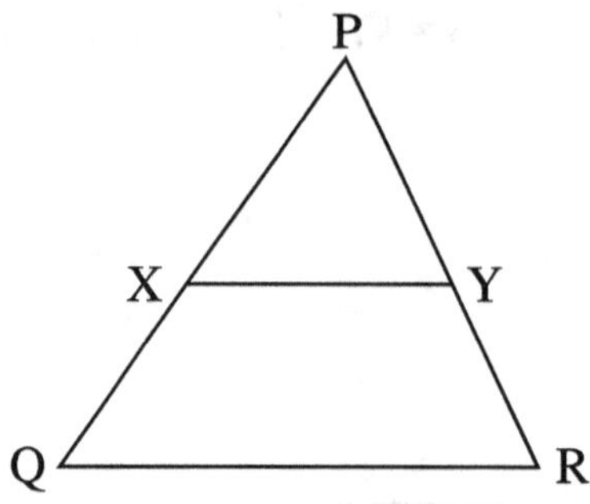

(a) $XY = Q.R$ (b) $XY = \dfrac{1}{3} Q.R$ (c) $XY^2 = Q.R^2$ (d) $XY = \dfrac{1}{2} Q.R$

Sol.

Q. 5. In the following figure, Q.A $\perp$ AB and PB $\perp$ AB, then AQ. is

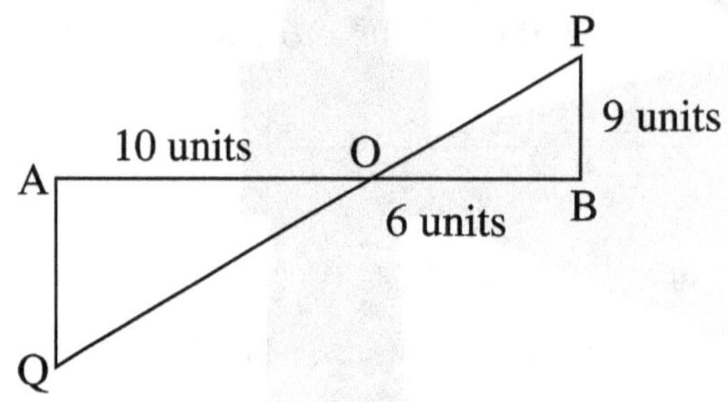

(a) 15 units   (b) 8 units   (c) 5 units   (d) 9 units

Sol.

Q. 6. If $\triangle ABC \sim \triangle EDF$ and $\triangle ABC$ is not similar to $\triangle DEF$, then which of the following not true? **(NCERT Exemplar)**

(a) $BC.EF = AC.FD$ (b) $AB.EF = AC.DE$

(c) $BC.DE = AB.EF$ (d) $BC.DE = AB.FD$

Sol.

Q. 7. Write the Statement of Basic Proportionality Theorem.

Sol.

Q. 8. In the given Fig., $\angle M = \angle N = 46°$, Express x in terms of a, b and c.

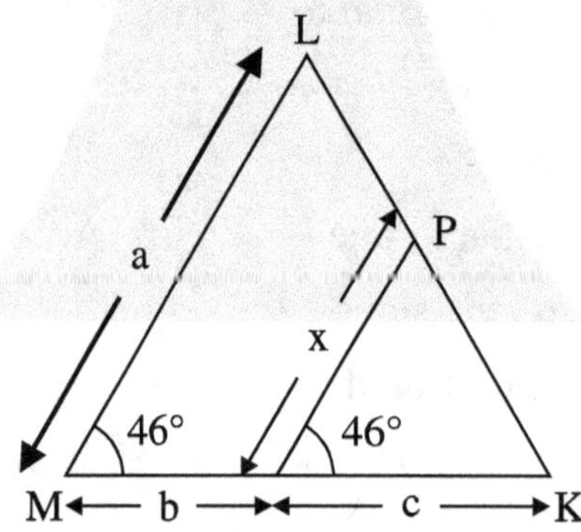

Sol.

Q. 9. In the given Fig. $\triangle AHK \sim \triangle ABC$. If AK = 10 cm, BC = 3.5 cm and HK = 7 cm find AC.

(CBSE 2010)

        Innovative Mathematics X-6

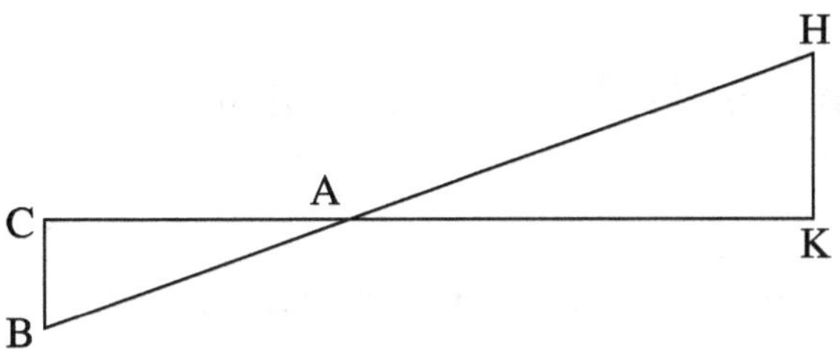

Sol.

Q. 10. It is given that $\triangle DEF \sim \triangle RPQ$. Is it true to say that $\angle D = \angle R$ and $\angle F = \angle P$?

Sol.

Q. 11. If the corresponding Medians of two similar triangles are in the ratio 5 : 7. Then find the ratio of their sides.

Sol.

Q. 12. In the given figure, if $\triangle ABC \sim \triangle PQ.R$, find the value of x?

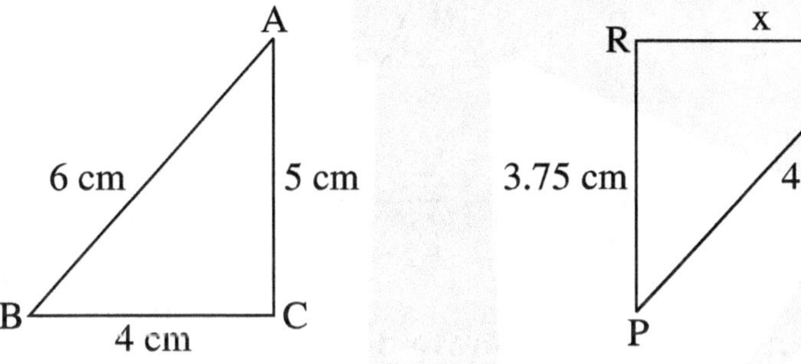

Sol.

Q. 13. In the given figure, XY || Q.R and $\dfrac{PX}{XQ} = \dfrac{PY}{YR} = \dfrac{1}{2}$, find XY : Q.R.

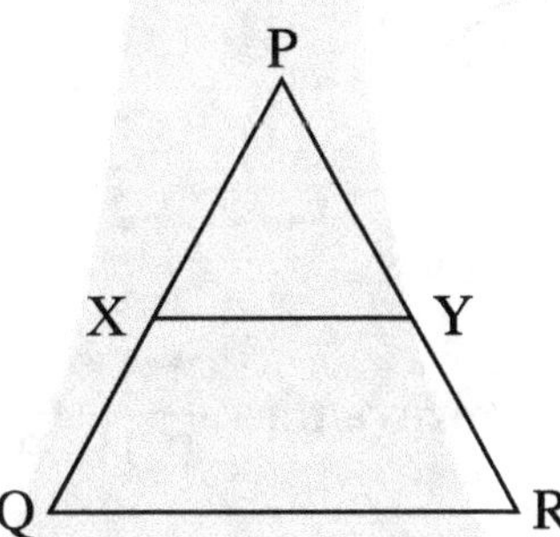

Sol.

Q. 14. In the given figure, find the value of x which will make DE || AB? **(NCERT Exemplar)**

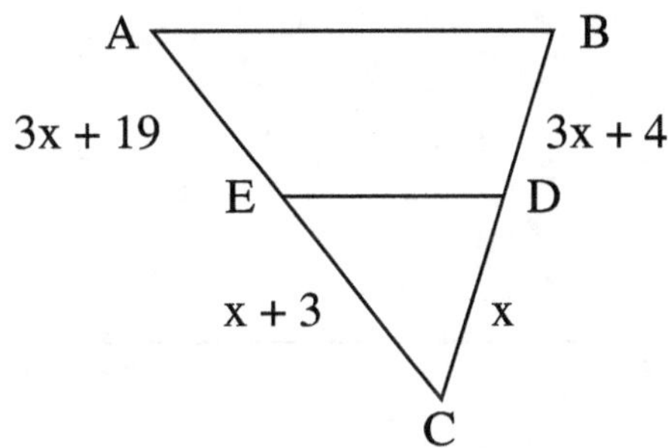

Sol.

Q. 15. If $\triangle ABC$ and $\triangle DEF$ are similar triangles such that $\angle A = 45°$ and $\angle F = 56°$, then find the value of $\angle C$.

Sol.

Q. 16. If the ratio of the corresponding sides of two similar triangles is 2 : 3, then find the ratio of their corresponding attitudes.

Sol.

SHORT ANSWER TYPE QUESTIONS-I

Q. 17. In the given fig. $\dfrac{BD}{AB} = \dfrac{CB}{AC}$, then prove that DE || BC

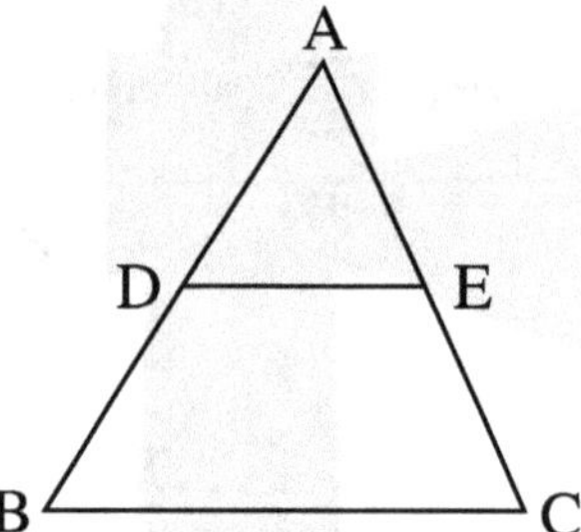

Sol.

Q. 18. In the given Fig., DE || AC and DC || AP Prove that $\dfrac{BE}{EC} = \dfrac{BC}{CP}$ (CBSE 2020)

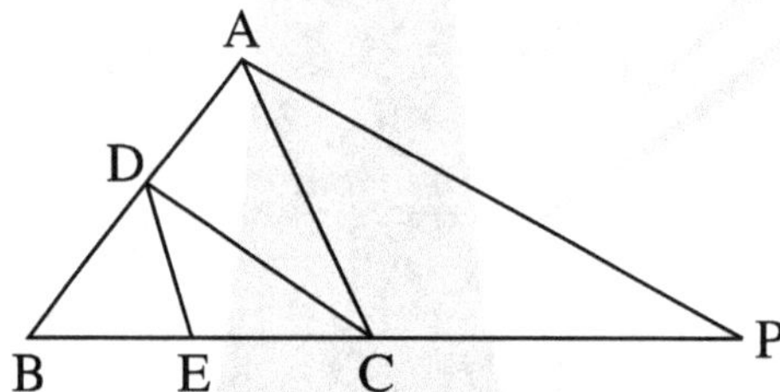

Sol.

Q. 19. In $\triangle PQ.R$, MN || Q.R, using B.P.T. prove that $\dfrac{PM}{PQ} = \dfrac{PN}{PR}$.

Sol.

Q. 20. In the given Fig., D and E are points on sides AB and CA of $\triangle ABC$ such that $\triangle B = \angle AED$. Show that $\triangle ABC \sim \triangle AED$.

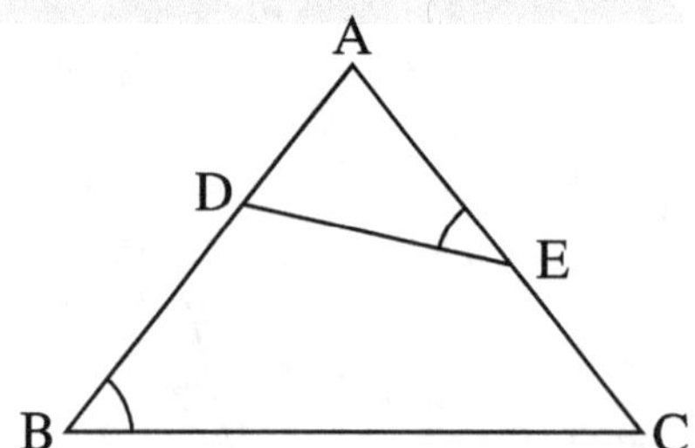

Sol.

Innovative Mathematics X-6

Q. 21. In the given fig., AB ∥ DC and diagonals AC and BD intersects at O. If OA = 3x − 1 and OB = 2x + 1, OC = 5x − 3 and OD = 6x − 5, find the value fo x.

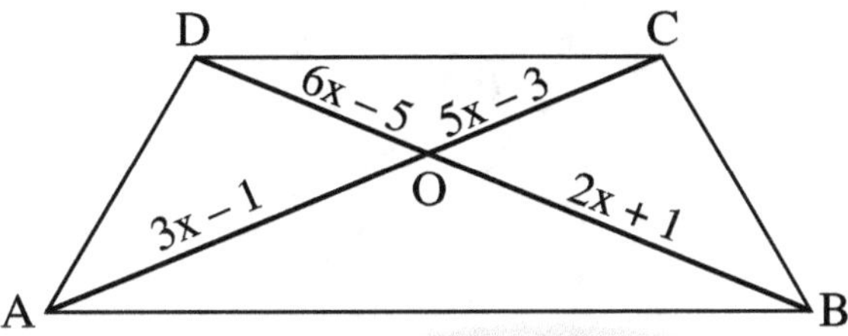

Sol.

Q. 22. In the given Fig. PQ.R is a triangle, right angled at Q. If XY ∥ Q.R, PQ. = 6 cm, PY = 4 cm and PX : XQ. = 1 : 2. Calculate the lengths of PR and Q.R.

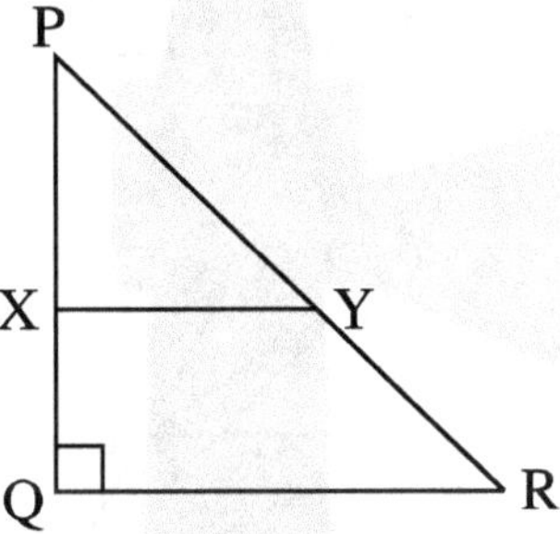

Sol.

Q. 23. In the given figure, AB ∥ DE. Find the length of CD.

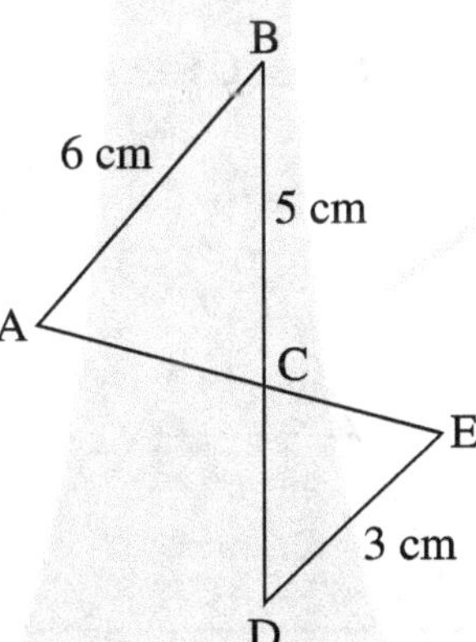

Sol.

Q. 24. In the given figure, ABCD is a parallelogram. AE divides the line segment BD in the ratio 1 : 2. If BE = 1.5 cm find BC.

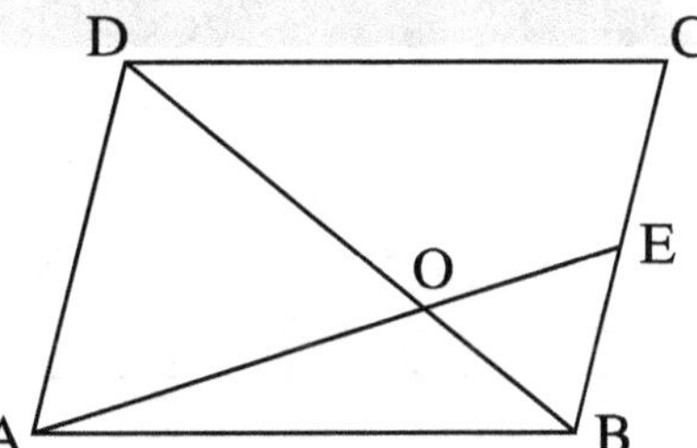

Sol.

Q. 25. In the given figure, $\triangle ODC \sim \triangle OBA$, $\angle BOC = 115°$ and $\angle CDO = 70°$. Find, (i) $\angle DOC$, (ii) $\angle DCO$, (iii) $\angle OAB$, (iv) $\angle OBA$.

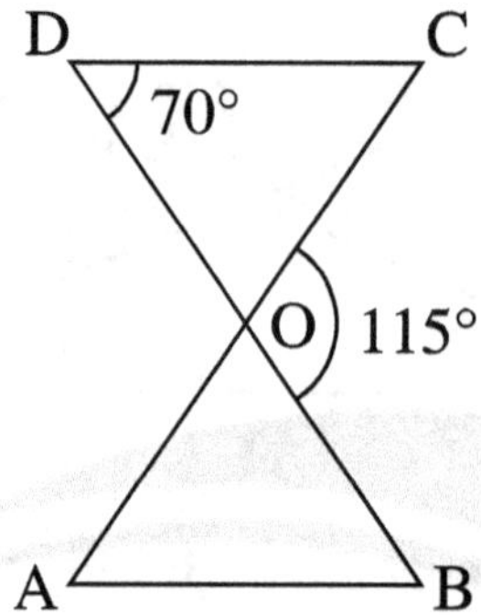

Sol.

Q. 26. In the given fig., AB || DE and BD || EF prove that $DC^2 = CF \times AC$

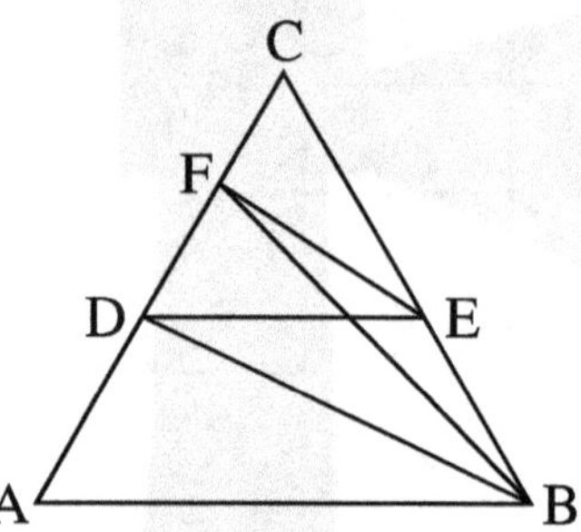

Sol.

Q. 27. In the given fig. $\dfrac{AD}{DC} = \dfrac{BE}{EC}$ and $\angle CDE = \angle CED$, prove that $\triangle CAB$ is isoceles.

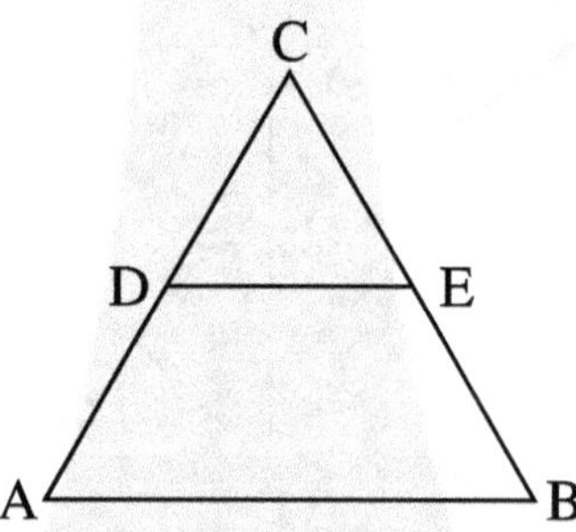

Sol.

Q. 28. In the given fig., Q.S || BA, Q.R || CA and PQ. = 10 cm. Find PB × PC.

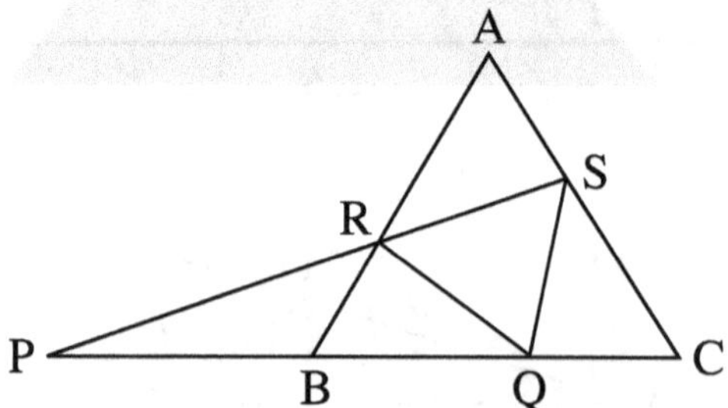

Sol.

Innovative Mathematics X-6

Q. 29. In the given fig., ΔFEC = ΔGBD and ∠1 = ∠2 prove that ΔADE = ΔABC.

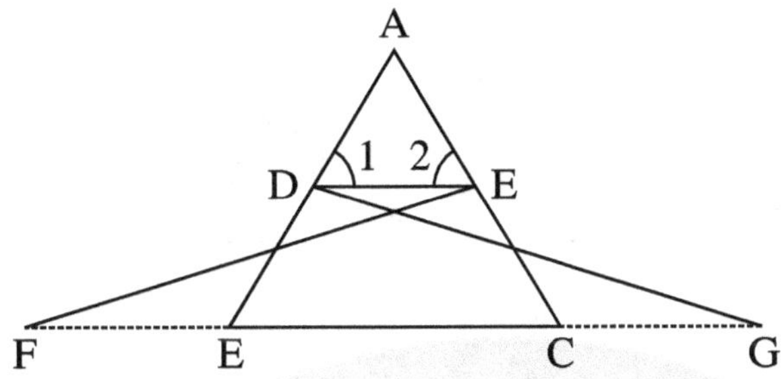

Sol.

SHORT ANSWER TYPE Q.UESTIONS-II

Q. 30. In the given figure, $\dfrac{QR}{QS} = \dfrac{QT}{PR}$ and ∠1 = ∠2 then prove that ΔPQ.S ~ ΔTQ.R.   **(NCERT)**

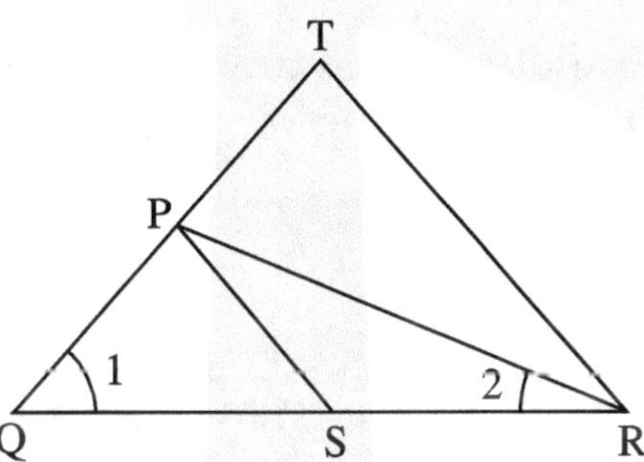

Sol.

Q. 31. In the adjoining figure ΔABC and ΔDBC are on the same base BC. AD and BC intersect at O.

Prove that $\dfrac{\text{area}\,(\Delta\,ABC)}{\text{area}\,(\Delta\,DBC)} = \dfrac{AO}{DO}$.   **(CBSE 2020)**

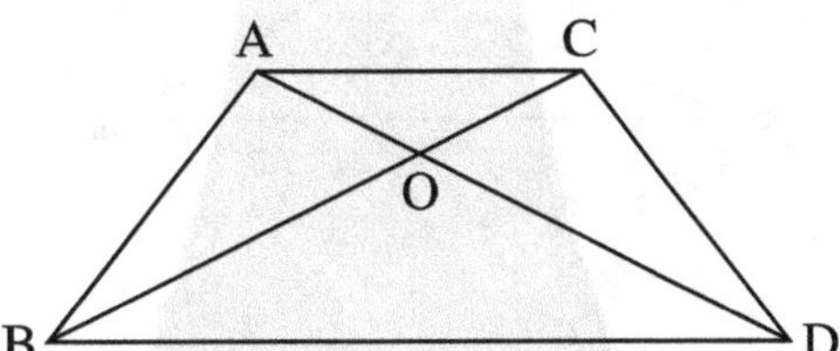

Sol.

Q. 32. If AD and PS are medians of ΔABC and ΔPQ.R respectively where ΔABC ~ ΔPQ.R, Prove that $\dfrac{AB}{PQ} = \dfrac{AD}{PS}$.

Sol.

Q. 33. In the given figure, DE ∥ AC. Which of the following is correct?

$$x = \frac{a+b}{ay} \qquad \text{or} \qquad x = \frac{ay}{a+b}$$

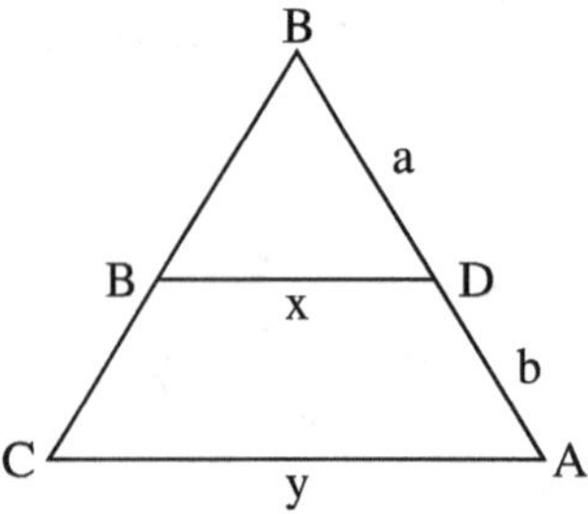

Sol.

Q. 34. If three parallel lines are intersected by two transversals, them prove that the intercepts made by them on the transversals are proportional.

Sol.

Q. 35. A street light bulb is fixed on a pole 6 m above the level of the street. If a woman of height 1.5 m casts a shadow of 3 m, find how far she is away from the base of the pole.

(NCERT Exemplar)

Sol.

Q. 36. Two poles of height a metrs and b metres are p metres apart. Prove that the height of the point of intersection of the lines joining the top of each pole to the foot of the opposite pole is given by $\dfrac{ab}{a+b}$ metres.

Sol.

Q. 37. In the given figure AB $\parallel$ PQ. $\parallel$ CD, AB = x, CD = y and PQ. = z. Prove that $\dfrac{1}{x}+\dfrac{1}{y}=\dfrac{1}{z}$.

(CBSE 2020)

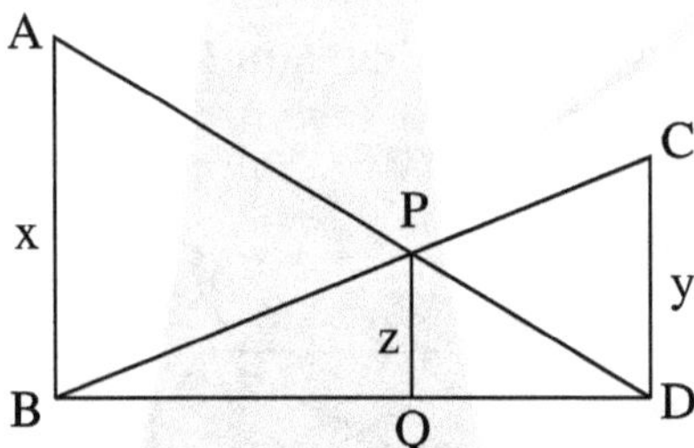

Sol.

Q. 38. In the given figure $\angle D = \angle E$ and $\dfrac{AD}{DB}=\dfrac{AE}{EC}$. Prove that $\triangle BAC$ is an isoscles triangle.

(CBSE 2020)

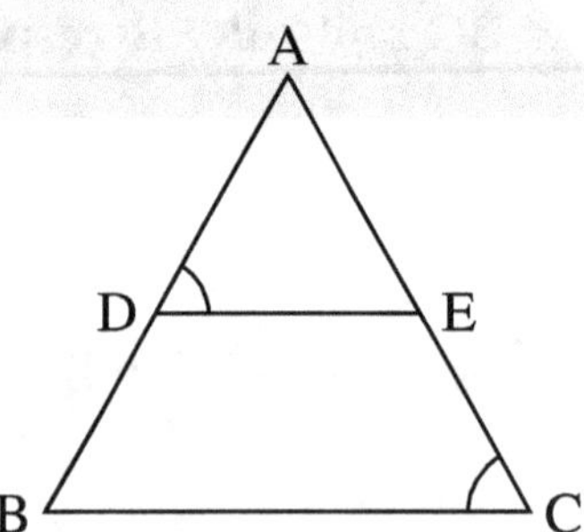

Sol.

Innovative Mathematics X-6

Q. 39. In the figure, a point O inside $\triangle$ABC is joined to its vertices. From a point D on AO, DE is drawn parallel to AB and from a point E on BO, EF is drawn parallel to BC. Prove that DF ∥ AC.

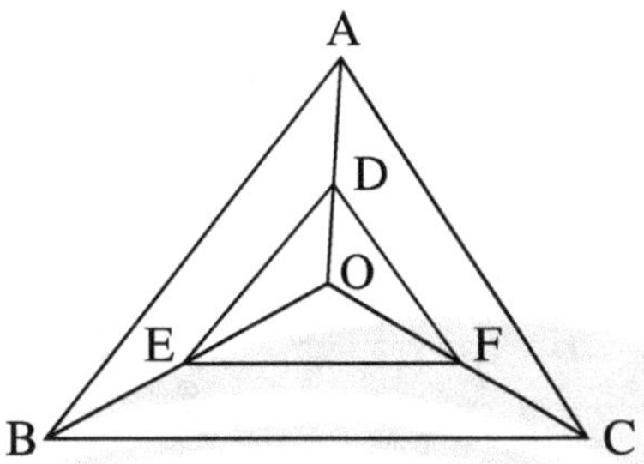

Sol.

Q. 40. Two triangles $\triangle$BAC and $\triangle$BDC, right angled at A and D respectively are drawn on the same base BC and on the same side of BC. If AC and DB intersect at P. Prove that AP × PC = DP × PB. **(CBSE 2019)**

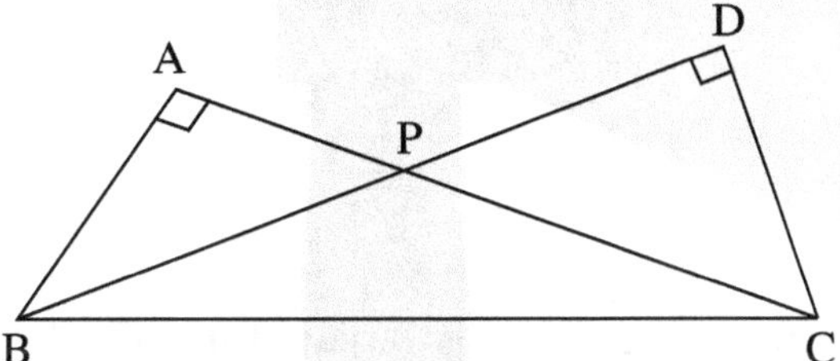

Sol.

Q. 41. In the given fig., P is the mid point of BC and Q. is the mid point of AP. If BQ. when produced meets AC at R, prove that RA = 1/3 CA. **(CBSE)**

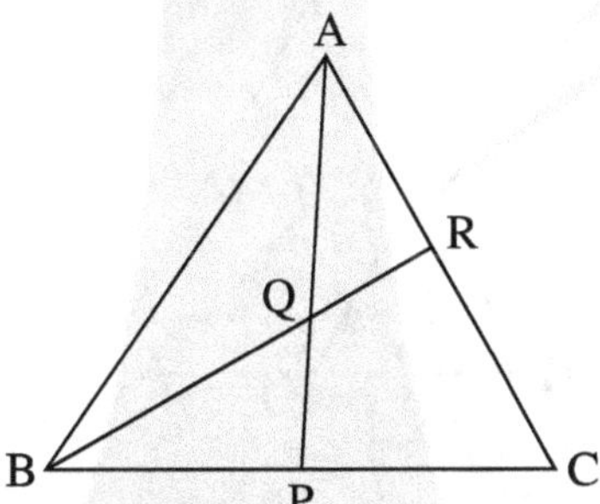

Sol.

Q. 42. In the given figure DE ∥ AC and $\dfrac{BE}{EC} = \dfrac{BC}{CP}$. Prove that DC ∥ AP.

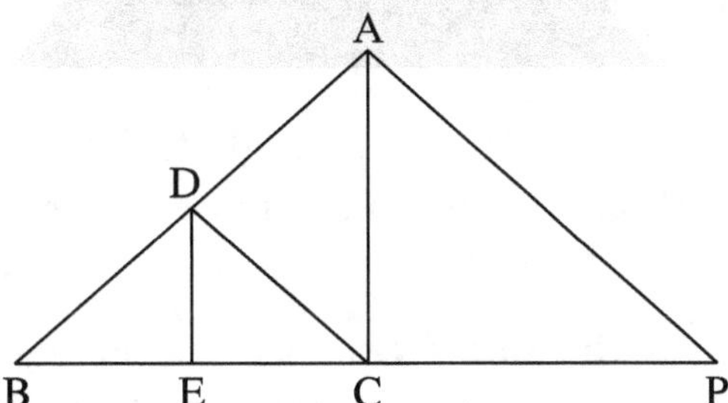

Sol.

Q. 43. In $\triangle ABC$, AD is a median, X is a point on AD such that AX : XD = 2 : 3. Ray B; intersects AC in Y. Prove that BX = 4XY.

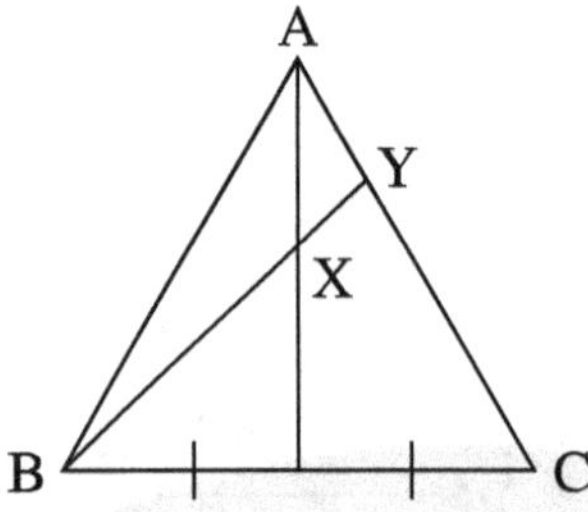

Sol.

Q. 44. In the given figure, DE ∥ BC, DE = 3 cm, BC = 9 cm and ar($\triangle ADE$) = 30 cm². Find ar(BCED).

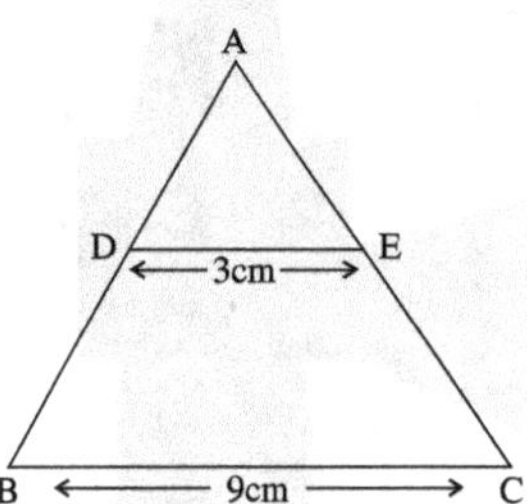

Sol.

Q. 45. In the given figure, the line segment XY is Parallel to AC of $\triangle ABC$ and it divides the triangle into two parts of equal areas. Prove that $\dfrac{AX}{AB} = \dfrac{\sqrt{2}-1}{\sqrt{2}}$.

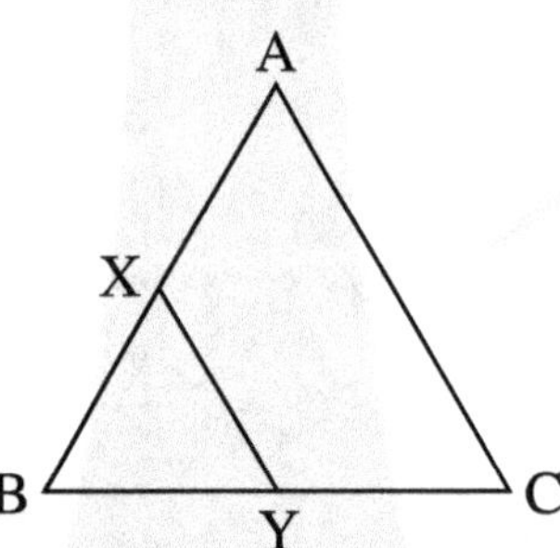

Sol.

Q. 46. Through the vertex D of a parallelogram ABCD, a line is drawn to intersect the side BA and BC produced at E and F respectively. Prove that $\dfrac{DA}{AE} = \dfrac{FB}{BE} = \dfrac{FC}{CD}$.

Sol.

Q. 47. If a line is drawn parallel to one side of a triangle to intersect the other two sides in distinct points, then prove that the other two sides are divided in the same ratio.   (CBSE 2019, 2020)

Sol.

Q. 48. Through the mid point M of the side CD of a parallelogram ABCD, the line BM is drawn intersecting AC in L and AD produced in E. Prove that EL = 2BL.

Sol.

Innovative Mathematics X-6

Q. 49. In the given figure, $\angle AEF = \angle AFE$ and E is the mid-point of CA. Prove that $\dfrac{BD}{CD} = \dfrac{BF}{CE}$.

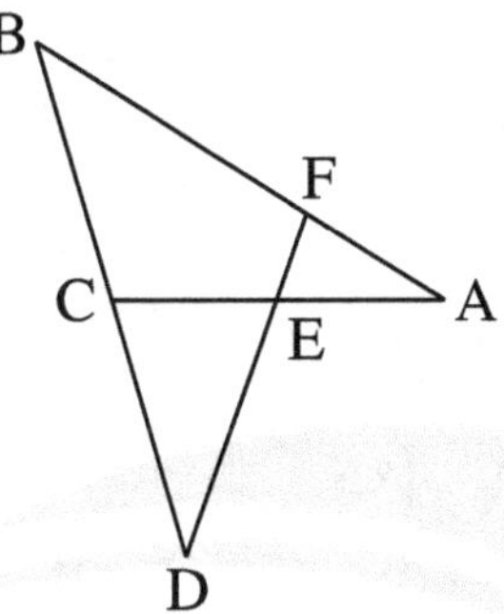

Sol.

Q. 50. Sides AB and AC and median AD of $\triangle ABC$ are respectively proportional to sides PQ. and PR and median PM of $\triangle PQ.R$. Show that $\triangle ABC \sim \triangle PQ.R$. **(CBSE 2020)**

Sol.

Q. 51. In figure if $\triangle ABC \sim \triangle DEF$ and their sides of lengths (in cm) are marked along them then find the lengths of sides of each triangle. **(CBSE 2020)**

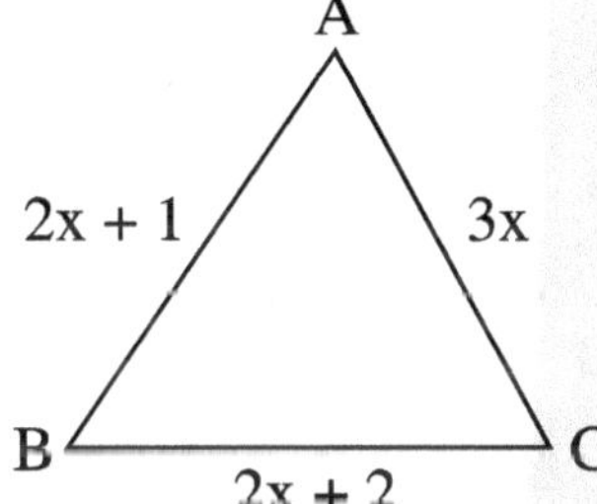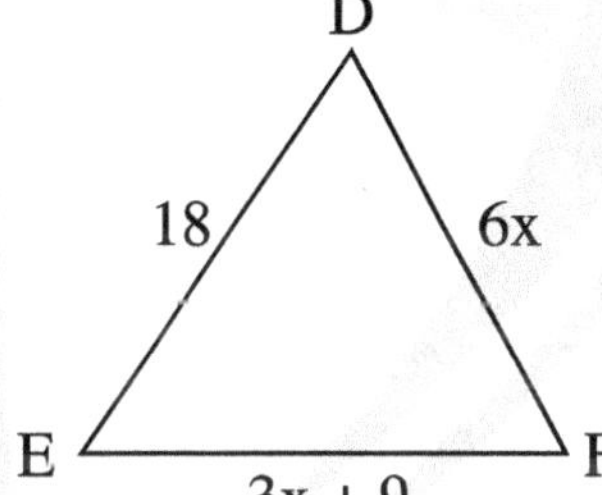

Sol.

Q. 52. The Perimeters of two similar triangles are 30 cm and 20 cm respectively. If one side of the first triangle is 9 cm long. Find the length of the corresponding side of the second triangle.

(CBSE 2020)

Sol.

Q. 53. If in $\triangle ABC$, D be a point on BC such that $\dfrac{BD}{DC} = \dfrac{AB}{AC}$, then show that AD is bisector of $\angle A$.

Sol.

1. (i) Similar   (ii) $\dfrac{AB}{FE} = \dfrac{BC}{ED}$ (iii) Congruent

   (iv) Same   (v) Proportional   (vi) $90°$

2. (i) False   (ii) False   (iii) True

   (iv) True   (v) False

3. (a) (iii) (AAA)   (b) (iv) (SSS)   (c) (i) (SAS)

4. (b)   5. (a)   6. (c)

9. $AC = 5$ cm   10. $\angle D = \angle R$ (True), $\angle F = \angle P$ (False)

11. $5 : 7$ 12. $x = 3$ cm   13. $XY : Q.R = 1 : 3$

14. $x = 2$ 15. $\angle F = \angle C = 56°$ 16. $2 : 3$

21. $x = 1/2$ or 2 22. $Q.R = 6\sqrt{3}$ cm   23. $CD = 2.5$ cm

24. $BC = DA = 3$ cm   25. (i) $65°$, (ii) $45°$, (iii) $45°$, (iv) $70°$

26. $DC^2 = CF \times AC$ 27. $\triangle ABC$ is Isoscles   28. $PB \times PC = 100$ cm^2

35. $BD = 9$ cm

SECTION-A

Q. 1. In the given fig., $\triangle ABC \sim \triangle PQ.R$, then find $(m + n)$

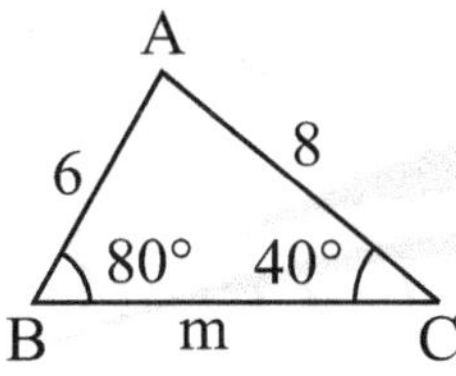 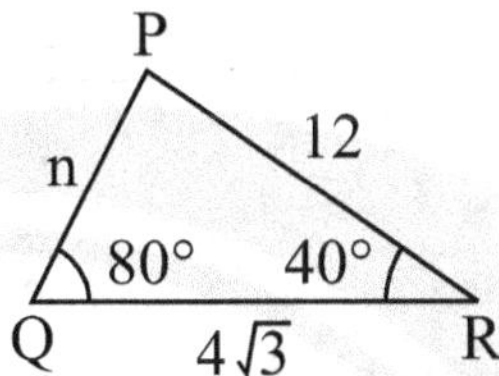

Sol.

Q. 2. In the given fig., DE || Q.R, PQ. = 5.6 cm and PD = 1.6 cm then find PE : ER.

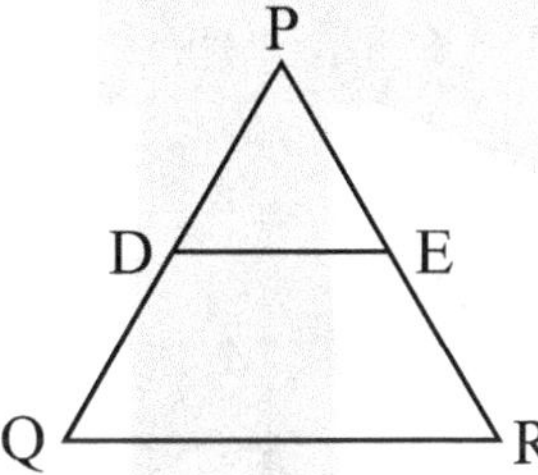

Sol.

Q. 3. $\triangle ABC$ is such that AB = 3 cm, BC = 2 cm and CA = 2.5 cm. In $\triangle PQ.R \sim \triangle ABC$ and Q.R = 6 cm, then perimeter of $\triangle DEF$ is __________. (1)

Sol.

Q. 4. If in two triangles ABC and DEF, $\dfrac{AB}{DE} = \dfrac{BC}{EF} = \dfrac{AC}{FD}$, then

(a) $\triangle BCA \sim \triangle FDE$ (b) $\triangle FDE \sim \triangle ABC$

(c) $\triangle CBA \sim \triangle FDE$ (d) $\triangle FDE \sim \triangle CAB$

Sol.

SECTION-B

Q. 5. In the given fig., Q.R || BC and Q.P || AC. If PB = 12 cm, PC = 20 cm and AR = BQ. = 15 cm, calculate AQ. and CR.

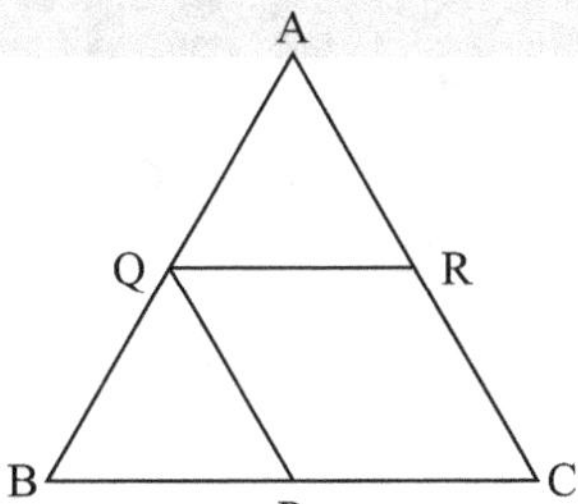

Sol.

Q. 6. In the given fig., BD $\perp$ AC and CE $\perp$ AB, prove that BP × PD = EP × PC.

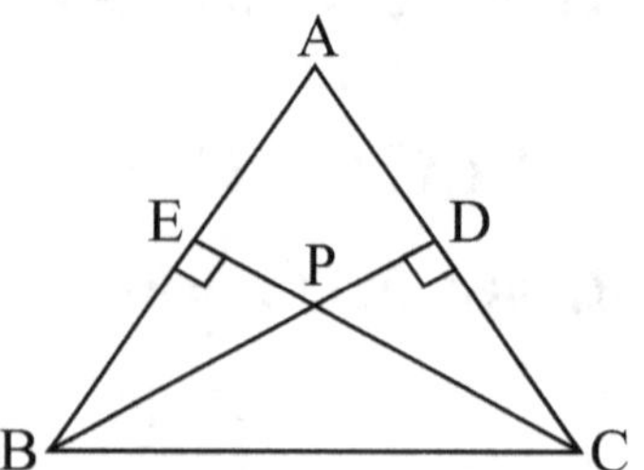

Sol.

Q. 7. If one diagonal of a trapezium divides the other diagonal in the ratio 1 : 3, prove that one of the parallel sideds is three times the other. (2)

Sol.

SECTION-C

Q. 8. In the given fig., if AB $\perp$ BC, PO $\perp$ AC and MN $\perp$ BC, prove that $\triangle$APQ. $\sim$ $\triangle$MCN.

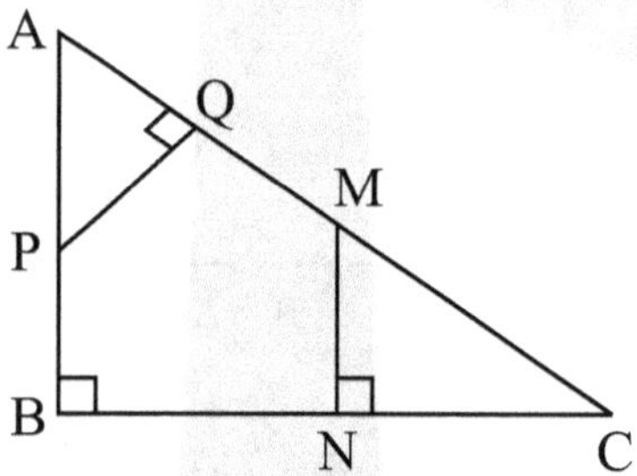

Sol.

Q. 9. E is a point on the side AD produced of a parallelogram ABCD and BE interects CD at F. Show that AB × BC = AE × CF. (3)

Sol.

SECTION-D

Q. 10. State and prove Basic Proportionality Theorem. (4)

Sol.

Innovative Mathematics X-6

CASE STUDY

CASE STUDY 1.

Vijay is trying to find the average height of a tower near his house. He is using the properties of similar triangles. The height of Vijay's house if 20m when Vijay's house casts a shadow 10m long on the ground. At the same time, the tower casts a shadow 50m long on the ground and the house of Ajay casts 20m shadow on the ground.

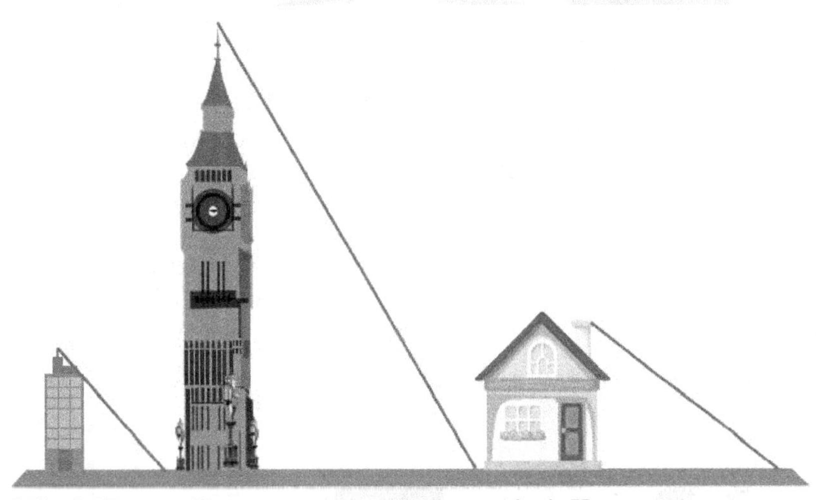

Q. 1. **What is the height of the tower?**

    (a) 20 m   (b) 50 m   (c) 100 m   (d) 200 m

Q. 2. **What will be the length of the shadow of the tower when Vijay's house casts a shadow of 12m?**

    (a) 75 m   (b) 50 m   (c) 45 m   (d) 60 m

Q. 3. **What is the height of Ajay's house?**

    (a) 30 m   (b) 40 m   (c) 50 m   (d) 20 m

Q. 4. **When the tower casts a shadow of 40m, same time what will be the length of the shadow of Ajay's house?**

    (a) 16 m   (b) 32 m   (c) 20 m   (d) 8 m

Q. 5. **When the tower casts a shadow of 40m, same time what will be the length of the shadow of Vijay's house?**

    (a) 15 m   (b) 32 m   (c) 16 m   (d) 8 m

ANSWERS

1. (c) 100 m   **2.** (d) 60 m   **3.** (b) 40 m

4. (a) 16 m   **5.** (d) 8 m

CASE STUDY 2.

Rohan wants to measure the distance of a pond during the visit to his native. He marks points A and B on the opposite edges of a pond as shown in the figure below. To find the distance between the points, he makes a right-angled triangle using rope connecting B with another point C are a distance of 12 m, connecting C to point D at a distance of 40 m from point C and the connecting D to the point A which is are a distance of 30 m from D such the ADC = 900.

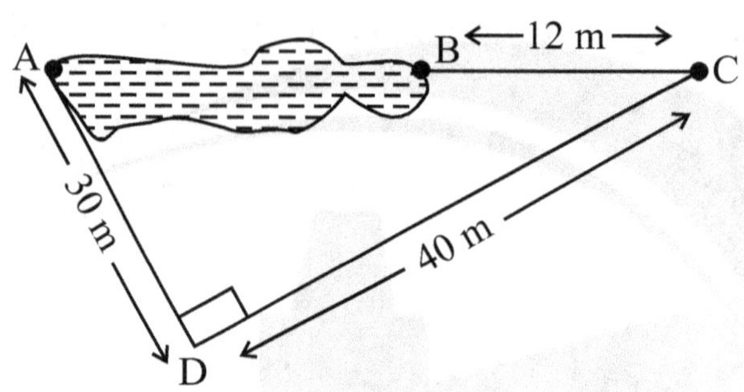

Q. 1. **Which property of geometry will be used to find the distance AC?**

    (a) Similarity of triangles   (b) Thales Theorem

    (c) Pythagoras Theorem   (d) Area of similar triangles

Q. 2. **What is the distance AC?**

    (a) 50 m   (b) 12 m   (c) 100 m   (d) 70 m

Q. 3. **Which is the following does not form a Pythagoras triplet?**

    (a) (7,24,25)   (b) (15,8,17)   (c) (5,12,13)   (d) (21,20,28)

Q. 4. **Find the length AB?**

    (a) 12 m   (b) 38 m   (c) 50 m   (d) 100 m

Q. 5. **Find the length of the rope used.**

    (a) 120 m   (b) 70 m   (c) 82 m   (d) 22 m

ANSWERS

1. (c) Pythagoras Theorem   **2.** (a) 50 m   **3.** (d) (21,20,28)

4. (b) 38 m   **5.** (c) 82 m

CASE STUDY 3.

A scale drawing of an object is the same shape at the object but a different size. The scale of a drawing is a comparison of the length used on a drawing to the length it represents. The scale is written as a ratio. The ratio of two corresponding sides in similar figures is called the scale factor.

Scale factor = length in image / corresponding length in object

If one shape can become another using revising, then the shapes are similar. Hence, two shapes are similar when one can become the other after a resize, flip, slide or turn. In the photograph below showing the side view of a train engine. Scale factor is 1:200

This means that a length of 1 cm on the photograph above corresponds to a length of 200cm or 2 m, of the actual engine. The scale can also be written as the ratio of two lengths.

Q. 1. **If the length of the model is 11cm, then the overall length of the engine in the photograph above, including the couplings(mechanism used to connect) is:**

    (a) 22 cm   (b) 220 cm   (c) 220 m   (d) 22 m

Q. 2. **What will affect the similarity of any two polygons?**

    (a) They are flipped horizontally   (b) They are dilated by a scale factor

    (c) They are translated down   (d) They are not the mirror image of one another.

Q. 3. **What is the actual width of the door if the width of the door in photograph is 0.35cm?**

    (a) 0.7 m   (b) 0.7 cm   (c) 0.07 cm   (d) 0.07 m

Q. 4. **If two similar triangles have a scale factor 5:3 which statement regarding the two triangles is true?**

    (a) The ratio of their perimeters is 15:1   (b) Their altitudes have a ratio 25:15

    (c) Their medians have a ratio 10:4   (d) Their angle bisectors have a ratio 11:5

Q. 5. **The length of AB in the given figure:**

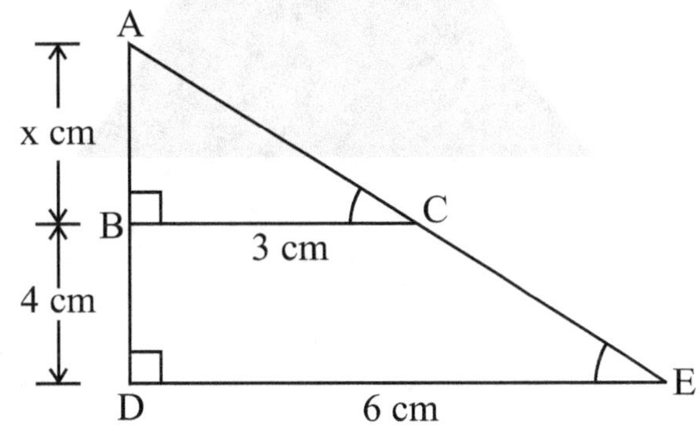

    (a) 8 cm   (b) 6 cm   (c) 4 cm   (d) 10 cm

1. (a) 22 m
2. (d) They are not the mirror image of one another
3. (a) 0.7 m
4. (b) Their altitudes have a ratio 25:15
5. (c) 4 cm

CHAPTER-7

COORDINATE GEOMETRY

FUNDAMENTALS

Coordinate Geometry (also known as Analytical Geometry) as a branch of mathematics where we make use of algebra in the study of geometry, has already been intoduced to you in class IX. Also, you have been introduced to some basic concepts and terms related to coordinate geometry like the coordinate system, the coordinate axes, the quadrants, the x- and y-coordinates (abscissa and ordinate), etc., and you know how to locate a point in the coordinate plane. In the present chapter, we extend our knowledge about coordinate geometry by learning how to find the distance between the two points whose coordinates are given, and to find the area of the triangle formed by three given points. We also learn about finding the coordinates of the point which divides a line segment joining two given points in a given ratio.

Distance Between Two Points (Distance Formula)

Using coordinate geometry, we can find out the distance between two points in terms of their coordinates.

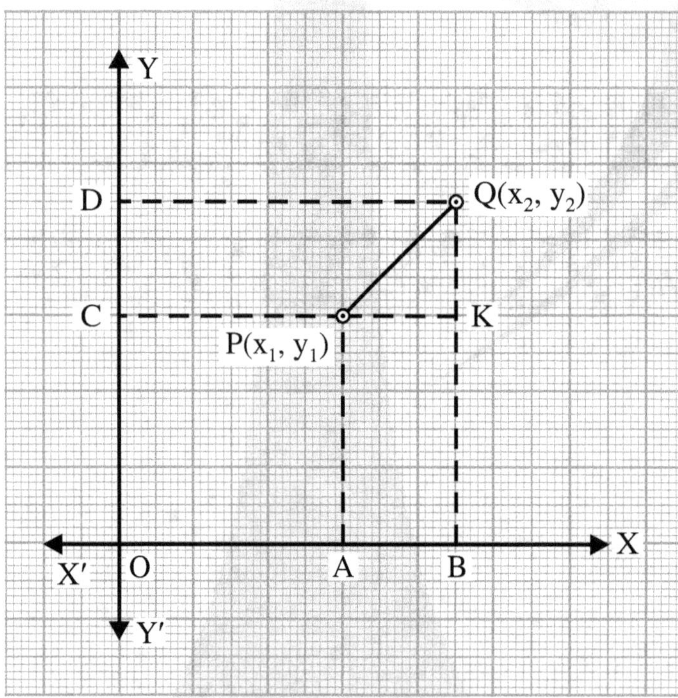

Let $P(x_1, y_1)$ and $Q. (x_2, y_2)$ be the given points. Draw PA, Q.B perpendicular to OX and PC, Q.D perpendicular to OY. Produce CP to meet Q.B in K.

Now $\qquad PK = AB = OB - OA = x_2 - x_1$

and $\qquad Q.K = Q.B - KB = Q.B - PA = y_2 - y_1$

Applying Pythagoras Theorem in right angled triangle PKQ., we get :

$$PQ.^2 = PK^2 + Q.K^2$$

or $\qquad PQ.^2 = (x^2 - x_1)^2 + (y_2 - y_1)^2$

$$\therefore \qquad PQ. = \sqrt{(x_2 - x_1)^2 + (y_2 - y_1)^2}$$

This is know as distance formula.

Corollary: The distance of any point P (x, y) from the origin O (0, 0) is given by:

$$OP = \sqrt{x^2 + y^2}$$

Section Formula

Section formula is used to find out the coordinates of the point which divides the line segment joining the two given points in a given ratio internally.

Let P (x, y) be the point dividing the line segment joining the points A (x_1, y_1) and B (x_1, y_2) internally in the ratio $m_1 : m_2$.

Draw AC, PQ. and BD perpendiculars to x-axis and AE and PR, both parallel to x-axis to meet PQ. and BD in E and R, respectively.

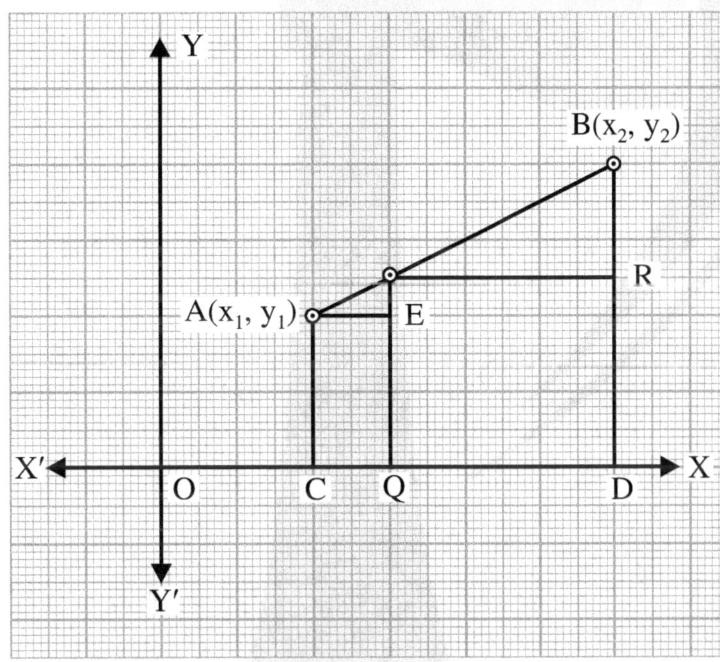

$$\Delta\, AEP \sim \Delta\, PRB \qquad\qquad \text{[AA similarity]}$$

$$\therefore \qquad \frac{AE}{PR} = \frac{EP}{RB} = \frac{AP}{PB} = \frac{m_1}{m_2}$$

Now
$$AE = CQ. = OQ. - OC = x - x_1$$
$$PR = Q.D = OD - OQ. = x_2 - x$$
$$EP = Q.P - Q.E = Q.P - CA = y - y1$$
and
$$RB = DB - DR = DB - Q.P = y_2 - y$$

Substituting the above in equaltion (1), we get:

$$\frac{x - x_1}{x_2 - x} = \frac{y - y_1}{y_2 - y} = \frac{m_1}{m_2}$$

Taking $\dfrac{x - x_1}{x_2 - x} = \dfrac{m_1}{m_2}$, we have

$$m_2 (x - x_1) = m_1 (x_2 - x)$$

or $$m_2 x - m_2 x_1 = m_1 x_2 - m_1 x$$

or $$x(m_1 + m_2) = m_1 x_2 + m_2 x_1$$

i.e., $$x = \dfrac{m_1 x_2 + m_2 x_1}{m_1 + m_2}$$

Similarly, $$y = \dfrac{m_1 y_2 + m_2 y_1}{m_1 + m_2}$$

Hence, the coordinates of point P, dividing the line segment joining the points $A(x_1, y_1)$ and $B(x_2, y_2)$ internally in ratio $m_1 : m_2$ are given by:

$$\left(\dfrac{m_1 x_2 + m_2 x_1}{m_1 + m_2}, \dfrac{m_1 y_2 + m_2 y_1}{m_1 + m_2} \right)$$

This is known as section formula.

Remarks:

1. In the case of mid point of the line segment joining $A(x_1, y_1)$ and $B(x_2, y_2)$ $m_1 = m_2 = 1$.

2. Coordinates of mid point are $\left(\dfrac{x_1 + x_2}{2}, \dfrac{y_1 + y_2}{2} \right)$

3. If the point P divides the line segment joining $A(x1, y1)$ and $B(x2, y2)$ in the ratio $k : 1$, its coordinates

   are given by : $\left(\dfrac{kx_2 + x_1}{k + 1}, \dfrac{ky_2 + y_1}{k + 1} \right)$

Q. 1. **Find the distance between the points P $(\cos \alpha, -\sin \alpha)$ and Q. $(-\cos \alpha, -\sin \alpha)$.**

Ans. Using distance formula:

$$PQ. = \sqrt{(-\cos \alpha - \cos \alpha)^2 + (\sin \alpha + \sin \alpha)^2}$$

$$= \sqrt{4 \cos^2 \alpha + 4 \sin^2 \alpha}$$

$$= \sqrt{4(\cos^2 \alpha + \sin^2 \alpha)}$$

$$= \sqrt{4(1)} \qquad\qquad [\sin^2\alpha + \cos^2 \alpha = 1]$$

$$= 2 \text{ units}$$

Q. 2. **Find the value(s) of x for which the distance between the points A $(x, 5)$ and B $(0, -3)$ is $4\sqrt{5}$ units.**

Ans. By distance formula, we get:

$$AB = \sqrt{(0 - x)^2 + (-3 - 5)^2}$$

$$\Rightarrow \qquad 4\sqrt{5} = \sqrt{x^2 + 64}$$

On squaring both sides, we get :

$$80 = x^2 + 64$$

$$\Rightarrow \qquad x^2 = 16$$

$$\therefore \qquad x = \pm 4$$

Q. 3. **Find the distance of the point P $(3, -4)$ from the origin.**

Ans. The given point is P $(3, -4)$ and the origin is given by O$(0, 0)$.

$\therefore$ By distance formula, we get :

$$OP = \sqrt{(3 - 0)^2 + (-4 - 0)^2}$$

$$= \sqrt{9 + 16}$$

$$= \sqrt{25}$$

$$= 5 \text{ units}$$

Q. 4. **The length of a line segment is 15 units. If one end point is $(-2, 5)$ and the ordinate of the other point is -4, find the abscissa of the other point.**

Ans. Let the abscissa of the other point be x.

$\therefore$ According to given:

$$\sqrt{(x + 2)^2 + (-4 - 5)^2} = 15$$

$$\sqrt{x^2 + 4x + 4 + 81} = 15$$

On squaring both sides, we get :

$$x^2 + 4x + 85 = 225$$

$$\Rightarrow \qquad x^2 + 4x - 140 = 0$$

$$\therefore \qquad x = \frac{-4 \pm \sqrt{16 - 4.1.(-140)}}{2}$$

$$= \frac{-4 \pm \sqrt{576}}{2}$$

$$= -14, 10$$

$\therefore$ The abscissa of the other point is -14 or 10.

Q. 5. **Find the values of x and y, if the point (x, y) is at a distance of 4 units from both the points (–3, 0) and (3, 0).**

Ans. Let the given points be : $A(x, y)$, $B(-3, 0)$ and $C(3, 0)$.

By distance formula, we get :

$$AB = \sqrt{(-3 - x)^2 + (0 - y)^2}$$

or

$$AB = \sqrt{x^2 + 6x + 9 + y^2}$$

and

$$AC = \sqrt{(3 - x)^2 + (0 - y)^2}$$

or

$$AC = \sqrt{x^2 - 6x + 9 + y^2}$$

But according to given $AB = AC$

$$\therefore \qquad \sqrt{x^2 + 6x + 9 + y^2} = \sqrt{x^2 - 6x + 9 + y^2}$$

On squaring both sides, we get,

$$x^2 + 6x + 9 + y^2 = x^2 - 6x + 9 + y^2$$

$$\Rightarrow \qquad 6x = -6x$$

or

$$12x = 0$$

$$\therefore \qquad x = 0$$

Also

$$AB = 4$$

$$\Rightarrow \qquad \sqrt{x^2 + 6x + 9 + y^2} = 4$$

On squaring both sides, we get :

$$x^2 + 6x + 9 + y^2 = 16$$

Putting $x = 0$, we get :

$$9 + y^2 = 16$$

or

$$y^2 = 7$$

$$\therefore \qquad y = \pm\sqrt{7}$$

Hence

$$x = 0 \text{ and } y = \pm\sqrt{7}$$

Q. 6. **Find the value of x such that PQ. = Q.R, where P, Q. and R are the points (2, 5), (x, –3) and (7, 9) respectively.**

Ans. Given that : $PQ. = Q.R$

$\therefore$ By distance formula, we get :

$$\sqrt{(x-2)^2+(-3-5)^2} = \sqrt{(7-x)^2+(9+3)^2}$$

$$\Rightarrow \quad \sqrt{x^2-4x+4+64} = \sqrt{49-14x+x^2+144}$$

$$\text{or} \quad x^2-4x+68 = x^2-14x+193$$

$$\Rightarrow \quad 10x = 125$$

$$\therefore \quad x = 12.5$$

Q. 7. **Determine if the points (1, 5), (2, 3) and (–2, –11) are collinear.**

Ans. Let the given points be A(1, 5), B(2, 3) and C(–2, –11).

By distance formula, we get :

$$AB = \sqrt{(2-1)^2+(3-5)^2}$$

$$= \sqrt{1+4}$$

$$= \sqrt{5}$$

$$BC = \sqrt{(-2-2)^2+(-11-3)^2}$$

$$= \sqrt{16+196}$$

$$= \sqrt{212} = 2\sqrt{53}$$

$$AC = \sqrt{(-2-1)^2+(-11-5)^2}$$

$$= \sqrt{9+256}$$

$$= \sqrt{256}$$

$$= \sqrt{5}\times\sqrt{53}$$

Since $AB + BC \neq AC$, the points A, B, C are not collinear.

Q. 8. **Using distance formula show that the points (8, 1), (3, –4) and (2, –5) are collinear.**

Ans. Let the given points (8, 1), (3, –4) and (2, –5) be denoted by A, B and C respectively.

By distance formula, we get :

$$AB = \sqrt{(3-8)^2+(-4-1)^2}$$

$$= \sqrt{25+25}$$

$$= \sqrt{50}$$

$$= 5\sqrt{2}$$

$$BC = \sqrt{(2-3)^2+(-5+4)^2}$$

$$= \sqrt{1+1}$$

$$= \sqrt{2}$$

$$AC = \sqrt{(2-8)^2 + (-5-1)^2}$$

$$= \sqrt{36+36}$$

$$= \sqrt{72}$$

$$= 6\sqrt{2}$$

Since $AB + BC = 5\sqrt{2} + \sqrt{2} = 6\sqrt{2} = AC$, the points A, B, C are collinear.

Q. 9. **Check whether (5, –2), (6, 4) and (7, –2) are the vertices of an isosceles triangle.**

Ans. Let the given points be A(5, –2), B(6, 4) and C(7, –2).

By distance formula, we get:

$$AB = \sqrt{(6-5)^2 + (4+2)^2}$$

$$= \sqrt{1+36}$$

$$= \sqrt{37}$$

$$BC = \sqrt{(7-6)^2 + (-2-4)^2}$$

$$= \sqrt{1+36}$$

$$= \sqrt{37}$$

Since AB = BC, $\triangle ABC$ is isosceles i.e., the given points are the vertices of an isosceles triangle.

Q. 10. **Show that the points P (4, 4), Q. (3, 5) and R (–1, 1) are the vertices of a right triangle.**

Ans. By distance formula, we get :

$$PQ. = \sqrt{(3-4)^2 + (5-4)^2}$$

$$= \sqrt{1+1} = \sqrt{2}$$

$$Q.R = \sqrt{(-1-3)^2 + (1-5)^2}$$

$$= \sqrt{16+16}$$

$$= \sqrt{32}$$

$$PR = \sqrt{(-1-4)^2 + (1-4)^2}$$

$$= \sqrt{25+9}$$

$$= \sqrt{34}$$

We observe that the longest side is PR and also that $PR^2 = PQ.^2 + Q.R^2$ $\qquad [34 = 2 + 32]$

$\therefore$ By converse of Pythagoras Theorem,

$\triangle$ PQ.R is a right angled $\triangle$ with right angle at Q..

Q. 11. **Show that the triangle whose vertices are P (2, –4), Q. (–4, 4) and R(4√3– 1, 3√4) is equilateral.**

Ans. By distance formula, we get :

$$PQ. = \sqrt{(-4-2)^2 + (4+4)^2}$$

$$= \sqrt{36 + 64}$$

$$= 10$$

$$Q.R = \sqrt{(4\sqrt{3} - 1 + 4)^2 + (3\sqrt{3} - 4)^2}$$

$$= \sqrt{48 + 9 + 24\sqrt{3} + 27 + 16 - 24\sqrt{3}}$$

$$= \sqrt{100}$$

$$= 10$$

and

$$PR = \sqrt{(4\sqrt{3} - 1 - 2)^2 + (3\sqrt{3} + 4)^2}$$

$$= \sqrt{48 + 9 - 24\sqrt{3} + 27 + 16 + 24\sqrt{3}}$$

$$= \sqrt{100}$$

$$= 10$$

We find that PQ. = Q.R = PR

∴   Δ PQ.R is equilateral.

Q. 12. **The vertices of a triangle are (–4, 0), (4, 6) and (2, –6). Show that it is a scalene triangle.**

Ans. Let P (–4, 0), Q. (4, 6) and R (2, –6) be the vertices of the given triangle.

By distance formula, we get :

$$PQ. = \sqrt{(4+4)^2 + (6-0)^2}$$

$$= \sqrt{64 + 36}$$

$$= \sqrt{100}$$

$$= 10$$

$$Q.R = \sqrt{(2-4)^2 + (-6-6)^2}$$

$$= \sqrt{4 + 144}$$

$$= \sqrt{148}$$

$$= 2\sqrt{37}$$

$$PR = \sqrt{(2+4)^2 + (-6-0)^2}$$

$$= \sqrt{36 + 36}$$

$$= \sqrt{72}$$

$$= 6\sqrt{2}$$

We find that PQ. = Q.R = PR

∴   PQ.R is a scalene triangle.

Q. 1. **Prove that the points (1, 1), (4, 4), (4, 8) and (1, 5) are the vertices of a parallelogram.**

Ans. Let A (1, 1), B (4, 4), C (4, 8) and D (1, 5) be the given vertices.

By distance formula, we get :

$$AB = \sqrt{(4-1)^2 + (4-1)^2}$$

$$= \sqrt{9+9}$$

$$= \sqrt{18}$$

$$= 3\sqrt{2}$$

$$BC = \sqrt{(4-4)^2 + (8-4)^2}$$

$$= \sqrt{16}$$

$$= 4$$

$$CD = \sqrt{(1-4)^2 + (5-8)^2}$$

$$= \sqrt{9+9}$$

$$= \sqrt{18}$$

$$= 3\sqrt{2}$$

and

$$DA = \sqrt{(1-1)^2 + (5-1)^2}$$

$$= \sqrt{16}$$

$$= 4$$

We find that $\qquad$ AB = CD

and $\qquad$ BC = DA

i.e., opposite sides of ABCD are equal.

$\therefore$ ABCD is a parallelogram.

Q. 2. **Show that the points (0, 1), (1, 4), (4, 3) and (3, 0) are the vertices of a square.**

Ans. Let A (0, 1), B (1, 4), C(4, 3) and D (3, 0) be the given vertices.

By distance formula, we get :

$$AB = \sqrt{(1-0)^2 + (4-1)^2}$$

$$= \sqrt{1+9}$$

$$= \sqrt{10}$$

$$BC = \sqrt{(4-1)^2 + (3-4)^2}$$

$$= \sqrt{9+1}$$

$$= \sqrt{10}$$

$$CD = \sqrt{(3-4)^2 + (0-3)^2}$$

$$= \sqrt{1+9}$$

$$= \sqrt{10}$$

$$DA = \sqrt{(3-0)^2 + (0-1)^2}$$

$$= \sqrt{9+1}$$

$$= \sqrt{10}$$

and

$$AC = \sqrt{(4-0)^2 + (3-1)^2}$$

$$= \sqrt{16+4}$$

$$= \sqrt{20}$$

We find that $\quad AB = BC = CD = DA$

and $\quad AB^2 + BC^2 = 10 + 10 = 20 = AC^2$

$\therefore \quad \angle B = 90°$ [by converse of Pythagoras Theorem]

Since all the four sides of equadrilateral ABCD are equal and $\angle B = 90°$

$\therefore$ ABCD is a square.

Q. 3. **Show that the points A (4, 2), B (7, 5) and C (9, 7) do not form a triangle.**

Ans. By distance formula, we get :

$$AB = \sqrt{(7-4)^2 + (5-2)^2}$$

$$= \sqrt{9+9}$$

$$= \sqrt{18}$$

$$= 3\sqrt{2}$$

$$BC = \sqrt{(9-7)^2 + (7-5)^2}$$

$$= \sqrt{4+4}$$

$$= \sqrt{8}$$

$$= 2\sqrt{2}$$

$$AC = \sqrt{(9-4)^2 + (7-2)^2}$$

$$= \sqrt{25+25}$$

$$= \sqrt{50}$$

$$= 5\sqrt{2}$$

We find that $\quad AB + BC = 3\sqrt{2} + 2\sqrt{2} = 5\sqrt{2} = AC$

$\therefore$ A, B, C are collinear and hence do not form a triangle.

Q. 4. **What point on y-axis is equidistant from the points (–1, 2) and (3, 4)?**

Ans. Let the point on the y-axis, P (0, y) be equidistant from the points A (–1, 2) and B (3, 4).

$$\therefore \qquad PA = PB$$

$$\text{or} \qquad PA^2 = PB^2$$

$$\therefore \qquad (-1-0)^2 + (2-y)^2 = (3-0)^2 + (4-y)^2$$

$$\text{or} \qquad 1 + 4 + y^2 - 4y = 9 + 16 + y^2 - 8y$$

$$\Rightarrow \qquad 4y = 20$$

$$\text{or} \qquad y = 5$$

$\therefore$ The required point is (0, 5).

Q. 5. **Find the values of y for which the distance between the points P(2, –3) and Q.(10, y) is 10 units.**

Ans. Distance between the points P(2, –3) and Q.(10, y) is given by,

$$PQ. = \sqrt{(10-2)^2 + (y+3)^2}$$

$$= \sqrt{64 + y^2 + 9 + 6y}$$

$$= \sqrt{y^2 + 6y + 73}$$

According to given:

$$\sqrt{y^2 + 6y + 73} = 10$$

$$\Rightarrow \qquad y^2 + 6y + 73 = 100$$

$$\text{or} \qquad y^2 + 6y - 27 = 0$$

$$\text{or} \qquad (y + 9)(y - 3) = 0$$

$$\Rightarrow \qquad y = -9 \quad \text{or} \quad y = 3$$

Q. 6. **Find the coordinates of a point which is at a distance of 10 units from (0, 1) and at a distance of 5 units from (3, 5).**

Ans. Let the coordinates of the point be (x, y).

Distance between the points (x, y) and (0, 1) is 10 units.

$$\Rightarrow \quad \sqrt{(x-0)^2 + (y-1)^2} = 10$$

On squaring both sides, we get:

$$x^2 + y^2 - 2y + 1 = 100$$

$$\text{or} \qquad x^2 + y^2 - 2y - 99 = 0$$

Distance between the points (x, y) and (3, 5) is 5 units

$$\Rightarrow \quad \sqrt{(x-3)^2 + (y-5)^2} = 5$$

On squaring both sides, we get:

$$x^2 - 6x + 9 + y^2 - 10y + 25 = 25$$

$$\text{or} \quad x^2 + y^2 - 6x - 10y + 9 = 0$$

Subtracting (2) from (1), we have:

Innovative Mathematics X-7

$$6x + 8y - 108 = 0$$

or
$$x = \dfrac{54 - 4y}{3}$$

Putting the value of x in (1), we get:

$$\left(\dfrac{54 - 4y}{3}\right)^2 + y^2 - 2y - 99 = 0$$

or $\quad 2916 - 432y + 16y^2 + 9y^2 - 18y - 891 = 0$

or $\qquad\qquad 25y^2 - 450y + 2025 = 0$

or $\qquad\qquad\qquad y^2 - 18y + 81 = 0$

or $\qquad\qquad\qquad\quad (y - 9)^2 = 0$

$\therefore \qquad\qquad\qquad\qquad\quad y = 9$

Putting the value of y in (3), we get:

$$x = \dfrac{54 - 4(9)}{3} = 6$$

$\therefore \quad$ The required point is (6, 9).

Q. 7. **Find a relation between x and y such that the point (x, y) is equidistant from the point (3, 6) and (–3, 4).**

Ans. Let the given points be A(x, y), B(3, 6) and C(–3, 4).

According to given:
$$AB = AC$$

$\Rightarrow \qquad\qquad\qquad AB^2 = AC^2$

$\Rightarrow \qquad (3 - x)^2 + (6 - y)^2 = (-3 - x)^2 + (4 - y)^2$

$\Rightarrow \quad 9 + x^2 - 6x + 36 + y^2 - 12y = 9 + x^2 + 6x + 16 + y^2 - 8y$

$\Rightarrow \qquad\qquad 12x + 4y - 20 = 0$

$\qquad\qquad\qquad 3x + y - 5 = 0$

which is the required relation.

Q. 8. **If the distances of P (x, y) from A (2, 1) and B (1, –2) are equal, prove that x + 3y = 0.**

Ans. P (x, y), A(2, 1) and B(1, –2) are the given points.

Given that : $\qquad\qquad AP = BP$

$\Rightarrow \qquad\qquad\qquad AP^2 = BP^2$

or $\qquad\qquad AP^2 - BP^2 = 0$

or $\quad [(2 - x)^2 + (1 - y)^2] - [(1 - x)^2 + (-2 - y)^2] = 0$

$\Rightarrow \quad 4 + x^2 - 4x + 1 + y^2 - 2y - 1 - x^2 + 2x - 4 - y^2 - 4y = 0$

or $\qquad\qquad -2x - 6y = 0$

$\Rightarrow \qquad\qquad\quad x + 3y = 0$

Q. 9. **Find the value of k, if the point P (0, 2) is equidistant from (3, k) and (k, 5).**

Ans. Given that : P (0, 2) is equidistant from A (3, k) and B (k, 5)

$$
\begin{aligned}
\therefore \qquad PA &= PB \\
\text{or} \qquad PA^2 &= PB^2 \\
\Rightarrow \qquad (3 - 0)^2 + (k - 2)^2 &= (k - 0)^2 + (5 - 2)^2 \\
\Rightarrow \qquad 9 + k^2 + 4 - 4k &= k^2 + 9 \\
\Rightarrow \qquad 4k &= 4 \\
\therefore \qquad k &= 1
\end{aligned}
$$

Q. 10. **Find the circumcentre of the triangle whose vertices are (3, 0), (–1, –6) and (4, –1). Also find its circumradius.**

Ans. Let P (x, y) be the circumcentre of the triangle having vertices A (3, 0), B (–1, –6) and C (4, –1).

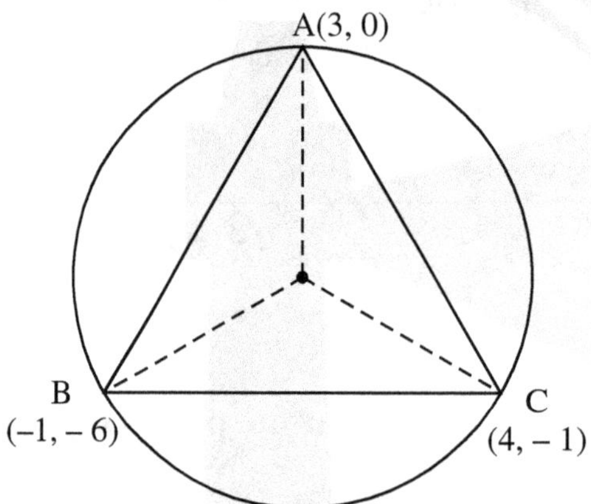

$\Rightarrow$ P is equidistant from A, B and C

$$
\begin{aligned}
\therefore \qquad PA &= PB = PC \\
\text{or} \qquad PA^2 &= PB^2 = PC^2 \\
PA^2 &= (x - 3)^2 + (y - 0)^2 \\
&= x^2 + y^2 - 6x + 9 \\
PB^2 &= (x + 1)^2 + (y + 6)^2 \\
&= x^2 + y^2 + 2x + 12y + 37 \\
PC^2 &= (x - 4)^2 + (y + 1)^2 \\
&= x^2 + y^2 - 8x + 2y + 17 \\
PA^2 &= PB^2 \\
\therefore \quad x^2 + y^2 - 6x + 9 &= x^2 + y^2 + 2x + 12y + 37 \\
\Rightarrow \qquad 8x + 12y &= -28 \\
\text{or} \qquad 2x + 3y &= -7 \\
PB^2 &= PC^2 \\
\therefore \quad x^2 + y^2 + 2x + 12y + 37 &= x^2 + y^2 - 8x + 2y + 17 \\
\Rightarrow \qquad 10x + 10y &= -20 \\
\text{or} \qquad x + y &= -2
\end{aligned}
$$

On solving (1) and (2) :

$$x = 1 \text{ and } y = -3$$

$\therefore$ The circumcentre is P $(1, -3)$

Also, circumradius $=$ PA (or PB or PC)

$$= \sqrt{(x-3)^2 + (y-0)^2}$$

$$= \sqrt{(1-3)^2 + (-3-0)^2}$$

$$= \sqrt{4+9}$$

$$= \sqrt{13}$$

Q. 11. **Find the coordinates of the point which divides the join of $(-1, 7)$ and $(4, -3)$ in the ratio $2 : 3$.**

Ans. Let P(x, y) be the required point.

Using section formula, we get:

$$x = \frac{2(4) + 3(-3)}{2+3} = 1$$

and

$$y = \frac{2(-3) + 3(7)}{2+3} = 3$$

$\therefore$ $(1, 3)$ is the required point.

Q. 12. **Find the coordinates of the points of trisection of the line segment joining $(4, -1)$ and $(-2, -3)$.**

A   P   Q   B

$(4, -1)$ $(-2, -3)$

Ans. Let the given points be A$(4, -1)$ and B$(-2, -3)$.

Also, let P and Q. be the points of trisection of AB, i.e., AP $=$ PQ. $=$ Q.B

$\Rightarrow$ P divides AB internally in the ratio $1 : 2$ and Q. divides AB internally in the ratio $2 : 1$.

Using section formula, coordinates of P are: $\left[\dfrac{1(-2) + 2(4)}{1+2}, \dfrac{1(-3) + 2(-1)}{1+2}\right]$ i.e., $\left(2, -\dfrac{5}{3}\right)$.

Similarly, coordinates of Q. are: $\left[\dfrac{2(-2) + 1(4)}{2+1}, \dfrac{2(-3) + 1(-1)}{2+1}\right]$ i.e., $\left(0, -\dfrac{7}{3}\right)$

Hence $\left(2, -\dfrac{5}{3}\right)$ and $\left(0, -\dfrac{7}{3}\right)$ are the required points.

Q. 13. **Find the point on the line through A $(5, -4)$ and B $(-3, 2)$ so that it is twice as far from A as from B.**

Ans. Let P (x, y) be the required point, so that it divides AB internally in the ratio $2 : 1$.

$2 : 1$
P

A   B

$(5, -4)$ $(-3, 2)$

$$\therefore \qquad x = \frac{2(-3) + 1 \times 5}{2+1} = \frac{-1}{3}$$

$$y = \frac{2 \times 2 + 1(-4)}{2+1}$$

$$= 0$$

$\therefore$ The required point is $\left(\dfrac{-1}{3}, 0\right)$

Q. 14. **Find the point which represents the three-fourths of the distance from (3, 2) to (5, 6).**

Ans. Let A (3, 2) and B (–5, 6) be the two given points.

Let C(x, y) be the point which represents three-fourths of the distance from A to B.

$$\therefore \qquad AC = \frac{3}{4} AB$$

$$\therefore \qquad BC = \frac{1}{4} AB$$

$$\Rightarrow \qquad AC : BC = 3 : 1$$

$\therefore$ Coordinates of C are $\left(\dfrac{3(-5) + 1(3)}{3+1}, \dfrac{3(6) + 1(2)}{3+1}\right)$

$\therefore$ The required point is (–3, 5)

Q. 15. **Find the ratio in which the segment joining the points of (–3, 10) and (6, –8) is divided by (–1, 6).**

Ans. Let (–1, 6) divide the join of (–3, 10) and (6, –8) in the ratio k : 1.

Using section formula, the coordinates of the dividing point are given by:

$$-1 = \frac{k(6) + 1(-3)}{k+1}$$

$$\Rightarrow \qquad -k - 1 = 6k - 3 \qquad \text{or} \qquad k = \frac{2}{7}$$

and

$$6 = \frac{k(-8) + 1(10)}{k+1}$$

$$\Rightarrow \qquad 6k + 6 = -8k + 10 \qquad \text{or} \qquad k = \frac{2}{7}$$

$\therefore$ The required ratio is $\dfrac{2}{7} : 1 \qquad \text{or} \qquad 2 : 7$

Q. 16. Find the ratio in which the line segment joining A(1, –5) and B(–4, 5) is divided by the x-axis. Also find the coordinates of the point of division.

Ans. Let the required ratio be k : 1.

Using section formula, the coordinates of the point of division are given by:

$$x = \frac{k(-4) + 1(1)}{k + 1} \text{ and } y = \frac{k(5) + 1(-5)}{k + 1}$$

Since it lies on x-axis,

∴ its y-coordinates is zero.

$$\Rightarrow \quad \frac{5k - 5}{k + 1} = 0$$

$$\Rightarrow \quad 5k - 5 = 0$$

or $\quad k = 1$

∴ The required ratio is 1 : 1

Substituting k = 1, the coordinates of the point of division are given by:

$$x = \frac{-4k + 1}{k + 1} = \frac{-4 + 1}{1 + 1} = \frac{-3}{2}$$

and $\quad y = 0$

∴ The required point is $\left(\dfrac{-3}{2}, 0\right)$.

Q. 17. In what ratio is the line segment joining the points (3, 5) and (–4, 2) divided by y-axis?

Ans. Let the required ratio be k : 1

∴ The coordinates of the required point on the y-axis are given by:

$$x = \frac{k(-4) + 3}{k + 1} \text{ and } y = \frac{k(2) + 5}{k + 1}$$

Since it lies on y-axis, ∴ its x-coordinate is zero

$$\therefore \quad \frac{-4k + 3}{k + 1} = 0$$

$$\Rightarrow \quad -4k + 3 = 0$$

or $\quad k = \dfrac{3}{4}$

∴ The required ratio is $\dfrac{3}{4}$: 1 or 3 : 4

Q. 18. Find the ratios in which the coordinate axes divide the join of (–2, –5) and (1, 9). Find also the coordinates of the points of division.

Ans. Let the join of A $(-2, -5)$ and B $(1, 9)$ be divided in the ratio $k : 1$ at P on the x-axis and at Q. on the y-axis.

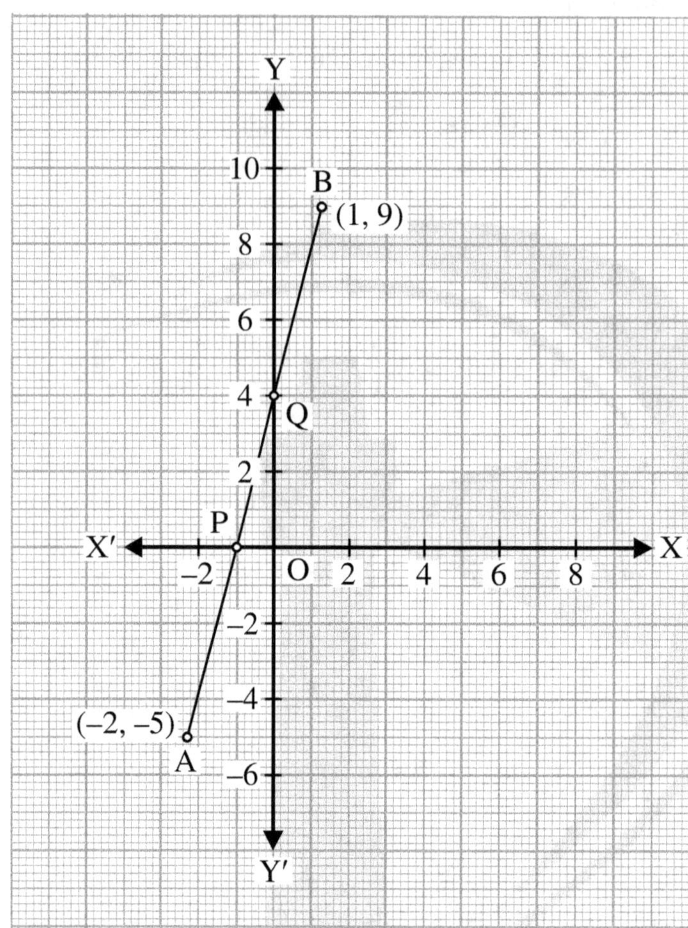

The coordinates of the dividing point are given by:

$$x = \frac{k \times 1 + (-2)}{k + 1}$$

$$= \frac{k - 2}{k + 1}$$

and

$$y = \frac{k \times 9 + (-5)}{k + 1}$$

$$= \frac{9k - 5}{k + 1}$$

(i) When the point (P) lies on x-axis, its y-coordinate $= 0$

$$\therefore \quad \frac{9k - 5}{k + 1} = 0$$

$$\Rightarrow \quad 9k - 5 = 0$$

or $\qquad k = 5/9$

$\therefore$ x-axis divides AB in the ratio 5/9 : 1 or 5 : 9

Putting k = 5/9, the coordinates of P are given by:

$$x = \frac{k-2}{k+1}$$

$$= \frac{\frac{5}{9}-2}{\frac{5}{9}+1}$$

$$= -\frac{13}{14}$$

and $\qquad\qquad y = 0$

$\therefore$ The point P is $\left(-\dfrac{13}{14}, 0\right)$

(ii) When the point (Q.) lies on y-axis, its x-coordinate = 0

$\therefore \qquad\qquad \dfrac{k-2}{k+1} = 0$

$\Rightarrow \qquad\qquad k-2 = 0$

or $\qquad\qquad k = 2$

$\therefore$ y-axis divides AB in the ratio 2 : 1

Putting k = 2, the coordinates of Q. are given by:

$$x = 0$$

and $\qquad\qquad y = \dfrac{9k-5}{k+1}$

$$= \frac{18-5}{2+1}$$

$$= \frac{13}{3}$$

$\therefore$ The point Q. is (0, 13/3)

Q. 19. **Find the ratio in which the line x – y – 2 = 0 divides the line segment joining the points (3, –1) and (8, 9).**

Ans. Let the required ratio be k : 1

$\therefore$ The point dividing the join of (3, –1) and (8, 9) in the ratio k : 1 is

$$\left(\frac{8k+3}{k+1}, \frac{9k-1}{k+1}\right)$$

Since, this point lies on the line x – y – 2 = 0, it should satisfy it.

i.e., $\dfrac{8k+3}{k+1}, \dfrac{9k-1}{k+1} - 2 = 0$

or $8k + 3 - 9k + 1 - 2k - 2 = 0$

$\Rightarrow \quad 3k = 2 \text{ or } k = \dfrac{2}{3}$

$\therefore$ The required ratio is 2 : 3.

Q. 20. **Find the distance of the point (1, 2) from the mid point of the line segment joining the points (6, 8) and (2, 4).**

Ans. Mid point of the line joining (6, 8) and (2, 4) is given by:

$$\dfrac{6+2}{2} \text{ and } \dfrac{8+4}{2}$$

i.e., $(4, 6)$

Now distance of the point (4, 6) from the point $(1, 2) = \sqrt{(1-4)^2 + (2-6)^2}$

$$= \sqrt{9+16}$$

$$= 5 \text{ units}$$

Q. 21. **Find the coordinate of a point A, where AB is the diameter of a circle whose centre is (2, –3) and B is (1, 4).**

Ans. Let (x, y) be the coordinates of A.

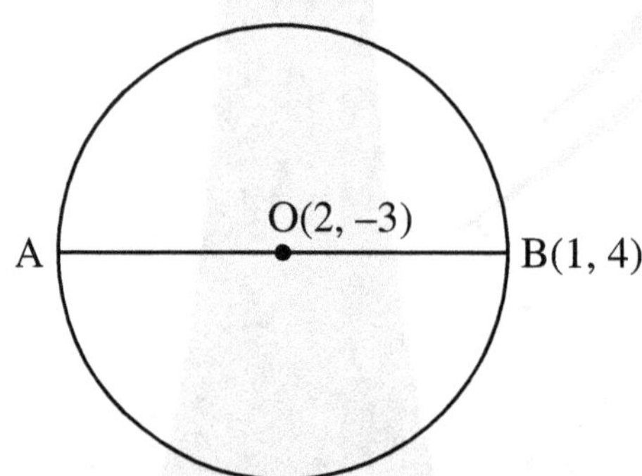

Since O(2, –3) is the mid point of the join of A(x, y) and B(1, 4).

$\therefore \qquad \dfrac{x+1}{2} = 2 \quad \Rightarrow \quad x = 3$

and $\qquad \dfrac{y+4}{2} = -3 \quad \Rightarrow \quad y = -10$

$\therefore$ Coordinates of A are (3, –10).

Q. 22. **Find the coordinates of the centre of circle, the coordinates of the end points of whose diameter are (–5, –2) and (7, –6). Also find the radius of the circle.**

Ans. Since A (–5, –2) and B (7, –6) are the end points of the diameter, so the centre of the circle is the mid point of AB.

Innovative Mathematics X-7

$\therefore$ Coordinates of the centre are $\left(\dfrac{-5+7}{2}, \dfrac{-2-6}{2}\right)$ or $(1, -4)$

$$\begin{aligned}
AB &= \sqrt{(7+5)^2 + (-6+2)^2} \\
&= \sqrt{144 + 16} \\
&= \sqrt{160} \\
&= 4\sqrt{10}
\end{aligned}$$

$\therefore$ Radius of the circle $= 2\sqrt{10}$ units

Q. 23. **If (1, 2), (4, y), (x, 6) and (3, 5) are the vertices of a parallelogram taken in order, find x and y.**

Ans. Let A(1, 2), B(4, y), C(x, 6) and D(3, 5) be the given vertices of the parallelogram.

Mid point of diagonal AC is $\left(\dfrac{1+x}{2}, \dfrac{2+6}{2}\right)$, i.e., $\dfrac{1+x}{2}, 4$

Mid point of diagonal BD is $\left(\dfrac{4+3}{2}, \dfrac{y+5}{2}\right)$, i.e., $\dfrac{7}{2}, \dfrac{y+5}{2}$

Since diagonals of a parallelogram bisect each other, the coordinates of their mid points are same.

$\Rightarrow \qquad \dfrac{1+x}{2} = \dfrac{7}{2}$ and $4 = \dfrac{y+5}{2}$

$\Rightarrow \qquad x = 6$ and $y = 3$

Q. 24. **If A (5, –1), B (–3, –2) and C(–1, 8) are the vertices of a triangle ABC, find the length of the median through A.**

Ans. Let AD be the median through A.

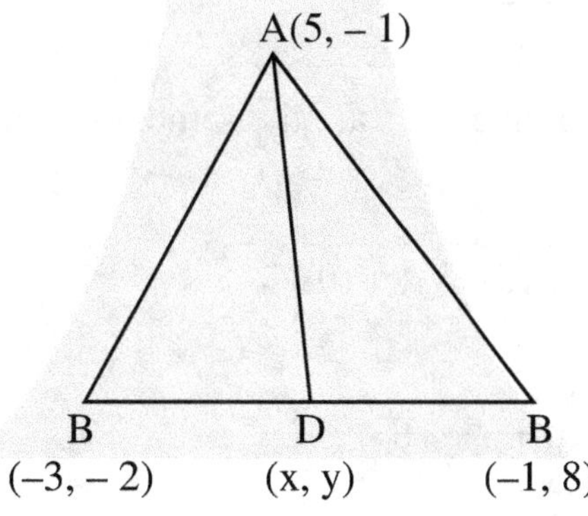

$\therefore$ D is the mid point of BC.

$\therefore$ Coordinates of D are given by :

$$x = \dfrac{-3-1}{2}$$

$$= -2$$

and
$$y = \frac{-2 + 8}{2}$$
$$= 3$$

$\therefore$ The point D is $(-2, 3)$

$\therefore$
$$AD = \sqrt{(-2 - 5)^2 + (3 + 1)^2}$$
$$= \sqrt{49 + 16}$$
$$= \sqrt{65} \text{ units}$$

Q. 25. **Prove that points $(-2, -1)$, $(1, 0)$, $(4, 3)$ and $(1, 2)$ form a parallelogram.**

Ans. Let $(-2, -1)$, $(1, 0)$, $(4, 3)$ and $(1, 2)$ be the coordinates of points A, B, C and D respectively.

Mid point of diagonal AC is $\left(\dfrac{-2 + 4}{2}, \dfrac{-1 + 3}{2} \right)$, i.e., $(1, 1)$

Mid point of diagonal BD is $\left(\dfrac{1 + 1}{2}, \dfrac{0 + 2}{2} \right)$, i.e., $(1, 1)$

Since the mid points of the two diagonals are the same, the diagonals bisect each other.

$\therefore$ ABCD is a parallelogram.

Q. 26. **Prove that points $(4, -1)$, $(6, 0)$, $(7, 2)$ and $(5, 1)$ are the vertices of a rhombus.**

Ans. Let $(4, -1)$, $(6, 0)$, $(7, 2)$ and $(5, 1)$ be the coordinates of points A, B, C and D respectively.

Mid point of diagonal AC is $\left(\dfrac{4 + 7}{2}, \dfrac{-1 + 2}{2} \right)$, i.e., $\left(\dfrac{11}{2}, \dfrac{1}{2} \right)$

Mid point of diagonal BD is $\left(\dfrac{6 + 5}{2}, \dfrac{0 + 1}{2} \right)$, i.e., $\left(\dfrac{11}{2}, \dfrac{1}{2} \right)$

Since the mid points of the two diagonals are the same, the diagonals bisect each other.

$\therefore$ ABCD is a parallelogram

Also,
$$AB = \sqrt{(6 - 4)^2 + (0 + 1)^2}$$
$$= \sqrt{5}$$

and
$$BC = \sqrt{(7 - 6)^2 + (2 - 0)^2}$$
$$= \sqrt{5}$$

$\therefore$ AB = BC, i.e., the adjacent sides of $\parallel$ gm ABCD are equal.

Hence, ABCD is rhombus.

Innovative Mathematics X-7

VERY SHORT ANSWER TYPE Q.UESTIONS

Q. 1.   **Fill in the blanks:**

    (i)   The distance of a point from the y-axis is called its x-coordinate or _______ .

    (ii)   The distance of a point from the x-axis is called its _________ or ordinate.

    (iii) The point (5, 0) lies on ______ axis.

    (iv) A point which lies on y-axis are of the form ________ .

    (v)   A linear equation of the form ax + by + c = 0 when represented graphically gives a ________ .

    (vi) The distance of a point P(x, y) from the origin is _________.

Sol.

MULTIPLE CHOICE QUESTIONS

Q. 2.   **P is a point on x-axis at a distance of 3 unit from y-axis to its left. The co-ordinates of P are:**

    (a) (3, 0)   (b) (0, 3)   (c) (– 3, 0)   (d) (0, – 3)

Sol.

Q. 3.   **The distance of P(3, – 2) from y-axis is**

    (a) 3 units   (b) 2 units   (c) – 2 units   (d) $\sqrt{13}$ units

Sol.

Q. 4.   **The co-ordinates of two points are (6, 0) and (0, – 8). The co-ordinates of the mid points are**

    (a) (3, 4)   (b) (3, – 4)   (c) (0, 0)   (d) (– 4, 3)

Sol.

Q. 5.   **If the distance between P(4, 0) and Q.(0, x) is 5 units, the value of x will be**

    (a) 2   (b) 3   (c) 4   (d) 5

Sol.

Q. 6.   **The co-ordinates of the point where line $\dfrac{x}{a}+\dfrac{y}{b}=7$ intersects y-axis are**

    (a) (a, 0)   (b) (0, b)   (c) (0, 7b)   (d) (2a, 0)

Sol.

Q. 7.   **The area of triangle OAB, the co-ordinates of whose vertices are A(4, 0), B(0, – 7) and O origin is:**

    (a) 11 sq. units   (b) 18 sq. units   (c) 28 sq. units   (d) 14 sq. units

Sol.

Q. 8.   **The distance between the points $P\left(-\dfrac{11}{3}, 5\right)$ and $Q\left(-\dfrac{2}{3}, 5\right)$ is**

    (a) 6 units   (b) 4 units   (c) 3 units   (d) 2 units

Sol.

Q. 9.   **The co-ordinate of the point which is reflection of point (–3, 5) in x axis are**

    (a) (3, 5)   (b) (3, –5)   (c) (–3, –5)   (d) (–3, 5)

Sol.

Q. 10. The co-ordinates of vertex A of $\triangle ABC$ are $(-4, 2)$ and a point D which is mid point of BC are) $(2, 5)$. The coordinates of centroid of $\triangle ABC$ are

(a) $(0, 4)$ (b) $\left(-1, \dfrac{7}{2}\right)$ (c) $\left(-2, \dfrac{7}{2}\right)$ (d) $(0, 2)$

Sol.

Q. 11. The distance between the line $2x + 4 = 0$ and $x - 5 = 0$ is

(a) 9 units   (b) 1 unit   (c) 5 units   (d) 7 units

Sol.

Q. 12. The perimeter of triangle formed by the points $(0, 0)$, $(2, 0)$ and $(0, 2)$ is

(a) 4 units   (b) 6 units   (c) $6\sqrt{2}$ units   (d) $4 + 2\sqrt{2}$ units

Sol.

Q. 13. If the centroid of the triangle formed by $(9, a)$, $(b, -4)$ and $(7, 8)$ is $(6, 8)$, then the value a and b are:

(a) $a = 4, b = 5$ (b) $a = 5, b = 4$ (c) $a = 5, b = 2$ (d) $a = 20, b = 2$

Sol.

Q. 14. The centre of circle having end points of its diameter as $(-4, 2)$ and $(4, -3)$ is

(a) $(2, -1)$ (b) $(0, -1)$ (c) $(0, -\dfrac{1}{2})$ (d) $(4, -\dfrac{5}{2})$ **(CBSE 2020)**

Sol.

Q. 15. The distance between the points $(0, 0)$ and $(a - b, a + b)$ is

(a) $2\sqrt{ab}$ (b) $\sqrt{2a^2 + ab}$ (c) $2\sqrt{a^2 + b^2}$ (d) $\sqrt{2a^2 + 2b^2}$ **(CBSE 2020)**

Sol.

SHORT ANSWER TYPE QUESTIONS-I

Q. 16. For what value of P, the points $(2, 1)$, $(p, -1)$ and $(-1, 3)$ are collinear.

Sol.

Q. 17. Three consecutive vertices of a parallelogram are $(-2, -1)$, $(1, 0)$ and $(4, 3)$. Find the co-ordinates of the fourth vertax.

Sol.

Q. 18. Find the point of trisection of the line segment joining the points $(1, -2)$ and $(-3, 4)$.

Sol.

Q. 19. The midpoints of the sides of a triangle are $(3, 4)$, $(4, 1)$ and $(2, 0)$. Find the vertices of the triangle.

Sol.

Q. 20. A circle has its centre at $(4, 4)$. If one end of a diameter is $(4, 0)$ then find the coordinates of the other end. **(CBSE 2020 Standard)**

Sol.

   Innovative Mathematics X-7

Q. 21. Find the ratio in which P(4, m) divides the line segment joining the points A(2, 3) and B(6, –3). Hence find m. **(CBSE 2018)**

Sol.

Q. 22. Show that the points (– 2, 3), (8, 3) and (6, 7) are the vertices of a right angle triangle.

Sol.

Q. 23. Find the point on y-axis which is equidistant from the points (5, –2) and (–3, 2). **(CBSE 2019)**

Sol.

Q. 24. Find the ratio in which y-axis divides the line segment joining the points A(5, – 6) and B(– 1, – 4).

Sol.

Q. 25. Find the co-ordinates of a centroid of a triangle whose vertices are (3, – 5), (– 7, 4) and (10, – 2).

Sol.

Q. 26. Find the relation between x and y such that the points (x, y) is equidistant from the points (7, 1) and (3, 5).

Sol.

Q. 27. Find the ratio in which the segment joining the points (1, –3) and (4, 5) is divided by x-axis. Also find the coordinates of the point on x-axis.

Sol.

Q. 28. What is the value of a if the points (3, 5) and (7, 1) are equidistant from the point (a, 0)?

Sol.

Q. 29. If the points A(4, 3) and B(x, 5) are on the circle with centre O(2, 3). Find the value of x.

Sol.

Q. 30. A(5, 1), B(1, 5) and C(–3, –1) are the vertices of $\triangle$ABC. Find the length of median passing through A. **(CBSE 2018)**

Sol.

Q. 31. Name the type of triangle formed by the points A(– 5, 6), B(– 4, – 2) and C(7, 5).

(NCERT Exempler)

Sol.

Q. 32. Find the points on the x-axis which are at a distance of $2\sqrt{5}$ from the point (7, – 4). How many such points are there? **(NCERT Exempler)**

Sol.

Q. 33. A line intersects the y-axis and x-axis at the point P and Q.. If (2, –5) is the midpoint of PQ. then find the co-ordinates of P and Q.. **(CBSE 2017)**

Sol.

Q. 34. If A(–2, 1), B(a, 0), C(4, b) and D(1, 2) are the vertices of a parallelogram ABCD, find the values of a and b. Hence find the lengths of its sides. **(CBSE 2018)**

Sol.

Q. 35. If P(x, y) is any point on the line joining A(a, 0) and B(0, b) then show that $\dfrac{x}{a} + \dfrac{y}{b} = 1$.
Sol.

Q. 36. Find the co-ordinates of the point which divides the line segment joining the points A(2, 6) and B(10, –10) in to 4 equal points. **(CBSE-2011)**

Sol.

Q. 37. Find the relation between x and y if A(x, y), B(– 2, 3) and C(2, 1) form an isosceles triangle with AB = AC.

Sol.

Q. 38. Prove that the point $\left(x, \sqrt{1-x^2}\right)$ is at a distance of 1 unit from the origin.

Sol.

Q. 39. Prove that the points (1, 2), (9, 3) and (17, 4) are collinear by section formula. **(CBSE 2017)**

Sol.

Q. 40. Determine the ratio in which the line $3x + y - 9 = 0$ divides the segment joining the points (1, 3) and (2, 7).

Sol.

Q. 41. Find the co-ordinate of the circumcenter of the triangle whose vertices are (3, 7), (0, 6) and (– 1, 5). Find the circumradius. **(HOTS)**

Sol.

Q. 42. In a triangle PQ.R, the co-ordinates of points P, Q. and R are (3, 2), (6, 4) and (9, 3) respectively. Find the co-ordinates of centroid G.

Sol.

Q. 43. If co-ordinates of two adjacent vertices of a parallelogram are (3, 2) and (1, 0) and diagonats bisect each other at (–2, 5). Find the co-ordinates of the other vertices.

Sol.

Q. 44. Determine if the points (1, 5), (2, 3) and (– 2, – 11) are collinear.

Sol.

Q. 45. Check whether (5, – 2), (6, 4) and (7, – 2) are the vertices of an isosceles triangle.

Sol.

Q. 46. Find the point on the x-axis which is equidistant from (2, – 5) and (–2, 9).

Sol.

Q. 47. Find the values of y for which the distance between the points P(2, – 3) and Q.(10, y) is 10 units.

Sol.

Q. 48. If Q.(0, 1) is equidistant from P(5, – 3) and R(x, 6), find the values of x. Also find the distances Q.R and PR.

Sol.

Q. 49. Find a relation between x and y such that the point (x, y) is equidistant from the point (3, 6) and (– 3, 4).

Sol.

Q. 50. Find the coordinates of the point which divides the join of (–1, 7) and (4, –3) in the ratio 2 : 3.

Sol.

Q. 51. Find the coordinates of the points of trisection of the line segment joining (4, – 1) and (–2, –3).

Sol.

Q. 52. Find the ratio in which the line segment joining the points $(-3, 10)$ and $(6, -8)$ is divided by $(-1, 6)$.

Sol.

Q. 53. Find the ratio in which the line segment joining $A(1, -5)$ and $B(-4, 5)$ is divided by the x-axis. Also find the coordinates of the point of division.

Sol.

Q. 54. If $(1, 2)$, $(4, y)$, $(x, 6)$ and $(3, 5)$ are the vertices of a parallelogram taken in order, find x and y.

Sol.

Q. 55. Find the coordinates of a point A, where AB is the diameter of a circle whose centre is $(2, -3)$ and B is $(1, 4)$.

Sol.

Q. 56. If A and B are $(-2, -2)$ and $(2, -4)$, respectively, find the coordinates of P such that $AP = \dfrac{3}{7} AB$ and P lies on the line segment AB.

Sol.

Q. 57. Find the coordinates of the points which divide the line segment joining $A(-2, 2)$ and $B(2, 8)$ into four equal parts.

Sol.

Q. 58. Find the area of a rhombus if its vertices are $(3, 0)$, $(4, 5)$, $(-1, 4)$ and $(-2, -1)$ taken in order.

Sol.

Q. 59. If the point $P(k - 1, 2)$ is equidistant from the points $A(3, k)$ and $B(k, 5)$ find the values of k.

Sol.

Q. 60. Find the relation between x and y such that the point $P(x, y)$ is equidistant from the points $A(1, 4)$ and $B(-1, 2)$.

Sol.

Q. 61. Find the point on x-axis which is equidistant from the points $(5, -2)$ and $(-3, 2)$.

Sol.

Q. 62. Find the point on y-axis which is equidistant from the points $(-5, 2)$ and $(9, -2)$.

Sol.

Q. 63. Show that the points (a, a), $(-a, -a)$ and $(-\sqrt{3}\, a, \sqrt{3}\, a)$ are the vertices of an equilateral triangle.

Sol.

Q. 64. Show that the points $(1, 1)$, $(-1, 5)$, $(7, 9)$ and $(9, 5)$ taken in that order are the vertices of a rectangle.

Sol.

Q. 65. Show that the points $A(3, 5)$, $B(6, 0)$, $C(1, -3)$ and $D(-2, 2)$ are the vertices of a square ABCD.

Sol.

Q. 66. If $A(2, -1)$, $B(3, 4)$, $C(-2, 3)$ and $D(-3, -2)$ be four points in a plane, show that ABCD is a rhombus but not a square.

Sol.

Q. 67. If the point $A(0, 2)$ is equidistant from the points $B(3, p)$ and $C(p, 5)$, find the value of p. Also, find the length of AB.

Sol.

Q. 68. If the distances of $P(x, y)$ from $A(5, 1)$ and $B(-1, 5)$ are equal then prove that $3x = 2y$.

Sol.

1. (i) abscissa   (ii) y-coordinate   (iii) x-axis   (iv) (0, y)

    (v) straight line   (vi) $\sqrt{x^2 + y^2}$

2. (c) $(-3, 0)$ **3.** (a) 3 units   **4.** (b) $(3, -4)$ **5.** (b) 3

6. (c) $(0, 7b)$ **7.** (d) 14 sq. units   **8.** (c) 3 units   **10.** (a) $(0, 4)$

11. (d) 7 units   **12.**   **13.**   **14.**

15.   **16.**   **17.**   **18.**

19.   **20.**   **21.**   **22.**

23.   **24.**   **25.**

26.   **28.**   **29.**   **30.**

31.   **32.**   **33.**   **34.**

35.   **37.**   **38.**   **39.**

40.   **41.**   **42.**   **43.**

44.   **45.**   **45.**   **48.**

49.   **50.**   **51.**   **52.**

53.   **54.**   **55.**   **56.**

57.   **58.**   **59.**   **60.**

61.   **62.** No   **62.** Yes   **63.**

64. $-9, 3$ **65.** $+4$, $Q.R = \sqrt{41}$, $PR = \sqrt{82}$, $9\sqrt{2}$ **66.** $3x + y - 5 = 0$

67. $(1, 3)$ **68.** $\left(2, -\dfrac{5}{3}\right); \left(2, -\dfrac{7}{3}\right)$ **69.** $2 : 7$ **70.** $1 : 1 ; \left(-\dfrac{3}{2}, 0\right)$

71. $x = 6, y = 3$ **72.** $(3, -10)$ **73.** $\left(-\dfrac{2}{7}, -\dfrac{20}{7}\right)$

74. $\left(-1, \dfrac{7}{2}\right), (0, 5)\left(1, \dfrac{13}{2}\right)$ **75.** 24 sq. units   **76.** 1 and 5.   **77.** $y = 3 - x$

78. $(1, 0)$.   **79.** $(0, -7)$ **80.** $p = 1$, $AB = 10$ units

PRACTICE-TEST

SECTION-A

Q. 1. x axis divides the line segment joining A(2, –3) and B(5, 6) in the ratio

(a) $2 : 3$ (b) $3 : 5$ (c) $1 : 3$ (d) $2 : 1$ **(1)**

Sol. ( )

Q. 2. What is the distance between the points A(c, 0) and B(0, –c)   **(1)**

Sol. ( )

Q. 3. The distance of point P(– 6, 8) from the origin is ______ .   **(1)**

Sol. ( )

Q. 4. Find the value of 'a' so that the point (3, a) lies on the line segment $2x - 3y = 5$.   **(1)**

Sol. ( )

SECTION-B

Q. 5. Find the point on y-axis which is equidistant from (–5, –2) and (3, 2)   **(2)**

Sol. ( )

Q. 6. If the points A(8, 6) and B(x, 10) lie on the circle whose centre is (4, 6) then find the value of x.   **(2)**

Sol. ( )

Q. 7. Find the perimeter of a triangle with vertices (0, 4), (0, 0) and (3, 0).   **(2)**

Sol. ( )

SECTION-C

Q. 8. Show that the points A(–3, 2), B(– 5, –5), C(2, –3) and D(4, 4) are the vertices of a rhombus.   **(3)**

Sol. ( )

Q. 9. Find the ratio in which the point (2, y) divides the line segment joining the points A(–2, 2) and B(3, 7). Also find the value of y.   **(3)**

Sol. ( )

SECTION-D

Q. 10. If the point P divides the line segment joining the points A(–2, –2) and B(2, –4) such that $\dfrac{AP}{AB} = \dfrac{3}{7}$, then find the coordinate of P.   **(4)**

Sol. ( )

CASE STUDY

CASE STUDY 1.

In order to conduct Sports Day activities in your School, lines have been drawn with chalk powder at a distance of 1 m each, in a rectangular shaped ground ABCD, 100 flowerpots have been placed at a distance of 1 m from each other along AD, as shown in given figure below. Niharika runs 1/4th the distance AD on the 2nd line and posts a green flag. Preet runs 1/5th distance AD on the eighth line and posts a red flag.

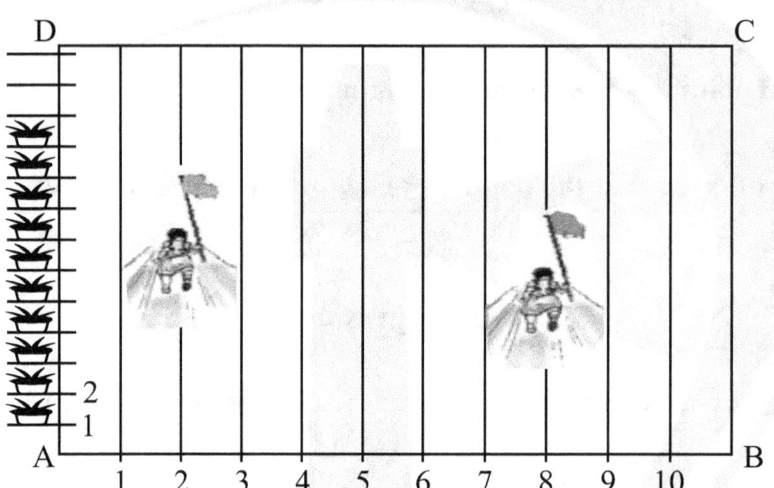

Q. 1. **Find the position of green flag**

    (a) (2, 25)   (b) (2, 0.25)   (c) (25, 2)   (d) (0, –25)

Q. 2. **Find the position of red flag**

    (a) (8, 0)   (b) (20, 8)   (c) (8, 20)   (d) (8, 0.2)

Q. 3. **What is the distance between both the flags?**

    (a) $\sqrt{41}$ (b) $\sqrt{11}$ (c) $\sqrt{61}$ (c) $\sqrt{51}$

Q. 4. **If Rashmi has to post a blue flag exactly halfway between the line segment joining the two flags, where should she post her flag?**

    (a) (5, 22.5)   (b) (10, 22)   (c) (2, 8.5)   (d) (2.5, 20)

Q. 5. **If Joy has to post a flag at one-fourth distance from green flag ,in the line segment joining the green and red flags, then where should he post his flag?**

    (a) (3.5, 24)   (b) (0.5, 12.5)   (c) (2.25, 8.5)   (d) (25,20)

ANSWERS

1. (a) (2, 25)   **2.** (c) (8, 20)   **3.** (c) $\sqrt{61}$

4. (a) (5, 22.5)   **5.** (a) (3.5, 24)

CASE STUDY 2.

The class X students school in krishnagar have been allotted a rectangular plot of land for their gardening activity. Saplings of Gulmohar are planted on the boundary at a distance of 1 m from each other. There is triangular grassy lawn in the plot as shown in the figure. The students are to sow seeds of flowering plants on the remaining area of the plot.

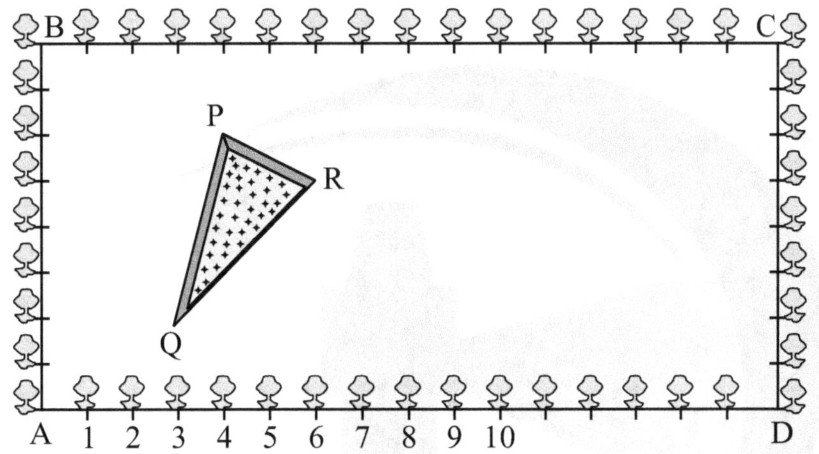

Q. 1. **Taking A as origin, find the coordinates of P**

(a) (4, 6) (b) (6, 4) (c) (0, 6) (d) (4, 0)

Q. 2. **What will be the coordinates of R, if C is the origin?**

(a) (8, 6) (b) (3, 10) (c) (10, 3) (d) (0,6)

Q. 3. **What will be the coordinates of Q., if C is the origin?**

(a) (6, 13) (b) (b) (–6, 13) (c) (–13, 6) (d) (13,6)

Q. 4. **Calculate the area of the triangles if A is the origin**

(a) 4.5 (b) 6 (c) 8 (d) 6.25

Q. 5. **Calculate the area of the triangles if C is the origin**

(a) 8 (b) 5 (c) 6.25 (d) 4.5

ANSWERS

1. (a) (4, 6)   **2.** (c) (10, 3)   **3.** (d) (13,6)

4. (a) 4.5   **5.** (d) 4.5

CHAPTER-8

INTRODUCTION TO TRIGONOMETRY

LEARNING OBJECTIVES

To find the distances and heights we can use the mathematical techniques, which come under the Trigonometry. It shows the relationship between the sides and the angles of the triangles. Generally, it is used in the case of a right angle triangle.

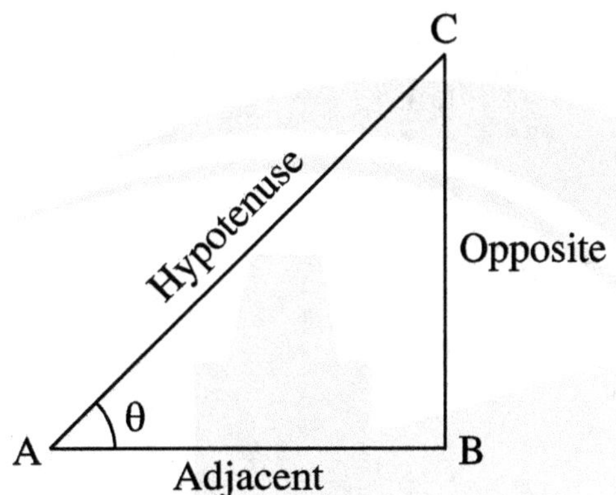

TRIGONOMETRY

It is made up of two greek word 'trigonon' means triangle and 'metron' means measurement.

Throughout history Trigonometry has been applied in areas such as surveying, celestial mechanics and navigation.

TRIGONOMETRIC RATIOS

In a right angle triangle, the ratio of its side and the acute angles is the trigonometric ratios of the angles.

From above right angle triangle $\angle B = 90°$. If $\angle A$ is an acute angle then then AB is the base, as the side adjacent to the acute angle. BC is the perpendicular as the side opposite to the acute angle. AC is the hypotenuse as the side opposite to the right angle.

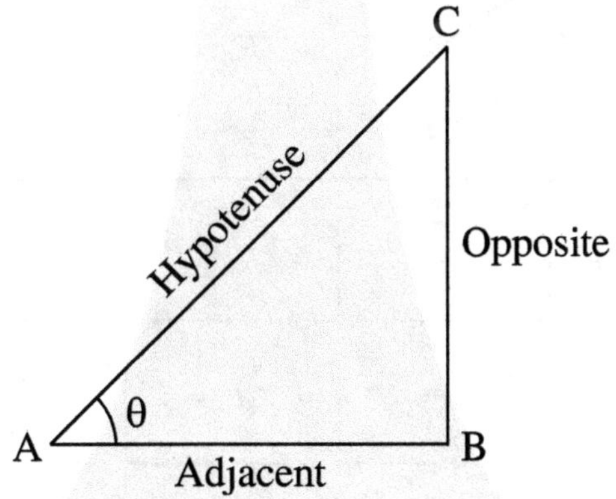

Trigonometric Ratios:-

sin A	P/H	BC/AC
cos B	B/H	AB/AC
tan A	P/B	BC/AB
cot A	H/P	AC/BC
sec A	H/B	AC/AB
cosec A	B/P	AB/BC

Reciprocal Relation between Trigonometric Rations:-

$$\sin \theta = \frac{1}{\text{cosec } \theta} \qquad \text{cosec } \theta = \frac{1}{\sin \theta}$$

$$\cos \theta = \frac{1}{\sec \theta} \qquad \sec \theta = \frac{1}{\cos \theta}$$

$$\tan \theta = \frac{1}{\cot \theta} \qquad \cot \theta = \frac{1}{\tan \theta}$$

- $\sin \theta \times \text{cosec } \theta = 1$
- $\cos \theta \times \sec \theta = 1$
- $\tan \theta \times \cot \theta = 1$

Q.UOTIENT RELATION

$$\tan \theta = \frac{\sin \theta}{\cos \theta} \qquad \cot \theta = \frac{\cos \theta}{\sin \theta}$$

TRIGONOMETRY TABLE

	0°	30°	45°	60°	90°
sin θ	0	$\frac{1}{2}$	$\frac{1}{\sqrt{2}}$	$\frac{\sqrt{3}}{2}$	1
cos θ	1	$\frac{\sqrt{3}}{2}$	$\frac{1}{\sqrt{2}}$	$\frac{1}{2}$	0
tan θ	0	$\frac{1}{\sqrt{3}}$	1	$\sqrt{3}$	N.D.
cot θ	N.D.	2	$\sqrt{2}$	$\frac{2}{\sqrt{3}}$	1
sec θ	1	$\frac{2}{\sqrt{3}}$	$\sqrt{2}$	2	N.D.
cosec θ	N.D.	$\sqrt{3}$	1	$\frac{1}{\sqrt{3}}$	0

TRIGONOMETRIC RATIOS OF COMPLEMENTRY ANGLES

If the sum of two angles is 90° then, it is called complementry angles in a right angle triangle one angle is 90°, so the sum of other two angles is also 90°. So the complementry angles of trigonometric ratios will be:-

$\sin (90° - A) = \cos A$

$\cos (90° - A) = \sin A$

$$\tan (90° - A) = \cot A$$
$$\cot (90° - A) = \tan A$$
$$\sec (90° - A) = \operatorname{cosec} A$$
$$\operatorname{cosec} (90° - A) = \sec A$$

Trigonometric Identities:-

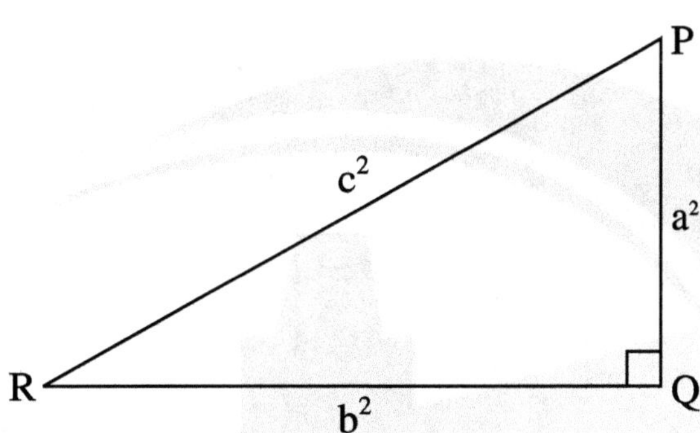

In Δ PQ.R, right angled at Q. and accourding to pythagoras theorem.

$$c^2 = a^2 + b^2 \qquad \text{(P.T.)}$$

Dividing all terms by c^2.

$$\frac{c^2}{c^2} = \frac{a^2}{c^2} + \frac{b^2}{c^2}$$

$$1 = \left(\frac{a}{c}\right)^2 + \left(\frac{b}{c}\right)^2$$

$$1 = \sin^2 \theta + \cos^2 \theta \qquad\qquad \textbf{IDENTITY NO. – 1}$$

$$\text{or}$$

$$1 - \sin^2 \theta = \cos^2 \theta$$

$$\text{or}$$

$$1 - \cos^2 \theta = \sin^2 \theta$$

Dividing all terms by a^2.

$$\frac{c^2}{a^2} = \frac{a^2}{a^2} + \frac{b^2}{a^2}$$

$$\left(\frac{c}{a}\right)^2 = 1 + \left(\frac{b}{a}\right)^2$$

$$\Rightarrow \quad \operatorname{cosec}^2 \theta = 1 + \cot^2 \theta$$

$$\text{or}$$

$$\operatorname{cosec}^2 \theta - \cot^2 \theta = 1$$

$$\text{or}$$

$$\cot^2 \theta = \mathrm{cosec}^2\, \theta - 1$$

Now, dividing all terms by b^2.

$$\frac{c^2}{b^2} = \frac{a^2}{b^2} + \frac{b^2}{b^2}$$

$$\Rightarrow \quad \left(\frac{c}{b}\right)^2 = \left(\frac{a}{b}\right)^2 + 1$$

$$\Rightarrow \quad \sec^2 \theta = \tan^2 \theta + 1$$

or

$$\sec^2 - 1 = \tan^2 \theta$$

or

$$\sec^2 \theta - \tan^2 \theta = 1$$

From R.P. Gupta Desk:-

T-values on your Finger tips:-

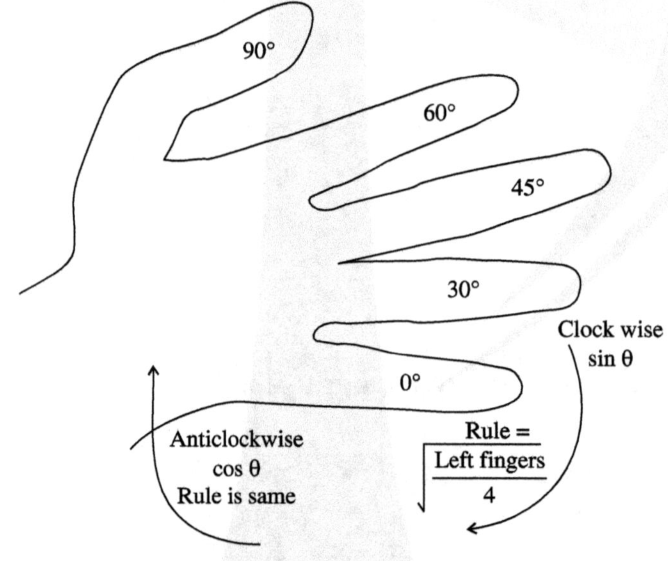

T-Ratios on your tongue.

Some	**C**urry	**T**o
People	**B**reak	**P**roper
Hane	**H**air	**B**rushing
Some	**C**urry	**T**o

or

Papa	**B**eer	**P**ears
Ha	**H**a	**B**eta.

WARM-UP – LEVEL-I

Q. 1. **If $\cos\theta = \dfrac{4}{5}$, find $\sin\theta$ and $\tan\theta$.**

Ans. Let us draw a right triangle ABC in which $\angle ABC = \theta$.

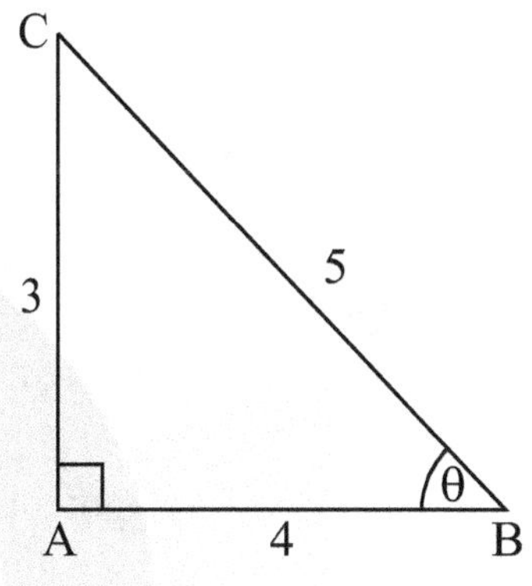

Now, since $\cos\theta = \dfrac{4}{5}$

$\therefore$ base $AB = 4$ units

and hypotenuse $BC = 5$ units

By Pythagoras Theorem

$$\text{perpendicular } AC = \sqrt{BC^2 - AB^2}$$
$$= \sqrt{25 - 16} = 3 \text{ units}$$

We know that $\sin\theta = \dfrac{\text{perpendicular}}{\text{hypotenuse}}$

$\therefore$ $\sin\theta = \dfrac{AC}{BC} = \dfrac{3}{5}$

and $\tan\theta = \dfrac{\text{perpendicular}}{\text{base}} = \dfrac{AC}{AB} = \dfrac{3}{4}$

Q. 2. **If $\operatorname{cosec} A = \sqrt{10}$, find other five t-ratios.**

Ans. $\operatorname{cosec} A = \sqrt{10}$

$\therefore$ $\sin A = \dfrac{1}{\operatorname{cosec} A} = \dfrac{1}{\sqrt{10}}$

Let us draw a right triangle ABC, right angled at A, in which perpendicular $AB = 1$ unit and hypotenuse $BC = \sqrt{10}$ units.

By Pythagoras Theorem

$$\text{base } AC = \sqrt{BC^2 - AB^2} = \sqrt{10 - 1} = 3 \text{ units}$$

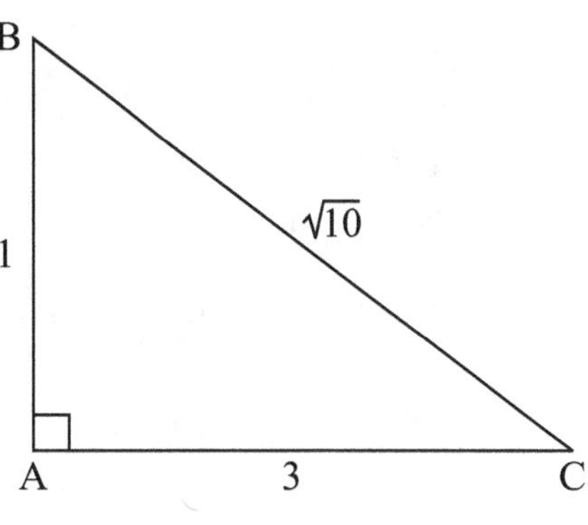

$\therefore$ $\cos A = \dfrac{\text{base}}{\text{hypotenuse}} = \dfrac{AC}{BC} = \dfrac{3}{\sqrt{10}}$

$\tan A = \dfrac{\text{perpendicular}}{\text{base}} = \dfrac{AB}{AC} = \dfrac{1}{3}$

$\cot A = \dfrac{1}{\cos A} = \dfrac{\sqrt{10}}{3}$

Q. 3. **If $\tan A = 1$ and $\tan B = \sqrt{3}$, evaluate $\cos A \cos B - \sin A \sin B$.**

Ans. Since $\tan A = 1$, we draw a right triangle ABC right angled at A, in which perpendicular $AB = 1$ unit

Innovative Mathematics X-8

and base AC = 1 unit

$$\therefore \quad \text{hypotenuse BC} = \sqrt{AB^2 + AC^2}$$

$$= \sqrt{1+1} = \sqrt{2} \text{ units}$$

$$\cos A = \frac{\text{perpendicular}}{\text{hypotenuse}} = \frac{AB}{BC} = \frac{1}{\sqrt{2}}$$

$$\sin A = \frac{\text{perpendicular}}{\text{hypotenuse}} = \frac{AB}{BC} = \frac{1}{\sqrt{2}}$$

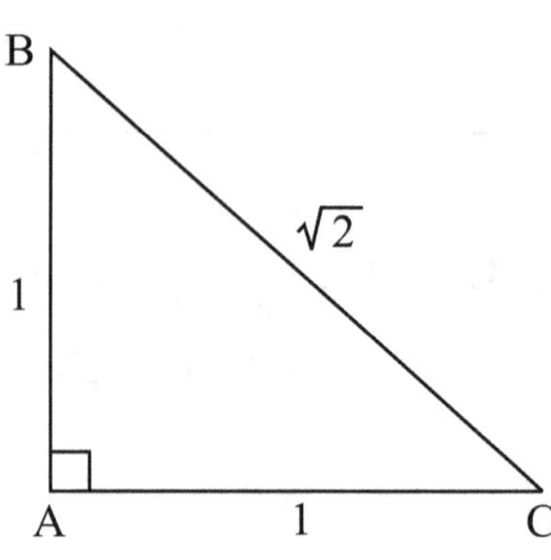

Also, since $\tan B = \sqrt{3}$, we draw a right triangle ABC, right angled at B, in which perpendicular BC = $\sqrt{3}$ units and base AB = 1 unit.

$$\therefore \quad \text{hypotenuse AC} = \sqrt{BC^2 + AB^2}$$

$$= \sqrt{3+1} = 2 \text{ units}$$

$$\cos B = \frac{\text{base}}{\text{hypotenuse}} = \frac{AB}{AC} = \frac{1}{2}$$

$$\sin B = \frac{\text{perpendicular}}{\text{hypotenuse}} = \frac{BC}{AC} = \frac{\sqrt{3}}{2}$$

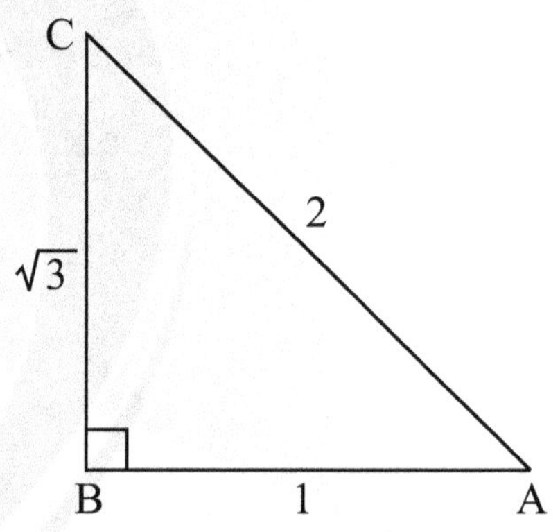

$$\therefore \quad \cos A \cos B - \sin A \sin B$$

$$= \frac{1}{\sqrt{2}} \times \frac{1}{2} - \frac{1}{\sqrt{2}} \times \frac{\sqrt{3}}{2}$$

$$= \frac{1}{2\sqrt{2}} - \frac{\sqrt{3}}{2\sqrt{2}} = \frac{1-\sqrt{3}}{2\sqrt{2}}$$

Q. 4. **The sine of an angle to its cosine is 8 : 15, find their values.**

Ans. Let the angle be θ

$$\therefore \quad \frac{\sin\theta}{\cos\theta} = \frac{8}{15} \Rightarrow \tan\theta = \frac{8}{15}$$

We draw a right triangle ABC, right angled at A, in which perpendicular AC = 8 units and base AB = 15 units.

$$\therefore \quad \text{hypotenuse BC} = \sqrt{AB^2 + AC^2}$$

$$= \sqrt{225 + 64} = \sqrt{289}$$

$$\sin\theta = \frac{\text{perpendicular}}{\text{hypotenuse}} = \frac{AC}{BC} = \frac{8}{17}$$

$$\cos\theta = \frac{\text{base}}{\text{hypotenuse}} = \frac{AB}{BC} = \frac{15}{17}$$

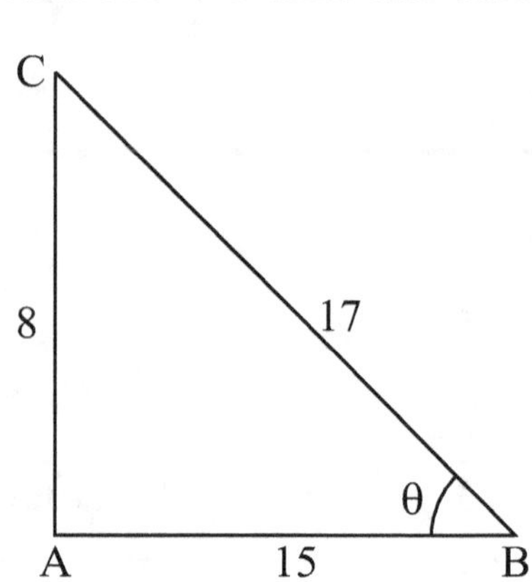

Q. 5. If $\cot \theta = \dfrac{7}{8}$, evaluate $\dfrac{(1 + \sin \theta)(1 - \sin \theta)}{(1 + \cos \theta)(1 - \cos \theta)}$.

Ans. Draw a right triangle ABC in which $\angle ABC = \theta$

Since, $\qquad \cot \theta = \dfrac{AB}{AC} = \dfrac{7}{8}$

$\therefore$ Let $\qquad AB = 7$ units and $AC = 8$ units

$\therefore \qquad BC = \sqrt{AB^2 + AC^2} \qquad$ (By Pythagoras Theorem)

$\qquad\qquad = \sqrt{7^2 + 8^2}$

$\qquad\qquad = \sqrt{49 + 64}$

$\qquad\qquad = \sqrt{113}$ units

$\therefore \qquad \sin \theta = \dfrac{AC}{BC} = \dfrac{8}{\sqrt{113}}$ and $\cos \theta = \dfrac{AB}{BC} = \dfrac{7}{\sqrt{113}}$

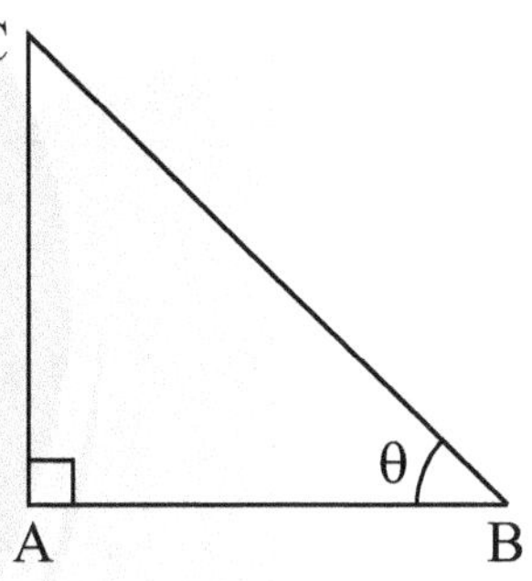

Now, $\dfrac{(1 + \sin \theta)(1 - \sin \theta)}{(1 + \cos \theta)(1 - \cos \theta)} = \dfrac{1 - \sin^2 \theta}{1 - \cos^2 \theta}$

$$= \dfrac{1 - \left(\dfrac{8}{\sqrt{113}}\right)^2}{1 - \left(\dfrac{7}{\sqrt{113}}\right)^2}$$

$$= \dfrac{1 - \dfrac{64}{113}}{1 - \dfrac{49}{113}} = \dfrac{133 - 64}{113 - 49} = \dfrac{49}{64}$$

Q. 6. In $\triangle PQ.R$, right angled at Q., $PR + Q.R = 25$ cm and $PQ. = 5$ cm. Determine the values of sin P, cos P and tan P.

Ans. $\qquad\qquad PR + Q.R = 25$ cm

$\therefore \qquad\qquad PR = (25 - Q.R)$

By Pythagoras Theorem:

$\qquad\qquad PR^2 = PQ.^2 + Q.R^2$

$\therefore \qquad (25 - Q.R)^2 = (5)^2 + Q.R^2$

or $\quad 625 + Q.R^2 - 50Q.R = 25 + Q.R^2$

or $\qquad\qquad 50Q.R = 600$

$\therefore \qquad\qquad Q.R = 12$ cm

and $\qquad\qquad PR = (25 - Q.R) = (25 - 12) = 13$ cm

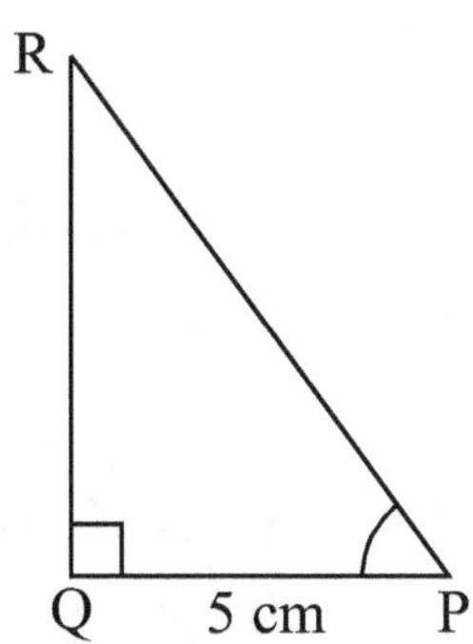

$\therefore \; \sin P = \dfrac{QR}{PR} = \dfrac{12}{13}$, $\cos P = \dfrac{PQ}{PR} = \dfrac{5}{13}$, and $\tan P = \dfrac{QR}{PQ} = \dfrac{12}{5}$

Q. 7. If **3 cot A = 4**, check whether $\dfrac{1-\tan^2 A}{1+\tan^2 A} = \cos^2 A - \sin^2 A$ or not.

Ans. Draw a right triangle ABC, right angled at B.

Since, $\quad 3\cot A = 4$

$\therefore \quad \cot A = \dfrac{4}{3} = \dfrac{AB}{BC}$

Let AB = 4 units and BC = 3 units

$\therefore \quad AC = \sqrt{AB^2 + BC^2}$ [By Pythagoras Theorem]

$\qquad\qquad = \sqrt{4^2 + 3^2}$

$\qquad\qquad = \sqrt{16 + 9}$

$\qquad\qquad = 5$ units

$\therefore \quad \tan A = \dfrac{BC}{AB} = \dfrac{3}{4}$

$\qquad \sin A = \dfrac{BC}{AC} = \dfrac{3}{5}$

and $\quad \cos A = \dfrac{AB}{AC} = \dfrac{4}{5}$

Now, $\quad \dfrac{1-\tan^2 A}{1+\tan^2 A} = \dfrac{1-\left(\dfrac{3}{4}\right)^2}{1+\left(\dfrac{3}{4}\right)^2} = \dfrac{1-\dfrac{9}{16}}{1+\dfrac{9}{16}} = \dfrac{16-9}{16+9} = \dfrac{7}{25}$

and, $\quad \cos^2 A - \sin^2 A = \left(\dfrac{4}{5}\right)^2 - \left(\dfrac{3}{5}\right)^2$

$\qquad\qquad\qquad\quad = \dfrac{16}{25} - \dfrac{9}{25}$

$\qquad\qquad\qquad\quad = \dfrac{7}{25}$

From (1) and (2):

$\dfrac{1-\tan^2 A}{1+\tan^2 A} = \cos^2 A - \sin^2 A.$

Q. 8. In $\triangle$ABC, right angled at B, if **AB = 12 cm** and **BC = 5 cm**, find

(i) sin A and tan A.   (ii) sin C and cot C.

Ans. Given, $\qquad AB = 12$ cm

$\qquad\qquad\qquad BC = 5$ cm

$\therefore \qquad AC = \sqrt{AB^2 + BC^2}$ [By Pythagoras Theorem]

$$= \sqrt{144 + 25}$$

$$= 13 \text{ cm}$$

(i)
$$\sin A = \dfrac{BC}{AC} = \dfrac{5}{13}$$

$$\tan A = \dfrac{BC}{AB} = \dfrac{5}{12}$$

(ii)
$$\sin C = \dfrac{AB}{AC} = \dfrac{12}{13}$$

and
$$\cot C = \dfrac{BC}{AB} = \dfrac{5}{12}$$

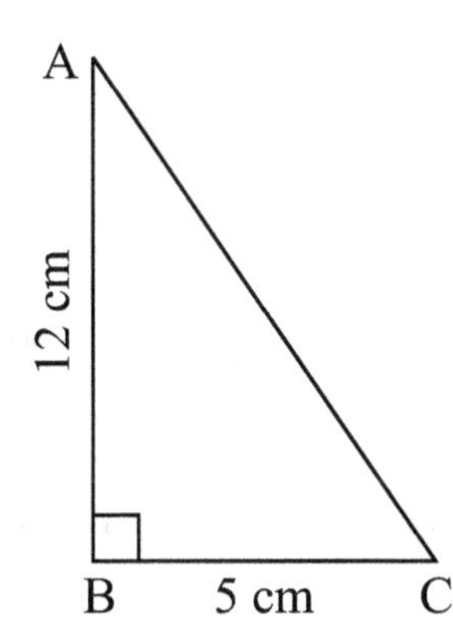

Q. 9. Given $\cos\theta = \dfrac{21}{29}$, determine the value of

$$\frac{\sec\theta}{\tan\theta - \sin\theta}$$

Ans. Let us draw a right $\triangle ABC$ in which $\angle ABC = \theta$

$$\text{base } AB = 21 \text{ units}$$

and $\quad$ hypotenuse $BC = 29$ units

$\therefore \quad$ perpendicular $AC = \sqrt{BC^2 - AB^2}$ $\qquad$ [By Pythagoras Theorem]

$$= \sqrt{841 - 441}$$

$$= 20 \text{ units}$$

$\therefore \quad \sec\theta = 1/\cos\theta = 29/21$

$\quad \sin\theta = AC/BC = 20/29$

and $\quad \tan\theta = AC/AB = 20/21$

Hence, $\quad \dfrac{\sec\theta}{\tan\theta - \sin\theta} = \dfrac{\dfrac{29}{21}}{\dfrac{20}{21} - \dfrac{20}{29}} = \dfrac{841}{160}$

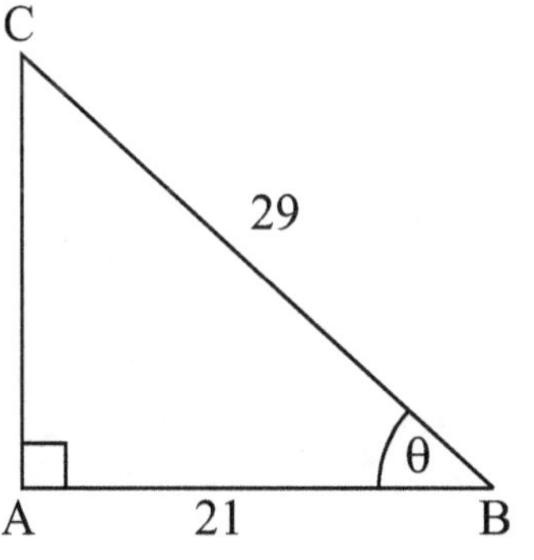

Q. 10. Evaluate $\dfrac{\operatorname{cosec}\theta - \cot\theta}{2\cot\theta}$, when $\sin\theta = \dfrac{3}{5}$.

Ans. Let us draw a right $\triangle ABC$ such that

$$\text{perpendicular } AC = 3 \text{ units}$$

and $\quad$ hypotenuse $BC = 5$ units

$\therefore \quad$ base $AB = \sqrt{BC^2 - AC^2}$ $\qquad$ [By Pythagoras Theorem]

$$= \sqrt{25 - 9}$$

$$= 4 \text{ units}$$

Innovative Mathematics X-8

$$\therefore \qquad \operatorname{cosec}\theta = \frac{1}{\sin\theta} = \frac{5}{3}$$

$$\cot\theta = \frac{\text{base}}{\text{perpendicular}} = \frac{4}{3}$$

Hence, $\quad \dfrac{\operatorname{cosec}\theta - \cot\theta}{2\cot\theta} = \dfrac{5/3 - 4/3}{2 \times 4/3} = \dfrac{1}{8}$

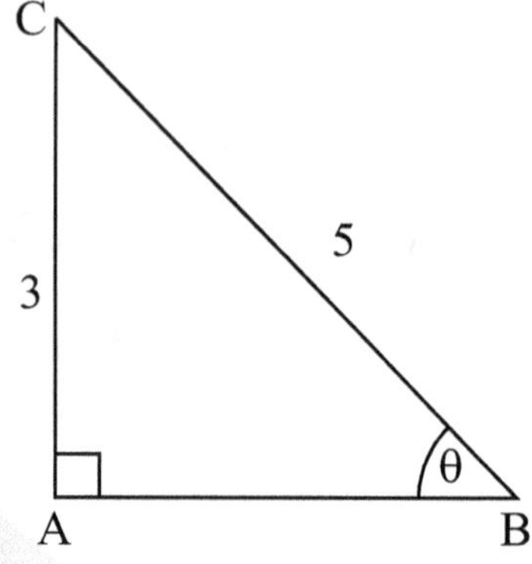

Q. 11. **Given that $\cos\theta = \dfrac{p}{q}$, find the value of $\tan\theta$.**

Ans. Let us draw a right $\triangle ABC$ such that

$$\text{base } AB = p \text{ units}$$

and $\qquad$ hypotenuse $BC = q$ units

$\therefore \qquad$ perpendicular $AC = \sqrt{BC^2 - AB^2} \qquad$ [By Pythagoras Theorem]

$$= \sqrt{q^2 - p^2}$$

$\therefore \qquad \tan\theta = \dfrac{\text{perpendicular}}{\text{base}} = \dfrac{AC}{AB} = \dfrac{\sqrt{q^2 - p^2}}{p}$

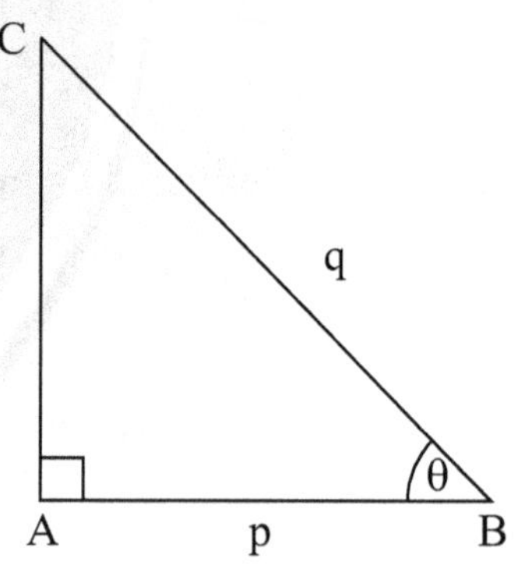

Q. 12. **If $\sin A = \dfrac{1}{3}$, evaluate $\cos A \operatorname{cosec} A + \tan A \sec A$.**

Ans. Let us draw a right $\triangle ABC$, such that

$$\text{perpendicular } BC = 1 \text{ unit}$$

and $\qquad$ hypotenuse $AC = 3$ units

$\therefore \qquad$ base $AB = \sqrt{AC^2 - BC^2} \qquad$ [By Pythagoras Theorem]

$$= \sqrt{9 - 1}$$

$$= 2\sqrt{2} \text{ units}$$

$\therefore \qquad \cos A = AB/AC = 2\sqrt{2}/3$

$$\tan A = BC/AB = 1/2\sqrt{2}$$

$$\sec A = 1/\cos A = 3/2\sqrt{2}$$

$$\operatorname{cosec} A = 1/\sin A = 3/1$$

Hence, $\cos A \operatorname{cosec} A + \tan A \sec A$

$$= \frac{2\sqrt{2}}{3} \times \frac{3}{1} + \frac{1}{2\sqrt{2}} \times \frac{3}{2\sqrt{2}}$$

$$= 2\sqrt{2} + \frac{3}{8}$$

$$= \frac{16\sqrt{2} + 3}{8}$$

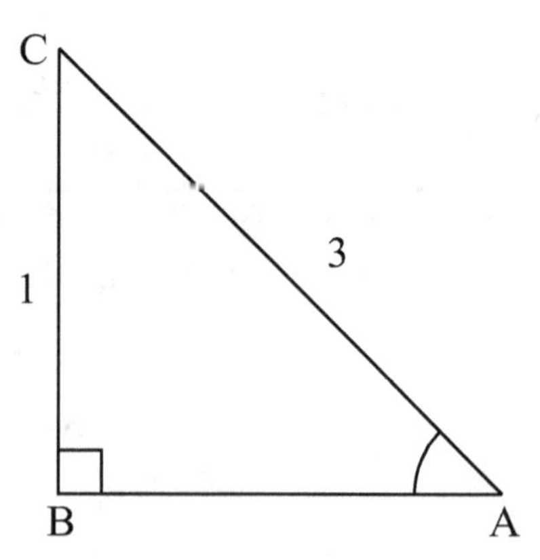

Q. 13. If cosec A = 2, find the value of $\cot A + \dfrac{\sin A}{1 + \cos A}$.

Ans. Let us draw a right $\triangle ABC$, such that cosec A = 2

(AC = 2 units and BC = 1 unit)

$$\therefore \qquad AB = \sqrt{AC^2 - BC^2} \qquad \text{[By Pythagoras Theorem]}$$

$$= \sqrt{4-1}$$

$$= \sqrt{3} \text{ units}$$

$$\therefore \qquad \cot A = AB/BC = \sqrt{3}/1$$

$$\sin A = 1/\text{cosec } A = 1/2$$

$$\cos A = AB/AC = \sqrt{3}/2$$

Hence, $\cot A + \dfrac{\sin A}{1 + \cos A} = \sqrt{3} + \dfrac{1/2}{1 + \sqrt{3}/2}$

$$= \sqrt{3} + \dfrac{1}{2} \times \dfrac{2}{2 + \sqrt{3}}$$

$$= \sqrt{3} + \dfrac{1}{2 + \sqrt{3}}$$

$$= \dfrac{2\sqrt{3} + 3 + 1}{2 + \sqrt{3}} = \dfrac{4 + 2\sqrt{3}}{2 + \sqrt{3}}$$

$$= \dfrac{4 + 2\sqrt{3}}{2 + \sqrt{3}} \times \dfrac{2 - \sqrt{3}}{2 - \sqrt{3}} = \dfrac{2}{1} = 2$$

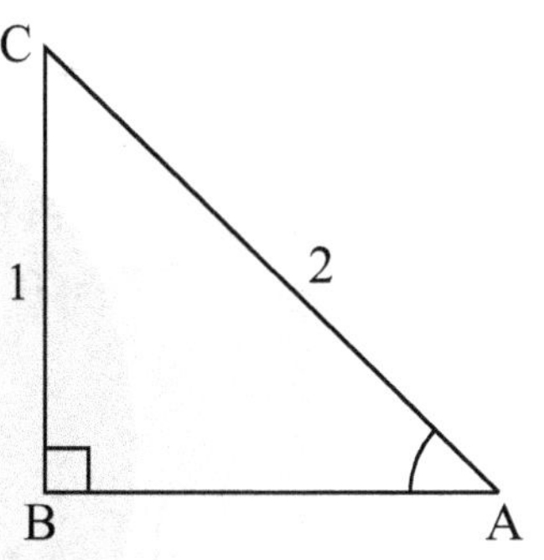

Q. 14. If $\sec \theta = \dfrac{5}{4}$, verify that

$$\dfrac{\tan \theta}{1 + \tan^2 \theta} = \dfrac{\sin \theta}{\sec \theta}$$

Ans. Let us draw a right $\triangle ABC$ such that $\angle ABC = \theta$

$$\text{base } AB = 4 \text{ units}$$

and $\qquad$ hypotenuse BC = 5 units

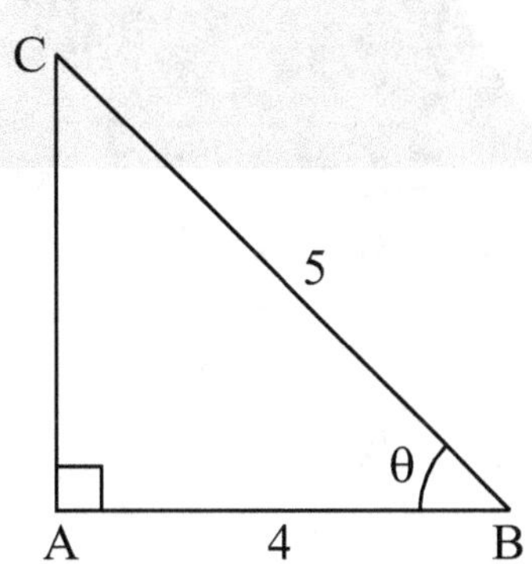

$$\therefore \qquad \text{perpendicular AC} = \sqrt{BC^2 - AB^2} \qquad \text{[By Pythagoras Theorem]}$$

$$= \sqrt{25 - 16}$$

$$= 3 \text{ units}$$

$$\therefore \qquad \sin\theta = AC/BC = 3/5$$

$$\tan\theta = AC/AB = 3/4$$

$$\sec\theta = 5/4$$

Q. 15. If $\tan\theta = 12/13$, find the value of $\dfrac{2\sin\theta\cos\theta}{\cos^2\theta - \sin^2\theta}$

Ans.
$$\frac{2\sin\theta\cos\theta}{\cos^2\theta - \sin^2\theta} = \frac{2\tan\theta}{1 - \tan^2\theta} \qquad \text{[Dividing numerator and denominator by } \cos^2\theta\text{]}$$

$$= \frac{2 \times \dfrac{12}{13}}{1 - \left(\dfrac{12}{13}\right)^2} \qquad [\tan\theta = 12/13]$$

$$= \frac{24}{13} \times \frac{169}{25} = \frac{312}{25}$$

Q. 16. Evaluate the following:

(i) $\sin 60° \cos 30° + \sin 30° \cos 60°$

(ii) $\dfrac{\sin 30° + \tan 45° - \operatorname{cosec} 60°}{\sec 30° + \cos 60° + \cot 45°}$

(iii) $\dfrac{5\cos^2 60° + 4\sec^2 30° - \tan^2 45°}{\sin^2 30° + \cos^2 30°}$

Ans. (i) $\sin 60° \cos 30° + \sin 30° \cos 60°$

$$= \frac{\sqrt{3}}{2} \times \frac{\sqrt{3}}{2} + \frac{1}{2} \times \frac{1}{2}$$

$$= \frac{3}{4} + \frac{1}{4}$$

$$= 1$$

(ii) $\dfrac{\sin 30° + \tan 45° - \operatorname{cosec} 60°}{\sec 30° + \cos 60° + \cot 45°}$

$$= \frac{\dfrac{1}{2} + 1 - \dfrac{2}{\sqrt{3}}}{\dfrac{2}{\sqrt{3}} + \dfrac{1}{2} + 1}$$

$$= \frac{\dfrac{\sqrt{3} + 2\sqrt{3} - 4}{2\sqrt{3}}}{\dfrac{4 + \sqrt{3} + 2\sqrt{3}}{2\sqrt{3}}}$$

$$= \frac{\sqrt{3} + 2\sqrt{3} - 4}{4 + \sqrt{3} + 2\sqrt{3}}$$

$$= \frac{3\sqrt{3} - 4}{4 + 3\sqrt{3}}$$

$$= \frac{3\sqrt{3} - 4}{4 + 3\sqrt{3}} \times \frac{4 - 3\sqrt{3}}{4 - 3\sqrt{3}}$$

$$= \frac{12\sqrt{3} - 27 - 16 + 12\sqrt{3}}{16 - 27}$$

$$= \frac{24\sqrt{3} - 43}{-11}$$

$$= \frac{43 - 24\sqrt{3}}{11}$$

(iii) $\dfrac{5\cos^2 60° + 4\sec^2 30° - \tan^2 45°}{\sin^2 30° + \cos^2 30°}$

$$= \frac{5\left(\dfrac{1}{2}\right)^2 + 4\left(\dfrac{2}{\sqrt{3}}\right)^2 - (1)^2}{\left(\dfrac{1}{2}\right)^2 + \left(\dfrac{\sqrt{3}}{2}\right)^2}$$

$$= \frac{\dfrac{5}{4} + \dfrac{16}{3} - 1}{\dfrac{1}{4} + \dfrac{3}{4}} = \frac{\dfrac{15 + 64 - 12}{12}}{\dfrac{4}{4}}$$

$$= \frac{67}{12}$$

Q. 17. **Given A = 30°, verify**

(i) $\sin 2A = 2 \sin A \cos A$ **(ii)** $\sin A = \sqrt{\dfrac{1 - \cos 2A}{2}}$

(iii) $\cos 3A = 4 \cos^3 A - 3 \cos A$ **(iv)** $\cos A = \dfrac{1}{\sqrt{1 + \tan^2 A}}$

(v) $\tan A = \dfrac{\sqrt{1 - \cos^2 A}}{\cos A}$

Ans.   **(i)**   L.H.S. $= \sin 2A = \sin 2\,(30°) = \dfrac{\sqrt{3}}{2}$

                 R.H.S. $= 2 \sin A \cos A = 2 \sin 30° \cos 30°$

$$= 2 \times \frac{1}{2} \times \frac{\sqrt{3}}{2} = \frac{\sqrt{3}}{2}$$

$\therefore$ L.H.S. = R.H.S.

(ii)   L.H.S. $= \sin A = \sin 30° = \dfrac{1}{2}$

R.H.S. $= \sqrt{\dfrac{1 - \cos 2A}{2}} = \sqrt{\dfrac{1 - \cos 60°}{2}}$

$$= \sqrt{\dfrac{1 - \dfrac{1}{2}}{2}} = \sqrt{\dfrac{2 - 1}{2 \times 2}} = \sqrt{\dfrac{1}{4}} = \dfrac{1}{2}$$

$\therefore$ L.H.S. = R.H.S.

(iii)   L.H.S. $= \cos 3A = \cos(3 \times 30°) = \cos 90° = 0$

R.H.S. $= 4\cos^3 A - 3\cos A = 4\cos^3 30° - 3\cos 30°$

$$= 4\left(\frac{\sqrt{3}}{2}\right)^3 - 3\left(\frac{\sqrt{3}}{2}\right) = 4 \times \frac{3\sqrt{3}}{8} - \frac{3\sqrt{3}}{2}$$

$$= \frac{3\sqrt{3}}{2} - \frac{3\sqrt{3}}{2} = 0$$

$\therefore$ L.H.S. = R.H.S.

(iv)   L.H.S. $= \cos A = \cos 30° = \dfrac{\sqrt{3}}{2}$

R.H.S. $= \dfrac{1}{\sqrt{1 + \tan^2 A}} = \dfrac{1}{\sqrt{1 + \tan^2 30°}} = \dfrac{1}{\sqrt{1 + \left(\dfrac{1}{\sqrt{3}}\right)^2}}$

$$= \dfrac{1}{\sqrt{1 + \dfrac{1}{2}}} = \dfrac{1}{\sqrt{\dfrac{3+1}{2}}} = \sqrt{\dfrac{3}{4}} = \dfrac{\sqrt{3}}{2}$$

$\therefore$ L.H.S. = R.H.S.

(v)   L.H.S. $= \tan A = \tan 30° = \dfrac{1}{\sqrt{3}}$

R.H.S. $= \dfrac{\sqrt{1 - \cos^2 A}}{\cos A} = \dfrac{\sqrt{1 - \cos^2 30°}}{\cos 30°}$

$$= \dfrac{\sqrt{1 - (\sqrt{3}/2)^2}}{\sqrt{3}/2} = \dfrac{\sqrt{1 - 3/4}}{\sqrt{3}/2}$$

$$= \dfrac{1/2}{\sqrt{3}/2} = \dfrac{1}{2} \times \dfrac{2}{\sqrt{3}} = \dfrac{1}{\sqrt{3}}$$

$$\therefore \qquad \text{L.H.S.} = \text{R.H.S.}$$

Q. 18. **If $\theta = 45°$, verify that : $\sin 2\theta = 2 \sin \theta \cos \theta$**

Ans.
$$\sin 2\theta = 2 \sin \theta \cos \theta$$

Putting
$$\theta = 45°$$
$$\sin (2 \times 45°) = 2 \sin 45° \cos 45°$$
$$\sin 90° = 2 \sin 45° \cos 45°$$
$$1 = 2 \times \frac{1}{\sqrt{2}} \times \frac{1}{\sqrt{2}}$$
$$1 = 2 \times \frac{1}{2} = 1$$

Hence Verified

Q. 19. **If $\tan (A + B) = \sqrt{3}$ and $\tan (A - B) = \dfrac{1}{\sqrt{3}}$; $0° < (A + B) < 90°$; $A > B$, find A and B.**

Ans.
$$\tan (A + B) = \sqrt{3} = \tan 60°$$
$$\therefore \qquad A + B = 60° \qquad (1)$$

Also,
$$\tan (A - B) = \frac{1}{\sqrt{3}} = \tan 30°$$
$$\therefore \qquad A - B = 30° \qquad (2)$$

From (1) and (2):
$$A = 45° \text{ and } B = 15°$$

Q. 20. **If $\sin (A + B) = \dfrac{\sqrt{3}}{2}$ and $\cos (A - B) = \dfrac{1}{\sqrt{2}}$, find A and B where $(A + B)$ and $(A - B)$ are acute angles.**

Ans.
$$\sin (A + B) = \frac{\sqrt{3}}{2}$$
$$\sin (A + B) = \sin 60°$$
$$\therefore \qquad A + B = 60° \qquad (1)$$

Now
$$\cos (A - B) = \frac{1}{\sqrt{2}}$$
$$\cos (A - B) = \cos 45°$$
$$\therefore \qquad A - B = 45° \qquad (2)$$

Adding (1) and (2), we get,
$$2A = 105°$$
$$A = 52.5°$$

Now
$$B = 60° - A$$
$$= 60 - 52.5° = 7.5°$$
$$\therefore \qquad A = 52.5° \text{ and } B = 7.5°$$

WARM-UP – LEVEL-II

Q. 1. Prove : $\cos^4\theta + \sin^4\theta + 2\sin^2\theta\cos^2\theta = 1$.

Ans.

$$
\begin{aligned}
\text{L.H.S.} &= \cos^4\theta + \sin^4\theta + 2\sin^2\theta\cos^2\theta \\
&= (\cos^2\theta)^2 + (\sin^2\theta)^2 + 2\sin^2\theta\cos^2\theta \\
&= (\cos^2\theta + \sin^2\theta)^2 \\
&= 1 \qquad\qquad\qquad [\text{using } \sin^2\theta + \cos^2\theta = 1] \\
&= \text{R.H.S.}
\end{aligned}
$$

Hence Proved.

Q. 2. Prove : $\dfrac{1 - \tan^2\theta}{\cot^2\theta - 1} = \tan^2\theta$.

Ans.

$$
\begin{aligned}
\text{L.H.S.} &= \frac{1 - \tan^2\theta}{\cot^2\theta - 1} \\[2mm]
&= \frac{1 - \dfrac{\sin^2\theta}{\cos^2\theta}}{\dfrac{\cos^2\theta}{\sin^2\theta} - 1} \\[2mm]
&= \frac{\dfrac{\cos^2\theta - \sin^2\theta}{\cos^2\theta}}{\dfrac{\cos^2\theta - \sin^2\theta}{\sin^2\theta}} \\[2mm]
&= \frac{\cos^2\theta - \sin^2\theta}{\cos^2\theta} \times \frac{\sin^2\theta}{\cos^2\theta - \sin^2\theta} \\[2mm]
&= \frac{\sin^2\theta}{\cos^2\theta} \\[2mm]
&= \tan^2\theta \\
&= \text{R.H.S.}
\end{aligned}
$$

Hence Proved.

Q. 3. If $x = r\cos\alpha\sin\beta$, $y = r\cos\alpha\cos\beta$ and $z = r\sin\alpha$, show that
$$x^2 + y^2 + z^2 = r^2$$

Ans.

$$
\begin{aligned}
\text{L.H.S.} &= x^2 + y^2 + z^2 \\
&= r^2\cos^2\alpha\sin^2\beta + r^2\cos^2\alpha\cos^2\beta + r^2\sin^2\alpha \\
&= r^2\cos^2\alpha(\sin^2\beta + \cos^2\beta) + r^2\sin^2\alpha \\
&= r^2\cos^2\alpha(1) + r^2\sin^2\alpha \\
&= r^2(\cos^2\alpha + \sin^2\alpha) \\
&= r^2(1)
\end{aligned}
$$

$$= r^2$$
$$= \text{R.H.S.}$$

Hence Proved.

Q. 4. **Prove :** $\dfrac{1}{1 - \sin^2 \theta} = 1 + \dfrac{1}{\dfrac{1}{\sin^2 \theta} - 1}.$

Ans.
$$\text{R.H.S.} = 1 + \dfrac{1}{\dfrac{1}{\sin^2 \theta} - 1}$$

$$= 1 + \dfrac{1}{\dfrac{1 - \sin^2 \theta}{\sin^2 \theta}}$$

$$= 1 + \dfrac{\sin^2 \theta}{1 - \sin^2 \theta}$$

$$= \dfrac{1 - \sin^2 \theta + \sin^2 \theta}{1 - \sin^2 \theta}$$

$$= \dfrac{1}{1 - \sin^2 \theta}$$

$$= \text{L.H.S.}$$

Hence Proved.

Q. 5. **Prove the following identities:**

(i) $\dfrac{\cos \theta + \sin \theta}{\sin \theta} = 1 + \cot \theta.$

(ii) $\dfrac{\sin^4 \theta - \cos^4 \theta}{\sin^2 \theta - \cos^2 \theta} = 1.$

Ans. (i)
$$\text{L.H.S.} = \dfrac{\cos \theta + \sin \theta}{\sin \theta}$$

$$= \dfrac{\cos \theta}{\sin \theta} + \dfrac{\sin \theta}{\sin \theta}$$

$$= \cot \theta + 1$$

$$= 1 + \cot \theta$$

$$= \text{R.H.S.}$$

Hence Proved.

(ii)
$$\text{L.H.S.} = \dfrac{\sin^4 \theta - \cos^4 \theta}{\sin^2 \theta - \cos^2 \theta}$$

$$= \dfrac{(\sin^2 0 + \cos^2 0)(\sin^2 0 - \cos^2 0)}{(\sin^2 \theta - \cos^2 \theta)} \qquad [\text{using } (a^2 - b^2) = (a + b)(a - b)]$$

$$= \frac{(1)(\sin^2\theta - \cos^2\theta)}{(\sin^2\theta - \cos^2\theta)} \qquad\qquad [\text{using } \sin^2\theta + \cos^2\theta = 1]$$

$$= 1$$

$$= \text{R.H.S.}$$

Hence Proved.

Q. 6. **Using trigonometric identities, write the following expressions as an integer:**

    (i) $3\cot^2 A - 3\csc^2 A$ **(ii)** $5\sin^2\theta + 5\cos^2\theta.$

Ans.   (i) $3\cot^2 A - 3\csc^2 A$

$$= 3\cot^2 A - 3(1 + \cot^2 A) \qquad\qquad [\text{using } \csc^2\theta = 1 + \cot^2\theta]$$

$$= 3\cot^2 A - 3 - 3\cot^2 A$$

$$= -3$$

    (ii) $5\sin^2\theta + 5\cos^2\theta$

$$= 5(\sin^2\theta + \cos^2\theta)$$

$$= 5.1 \qquad\qquad [\text{using } \sin^2\theta + \cos^2\theta = 1]$$

$$= 5$$

Q. 7. **Prove that $(\csc\theta + \cot\theta)(1 - \cos\theta) = \sin\theta.$**

Ans.   L.H.S. $= (\csc\theta + \cot\theta)(1 - \cos\theta)$

$$= \left(\frac{1}{\sin\theta} + \frac{\cos\theta}{\sin\theta}\right)(1 - \cos\theta)$$

$$= \left(\frac{1 + \cos\theta}{\sin\theta}\right)(1 - \cos\theta)$$

$$= \frac{1 - \cos^2\theta}{\sin\theta}$$

$$= \frac{\sin^2\theta}{\sin\theta} \qquad\qquad [\text{using } \sin^2\theta = 1 - \cos^2\theta]$$

$$= \sin\theta$$

$$= \text{R.H.S.}$$

Hence Proved.

Q. 8. **Prove that $\dfrac{\tan^2\theta}{\sec\theta + 1} = \sec\theta - 1.$**

Ans.   L.H.S. $= \dfrac{\tan^2\theta}{\sec\theta + 1}$

$$= \frac{\sec^2\theta - 1}{\sec\theta + 1} \qquad\qquad [\text{using } \tan^2\theta = \sec^2\theta - 1]$$

$$= \frac{(\sec\theta + 1)(\sec\theta - 1)}{(\sec\theta + 1)}$$

$$= \sec \theta - 1$$
$$= \text{R.H.S.}$$

Hence Proved.

Q. 9. **Prove that following identities:**

(i) $\dfrac{1 - \sin A}{\cos A} = \dfrac{\cos A}{1 + \sin A}$

(ii) $\dfrac{1 - \cos \theta}{1 + \cos \theta} = (\text{cosec } \theta - \cot \theta)^2$

Ans. **(i)**

$$\text{R.H.S.} = \frac{\cos A}{1 + \sin A}$$

$$= \frac{\cos A}{1 + \sin A} \times \frac{1 - \sin A}{1 - \sin A}$$

$$= \frac{\cos A(1 - \sin A)}{1 - \sin^2 A}$$

$$= \frac{\cos A(1 - \sin A)}{\cos^2 A}$$

$$= \frac{1 - \sin A}{\cos A}$$

$$= \text{L.H.S.}$$

Hence Proved.

(ii)

$$\text{L.H.S.} = \frac{1 - \cos \theta}{1 + \cos \theta}$$

$$= \frac{1 - \cos \theta}{1 + \cos \theta} \times \frac{1 - \cos \theta}{1 - \cos \theta}$$

$$= \frac{(1 - \cos \theta)^2}{1 - \cos^2 \theta}$$

$$= \frac{(1 - \cos \theta)^2}{\sin^2 \theta} \qquad [\text{using } \sin^2 \theta = 1 - \cos^2 \theta]$$

$$= \left(\frac{1 - \cos \theta}{\sin \theta} \right)^2$$

$$= \left(\frac{1}{\sin \theta} - \frac{\cos \theta}{\sin \theta} \right)^2$$

$$= (\text{cosec } \theta - \cot \theta)^2$$

$$= \text{R.H.S.}$$

Hence Proved.

Q. 10. **Prove the following identities:**

(i) $\sqrt{\dfrac{1 + \sin A}{1 - \sin A}} = \sec A + \tan A$

Ans. (i)
$$\text{L.H.S.} = \sqrt{\frac{1 + \sin A}{1 - \sin A}}$$

$$= \sqrt{\frac{1 + \sin A}{1 - \sin A} \times \frac{1 + \cos A}{1 + \cos A}}$$

$$= \sqrt{\frac{(1 + \cos A)^2}{1 - \cos^2 A}}$$

$$= \sqrt{\frac{(1 + \cos A)^2}{\sin^2 A}}$$

$$= \frac{1 + \cos A}{\sin A}$$

$$= \text{R.H.S.}$$

Q. 11. **Prove the following identities:**

(i) $\dfrac{\sin \theta + \cos \theta}{\sin \theta - \cos \theta} + \dfrac{\sin \theta - \cos \theta}{\sin \theta + \cos \theta} = \dfrac{2}{1 - 2 \cos^2 \theta} = \dfrac{2}{2 \sin^2 \theta - 1}$

(ii) $(\operatorname{cosec} A - \sin A)(\sec A - \cos A)(\tan A + \cot A) = 1.$

Ans. (i)
$$\text{L.H.S.} = \frac{\sin \theta + \cos \theta}{\sin \theta - \cos \theta} + \frac{\sin \theta - \cos \theta}{\sin \theta + \cos \theta}$$

$$= \frac{(\sin \theta + \cos \theta)^2 + (\sin \theta - \cos \theta)^2}{(\sin \theta - \cos \theta)(\sin \theta + \cos \theta)}$$

$$= \frac{\sin^2 \theta + \cos^2 \theta + 2 \sin \theta \cos \theta + \sin^2 \theta + \cos^2 \theta - 2 \sin \theta \cos \theta}{\sin^2 \theta - \cos^2 \theta}$$

$$= \frac{(\sin^2 \theta + \cos^2 \theta) + (\sin^2 \theta + \cos^2 \theta)}{\sin^2 \theta - \cos^2 \theta}$$

$$= \frac{1 + 1}{\sin^2 \theta - \cos^2 \theta} \qquad [\text{using } \sin^2 \theta + \cos^2 \theta = 1]$$

$$= \frac{2}{(1 - \cos^2 \theta) - \cos^2 \theta} \qquad [\text{using } \sin^2 \theta = 1 - \cos^2 \theta]$$

$$= \frac{2}{1 - 2 \cos^2 \theta}$$

$$= \text{R.H.S.}$$

or
$$= \frac{2}{1 - 2(1 - \sin^2 \theta)} \qquad [\text{using } \cos^2 \theta = 1 - \sin^2 \theta]$$

$$= \frac{2}{1 - 2 + 2 \sin^2 \theta}$$

$$= \frac{2}{2\sin^2\theta - 1}$$

$$= \text{R.H.S.}$$

Henced Proved.

(ii) $\qquad$ L.H.S. $= (\operatorname{cosec} A - \sin A)(\sec A - \cos A)(\tan A + \cot A)$

$$= \left(\frac{1}{\sin A} - \sin A\right)\left(\frac{1}{\cos A} - \cos A\right)\left(\frac{1}{\cot A} + \cot A\right)$$

$$= \left(\frac{1 - \sin^2 A}{\sin A}\right)\left(\frac{1 - \cos^2 A}{\cos A}\right)\left(\frac{1 + \cot^2 A}{\cot A}\right)$$

$$= \frac{\cos^2 A}{\sin A} \times \frac{\sin^2 A}{\cos A} \times \frac{\operatorname{cosec}^2 A}{\cot A}$$

$$= \frac{\cos^2 A}{\sin A} \times \frac{\sin^2 A}{\cos A} \times \frac{1}{\sin^2 A} \times \frac{\sin A}{\cos A} \qquad \begin{bmatrix} \text{using } \cos^2 A = 1 - \sin^2 A \\ \sin^2 A = 1 - \cos^2 A \text{ and} \\ \operatorname{cosec}^2 A = 1 + \cot^2 A \end{bmatrix}$$

$$= 1$$

$$= \text{R.H.S.}$$

Hence Proved.

Q. 12. Prove that: $\dfrac{1}{\operatorname{cosec}\theta - \cot\theta} - \dfrac{1}{\sin\theta} = \dfrac{1}{\sin\theta} - \dfrac{1}{\operatorname{cosec}\theta + \cot\theta}$

Ans. We can write the given identity as :

$$\frac{1}{\operatorname{cosec}\theta - \cot\theta} + \frac{1}{\operatorname{cosec}\theta + \cot\theta} = \frac{1}{\sin\theta} + \frac{1}{\sin\theta} = \frac{2}{\sin\theta}$$

$$\text{L.H.S.} = \frac{1}{\operatorname{cosec}\theta - \cot\theta} + \frac{1}{\operatorname{cosec}\theta + \cot\theta}$$

$$= \frac{\operatorname{cosec}\theta + \cot\theta + \operatorname{cosec}\theta - \cot\theta}{\operatorname{cosec}^2\theta - \cot^2\theta}$$

$$= \frac{2\operatorname{cosec}\theta}{1} \qquad [\text{using } \operatorname{cosec}^2\theta - \cot^2\theta = 1]$$

$$= \frac{2}{\sin\theta}$$

$$= \text{R.H.S.}$$

Hence Proved.

Q. 13. Prove that: $\tan^2 A - \tan^2 B = \dfrac{\sin^2 A - \sin^2 B}{\cos^2 A \cos^2 B}$

Ans. $\qquad$ L.H.S. $= \tan^2 A - \tan^2 B$

$$= \frac{\sin^2 A}{\cos^2 A} - \frac{\sin^2 B}{\cos^2 B}$$

$$= \frac{\sin^2 A \cos^2 B - \cos^2 A \sin^2 B}{\cos^2 A \cos^2 B}$$

$$= \frac{\sin^2 A(1 - \sin^2 B) - (1 - \sin^2 A)\sin^2 B}{\cos^2 A \cos^2 B}$$

$$= \frac{\sin^2 A - \sin^2 A \sin^2 B - \sin^2 B + \sin^2 A \sin^2 B}{\cos^2 A \cos^2 B}$$

$$= \frac{\sin^2 A - \sin^2 B}{\cos^2 A \cos^2 B}$$

$$= \text{R.H.S.}$$

Hence Proved.

Q. 14. **Prove the following identities:**

(i) $\dfrac{1 + \sec A}{\sec A} = \dfrac{\sin^2 A}{1 - \cos A}$

Ans. (i)
$$\text{L.H.S.} = \frac{1 + \sec A}{\sec A}$$

$$= \frac{1 + \dfrac{1}{\cos A}}{\dfrac{1}{\cos A}}$$

$$= \frac{\cos A + 1}{\cos A} \times \frac{\cos A}{1}$$

$$= (\cos A + 1)$$

$$= (1 + \cos A) \times \frac{1 - \cos A}{1 - \cos A}$$

$$= \frac{1 - \cos^2 A}{1 - \cos A}$$

$$= \frac{\sin^2 A}{1 - \cos A}$$

$$= \text{R.H.S.}$$

Hence Proved.

Q. 15. **If** $\cos \theta + \sin \theta = \sqrt{2} \cos \theta$**, then show that**
$$\cos \theta - \sin \theta = \sqrt{2} \sin \theta.$$

Ans.
$$\cos \theta + \sin \theta = \sqrt{2} \cos \theta$$

$$\sqrt{2}\,\cos\theta - \cos\theta \ = \ \sin\theta$$

$$\cos\theta\,(\sqrt{2}-1) \ = \ \sin\theta$$

$$\cos\theta \ = \ \frac{\sin\theta}{\sqrt{2}-1}$$

$$= \ \frac{\sin\theta}{\sqrt{2}-1}\times\frac{\sqrt{2}+1}{\sqrt{2}+1}$$

$$= \ \sin\theta\,(\sqrt{2}+1)$$

or $$\cos\theta \ = \ \sqrt{2}\,\sin\theta + \sin\theta$$

$$\therefore \qquad \cos\theta - \sin\theta \ = \ \sqrt{2}\,\sin\theta$$

Hence Proved.

Q. 16. **If m = tan θ + sin θ and n = tan θ – sin θ, then prove that $m^2 - n^2 = 4\sqrt{mn}$.**

Ans.

$$\text{L.H.S.} \ = \ m^2 - n^2$$

$$= \ (\tan\theta + \sin\theta)^2 - (\tan\theta - \sin\theta)^2$$

$$= \ (\tan^2\theta + \sin^2\theta + 2\tan\theta\sin\theta) - (\tan^2\theta + \sin^2\theta - 2\tan\theta\sin\theta)$$

$$= \ 4\tan\theta\sin\theta \qquad\qquad (1)$$

$$\text{R.H.S.} \ = \ 4\sqrt{mn}$$

$$= \ 4\sqrt{(\tan\theta + \sin\theta)(\tan\theta - \sin\theta)}$$

$$= \ 4\sqrt{\tan^2\theta - \sin^2\theta}$$

$$= \ 4\sqrt{\frac{\sin^2\theta}{\cos^2\theta} - \sin^2\theta}$$

$$= \ 4\sqrt{\frac{\sin^2\theta - \cos^2\theta\sin^2\theta}{\cos^2\theta}}$$

$$= \ 4\sqrt{\frac{\sin^2\theta(1 - \cos^2\theta)}{\cos^2\theta}}$$

$$= \ 4\sqrt{\frac{\sin^2\theta.\sin^2\theta}{\cos^2\theta}}$$

$$= \ 4.\frac{\sin^2\theta}{\cos\theta}$$

$$= \ 4.\sin\theta.\frac{\sin\theta}{\cos\theta}$$

$$= \ 4\tan\theta\sin\theta \qquad\qquad (2)$$

From (1) and (2), we get :

$$\text{L.H.S.} \ = \ \text{R.H.S.}$$

Hence Proved.

Q. 17. If $a \cos \theta - b \sin \theta = c$, prove that $a \sin \theta + b \cos \theta =$

Ans. Given that

$$a \cos \theta - b \sin \theta = c$$
$$\Rightarrow \qquad (a \cos \theta - b \sin \theta)^2 = c^2$$
$$\Rightarrow \qquad a^2 \cos^2 \theta + b^2 \sin^2 \theta - 2ab \sin \theta \cos \theta = c^2$$
$$\Rightarrow \qquad a^2(1 - \sin^2 \theta) + b^2(1 - \cos^2 \theta) - 2ab \sin \theta \cos \theta = c^2$$
$$\Rightarrow \qquad a^2 - a^2 \sin^2 \theta + b^2 - b^2 \cos^2 \theta - 2ab \sin \theta \cos \theta = c^2$$
$$\Rightarrow \qquad a^2 + b^2 - c^2 = a^2 \sin^2 \theta + b^2 \cos^2 \theta + 2ab \sin \theta \cos$$
$$\Rightarrow \qquad a^2 + b^2 - c^2 = (a \sin \theta + b \cos \theta)^2$$
$$\text{or} \qquad (a \sin \theta + b \cos \theta)^2 = a^2 + b^2 - c^2$$
$$\Rightarrow \qquad a \sin \theta + b \cos \theta =$$

Henced Proved.

Q. 18. If $\sin \theta + \sin^2 \theta = 1$, prove that $\cos^2 \theta + \cos^4 \theta = 1$.

Ans. Given that

$$\sin \theta + \sin^2 \theta = 1$$
$$\Rightarrow \qquad \sin \theta = 1 - \sin^2 \theta$$
$$\Rightarrow \qquad \sin \theta = \cos^2 \theta \qquad\qquad\qquad (1)$$

Now,
$$\text{L.H.S.} = \cos^2 \theta + \cos^4 \theta$$
$$= \cos^2 \theta + (\cos^2 \theta)^2$$
$$= \cos^2 \theta + (\sin \theta)^2 \qquad\qquad [\text{From (1)} : \cos^2 \theta = \sin \theta]$$
$$= \cos^2 \theta + \sin^2 \theta$$
$$= \cos^2 \theta + \sin^2 \theta$$
$$= 1$$
$$= \text{R.H.S.}$$

Hence Proved.

VERY SHORT ANSWER TYPE Q.UESTIONS

Q. 1.   If $\sin \theta = \cos \theta$, find the value of θ

Sol.

Q. 2.   Find the value of $\tan^4\theta + \cot^4\theta$, if $\sin\theta - \cos\theta = 0$

Sol.

Q. 3.   Find the value of $\tan\theta + \cot\theta$, if $\tan^2\theta - 3\tan\theta + 1 = 0$

Sol.

Q. 4.   If $\tan \theta = \dfrac{4}{3}$ then find the value of $\dfrac{\sin\theta + \cos\theta}{\sin\theta - \cos\theta}$

Sol.

Q. 5.   If $3x = \operatorname{cosec}\theta$ and $\dfrac{3}{x} = \cot\theta$ then find $3\left(x^2 - \dfrac{1}{x^2}\right)$

Sol.

Q. 6.   If $x = a\sin\theta$ and $y = a\cos\theta$ than find the value of $x^2 + y^2$

Sol.

Q. 7.   If $\cos A = \dfrac{3}{5}$, find the value of $4 + 4\tan^2 A$

Sol.

Q. 8.   Find the value of $9\sec^2 A - 9\tan^2 A$

Sol.

Q. 9.   Express $\sec\theta$ in terms of $\cot\theta$

Sol.

Q. 10.   If $x = a\sec\theta$, $y = b\tan\theta$, tan find the value of $b^2x^2 - a^2y^2$.

Sol.

Q. 11.   Find the value of $\dfrac{1 + \tan^2\theta}{1 + \cot^2\theta}$, if $\tan\theta = \dfrac{4}{3}$.

Sol.

Q. 12.   Find the value of $\dfrac{1 + \tan^2\theta}{1 + \cot^2\theta}$

Sol.

Q. 13.   Given $\tan\theta = \dfrac{1}{\sqrt{3}}$, find the value of $\dfrac{\operatorname{cosec}^2\theta - \sec^2\theta}{\operatorname{cosec}^2\theta + \sec^2\theta}$.   (CBSE 2010)

Sol.

Q. 14.   If $\sqrt{3}\cot^2\theta - 4\cot\theta + \sqrt{3} = 0$, then find the value of $\tan^2\theta + \cot^2\theta$.

Sol.

Q. 15.   If $5\tan\theta - 4 = 0$, then value of $\dfrac{5\sin\theta - 4\cos\theta}{5\sin\theta + 4\cos\theta}$ is

(a) $\dfrac{5}{3}$ (b) $\dfrac{5}{6}$ (c) 0 (d) $\dfrac{1}{6}$

Sol.

Innovative Mathematics X-8

Q. 16. $3 \tan^2 \theta - 3 \sec^2 \theta + 4$ is equal to

(a) 3   (b) 2   (c) 1   (d) 0

Sol.

Q. 17. In Fig. if AD = 4 cm, BD = 3 cm and CB = 12 cm, then cot θ =

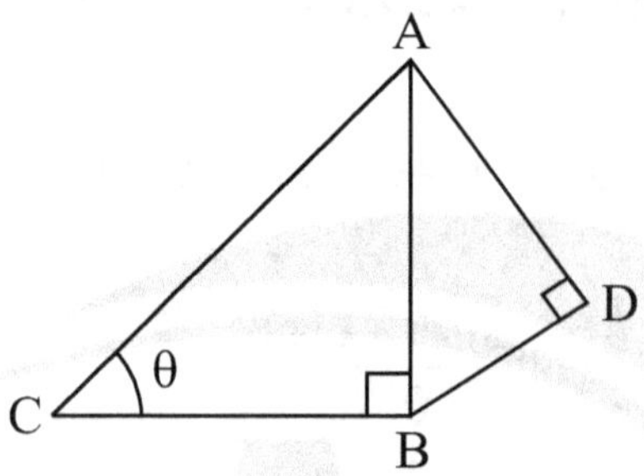

(a) $\dfrac{12}{5}$ (b) $\dfrac{5}{12}$ (c) $\dfrac{13}{12}$ (d) $\dfrac{12}{13}$

Sol.

Q. 18. If $x = 3 \sin \theta + 4 \cos \theta$ and $y = 3 \cos \theta - 4 \sin \theta$ then $x^2 + y^2$ is

(a) 25   (b) 45   (c) 7   (d) 49

Sol.

Q. 19. If $\sin \theta = \dfrac{a}{b}$, then the value of $\sec \theta + \tan \theta$ is

(a) $\sqrt{\dfrac{a+b}{a-b}}$ (b) $\dfrac{a+b}{a-b}$ (c) $\sqrt{\dfrac{b+a}{b-a}}$ (d) $\dfrac{b+a}{b-a}$

Sol.

SHORT ANSWER TYPE Q.UESTIONS (1)

Q. 20. **Prove that:** $\sec^4 \theta - \sec^2 \theta = \tan^4 \theta + \tan^2 \theta$

Sol.

Q. 21. $\sqrt{\dfrac{1+\sin \theta}{1-\sin \theta}} = \textbf{tan } \theta + \textbf{Sec } \theta$

Sol.

Q. 22. If $x = p \sec \theta + q \tan \theta$ & $y = p \tan \theta + q \sec \theta$ then prove that $x^2 - y^2 = p^2 - q^2$

Sol.

Q. 23. If $7 \sin^2 \theta + 3 \cos^2 \theta = 4$ then show that $\tan \theta = \dfrac{1}{\sqrt{3}}$

Sol.

Q. 24. If $\sin (A - B) = \dfrac{1}{2}$, $\cos (A + B) = \dfrac{1}{2}$ then find the value of A and B.

Sol.

Q. 25. Find the value of $\cos\theta$, if $\sec\theta + \tan\theta = 5$

Sol.

Q. 26. If $3 \cot A = 4$, find the value of $\dfrac{\operatorname{cosec}^2 A + 1}{\operatorname{cosec}^2 A - 1}$.

Sol.

Q. 27. Find the value of $\tan^3 \theta + \cot^3 \theta$, if $\tan \theta + \cot \theta = 2$.

Sol.

Q. 28. Find the value of $\tan \theta$, if $\sin \theta + \cos \theta = \sqrt{2} \, \cos \theta$.   (CBSE 2011)

Sol.

Q. 29. In $\triangle ABC$, right angled at B, $AB = 5$ cm and $\angle ACB = 30°$. Find BC and AC.

Sol.

Q. 30. Show that: $\dfrac{1 - \sin 60°}{\cos 60°} = 2 - \sqrt{3}$ (CBSE 2014)

Sol.

Q. 31. Find the value of θ, if $\dfrac{\cos \theta}{1 - \sin \theta} + \dfrac{\cos \theta}{1 + \sin \theta} = 4, \theta \leq 90°$.   (CBSE 2014)

Sol.

SHORT ANSWER TYPE Q.UESTIONS

Q. 32. Prove that: $\dfrac{\tan A + \operatorname{Sec} A - 1}{\tan A - \operatorname{Sec} A + 1} = \dfrac{1 + \operatorname{Sin} A}{\operatorname{Cos} A}$

Sol.

Q. 33. $\dfrac{1}{\sec x - \tan x} - \dfrac{1}{\cos x} = \dfrac{1}{\cos x} - \dfrac{1}{\sec x + \tan x}$

Sol.

Q. 34. $\dfrac{\tan \theta}{1 - \cot \theta} + \dfrac{\cot \theta}{1 - \tan \theta} = 1 + \tan \theta + \cot \theta = \sec \theta \operatorname{cosec} \theta + 1$ (CBSE 2019)

Sol.

Q. 35. $(\sin \theta + \operatorname{cosec} \theta)^2 + (\cos \theta + \sec \theta)^2 = 7 + \tan^2 \theta + \cot^2 \theta$

Sol.

Q. 36. $\sec A(1 - \sin A)(\sec A + \tan A) = 1$

Sol.

Q. 37. If $\sec \theta = x + \dfrac{1}{4x}$, prove that $\sec \theta + \tan \theta = 2x$ or $\dfrac{1}{2x}$

Sol.

Q. 38. If $\sin \theta + \sin^2 \theta = 1$, prove that $\cos^2 \theta + \cos^4 \theta = 1$

Sol.

Q. 39. Prove that $\cos \theta = \dfrac{p^2 - 1}{p^2 + 1}$, if $p = \operatorname{cosec} \theta + \cot \theta$.

Sol.

Q. 40. Show that: $x^2 + y^2 + z^2 = r^2$

if $x = r \cos \alpha \sin \beta$, $y = r \cos \alpha \cos \beta$ and $z = r \sin \alpha$ prove that

Sol.

Q. 41. Find the value of $\sin^{10} \theta + \operatorname{cosec}^{19}\theta$, if $\sin \theta + \operatorname{cosec} \theta = 2$.

Sol.

Q. 42. Prove that: $2 \sec^2 x - \sec^4 x - 2\operatorname{cosec}^2 x + \operatorname{cosec}^4 x = \cot^4 x - \tan^4 x$

Sol.

Q. 43. Find the value of $\operatorname{cosec} \theta$, if $\operatorname{cosec}\theta - \cot\theta = \dfrac{1}{3}$

Sol.

Q. 44. If $\cos \theta + \sin \theta = \sqrt{2} \, \cos \theta$, then show that $\cos \theta - \sin \theta = \sqrt{2} \, \sin \theta$.

Sol.

Q. 45. Evaluate: $\dfrac{\tan^2 60° + 4 \cos^2 45° + 3 \sec^2 30° + 5 \cos^2 90°}{\operatorname{cosec} 30° + \sec 60° - \cot^2 30°}$

Sol.

Q. 46. If $a \cos \theta + b \sin \theta = m$ and $a \sin \theta - b \cos \theta = n$ (CBSE 2001 C)

Prove that: $a^2 + b^2 = m^2 + n^2$

Sol.

LONG ANSWER TYPE Q.UESTIONS

Q. 47. Prove that: $\left(1 + \dfrac{1}{\tan^2 \theta}\right)\left(1 + \dfrac{1}{\cot^2 \theta}\right) = \dfrac{1}{\sin^2 \theta - \sin^4 \theta}$

Sol.

Q. 48. $2 (\sin^6 \theta + \cos^6 \theta) - 3(\sin^4 \theta + \cos^4 \theta) + 1 = 0$

Sol.

Q. 49. $(1 + \cot A + \tan A) (\sin A - \cos A) = \sin A \tan A - \cot A \cos A$

Sol.

Q. 50. If $\sin \theta + \cos \theta = m$ and $\sec \theta + \operatorname{cosec} \theta = n$ then show that $n(m^2 - 1) = 2m$

Sol.

Q. 51. Prove that: $\sqrt{\dfrac{\sec \theta - 1}{\sec \theta + 1}} + \sqrt{\dfrac{\sec \theta + 1}{\sec \theta - 1}} = 2 \operatorname{cosec} \theta$

Sol.

Q. 52. Prove that: $\dfrac{1}{\operatorname{cosec} \theta + \cot \theta} - \dfrac{1}{\sin \theta} = \dfrac{1}{\sin \theta} - \dfrac{1}{\operatorname{cosec} \theta - \cot \theta}$

Sol.

Q. 53. If $\dfrac{\cos \alpha}{\cos \beta} = m$ and $\dfrac{\cos \alpha}{\sin \beta} = n$, then prove that $(m^2 + n^2)\, \mathrm{Cos}^2\, \beta = n^2$

Sol.

Q. 54. Prove that:

$$\sec^2 \theta - \frac{\sin^2 \theta - 2 \sin^4 \theta}{2 \cos^4 \theta - \cos^2 \theta} = 1$$

Sol.

Q. 55. **Prove that:** $\sin^6 \theta + \cos^6 \theta = 1 - 3 \sin^2 \theta \cos^2 \theta$

Sol.

Q. 56. **Prove that:** $\dfrac{\cot \theta + \operatorname{cosec} \theta - 1}{\cot \theta - \operatorname{cosec} \theta + 1} = \dfrac{\sin \theta}{1 - \cos \theta}$

Sol.

Q. 57. **If** $\sin \theta + \cos \theta = \sqrt{3}$, **then prove that** $\tan \theta + \cot \theta = 1$ **(CBSE 2020)**

Sol.

Q. 58. **Prove** $\dfrac{\cot A - \cos A}{\cot A + \cos A} = \sec^2 A + \tan^2 A - 2\sec A \tan A$ **(CBSE 2020 Basic)**

Sol.

Q. 59. **Prove** $\dfrac{\sin \theta - 2 \sin^3 \theta}{2 \cos^3 \theta - \cos \theta} = \tan \theta$ **(CBSE 2020 Basic)**

Sol.

Q. 60. **If** $\cos(A + B) = \sin(A - B) = \dfrac{1}{2}$, $0 < A + B < 90°$ **and** $A > B$ **then find the value of A and B.**

(CBSE 2020 Basic)

Sol.

Q. 61. **If** $\tan \theta + \sin \theta = m$, $\tan \theta - \sin \theta = n$, **then prove that** $m^2 - n^2 = 4\sqrt{mn}$. **(CBSE 2020 Standard)**

Sol.

Q. 62. **Prove that:** $l^2 m^2 (l^2 + m^2 + 3) = 1$

If $l = \operatorname{cosec} x - \sin x$, $m = \sec x - \cos x$ **(CBSE 2020 Standard)**

Sol.

Q. 63. **Prove** $\dfrac{1 + \sec \theta - \tan \theta}{1 + \sec \theta + \tan \theta} = \dfrac{1 - \sin \theta}{1 + \cos x}$ **(CBSE 2020 Standard)**

Sol.

Q. 64. **Prove that** $\dfrac{(1 + \sin x - \cos x)^2}{(1 + \sin x + \cos x)^2} = \dfrac{1 - \cos x}{1 + \cos x}$ **(CBSE 2019)**

Sol.

Q. 65. **Prove that** $\dfrac{\sin\theta}{\cot\theta+\operatorname{cosec}\theta}=2+\dfrac{\sin\theta}{\cot\theta-\operatorname{cosec}\theta}$ **(CBSE 2019)**

Sol.

Q. 66. **If 4 tan θ = 3 then find the value of** $\dfrac{4\sin\theta-\cos\theta+1}{4\sin\theta+\cos\theta-1}$ **(CBSE 2018)**

Sol.

Q. 67. **Prove that** $\dfrac{\tan\theta+\sec\theta-1}{\tan\theta-\sec\theta+1}=\sec\theta+\tan\theta$ **(CBSE 2018)**

Sol.

Q. 68. **Prove that** $\dfrac{1}{1+\sin^2\theta}+\dfrac{1}{1+\cos^2\theta}+\dfrac{1}{1+\sec^2\theta}+\dfrac{1}{1+\operatorname{cosec}^2\theta}=2$

Sol.

Q. 69. **Prove that** $\dfrac{\tan^3\theta}{1+\tan^2\theta}+\dfrac{\cot^3\theta}{1+\cot^2\theta}=\sec\theta\operatorname{cosec}\theta-2\sin\theta\cos\theta$

Sol.

Q. 70. **If cosec $=4x+\dfrac{1}{16x}$, prove that cosec $\theta+\cot\theta=8x$ or $\dfrac{1}{8x}$**

Sol.

1. $45°$ 2. 2 3. 3

4. 7 5. $\dfrac{1}{3}$ 6. a^2

7. $\dfrac{100}{9}$ 8. a^2b^2 9. $\sqrt{\dfrac{1+\cot^2\theta}{\cot\theta}}$

10. $0°$ 11. $\dfrac{16}{9}$ 12. $\tan^2\theta$

13. $\dfrac{1}{2}$ 14. $10/3$ 15. (c)

16. (iii) -1 17. (a) 18. (a)

19. (iii) $\sqrt{\dfrac{b+a}{b-a}}$

20. LHS $= \sec^2\theta\,(\sec^2\theta - 1)$

   RHS $= \tan^2\theta\,(\tan^2\theta + 1)$

   Use $1 + \tan^2\theta = \sec^2\theta$

21. Relationalise and proceed in LHS

22. Squaring both sides of x and y and subtracting.

23. Divide both sides by $\cos^2\theta$

24. $A = 45°$, $B = 15°$

25. $\cos\theta = 5/13$

26. $\dfrac{17}{8}$

27. 2

28. $\sqrt{2}-1$

29. $AC = 10$, $BC = 5\sqrt{3}$, use Pythagoras theorem

30. Substitute values of $\sin 60°$ and $\cos 60°$ and solve

31. $60°$

Note : 32 to 38 use trigonometric identities and prove (based on Ex. 8.4 of NCERT)

39. $\sqrt{3}$

41. -1

42. 2

43. $\operatorname{cosec}\theta = \dfrac{5}{3}$

44. $\cos \theta + \sin \theta = \sqrt{2} \cos \theta$

Square both sides and get $1 + 2 \cos \theta \sin \theta = 2 \cos^2 \theta$

$\Rightarrow 2 \cos \theta \sin \theta = 2 \cos^2 \theta - 1$ ...(1)

Now square $(\cos \theta - \sin \theta)^2$ and get

$(\cos \theta - \sin \theta)^2 = 1 - 2 \cos \theta \sin \theta$ ...(2)

Substitute (1) in (2)

45. 9.

46. Find m^2 and n^2 and add

Note : Q.47 to Q.50 Use identities to prove

51. 0

52. Rationalise $\dfrac{1}{\cosec \theta + \cot \theta}$ in LHS and proceed, use $\dfrac{1}{\sin \theta} = \cosec \theta$.

Rationalise $\dfrac{1}{\cosec \theta - \cot \theta}$ in RHS and proceed, use $\dfrac{1}{\sin \theta} = \cosec \theta$.

53. Find m^2 and n^2 and substitute in LHS.

54. Take common $\sin^2\theta$ in Numerator and $\cos^2\theta$ in Denominator of 2nd term on LHS and replace 1 by $\sin^2 \theta + \cos^2 \theta$.

55. 0

56. $\dfrac{2}{3}$

57. $(\sin \theta + \cos \theta) = \sqrt{3}$

square both sides and get value of $\dfrac{1}{\sin \theta \times \cos \theta}$

Change $\tan \theta + \cot \theta$ into $\sin \theta$ and $\cos \theta$ proceed.

58. Change $\cot A = \dfrac{\cos A}{\sin A}$, take cos A common from Numerator and Denominator, Rationalise remaining term and change into sec A and tan A.

59. LHS $= \dfrac{\sin \theta(1 - 2 \sin^2 \theta)}{\cos \theta(2 \cos^2 \theta - 1)}$, write $1 = \sin^2 \theta + \cos^2 \theta$ and proceed.

60. $\cos (A + B) = \dfrac{1}{2} = \cos 60°$

$\Rightarrow \left.\begin{array}{l} A + B = 60° \\ A - B = 30° \end{array}\right]$ Solve these equations

$\sin (A - B) = \dfrac{1}{2} = \sin 30°$

$A = 45°, B = 15°$

61. Find m^2 and n^2 substitute in $m^2 - n^2$ and substitute m and n in $4\sqrt{mn}$

62. $\dfrac{2}{3}$

65. Convert cot θ and cosec θ into sin θ and cos θ

and use $\sin^2\theta = 1 - \cos^2\theta$

66. Divide Numerator and Denominator by cos θ, and use sec $\theta = \sqrt{1 + \tan^2\theta}$ or use pythagoras theorem and trigonometeric ratios,

Ans. $\dfrac{13}{11}$

67. Same as Q. 59.

PRACTICE-TEST

SECTION-A

Q. 1. If $\sin \theta = \dfrac{4}{5}$ what is the value of $\cos \theta$. (1)

Sol.

Q. 2. Find the value of $\tan^4 \theta + \cot^4 \theta$, if $\tan \theta + \cot \theta = 2$ (1)

Sol.

Q. 3. If $5x = \sec \theta$ and $\dfrac{5}{x} = \tan \theta$ then find the value of $5\left(x^2 - \dfrac{1}{x^2} \right)$

Sol.

Q. 4. If $\sin A + \sin^2 A = 1$, then the value of $(\cos^2 A + \cos^4 A)$ is

(a) 1 (b) $\dfrac{1}{2}$ (c) 2 (d) 3

Sol.

SECTION-B

Q. 5. If $5 \tan \theta = 4$ then find the value of $\dfrac{5 \sin \theta - 3 \cos \theta}{5 \sin \theta + 2 \cos \theta}$ (2)

Sol.

Q. 6. Find the value of $5 \sin \theta - 3 \cos \theta$ if $3 \sin \theta + 5 \cos \theta = 5$ (2)

Sol.

Q. 7. Prove that $(\sin \alpha + \cos \alpha)(\tan \alpha + \cot \alpha) = \sec \alpha + \operatorname{cosec} \alpha$ (2)

Sol.

SECTION-C

Q. 8. Prove that $\dfrac{\sin \theta}{1 + \cos \theta} + \dfrac{1 + \cos \theta}{\sin \theta} = 2 \operatorname{cosec} \theta$ (3)

Sol.

Q. 9. Prove that $\dfrac{\cos A}{1 - \tan A} - \dfrac{\sin^2 A}{\cos A - \sin A} = \sin A + \cos A$ (3)

Sol.

SECTION-D

Q. 10. Prove that $\dfrac{\tan \theta + \sec \theta - 1}{\tan \theta - \sec \theta + 1} = \dfrac{\cos \theta}{1 - \sin \theta}$.

Sol.

CHAPTER-9

SOME APPLICATIONS OF TRIGONOMETRY

LEARNING OBJECTIVES

NOTES

We have so far learnt to define trigonometric ratios of an angle. Also we have learnt to evaluate the trigonometric angles of 30°, 45°, 60°, we also know those trigonometric ratios for angles of 0° and 90° which are well defined. In this chapter we will learn trigonometry can be used to determine the distance between the objects or the distance between the objects or the heights of objects by taking examples from day to day life, that's why this topic is commonly known as heights and distances.

Angle of Elevation: When the observer is looking at an object (P) which is at a greater height than the observer (A), he has to lift his eyes to see the objects and the angle of elevation is formed between the line of sight joining the observer's eye to the object and the horizontal line.

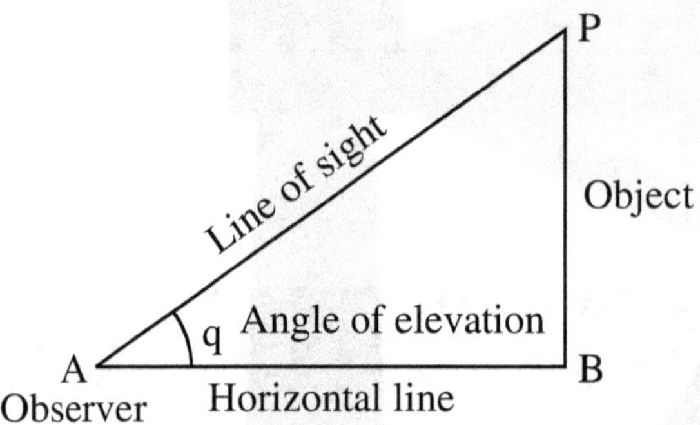

Angle of Depression: When the observer (A) is looking at an object, the angle formed between the line of sight and the horizontal line is called an angle of depression.

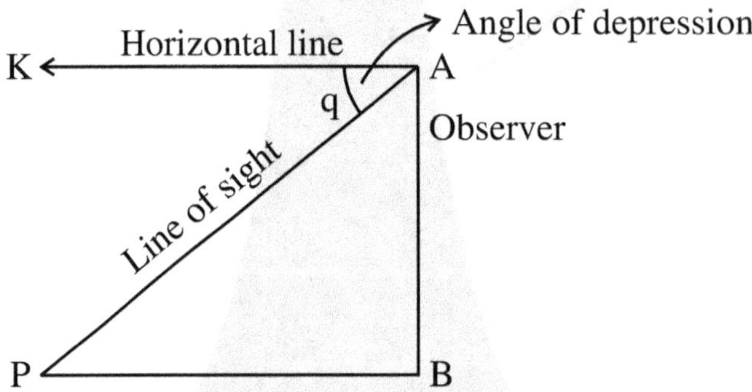

NOTE: Remember from the above diagram KA ∥ PB and PA is transversal, therefore we say that angle of elevation is always eQ.ual to angle of depression.

T-Ratios from $0° - 90°$:-

$$\sin Q. = \frac{P}{H} = \frac{Opp}{Hyp} = \frac{1}{\csc \theta}$$

$$\cos Q. = \frac{B}{H} = \frac{Adj}{Hyp} = \frac{1}{\sec \theta}$$

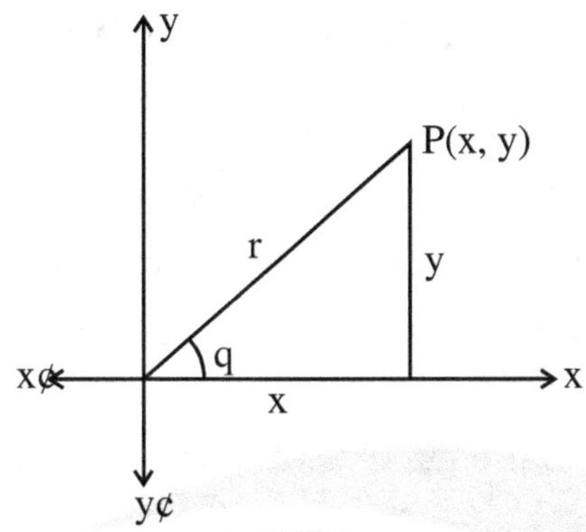

$$\tan \Theta. = \frac{P}{B} = \frac{Opp}{Adj} = \frac{\sin \theta}{\cos \theta} = \frac{1}{\cot \theta} \qquad \cot \Theta. = \frac{B}{P} = \frac{Adj}{Opp} = \frac{\cos \theta}{\sin \theta} = \frac{1}{\tan \theta}$$

$$\sec \Theta. = \frac{H}{B} = \frac{Hyp}{Adj} = \frac{1}{\cos \theta} \qquad \csc \Theta. = \frac{H}{P} = \frac{Hyp}{Opp} = \frac{1}{\sin \theta}$$

T–Ratios of Complementry Angles:-

$$\sin (90° - \Theta.) = \cos \Theta. \qquad \cos (90° - \Theta.) = \sin \Theta.$$

$$\tan (90° - \Theta.) = \cot \Theta. \qquad \cot (90° - \Theta.) = \tan \Theta.$$

$$\sec (90° - \Theta.) = \csc \Theta. \qquad \csc (90° - \Theta.) = \sec \Theta.$$

T–Ratios for Angle of Measure:

$0°, 30°, 45°, 60°,$ and $90°$

∠A	0°	30°	45°	60°	90°
sin A	0	$\frac{1}{2}$	$\frac{1}{\sqrt{2}}$	$\frac{\sqrt{3}}{2}$	1
cos A	1	$\frac{\sqrt{3}}{2}$	$\frac{1}{\sqrt{2}}$	$\frac{1}{2}$	0
tan A	0	$\frac{1}{\sqrt{3}}$	1	$\sqrt{3}$	N.D.
cosec A	N.D.	2	$\sqrt{2}$	$\frac{2}{\sqrt{3}}$	1
sec A	1	$\frac{2}{\sqrt{3}}$	$\sqrt{2}$	2	N.D.
cot A	N.D.	$\sqrt{3}$	1	$\frac{1}{\sqrt{3}}$	0

Innovative Mathematics X-9

BY R.P. GUPTA DESK:-

1. To calculate Heights and Distances follow the steps:

   Step 1:- Draw the line diagram corresponding to the problem.

   Step 2:- Mark all unknown values as x, y (Distances) and (h, H) as Heights.

   Step 3:- Use the values of various trigonometric ratios of the angles to obtain the unknown lengths from the known lengths.

2. Remember in Most of the sums we use tan Θ. as trigonometric ratio of the angle only some where sin Θ. and cos Θ..

3. In this chapter we only use trigonometric values from 30° to 60°. i.e., 30° 45°, 60°.

WARM-UP – LEVEL-I

Q. 1. **A ladder against a vertical wall makes an angle of 45° with the ground. The foot of the ladder is 3 m from the wall. Find the length of the ladder.**

Ans. Let AB be the wall and CB, the ladder.

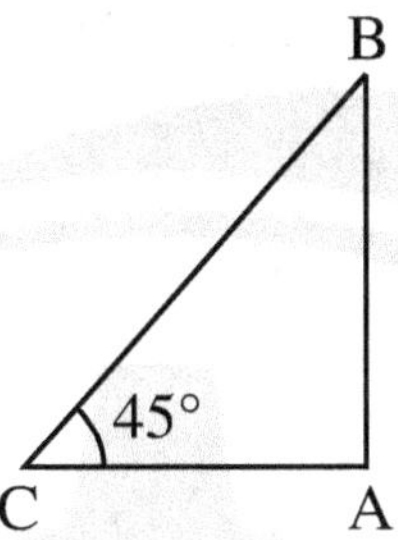

Then, AC = 3 m and $\angle$ACB = 45°

Now, $\dfrac{CB}{AC}$ = sec 45° = $\sqrt{2}$ $\Rightarrow$ $\dfrac{CB}{AC}$ = $\sqrt{2}$

$\therefore$ Length of the ladder = CB = $3\sqrt{2}$

= (3×1.41) = 4.23 m

Q. 2. **A balloon is connected to a meteorological station by a cable of length 200 m, inclined at 60° to the horizontal. Find the height of the balloon from the ground. Assume that there is no slack in the cable.**

Ans. Let B be the balloon and AB be the vertical height. Let C be the meteorological station and CB be the cable.

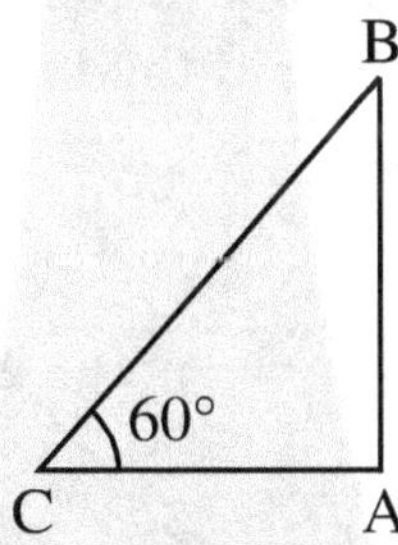

Then, BC = 200 m and $\angle$ACB = 60°

Then, $\dfrac{AB}{BC}$ = sin 60° = $\dfrac{\sqrt{3}}{2}$

$\Rightarrow$ $\dfrac{AB}{200} = \dfrac{\sqrt{3}}{2}$

$\Rightarrow$ $AB = \left(\dfrac{200 \times \sqrt{3}}{2}\right)$ = 173.2 m.

Q. 3. **The pilot of a helicopter, at an altitude of 1200m finds that the two ships are sailing towards it in the same direction. The angle of depression of the ships as observed from the helicopter are 60° and 45° respectively. Find the distance between the two ships.**

Innovative Mathematics X-9

Ans. Let B the position of the helicopter and let C, D be the ships. Let AB be the vertical height.

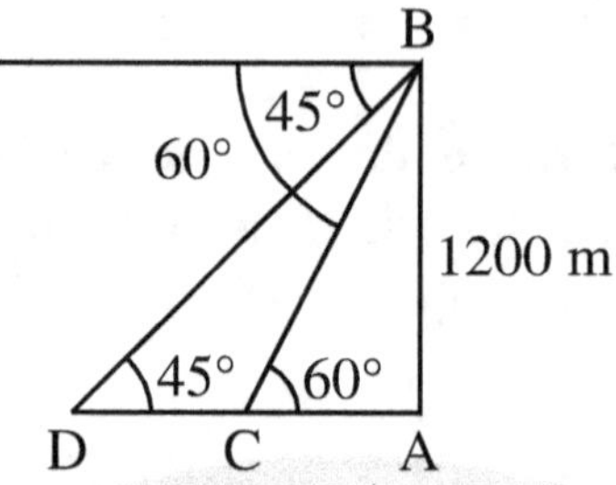

Then, AB = 1200 m,

$\angle ACB = 60°$ and $\angle ADB = 45°$.

Then, $\dfrac{AD}{AB} = \cot 45° = 1$

$\therefore \quad \dfrac{AC}{1200} = 1 \Rightarrow AD = 1200$ m

And, $\dfrac{AC}{AB} = \cot 60° = \dfrac{1}{\sqrt{3}}$

$\Rightarrow \qquad \dfrac{AC}{1200} = \dfrac{1}{\sqrt{3}}$

$\Rightarrow \qquad AC = \dfrac{1200}{\sqrt{3}} = 400\sqrt{3}$ m

Q. 4. **A vertical tower stands on a horizontal plane and is surmounted by a flagstaff of height 7 m. At a point on the plane, the angle of elevation of the bottom of the flagstaff is 30° and that of the top of the flagstaff is 45°. Find the height of the tower.**

Ans. Let AB be the tower and BC be the flagstaff.

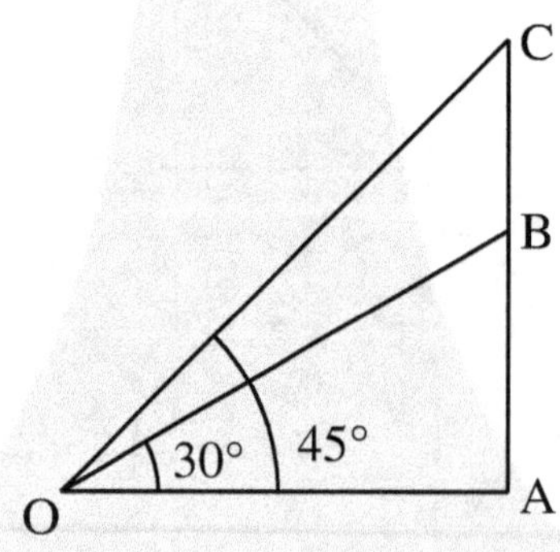

Then, BC = 7 m. Let AB = h.

Let O be the point of observation.

Then, $\angle AOB = 30°$ amd $\angle AOC = 45°$.

Now, $\dfrac{OA}{AC} = \cot 45° = 1$

$\Rightarrow \quad OA = AC = h + 7.$

And, $\dfrac{OA}{AB} = \cot 30° = \sqrt{3}$

$\Rightarrow \quad \dfrac{OA}{h} = \sqrt{3} \Rightarrow OA = h\sqrt{3}$

$\therefore \quad h + 7 = h\sqrt{3}$

$\Rightarrow \quad h = \dfrac{7}{\sqrt{3}-1} \times \dfrac{\sqrt{3}+1}{\sqrt{3}+1} = \dfrac{7(\sqrt{3}+1)}{2} = 9.562 \text{ m}$

Q. 5. **From the top of a building 30 m high, the top and bottom of a tower are observed to have angles of depression 30° and 45° respectively. The height of the tower is :**

(a) $15(1 + \sqrt{3})$m

(b) $30(\sqrt{3} - 1)$m

(b) $30\left(1 + \dfrac{1}{\sqrt{3}}\right)$m

(c) $30\left(1 - \dfrac{1}{\sqrt{3}}\right)$m

Ans. Let AB be the building and CD be the tower.

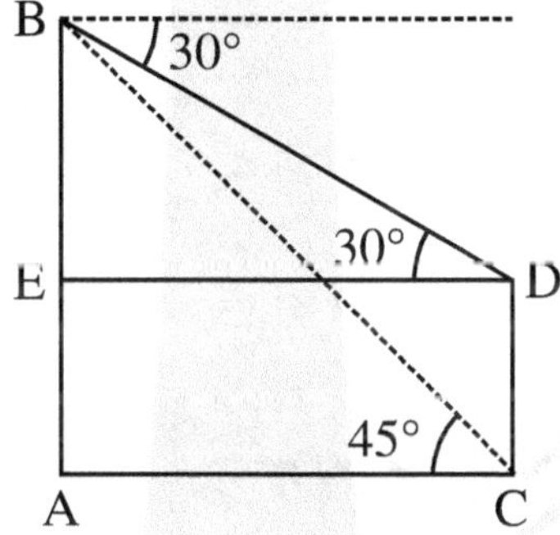

Then, AB = 30 m. Let DC = x.

Draw DE $\perp$ AB. Then AE = CD = x.

$\therefore \quad$ BE = (30 – x) m.

Now, $\dfrac{AC}{AB} = \cot 45° = 1$

$\Rightarrow \quad \dfrac{AC}{30} = 1 \Rightarrow AC = 30$ m.

$\therefore \quad$ DE = AC = 30 m.

$\dfrac{BE}{DE} = \tan 30° = \dfrac{1}{\sqrt{3}} \Rightarrow \dfrac{BE}{30} = \dfrac{1}{\sqrt{3}}$

$\Rightarrow \quad BE = \dfrac{30}{\sqrt{3}}$

$\therefore \quad CD = AE = AB = \left(30 - \dfrac{30}{\sqrt{3}}\right)$

$$= 30\left(1 - \frac{1}{\sqrt{3}}\right) m$$

Q. 6. **From the top of a cliff 25 m high the angle of elevation of a tower is found to be eQ.ual to the angle of depression of the foot of the tower. Find the height of the tower.**

Ans. Let AB be the cliff and CD be the tower.

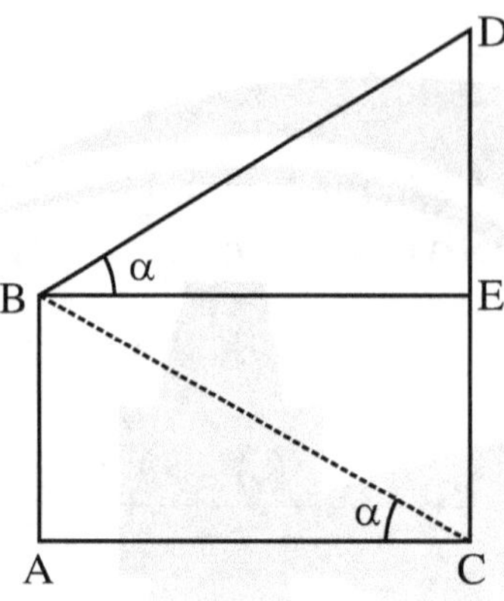

Then, AB = 25 m. From B draw BE ⊥ CD.

Let ∠EBD = ∠ACB = α.

Now, $\dfrac{DE}{BE} = \tan\alpha$ and $\dfrac{AB}{AC} = \tan\alpha$

∴ $\dfrac{DE}{BE} = \dfrac{AB}{AC}$. So, DE = AB   [Θ BE = AC]

∴ CD = CE + DE = AB + AB = 2AB = 50 m

Q. 7. **The altitude of the sun at any instant is 60°. The height of the vertical pole that will cast a shadow of 30 m is**

(a) $30\sqrt{3}$ m   (b) 15 m   (c) $\dfrac{30}{\sqrt{3}}$ m   (d) $15\sqrt{2}$ m

Ans. Let AB be the pole and AC be its shadow.

Then, Θ. = 60° and AC = 30 m.

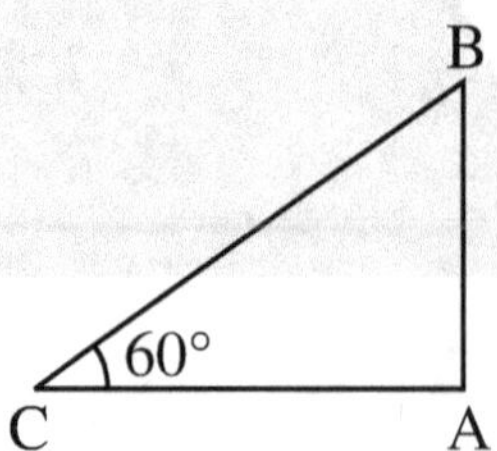

∴ $\dfrac{AB}{AC} = \tan 60°$

⇒ $\dfrac{AB}{30} = \sqrt{3}$ ⇒ AB = $30\sqrt{3}$ m

Q. 8. **If the angle of elevation of a cloud from a point h metres above a lake is α and the angle of depression of its reflection in the lake is β, prove that the height of the cloud is $\dfrac{h(\tan\beta + \tan\alpha)}{\tan\beta - \tan\alpha}$.**

Ans. Let AB be the surface of the lake and let P be a point of observation such that AP = h metres. Let C be the position of the cloud and C′ be its reflection in the lake. Then, CB = C′B. Let PM be perpendicular from P on CB. Then, $\angle$CPM = a and $\angle$MPC′ = β. Let CM = x.

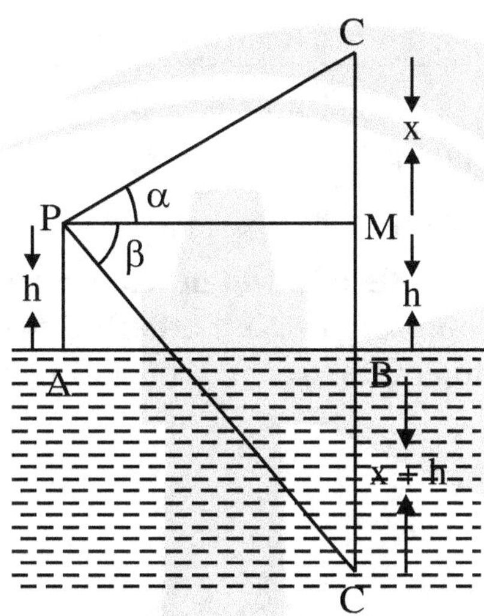

Then, CB = CM + MB = CM + PA = x + h.

In $\triangle$CPM, we have

$$\tan\alpha = \frac{CM}{PM}$$

$\Rightarrow \qquad \tan\alpha = \dfrac{x}{AB}$ $\qquad\qquad\qquad$ [Θ PM = AB]

$\Rightarrow \qquad\qquad AB = x\cot\alpha$ $\qquad\qquad\qquad$ ...(i)

In $\triangle$ PMC′, we have

$$\tan\alpha = \frac{C'M}{PM}$$

$\Rightarrow \qquad \tan\beta = \dfrac{x + 2h}{AB}$ $\qquad\qquad\qquad$ [Θ C′m = C′B + BM = x + h + h]

$\Rightarrow \qquad\qquad AB = (x + 2h)\cot\beta$ $\qquad\qquad\qquad$ ...(ii)

From (i) and (ii), we have

$$x\cot\alpha = (x + 2h)\cot\beta$$

$\Rightarrow \qquad x(\cot\alpha - \cot\beta) = 2h\cot\beta$

$\Rightarrow \qquad x\left(\dfrac{1}{\tan\alpha} - \dfrac{1}{\tan\alpha}\right) = \dfrac{2h}{\tan\beta}$

$\Rightarrow \qquad x\left(\dfrac{\tan\beta - \tan\alpha}{\tan\alpha\,\tan\beta}\right) = \dfrac{2h}{\tan\beta}$

Innovative Mathematics X-9

$$\Rightarrow \qquad x = \frac{2h \tan \alpha}{\tan \beta - \tan \alpha}$$

Hence,

Height of the cloud $= x + h$

$$= \frac{2h \tan \alpha}{\tan \beta - \tan \alpha} + h$$

$$= \frac{2h \tan \alpha + h \tan \beta - h \tan \alpha}{\tan \beta - \tan \alpha}$$

$$= \frac{h(\tan \alpha + \tan \beta)}{\tan \beta - \tan \alpha}$$

Q. 9. There is a small island in the middle of a 100 m wide river and a tall tree stands on the island. P and Q. are points directlly opposite to each other on two banks, and in line with the tree. If the angles of elevation of the top of the tree from P and Q. are respectively 30° and 45°, find the height of the tree.

Ans. Let OA be the tree of height h metre.

In triangles POA and Q.OA, we have

$$\tan 30° = \frac{OA}{OP} \text{ and } \tan 45° = \frac{OA}{OQ}$$

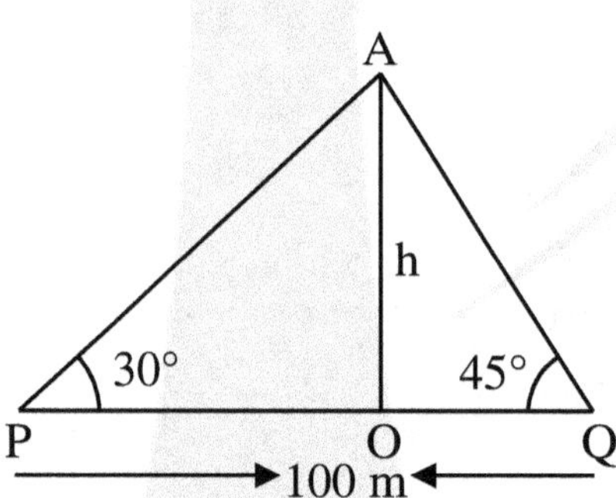

$$\Rightarrow \qquad \frac{1}{\sqrt{3}} = \frac{h}{OP} \text{ and } 1 = \frac{h}{OQ}$$

$$\Rightarrow \qquad OP = \sqrt{3}\, h \text{ and } OQ. = h$$

$$\Rightarrow \qquad OP + OQ. = \sqrt{3}\, h + h \Rightarrow PQ = (\sqrt{3} + 1)h$$

$$\Rightarrow \qquad 100 = (\sqrt{3} + 1)h \qquad\qquad [\odot PQ. = 100 \text{ m}]$$

$$\Rightarrow \qquad h = \frac{100}{\sqrt{3} + 1} m \Rightarrow h = \frac{100(\sqrt{3} - 1)}{2} m$$

$$\Rightarrow \qquad h = 50(1.732 - 1) m = 36.6 \text{ m}$$

Hence, the height of the tree is 36.6 m

Q. 10. The angle of elevation of a cliff from a fixed point is Θ.. After going up a distance of k metres towards the top of cliff at an angle of ϕ, it is found that the angle of elevation is α. Show that the height of the cliff is $\dfrac{k(\cos\phi - \sin\phi\cot\alpha)}{\cot\theta - \cot\alpha}$ metres

Ans. Let AB be the cliff and O be the fixed point such that the angle of elevation of the cliff from O is Θ. i.e. $\angle AOB = \Theta$.. Let $\angle AOC = \phi$ and OC = k metres. From C draw CD and CE perpendiculars on AB and OA respectively.

Then, $\angle DCB = \alpha$.

Let h be the height of the cliff AB.

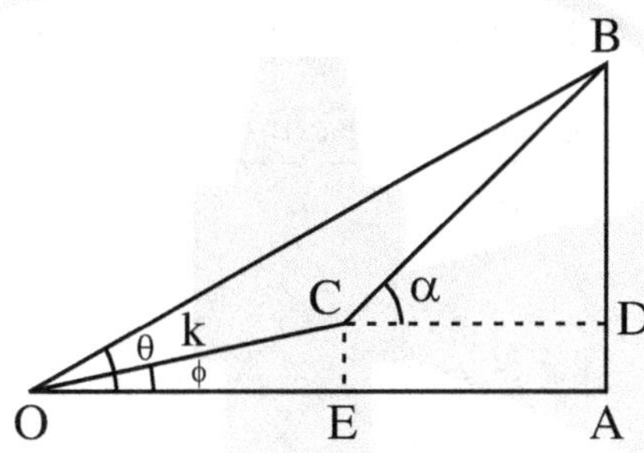

In $\triangle OCE$, we have

$$\sin\phi = \frac{CE}{OC}$$

$$\Rightarrow \quad \sin\phi = \frac{CE}{k}$$

$$\Rightarrow \quad CE = k\sin\phi \qquad \text{...(i)} \qquad\qquad [\because CE = AD]$$

$$\Rightarrow \quad AD = k\sin\phi$$

And, $\quad \cos\phi = \dfrac{OE}{OC}$

$$\Rightarrow \quad \cos\phi = \frac{OE}{k}$$

$$\Rightarrow \quad OE = k\cos\phi \qquad \text{...(ii)}$$

In $\triangle OAB$, we have

$$\tan\Theta. = \frac{AB}{OA}$$

$$\Rightarrow \quad \tan\Theta. = \frac{h}{OA}$$

$$\Rightarrow \quad OA = h\cot\Theta. \qquad \text{...(iii)}$$

$$\therefore \quad CD = EA = OA - OE$$

$$= h\cot\Theta. \quad k\cos\phi \qquad \text{...(iv)} \qquad [\text{Using eQ.s.(ii) and (iii)}]$$

and, $\quad BD = AB - AD = AB - CE$

$$= (h - k\sin\phi) \qquad \text{...(v)} \qquad [\text{Using eQ.uation (i)}]$$

In ΔBCD, we have

$$\tan \alpha \;=\; \frac{BD}{CD}$$

$\Rightarrow \qquad \tan \alpha \;=\; \dfrac{h - k \sin \phi}{h \cot \theta - k \cos \phi}$ $\qquad$ [Using eQ.uations (iv) and (v)]

$\Rightarrow \qquad \dfrac{1}{\cot \alpha} \;=\; \dfrac{h - k \sin \phi}{h \cot \theta - k \cos \phi}$

$\Rightarrow \qquad h \cot \alpha - k \sin \phi \cot \alpha = h \cot \Theta. - k \cos \phi$

$\Rightarrow \qquad h(\cot \Theta. - \cot \alpha) = k(\cos \phi - \sin \phi \cot \alpha)$

$\Rightarrow \qquad h \;=\; \dfrac{k(\cos \phi - \sin \phi \cot \alpha)}{\cot \theta - \cot \alpha}$

<h1 align="center">WARM-UP – LEVEL-II</h1>

Q. 1. The elevation of a tower at a station A due north of it is a and at a station B due west of A is β. Prove that the height of the tower is $\dfrac{AB \sin \alpha \sin \beta}{\sqrt{\sin^2 \alpha - \sin^2 \beta}}$.

Ans. Let OP be the tower and let A be a point due north of the tower OP and let B be the point due west of A. Such that $\angle OAP = \alpha$ and $\angle OBP = \beta$. Let h be the height of the tower.

In right angled triangles OAP and OBP, we have

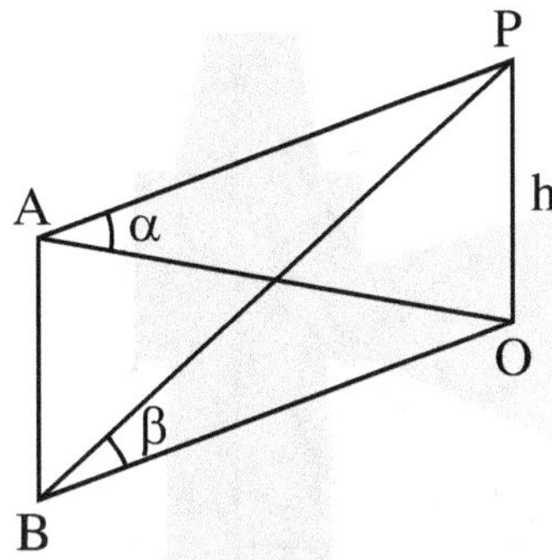

$$\tan \alpha = \frac{h}{OA} \text{ and } \tan \beta = \frac{h}{OB}$$

$$\Rightarrow \quad OA = h \cot \alpha \text{ and } OB = h \cot \beta.$$

In ΔOAB, we have

$$OB^2 = OA^2 + AB^2$$
$$\Rightarrow \quad AB^2 = OB^2 - OA^2$$
$$\Rightarrow \quad AB^2 = h^2 \cot^2 \beta - h^2 \cot^2 \alpha$$
$$\Rightarrow \quad AB^2 = h^2[\cot^2 \beta - \cot^2 \alpha]$$
$$\Rightarrow \quad AB^2 = h^2[(\operatorname{cosec}^2 \beta - 1) - (\operatorname{cosec}^2 \alpha - 1)]$$
$$\Rightarrow \quad AB^2 = h^2(\operatorname{cosec}^2 \beta - \operatorname{cosec}^2 \alpha)$$
$$\Rightarrow \quad AB^2 = h^2\left(\frac{\sin^2 \alpha - \sin^2 \beta}{\sin^2 \alpha \sin^2 \beta}\right)$$
$$\Rightarrow \quad h = \frac{AB \sin \alpha \sin \beta}{\sqrt{\sin^2 \alpha - \sin^2 \beta}}$$

Q. 2. An aeroplane when flying at a height of 4000m from the ground passes vertically above another aeroplane at an instant when the angles of the elevation of the two planes from the same point on the ground are 60° and 45° respectively. Find the vertical distance between the aeroplanes at the instant.

Ans. Let P and Q. be the positions of two aeroplanes when Q. is vertically below P and OP = 4000 m. Let the angles of elevation of P and Q. at a point A on the ground be 60° and 45° respectively.

Innovative Mathematics X-9

In triangles AOP and AOQ., we have

$$\tan 60° = \frac{OP}{OA} \text{ and } \tan 45° = \frac{OQ}{OA}$$

$$\Rightarrow \quad \sqrt{3} = \frac{4000}{OA} \text{ and } 1 = \frac{OQ}{OA}$$

$$\Rightarrow \quad OA = \frac{4000}{OA} \text{ and } OQ = OA$$

$$\Rightarrow \quad OQ. = \frac{4000}{\sqrt{3}} \text{ m}$$

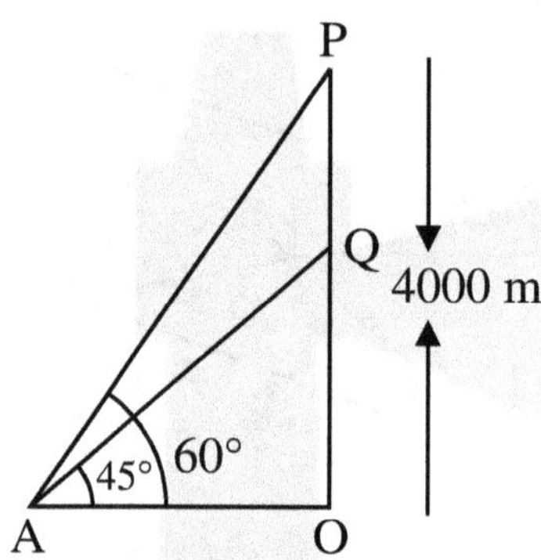

∴ Vertical distance between the aeroplanes

$$= PQ. = OP - OQ.$$

$$= \left(4000 - \frac{4000}{\sqrt{3}}\right) m = 4000 \frac{(\sqrt{3}-1)}{\sqrt{3}} m$$

$$= 1690.53 \text{ m}$$

Q. 3. **When the sun is 30° above the horizontal, the length of shadow cast by a building 50m high is-**

(a) $\dfrac{50}{\sqrt{3}}$ m (b) $50\sqrt{3}$ m (c) 25 m (d) $25\sqrt{3}$ m

Ans. Let AB be the building and AC be its shadow.

Then, AB = 50 m and $\Theta. = 30°$.

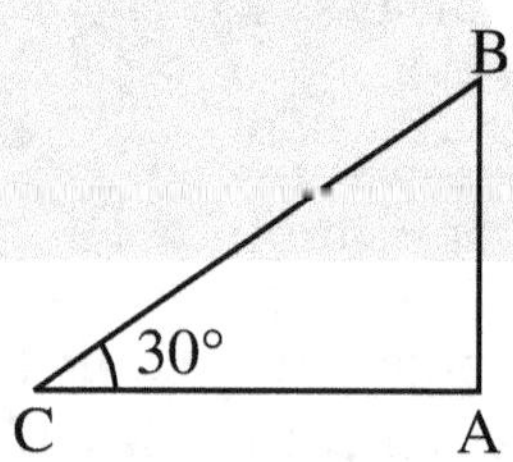

∴ $$\frac{AC}{AB} = \cot 30° = \sqrt{3}$$

$$\Rightarrow \quad \frac{AC}{50} = \sqrt{3}$$

$$\Rightarrow \qquad\qquad AC = 50\sqrt{3} \text{ cm}$$

Q. 4. If the elevation of the sun changed from 30° to 60°, then the difference between the lengths of shadows of a pole 15 m high, made at these two positions, is–

(a) 7.5 m (b) 15 m (c) $10\sqrt{3}$ m (d) $\dfrac{15}{\sqrt{3}}$ m

Ans. When $AB = 15$m, $\Theta. = 30°$, then $\dfrac{AC}{AB} = \tan 30°$

$$\Rightarrow \qquad\qquad AC = \dfrac{15}{\sqrt{3}} \text{ m}$$

When $AB = 15$m, $\Theta. = 60°$, then $\dfrac{AC}{AB} = \tan 60°$

$$\Rightarrow \qquad\qquad AC = 15\sqrt{3} \text{ m}$$

$\therefore$ Diff. in lengths of shadows

$$= \left(15\sqrt{3} - \dfrac{15}{\sqrt{3}}\right)$$

$$= \dfrac{30}{\sqrt{3}} = 10\sqrt{3} \text{ m}$$

Q. 5. The heights of two poles are 80 m and 62.5 m. If the line joining their tops makes an angle of 45° with the horizontal, then the distance between the poles, is-

(a) 17.5 m (b) 56.4 m (c) 12.33 m (d) 44 m

Ans. Let AB and CD be the poles such that $AB = 80$ m and $CD = 62.5$ m.

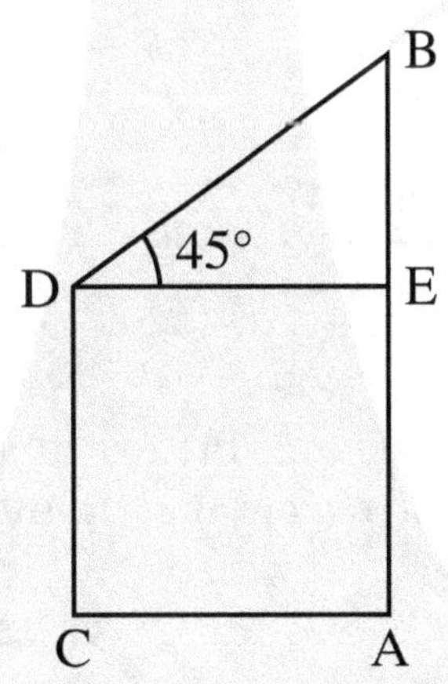

Draw $DE \perp AB$. Then,

$$\angle EDB = 45°$$

Now, $BE = AB - AE = AB - CD = 17.5$

$$\dfrac{DE}{BE} = \cot 45° = 1$$

$$\Rightarrow \qquad\qquad DE = BE = 17.5 \text{ m.}$$

Innovative Mathematics X-9

Q. 6. If the angle of elevation of cloud from a point 200 m above a lake is 30° and the angle of depression of its reflection in the lake is 60°, then the height of the cloud above the lake, is

   (a) 200 m   (b) 500 m

   (c) 30 m   (d) None of these

Ans. Let C be the cloud and C′ be its reflection in the lake.

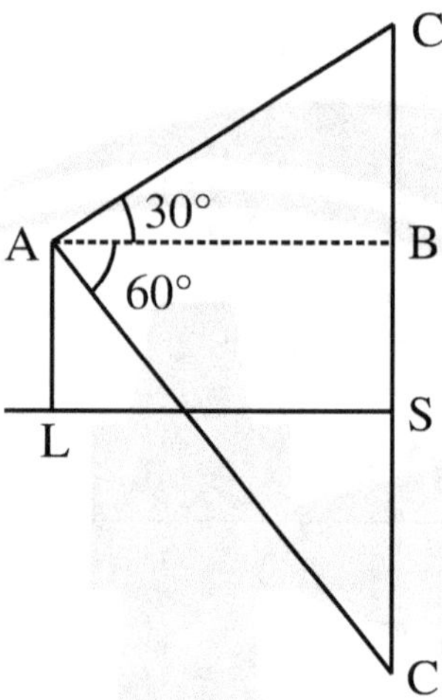

Let $CS = C'S = x$.

Now, $\qquad \dfrac{BC}{AB} = \tan 30° = \dfrac{1}{\sqrt{3}}$

$\Rightarrow \qquad x - 200 = \dfrac{AB}{\sqrt{3}}$

Also, $\qquad \dfrac{BC}{AB} = \tan 60° = \sqrt{3}$

$\Rightarrow \qquad x + 200 = (AB)\sqrt{3}$

$\therefore \qquad \sqrt{3}\,(x - 200) = \dfrac{x + 200}{\sqrt{3}} \text{ or } x = 400$

$\therefore \qquad CS = 400 \text{ m}$

Q. 7. A balloon of radius γ makes an angle α at the eye of an observer and the angle of elevation of its centre is β. The height of its centre from the ground level is given by:

   (a) $\gamma \cos \dfrac{\beta}{2} \sec \alpha$ (b) $\gamma \cos \beta \sec \dfrac{\alpha}{2}$

   (c) $\gamma \sin \dfrac{\beta}{2} \operatorname{cosec} \alpha$ (d) $\gamma \sin \beta \operatorname{cosec} \dfrac{\alpha}{2}$

Ans. Let C be the centre of the balloon and O be the position of the observer at the horizontal line OX. Let OA and OB be the tangents to the balloon so that $\angle AOB = \alpha$, $\angle XOC = \beta$ and $CA = CB = \gamma$.

Clearly, right angled triangles OAC and OBC are congruent.

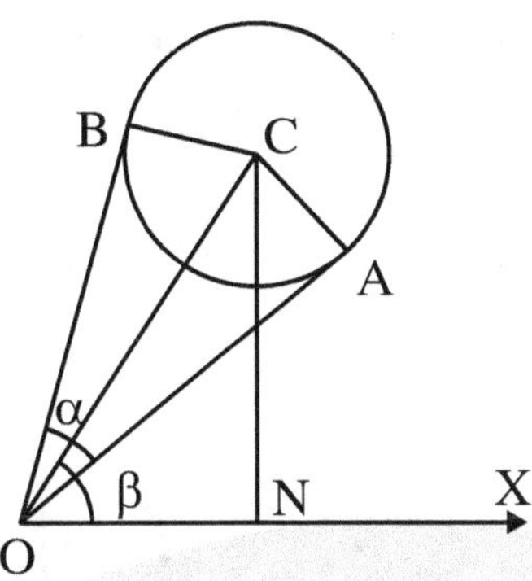

$$\therefore \qquad \angle AOC \;=\; \angle BOC = \frac{\alpha}{2}$$

Let $CN \perp OX$.

$$\text{Now,} \qquad \frac{OC}{CA} \;=\; \operatorname{cosec} \frac{\alpha}{2}$$

$$\Rightarrow \qquad OC \;=\; \gamma \operatorname{cosec} \frac{\alpha}{2} \qquad\qquad \text{...(i)}$$

$$\text{Also,} \qquad \frac{CN}{OC} \;=\; \sin \beta$$

$$\therefore \qquad CN = OC \sin\beta \;=\; \gamma \operatorname{cosec} \frac{\alpha}{2} \sin \beta \qquad\qquad [\text{Using (i)}]$$

Q. 8. **The banks of a river are parallel. A swimmer starts from a point on one of the banks and swims in a straight line inclined to the bank at 45° and reaches the opposite bank at a point 20 m from the point opposite to the starting point. The breadth of the river is -**

(a) 20 m (b) 28.28 m (c) 14.14 m (d) 40 m

Ans. Let A be the starting point and B, the end point of the swimmer. Then AB = 20 m and $\angle BAC = 45°$.

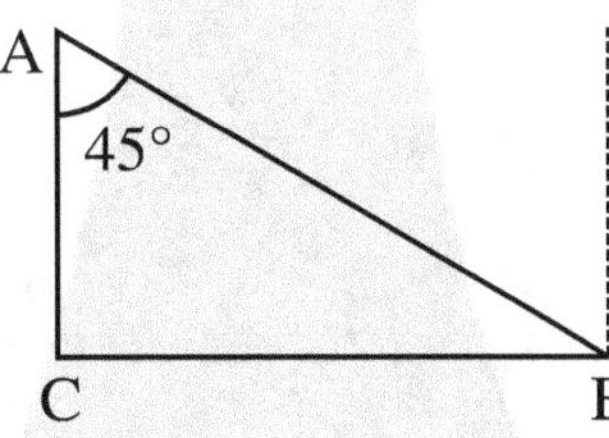

$$\text{Now,} \qquad \frac{BC}{AB} \;=\; \sin 45° = \frac{1}{\sqrt{2}}$$

$$\Rightarrow \qquad \frac{BC}{20} \;=\; \frac{1}{\sqrt{2}}$$

$$\Rightarrow \qquad BC \;=\; \frac{20 \times \sqrt{2}}{2} = 14.14 \text{ m.}$$

Q. 11. **At the foot of a mountain the elevation of its summit is 45°, after ascending 1000 m towards the mountain up a slope of 30° inclination is found to be 60°. Find the height of the mountain.**

Ans. Let F be the foot and S be the summit of the mountain FOS. Then, $\angle OFS = 45°$ and therefore $\angle OSF = 45°$. ConseQ.uently.

OF = OS = h km (say).

Let FP = 1000 m = 1 km be the slope so that $\angle OFP = 30°$. Draw PM $\perp$ OS and PL $\perp$ OF.

Join PS. It is given that $\angle MPS = 60°$.

In ΔFPL, We have

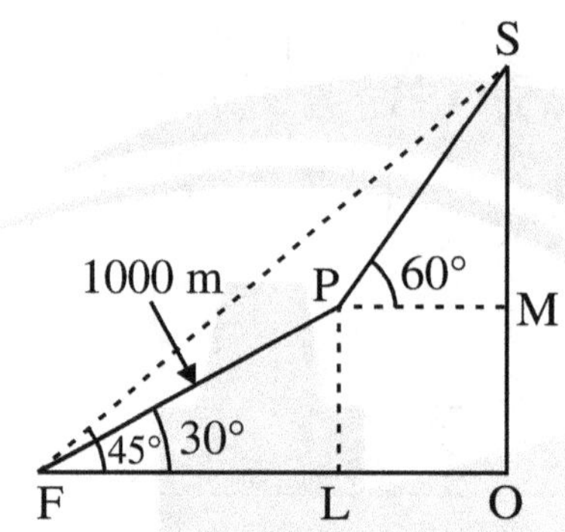

$$\sin 30° = \frac{PL}{PF}$$

$$\Rightarrow \quad PL = PF \sin 30° = \left(1 \times \frac{1}{2}\right) km. = \frac{1}{2} km.$$

$$\therefore \quad OM = PL = \frac{1}{2} km$$

$$\Rightarrow \quad MS = OS - OM = \left(h - \frac{1}{2}\right) km \qquad \qquad ...(i)$$

Also, $$\cos 30° = \frac{FL}{PF}$$

$$\Rightarrow \quad FL = PF \cos 30° = \left(1 \times \frac{\sqrt{3}}{2}\right) km = \frac{\sqrt{3}}{2} km$$

Now, $$h = OS = OF = OL + LF$$

$$\Rightarrow \quad h = OL + \frac{\sqrt{3}}{2}$$

$$\Rightarrow \quad OL = \left(h - \frac{\sqrt{3}}{2}\right) km$$

$$\Rightarrow \quad PM = \left(h - \frac{\sqrt{3}}{2}\right) km$$

In ΔPSM, we have

$$\tan 60° = \frac{SM}{PM}$$

$$\Rightarrow \qquad SM = PM.\tan 60° \qquad\qquad \text{...(ii)}$$

$$\Rightarrow \qquad \left(h - \frac{1}{2}\right) = \left(h - \frac{\sqrt{3}}{2}\right)\sqrt{3} \qquad\qquad \text{[Using eQ.uations (i) and (ii)]}$$

$$\Rightarrow \qquad h - \frac{1}{2} = h\sqrt{3} - \frac{3}{2}$$

$$\Rightarrow \qquad h(\sqrt{3} - 1) = 1$$

$$\Rightarrow \qquad h = \frac{1}{\sqrt{3} - 1}$$

$$\Rightarrow \qquad h = \frac{\sqrt{3} + 1}{(\sqrt{3} - 1)(\sqrt{3} + 1)}$$

$$= \frac{\sqrt{3} + 1}{2}$$

$$= \frac{2.732}{2} = 1.336 \text{ km}$$

Hence, the height of the mountain is 1.366 km.

Q. 12. **The angle of elevation of the top of a tower from a point A due south of the tower is a and from B due east of the tower is β. If AB = d, show that the height of the tower is $\dfrac{d}{\sqrt{\cot^2 \alpha + \cot^2 \beta}}$.**

Ans. Let OP be the tower and let A and B be two points due south and east respectively of the tower such that $\angle OAP = a$ and $\angle OBP = h$. Let OP = h. In $\triangle OAP$, we have

$$\tan \alpha = \frac{h}{OA}$$

$$\Rightarrow \qquad OA = h \cot \alpha \qquad\qquad \text{...(i)}$$

In $\triangle$ OBP, we have

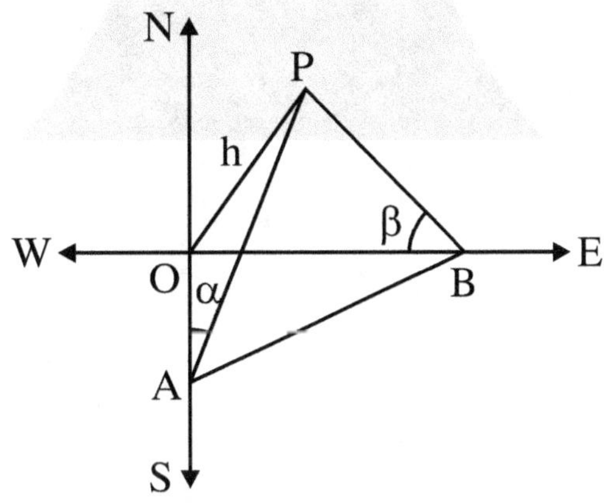

$$\tan \beta = \frac{h}{OB}$$

$$\Rightarrow \qquad OB = h \cot \beta. \qquad\qquad\qquad ...(ii)$$

Since OAB is a right angled triangle. Therefore,

$$AB^2 = OA^2 + OB^2$$

$$\Rightarrow \qquad d^2 = h^2 \cot^2 \alpha + h^2 \cot^2 \beta$$

$$\Rightarrow \qquad h = \frac{d}{\sqrt{\cot^2 \alpha + \cot^2 \beta}} \qquad\qquad \text{[Using (i) and (ii)]}$$

VERY SHORT ANSWER TYPE Q.UESTIONS

Q. 1. The length of the shadow of a tower on the plane ground is $\sqrt{3}$ times the height of the tower. The angle of elevation of sun is:

(a) 45° (b) 30° (c) 60° (d) 90°

Sol.

Q. 2. The tops of the poles of height 16 m and 10 m are connected by a wire of length l metres. If the wire makes an angle of 30° with the horizontal, then $l =$

(a) 26 m (b) 16 m (c) 12 m (d) 10 m

Sol.

Q. 3. A pole of height 6 m casts a shadow $2\sqrt{3}$ m long on the ground. the angle of elevation of the sun is **(CBSE 2017)**

(a) 30° (b) 60° (c) 45° (d) 90°

Sol.

Q. 4. A ladder leaning aginast a wall makes an angle of 60° with the horizontal. If the foot of the ladder is 2.5 m away from the wall, then the length of the ladder is— **(CBSE 2016)**

(a) 3 m (b) 4 m (c) 5 m (d) 6 m

Sol.

Q. 5. If a tower is 30 m high, casts a shadow $10\sqrt{3}$ m long on the ground, then the angle of elevation of the sun is: **(CBSE 2017)**

(a) 30° (b) 45° (c) 60° (d) 90°

Sol.

Q. 6. A tower is 50 m high. When the sun's altitude is 45° then what will be the length of its shadow?

Sol.

Q. 7. The length of shadow of a pole 50 m high is $\dfrac{50}{\sqrt{3}}$ m. find the sun's altitude.

Sol.

Q. 8. Find the angle of elevation of a point which is at a distance of 30 m from the base of a tower $10\sqrt{3}$ m high.

Sol.

Q. 9. A kite is flying at a height of $\dfrac{50}{\sqrt{3}}$ m from the horizontal. It is attached with a string and makes an angle 60° with the horizontal. Find the length of the string.

Sol.

Q. 10. In the given figure find the perimeter of rectangle ABCD.

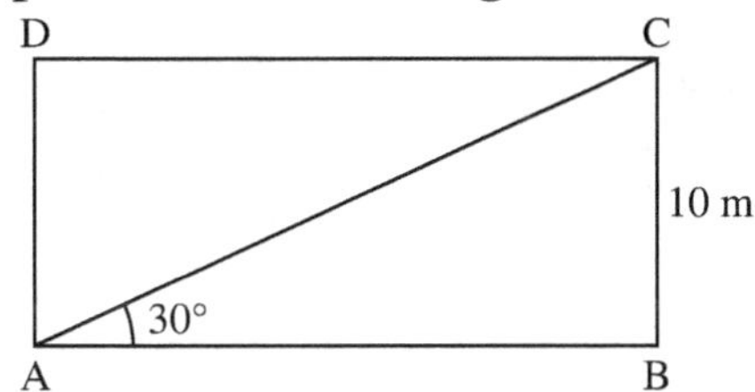

    Innovative Mathematics X-9

SHORT ANSWER TYPE Q.UESTIONS

Q. 11.   In the figure, find the value of BC.

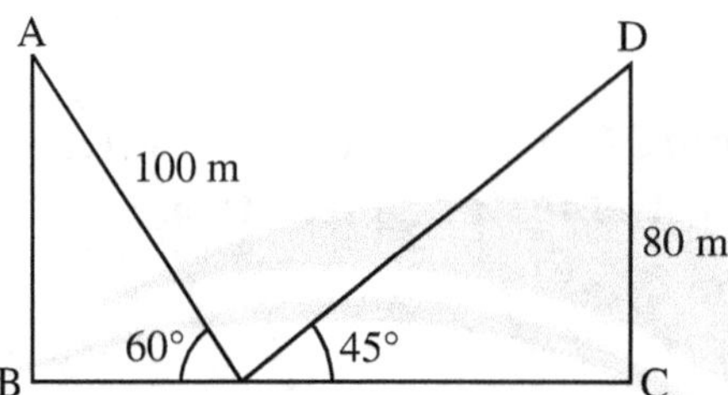

Sol.

Q. 12.   In the figure, two persons are standing at the opposite direction P & Q. of the tower. If the height of the tower is 60 m then find the distance hetween the two persons.

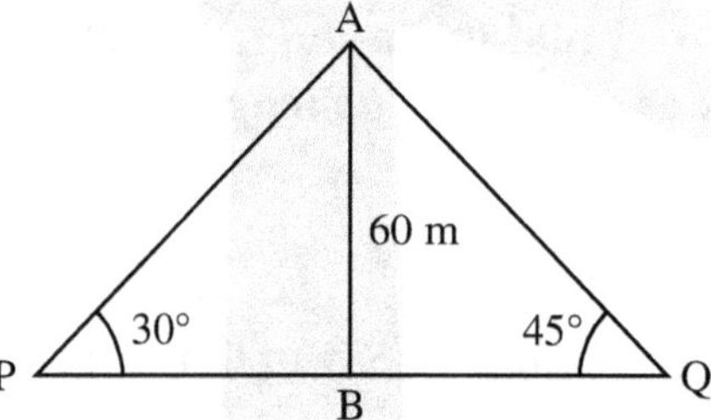

Sol.

Q. 13.   In the figure, find the value of AB.

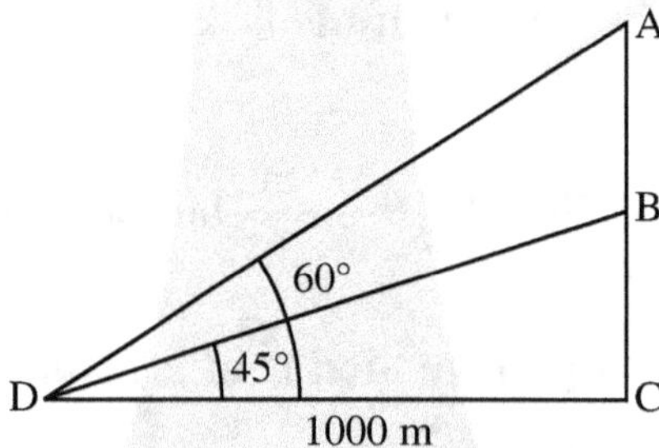

Sol.

Q. 14.   In the figure, find the value of CF.

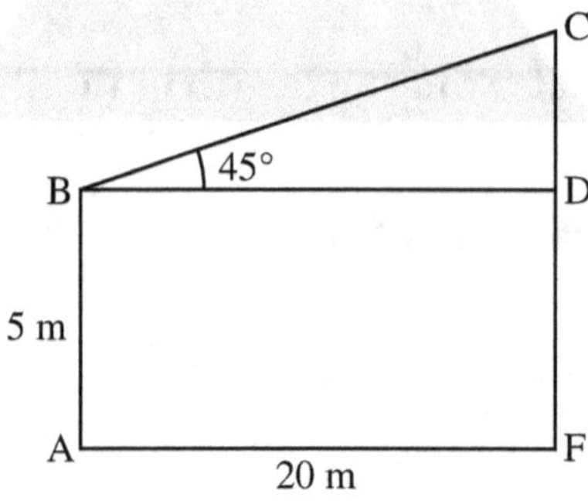

Sol.

Q. 15. If the horizontal distance of the boat from the bridge is 25 m and the height of the bridge is 25 m, then find the angle of depression of the boat from the bridge.

Sol.

Q. 16. If the length of the shadow of a tower is increasing, then the angle of elevation of the sun is also increasing. **(True / False)**

Sol.

Q. 17. If a man standing on the deck of a ship 3 m above the surface of sea observes a cloud and its reflection in the sea, then the angle of elevation of the cloud is eQ.ual to the angle of depression of its reflection. **(True / False)**

Sol.

Q. 18. The string of a kite is 150 m long and it makes an angle $60°$ with the horizontal. Find the height of the kite above the ground. (Assume string to be tight)

Sol.

Q. 19. The shadow of a vertical tower on level ground increases by 10 m when the altitude of the sun changes from $45°$ to $30°$. Find the height of the tower. (Use $\sqrt{3} = 1.73$)

Sol.

Q. 20. An acroplane at an altitude of 200 m observes angles of depression of opposite points on the two banks of the river to be $45°$ and $60°$, find the width of the river. (Use $\sqrt{3} = 1.732$)

Sol.

Q. 21. The angle of elevation of a tower at a point is $45°$. After going 40 m towards the foot of the tower, the angle of elevation of the tower becomes $60°$. Find the height of the tower. (Use $\sqrt{3} = 1.732$)

Sol.

Q. 22. The upper part of a tree broken over by the wind makes an angle of $30°$ with the ground and the distance of the foot of the tree from the point where the top touches the ground is 25 m. What was the total height of the tree?

Sol.

Q. 23. A vertical flagstaff stands on a horizontal plane. From a point 100 m from its foot, the angle of elevation of its top is found to be $45°$. Find the height of the flagstaff.

Sol.

Q. 24. The length of a string between kite and a point on the ground is 90 m. If the string makes an angle α with the level ground and $\sin \alpha = \dfrac{3}{5}$ Find the height of the kite. There is no slack in the string.

Sol.

Q. 25. An aeroplane, when 3000 m high, passes vertically above another plane at an instant when the angle of elevation of two aeroplanes from the same point on the ground are $60°$ and $45°$ respectively. Find the vertical distance between the two planes. (Use $\sqrt{3} = 1.732$)

Sol.

Q. 26. A 7m long flagstaffis fixed on the top of a tower on the horizontal plane. From a point on the ground, the angle of elevation of the top and the bottom of the flagstaff are $45°$ and $30°$ respectively. Find the height of the tower. (Use $\sqrt{3} = 1.732$)

Sol.

   Innovative Mathematics X-9

Q. 27. From the top of a 7 m high building, the angle of elevation of the top of the tower is 60° and the angle of depression of the foot of the tower is 45°. Find the height of the tower.   (CBSE 2020)

Sol.

Q. 28. Anand is watching a circus artist climbing a 20m long rope which is tightly stretched and tied from the top of vertical pole to the ground. Find the height of the pole if the angle made by the rope with the ground level is 30°.

Sol.

LONG ANSWER TYPE Q.UESTIONS

Q. 29. A statue 1.6 m tall, stands on the top of a pedestal. From a point on the ground, the angle of elevation of the top of the statue is 60° and from the same point the angle of elevation of the top of the pedestal is 45°. Find the height of the pedestal.   (Use $\sqrt{3} = 1.73$)(CBSE 2020)

Sol.

Q. 30. A man standing on the deck of a ship, 10 m above the water level observes the angle of elevation of the top of a hill as 60° and angle of depression of the bottom of the hill as 30°. Find the distance of the hill from the ship and height of the hill.

Sol.

Q. 31. From a window 60 m high above the ground of a house in a street, the angle of elevation and depression of the top and the foot of another house on the opposite side of the street are 60° and 45° respectively. Show that the height of opposite house is $60(1 + \sqrt{3})$ metres.

Sol.

Q. 32. The angle of elevation of an aeroplane from a point A on the ground is 60°. After a flight of 30 seconds, the angle of elevation changes to 30°. If the plane is flying at a constant height of 3600 $\sqrt{3}$ m, find the speed in km/hour of the plane.

Sol.

Q. 33. A bird is sitting on the top of a tree, which is 80 m high. The angle of elevation of the bird, from a point on the ground is 45°. The bird flies away from the point of observation horizontally and remains at a constant height. After 2 seconds, the angle of elevation of the bird from the point of observation becomes 30°. Find the speed of flying of the bird.   (Use $\sqrt{3} = 1.732$)

Sol.

Q. 34. The shadow of a tower standing on a level ground is found to be 30 m longer when the sun altitude is 30° longer when the sun altitude is 30° than when it is 60°. Find the height of the tower.

Sol.

Q. 35. The angle of elevation of the top of a building from the foot of a tower is 30°. The angle of elevation of the top of the tower from the foot of the building is 60°. If the tower is 60 m high, find the height ofthe building.   (CBSE 2020)

Sol.

Q. 36. An observer from the top of a light house, 100 m high above sea level, observes the angle of depression of a ship, sailing directly towards him, changes from 30° to 60°. Determine the distance travelled by the ship during the period of observation.   (Use $\sqrt{3} = 1.732$)

Sol.

Q. 37. The angles of elevation and depression of the top and bottom of a light house from the top of a building 60 m high are 30° and 60° respectively. Find

(i) The difference between the height of the light house and the building.

(ii) distance between the light house and the building.

Sol.

Q. 38. A fire in a building 'B' is reported on telephone in two fire stations P an Q., 20 km apart from each other on a straight road. P observes that the fire is at an angle of 60° to the road, and Q. observes, that it is at an angle of 45° to the road. Which station should send its team to start the work at the earliest and how much distance will this team has to travel?

Sol.

Q. 39. A 1.2 m tall girl spots a balloon on the eve of Independence Day, moving with the wind in a horizontal line at a height of 88.2 m from the ground. The angle of elevation of the balloon from the girl at an instant is 60°. After some time, the angle of elevation reduces to 30°, Find the distance travelled by the balloon.

Sol.

Q. 40. The angle of elevation of the cloud from a point 10 m above a lake is 30° and the angle of depression of the reflection of the cloud in the lake is 60°. Find the height of the cloud from the surface of lake. *(CBSE 2020)*

Sol.

Q. 41. Two pillars of eQ.ual heights stand on either side of a roadway 150 m wide. From a point on the roadway between the pillars, the angles of elevation of the top of the pillars are 60° and 30°. Find the height of pillars and the position of the point. *(CBSE, 2011)*

Sol.

Q. 42. The angle of elevation of the top of tower from certain point is 30°. If the observer moves 20 m towards the tower the angle of elevation of the top increases by 15°. Find the height of the tower.

Sol.

Q. 43. A moving boat is observed from the top of a 1 50 m high cliff moving away form the cliff. The angle of depression of the boat changes form 60° to 45° in 2 minutes. Find the speed of the boat in m/h. *(Take $\sqrt{3} = 1.732$)*

Sol.

Q. 44. From the top of a 120 m high tower a man observes two cars on the opposite sides of the tower and in straight line with the base of tower with angles of depression as 60° and 45°. Find the distance between the cars. *(Use $\sqrt{3} = 1.732$)*

Sol.

Q. 45. A vertical tower of height 20 m stands on a horizontal plane and is surmounted by a vertical flag-staff of height h. Ata point on the plane, the angle of elevation of the bottom and top of the flag staff are 45° and 60° respectively. Find the value of h. *(CBSE 2020)*

Sol.

Q. 46. The rod AC of a TV disc antenna is fixed at right angles to the wall AB and a rod CD is supporting the disc as shown in the figure. If AC = 1.5 m long and CD =3 m, find (i) tan Θ. (ii) sec Θ. + cosec Θ.. *(CBSE 2020)*

Sol.

 Innovative Mathematics X-9

Q. 47. At a point on level ground, the angle of elevation of a vertical tower is found to be α such that $\tan \alpha = \dfrac{1}{3}$. After walking 200 m towards the tower, then angle of elevation B becomes such that $\tan \beta = \dfrac{3}{4}$ find the height of the tower.

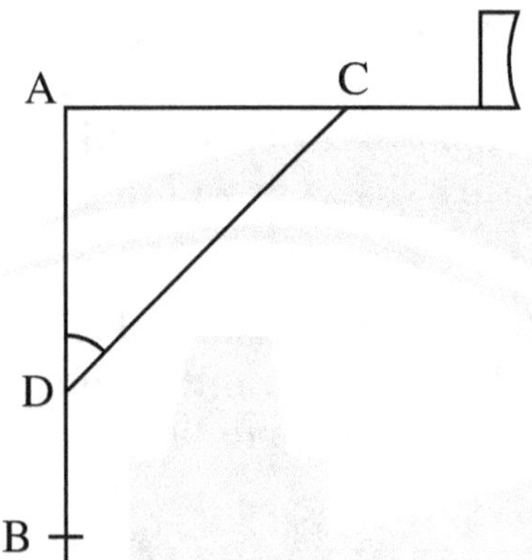

Sol.

Q. 48. A vertically straight tree, 20 m high, is broken by the wind in such a way that its top Just touches the ground and makes an angle of 60° with the ground. At what height of the ground did the tree break?

Sol.

Q. 49. If the angle of elevations of a cloud from a point 1 meters above a lake be 30° and the angle of depression of its reflection in the lake be 60°. Prove that the height of cloud is 2h, also find the distance of observer from cloud.

Sol.

Q. 50. The angles of elevation of the top of a tower of height h meter from two points P and Q. at a distance of x m and y m from the base of the tower respectively and in the same straight line with it, are 60° and 30°, respectively prove that height of tower be $\sqrt{x}$ m.

Sol.

Q. 51. Two poles of heights 18 m and 30 m stand vertically on the ground. The tops of two poles are connected by a wire, which is inclined to the horizontal at an angle of 60°. Find the length of wire and the distance between the poles.

Sol.

Q. 52. The angles of depression of the top and bottom of a 10 m tall pole from the top of a transimission tower are 45° and 60° respectively. Find the height of the transmission tower and the distance between the pole and tower.

Sol.

Q. 53. A tree breaks due to storm and the broken part bends so that the top of the tree touches the ground making and angle of 30° with it. The height of the breaking point from the ground is 10 m. Find the total height of the tree.

Sol.

1. (b) 2. (c) 3. (b) 4. (c)

5. (c) 6. 50 m 7. 60° 8. 30°

9. 100 m 10. $20\left(\sqrt{3}+1\right)$m 11. 130 m 12. $60\left(\sqrt{3}+1\right)$m

13. $1000\left(\sqrt{3}-1\right)$m 14. 25 m 15. 45° 16. False

17. False 18. $75\sqrt{3}$ m 19. 13.65 m 20. 315.46 m

21. 94.64 m 22. $25\sqrt{3}$ m 23. 100 m 24. 54 m

25. 1268 m 26. 9.562 m 27. $7(\sqrt{3}+1)$m 28. 10 m

29. 2.184 m 30. $10\sqrt{3}$ m, 40 m 31. ? 32. 864 km/hr

33. 2928 m 34. $15\sqrt{3}$ m 35. 20 m 36. 115.46 m

37. 20 m, $20\sqrt{3}$ m 38. Station P, 7.4 km (approx) 39. $58\sqrt{3}$m 40. 20 m

41. height = 64.95 m, distance (Position) = 112.5 m from the pillar having angle of elevation 60°

42. $10(\sqrt{3}+1)$m 43. 1902 m/h (approx.) 44. 189.28 m 45. h = $20(\sqrt{3}-1)$m

46. (i) $\tan\Theta. = \dfrac{1}{\sqrt{3}}$ 47. $\sec\Theta. + \csc\Theta. = \dfrac{2}{\sqrt{3}}+2$ 48. h = 120 m

49. $40(2-\sqrt{3})$m 50. 2 h 51. Length of wire = $8\sqrt{3}$ m distance $-$ $4\sqrt{3}$ m

52. h = $5\sqrt{3}(\sqrt{3}+1)$m 53. Height of tree = 30 m

SOME APPLICATIONS OF TRIGONOMETRY

SECTION-A

Q. 1. A pole which is 6 m high cast a shadow $2\sqrt{3}$ onthe ground. What is the sun's angle of elevation. **(1)**

Sol.

Q. 2. The height ofa tower is 100 m. When the angle of elevation of sun is 30°, then what is the shadow of the tower? **(1)**

Sol.

Q. 3. The angle of elevation of the sun, when the shadow of a pole h meters high is $\sqrt{3}$ h is. **(1)**

    (a) 30°   (b) 45°   (c) 60°   (d) 90°

Sol.

Q. 4. An observer 1.5 metre tall is 20.5 metre away from a tower 22 metres high. The angle of elevation of the top of the tower from the eye of the observer is, **(1)**

    (a) 30°   (b) 45°   (c) 60°   (d) 90°

Sol.

SECTION-B

Q. 5. From a point on the ground 20 m away from the foot of a tower the angle of elevation is 60°. What is the height of tower? **(2)**

Sol.

Q. 6. The ratio of height and shadow of a tower is $1 : \dfrac{1}{\sqrt{3}}$. What is the angle of elevation of the sun? **(2)**

Sol.

Q. 7. The angle of elevation of the top of a tower is 30°. If the height of the tower is tripled, then prove that the angle of elevation would be doubled. **(2)**

Sol.

SECTION-C

Q. 8. The tops of the two towers of height x and y standing on level ground, subtend angles of 30° and 60° respectively at the centre of the line joining their feet, then find x : y. **(3)**

Sol.

Q. 9. The angle of elevation of the top of a rock from the top and foot of a 100 m high tower are 30° and 45° respectively. Find the height of the rock. (3)

Sol.

SECTION-D

Q. 10. A man standing on the deck of a ship, 10 m above the water level observes the angle of elevation of the top of a hill as 60° and angle of depression of the base of the hill as 30°. Find the distance of the hill from the ship and height of the hill.

Sol.

CASE STUDY

CASE STUDY 1.

A group of students of class X visited India Gate on an education trip. The teacher and students had interest in history as well. The teacher narrated that **India Gate**, official name **Delhi Memorial**, originally called **All-India War Memorial**, monumental sandstone <u>arch</u> in <u>New Delhi</u>, dedicated to the troops of British <u>India</u> who died in wars fought between 1914 and 1919. The teacher also said that India Gate, which is located at the eastern end of the Rajpath (formerly called the Kingsway), is about 138 feet (42 metres) in height.

Q.. 1. What is the angle of elevation if they are standing at a distance of 42 m away from the monument?

(a) 30   (b) 45   (c) 60   (d) 0

Q.. 2. They want to see the tower at an angle of 60°. So, they want to know the distance where they should stand and hence find the distance.

(a) 25.24 m   (b) 20.12 m   (c) 42 m   (d) 24.64 m

Q.. 3. If the altitude of the Sun is at 60°, then the height of the vertical tower that will cast a shadow of length 20 m is

(a) $20\sqrt{3}$ m   (b) $\dfrac{20}{\sqrt{3}}$ m   (c) $\dfrac{15}{\sqrt{3}}$ m   (d) $15\sqrt{3}$ m

Q.. 4. The ratio of the length of a rod and its shadow is 1:1 . The angle of elevation of the Sun is

(a) 30°   (b) 45°   (c) 60°   (d) 90

Q.. 5. The angle formed by the line of sight with the horizontal when the object viewd is below the horizontal level is

(a) corresponding angle   (b) angle of elevation

(c) angle of depression   (d) complete angle

1. (b) 45°   2. (a) 25.24 m   3. (a) $20\sqrt{3}$ m
4. (b) 45°   5. (a) corresponding angle

CASE STUDY 2.

A Satellite flying at height h is watching the top of the two tallest mountains in Uttarakhand and Karnataka, them being Nanda Devi(height 7,816m) and Mullayanagiri (height 1,930 m). The angles of depression from the satellite, to the top of Nanda Devi and Mullayanagiri are 30° and 60° respectively. If the distance between the peaks of two mountains is 1937 km, and the satellite is vertically above the midpoint of the distance between the two mountains.

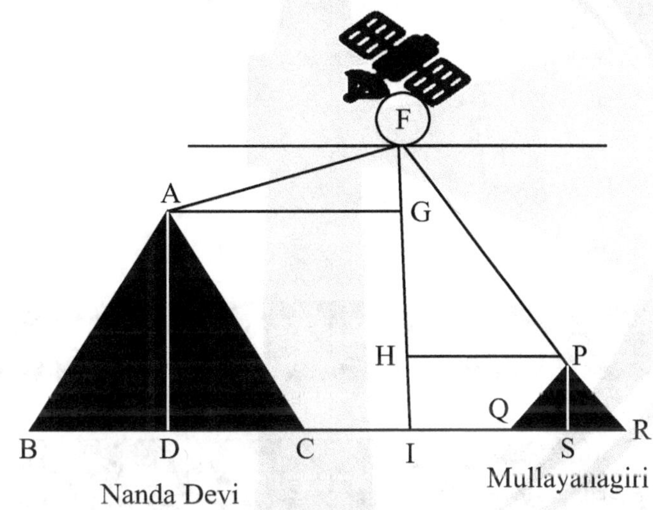

Q.. 1. **The distance of the satellite from the top of Nanda Devi is**

(a) 1139.4 km   (b) 577.52 km   (c) 1937 km   (d) 1025.36 km

Q.. 2. **The distance of the satellite from the top of Mullayanagiri is**

(a) 1139.4 km   (b) 577.52 km   (c) 1937 km   (d) 1025.36 km

Q.. 3. **The distance of the satellite from the ground is**

(a) 1139.4 km   (b) 577.52 km   (c) 1937 km   (d) 1025.36 km

Q.. 4. **What is the angle of elevation if a man is standing at a distance of 7816 m from Nanda Devi?**

(a) 30   (b) 45   (c) 60   (d) 0

Q.. 5. **If a mile stone very far away from, makes 45 to the top of Mullanyangiri montain. So, find the distance of this mile stone form the mountain.**

(a) 1118.327 km   (b) 566.976 km   (c) 1937 km   (d) 1025.36 km

1. (a) 1139.4 km   2. (c) 1937 km   3. (b) 577.52 km
4. (b) 45°   5. (c) 1937 km

CHAPTER-10

CIRCLES

LEARNING OBJECTIVES

If you consider a wheel or a coin as a circle and the touching surface as a line this illustrations show that a line touches a circle. In this topic we shall study about the possible contacts that a line and a circle can have and try to study their properties.

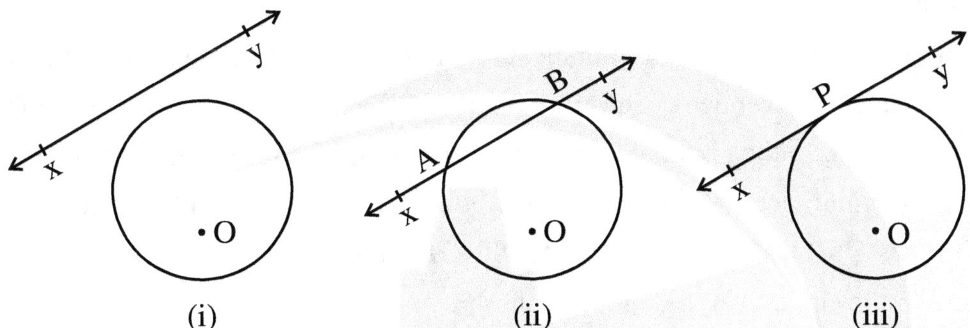

There are three diagrams above which describe three possibilities which are:

(i)   The line does not intersect circle at all i.e. the line is in the exterior of the circle.

(ii) The line intersects the circle at two distinct points. In that case a part of the line lies in the interior of the interior of the circle, the two points of intersection lie on the circle and the remaining portion of the line lies in the exterior of a circle.

(iii) The line touches the circle in exactly one point, we therefore define some specific terms.

Tangent: A line which touches a circle at exactly one point is called a tangent line and the point where it touches the circle is called the point of contact.

TANGENTS FROM A POINT OUTSIDE THE CIRCLE:-

Given: Tangents PT and PQ. are drawn from an external point P.

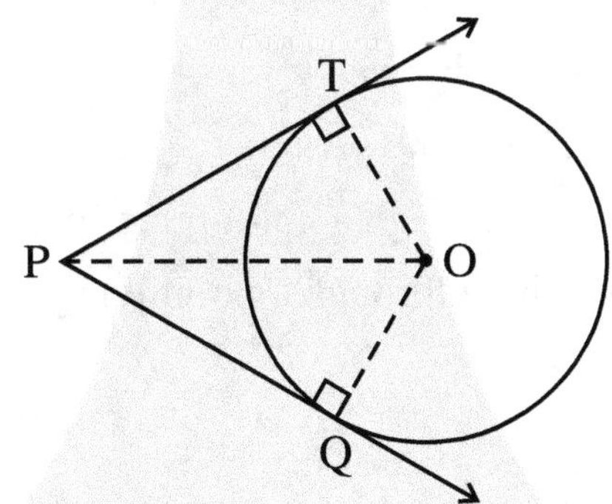

To prove: TP = Q.P

Const: Join to, Q.O and PO

Proof: In $\triangle$ PTO and $\triangle$ PQ.O

$\qquad \angle T = \angle Q.$ (90°)

$\qquad$ PO = PO (Common)

$\qquad$ OT = OQ. (Equal radii)

$\therefore \qquad \triangle$ PTO $\cong \triangle$ PQ.O

∴   PT = PQ. (C.P.C.T.)

So, from an external point, two tangents can be drawn to a circle.

∴   OP is less than each of OQ. OR, OS and OT.

As OP is the shortest distance from O to the line XY.

∴   OP ⊥ XY

Thus we can say that

A radius through the point of contact of tangent to a circle is perpendicular to the tangent at that point.

The above result can also be verified by measuring angles OPX and OPY and finding each of them equal to 90°.

Secant: A line which intersects the circle in two distinct points is called a secant line or secant.

NOTE: A tangent is the limiting position of a secant when the two points of intersection coincide.

TANGENT AND RADIUS THROUGH THE POINT OF CONTACT:-

Let XY be a tangent to the circle O at the point P, join OP.

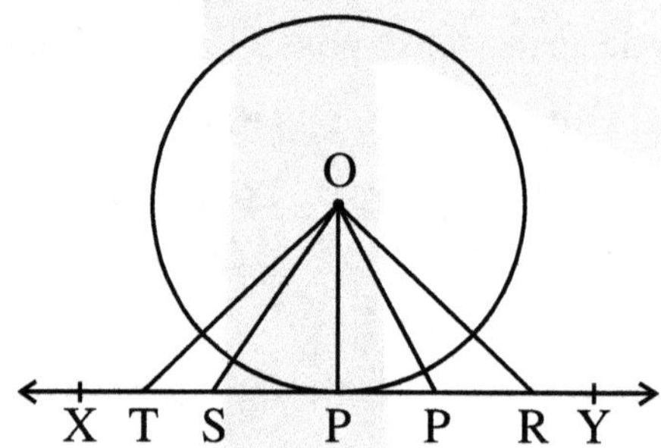

Take points Q., R, S and T on the tangent XY and Join OQ., OR, OS and OT As Q., R, S, T are points in the exterior of the circle.

From R.P. GUPTA DESK:-

(i)   From one point infinite tangents can be drawn.

(ii)   From two points only one secant can be drawn.

(iii)   The tangent to a circle is a special case of the secant.

(iv)   These is no tangent to a circle passing through a point lying inside the circle.

(v)   These are exactly two tangents to a circle through a point lying outside the circles.

Q. 1. In the given figure, PQ. is a chord of length 8 cm of a circle of radius 5 cm. The tangents at P and Q. intersect at a point T. Find the length TP.

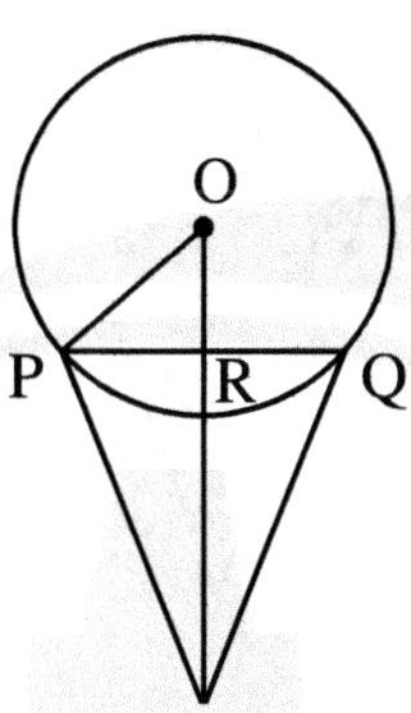

Ans. Join OP and OT Let OT intersect PQ. at a point R.

Then, TP = TQ. and $\angle$PTR = $\angle$Q.TR.

$\therefore$ TR $\perp$ PQ. and TR bisects PQ..

$\therefore$ PR = RQ. = 4 cm.

Also, OR = $\sqrt{OP^2 - PR^2}$ = $\sqrt{5^2 - 4^2}$ cm

$\qquad$ = $\sqrt{25 - 16}$ cm $-\sqrt{9}$ cm = 3 cm.

Let TP = x cm and TR = y cm.

From right $\triangle$TRP, we get

$\qquad$ TP^2 = TR^2 + PR^2

$\Rightarrow$ x^2 = y^2 + 16 $\Rightarrow$ x^2 – y^2 = 16 $\qquad\qquad$ ...(i)

From right $\triangle$OPT, we get

TP^2 + OP^2 = OT^2

$\Rightarrow$ x^2 + 5^2 = (y + 3)^2 [Θ OT^2 = (OR + RT)^2]

$\Rightarrow$ x^2 – y^2 = 6y – 16

From (i) and (ii), we get

$\qquad$ 6y – 16 = 16 $\Rightarrow$ 6y = 32 $\Rightarrow$ y = $\dfrac{16}{3}$

Putting y = $\dfrac{16}{3}$ in (i), we get

$\qquad$ x^2 = $16 + \left(\dfrac{16}{3}\right)^2 = \left(\dfrac{256}{9} + 16\right) = \dfrac{400}{9}$

$\Rightarrow$ x = $\sqrt{\dfrac{400}{9}} = \dfrac{20}{3}$.

Hence, length TP = x cm = $\left(\dfrac{20}{3}\right)$ cm

Q. 2. Two tangents TP and TQ. are drawn to a circle with centre O from an external point T. Proe that ∠PTQ. = 2∠OPQ..

Ans. **Given :** A circle with centre O and an external point T from which tangents TP and TQ. are drawn to touch the circle at P and Q..

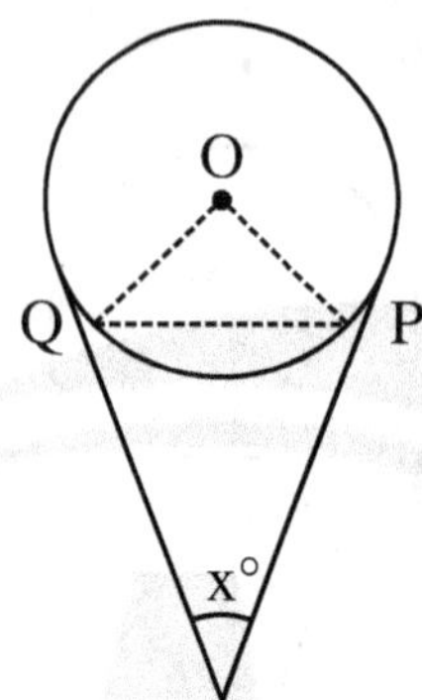

To prove : ∠PTQ. = 2∠OPQ..
Proof : Let ∠PTQ. = x°. Then,
$$\angle TQ.P + \angle TPQ. + \angle PTQ. = 180°$$
[⊙ sum of the ∠s of a triangle is 180°]
$$\Rightarrow \quad \angle TQ.P + \angle TPQ. = (180° - x) \qquad \qquad ...(i)$$
We know that the lengths of tangent drawn from an external point to a circle are equal.
So, TP = TQ..
Now, TP = TQ.
$$\Rightarrow \quad \angle TQ.P = \angle TPQ.$$

$$= \frac{1}{2}(180° - x) = \left(90° - \frac{x}{2}\right)$$

$$\therefore \quad \angle OPQ. = (\angle OPT - \angle TPQ.)$$

$$= 90° - \left(90° - \frac{x}{2}\right) = \frac{x}{2}$$

$$\Rightarrow \quad \angle OPQ. = \frac{1}{2} \angle PTQ.$$

$$\Rightarrow \quad \angle PTQ. = 2\angle OPQ..$$

Q. 3. Prove that in two concentric circles, the chord of the larger circle which touches the smaller circle, is bisected at the point of contact.

Ans. **Given :** Two circles with the same centre O and AB is a chord of the larger circle which touches the smaller circle at P.

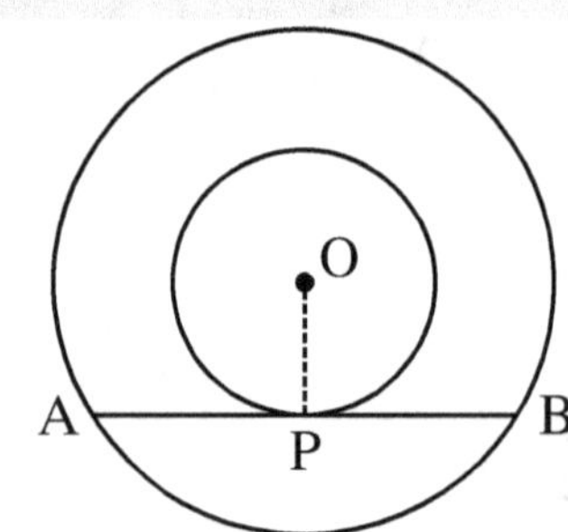

Innovative Mathematics X-10

To prove : AP = BP.

Construction : Join OP.

Proof : AB is a tangent to the smaller circle at the point P and OP is the radius through P.

∴   OP ⊥ AB.

But, the perpendicular drawn from the centre of a circle to a chord bisects the chord.

∴   OP bisects AB.   Hence, AP = BP.

Q. 4. **Prove that the tangents drawn at the ends of a diameter of a circle are parallel.**

Ans. **Given :** CD and EF are the tangents at the end points A and B of the diameter AB of a circle with centre O.

To prove : CD ∥ EF.

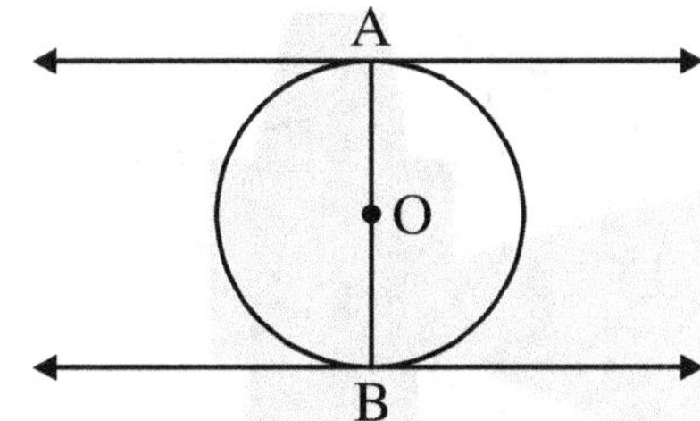

Proof : CD is the tangent to the circle at the point A.

∴   ∠BAD = 90°

EF is the tangent to the circle at the point B.

∴   ∠ABE = 90°

Thus, ∠BAD = ∠ABE (each equal to 90°).

But these are alternate interior angles.

∴   CD ∥ EF

Q. 5. **Prove that the line segment joining the point of contact of two parallel tangents to a circle is a diameter of the circle.**

Ans. **Given :** CD and EF are two parallel tangents at the points A and B of a circle with centre O.

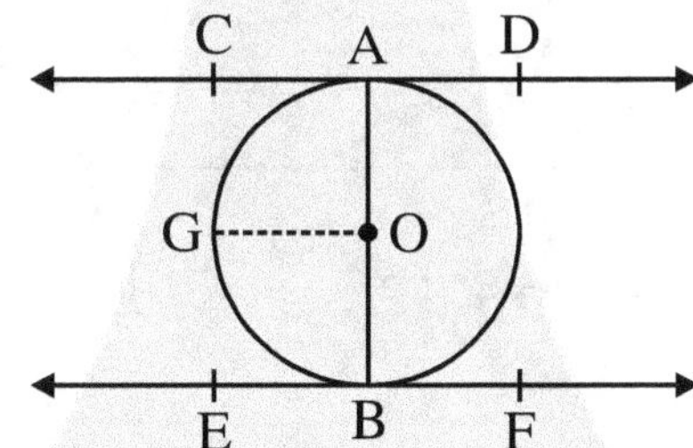

To prove : AOB is a diameter of the circle.

Construction : Join OA and OB.

Draw OG ∥ CD

Proof : OG ∥ CD and AO cuts them.

∴   ∠CAO + ∠GOA = 180°

⇒   90° + ∠GOA = 180° [OA ⊥ CD]

⇒   ∠GOA = 90°

Similarly, ∠GOB = 90°

$\therefore \quad \angle GOA + \angle GOB = (90° + 90°) = 180°$

$\Rightarrow \quad$ AOB is a straight line

Hence, AOB is a diameter of the circle with centrre O.

Q. 6. **Prove that the angle between the two tangents drawn from an external point to a circle is supplementary to the angle subtended by the line segments joining the pointsof contact to the centre.**

Ans. **Given :** PA and PB are the tangent drawn from a point P to a circle with centre O. Also, the line segments OA and OB are drawn.

To Prove : $\angle APB + \angle AOB = 180°$

Proof : We know that the tangent to a circle is perpendicular to the radius through the point of contact.

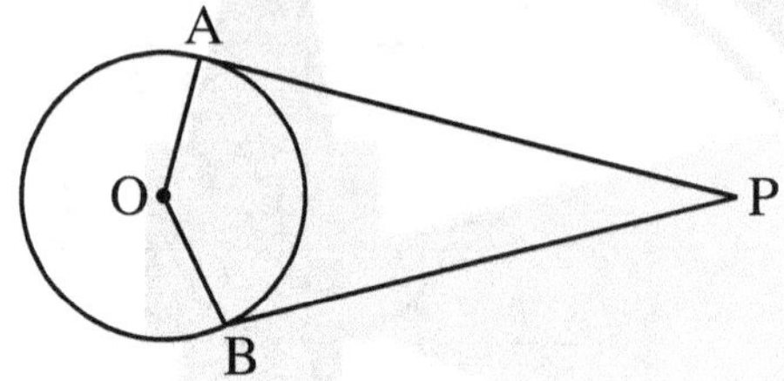

$\therefore \quad PA \perp OA \Rightarrow \angle OAP = 90°$, and

$PB \perp OB \Rightarrow \angle OBP = 90°$.

$\therefore \quad \angle OAP + \angle OBP = 90°$.

Hence, $\angle APB + \angle AOB = 180°$

[$\ominus$ sum of the all the angles of a quadrilateral is 360°]

Q. 7. **In the given figure, the incircle of $\triangle ABC$ touches the sides BC, CA and AB at D, E, F respectively.**

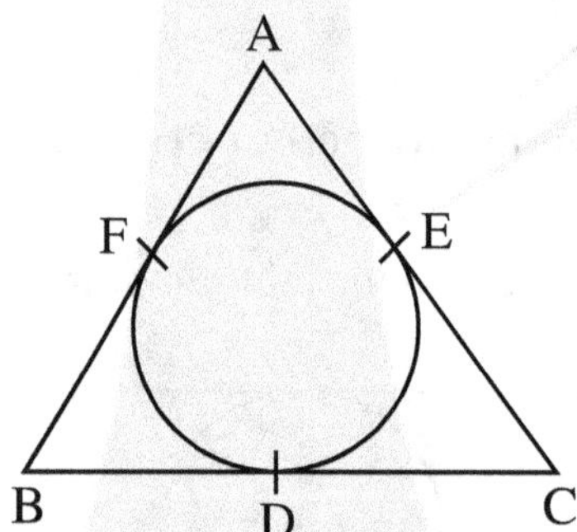

Prove that AF + BD + CE = AE + CD + BF

$$= \frac{1}{2} \text{ (perimeter of } \triangle ABC)$$

Ans. We know that the lengths of tangents from an exterior point to a circle are equal.

$\therefore \quad AF = AE \quad$ .... (i) [tangents from A]

$BD = BF \quad$ ..... (ii) [tangents from B]

$CE = CD \quad$ .... (iii) [tangents from C]

Adding (i), (ii) and (iii), we get

$(AF + BD + CE) = (AE + BF + CD) = k$ (say)

Perimeter of $\triangle ABC = (AF + BD + CE) + (AE + BF + CD)$

$$= (k + k) = 2k$$

$$\therefore \quad k = \frac{1}{2} \text{ (perimeter of } \Delta ABC).$$

Hence $AF + BD + CE = AE + CD + BF$

$$= \frac{1}{2} \text{ (perimeter of } \Delta ABC).$$

Q. 8. **A circle touches the side BC of a ΔABC at P, and touches AB and AC produced at Q. and R respectively, as shown in the figure.**

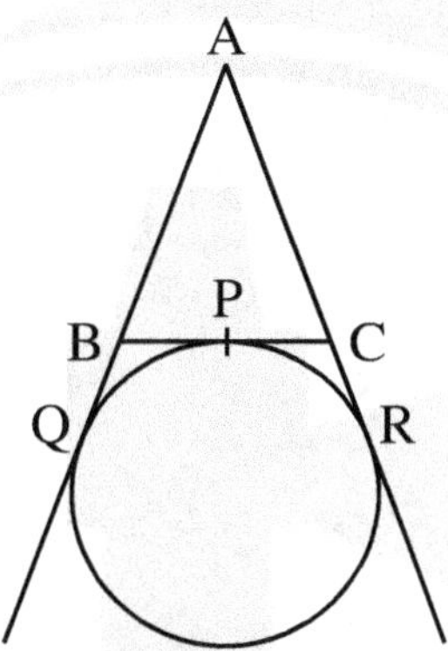

Show that AQ. $= \dfrac{1}{2}$ (perimeter of ΔABC)

Ans. We know that the lengths of tangents drawn from an exterior point to a circle are equal.

$\therefore \quad$ AQ. = AR $\hspace{5cm}$ ....(i) [tangents from A]

$\quad\quad$ BP = BQ. $\hspace{5.2cm}$ ....(ii) [tangents from B]

$\quad\quad$ CP = CR $\hspace{5.3cm}$ ....(iii) [tangents from C]

Perimeter of ΔABC

= AB + BC + AC

= AB + BP + CP + AC

= AB + BQ. + CR + AC [using (ii) and (iii)]

= AQ. + AR

= 2AQ. [using (i)].

Hence, AQ. $= \dfrac{1}{2}$ (perimeter of ΔABC)

Q. 9. **Prove that there is one and only one tangent at any point on the circumference of a circle.**

Ans. Let P be a point on the circumference of a circle with centre O. If possible, Let PT and PT be two tangents at a point P of the circle.

Now, the tangent at any point of a circle is perpendicular to the radius through the point of contact.

$\therefore \quad$ OP $\perp$ PT and similarly, OP $\perp$ PT

$\Rightarrow \quad \angle OPT = 90°$ and $\angle OPT' = 90°$

$\Rightarrow \quad \angle OPT = \angle OPT'$

This is possible only when PT and PT′ coincide. Hence, there is one and only one tangent at any point on the circumference of a circle.

Q. 10. **A quadrilateral ABCD is drawn to circumscribe a circle, as shown in the figure.**

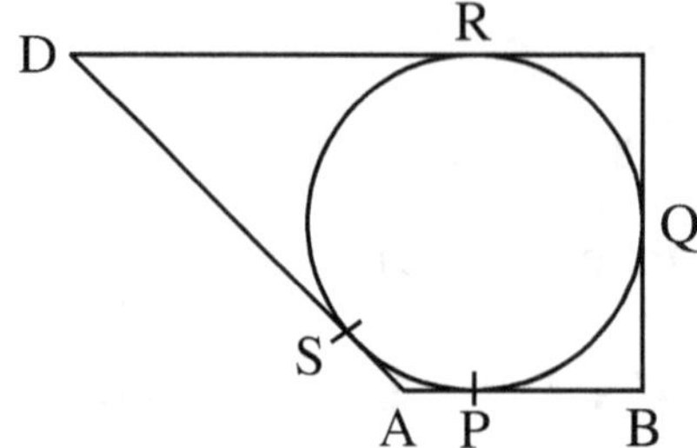

Prove that AB + CD = AD + BC

Ans. We known that the lengths of tangents drawn from an exterior point to a circle are equal.

∴ AP = AS   ....(i) [tangents from A]

BP = BQ.   ....(ii) [tangents from C]

DR = DS   ....(iii) [tangents from D]

∴ AB + CD = (AP + BP) + (CR + DS)

= (AS + BQ.) + (CQ. + DS)   [using (i), (ii), (iii), (iv)]

= (AS + DS) + (BQ. + CQ.)

= (AD + BC).

Hence, (AB + CD) = (AD + BC)

Q. 11. **Prove that the parallelogram circumscribing a circle, is a rhombus.**

Ans. **Given :** A parallelogram ABCD circumsribes a circle with centre O.

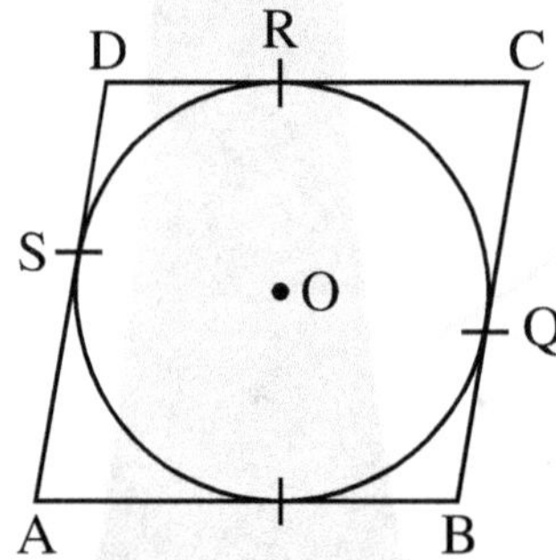

To prove : AB = BC = CD = AD

Proof : we know that the lengths of tangents drawn from an exterior point to a circle are equal.

∴ AP = AS   ....(i) [tangents from A]

BP = BQ.   ....(ii) [tangents from B]

CR = CQ.   ....(iii) [tangents from C]

DR = DS   ....(iv) [tangents from D]

∴ AB + CD = AP + BP + CR + DR

= AS + BQ. + CQ. + DS   [From (i), (ii), (iii), (iv)]

= AD + BC

Hence, (AB + CD) = (AD + BC)

⇒ 2AB = 2AD

[Θ opposite sides of a parallelogram are equal]

$\Rightarrow \quad AB = AD$

$\therefore \quad CD = AB = AD = BC$

Hence, ABCD is a rhombus

Q. 12. **Prove that the opposite sides of a quadrilateral circumscribing a circle subtend supplementary angles at the centre of the circle.**

Ans. **Given :** A quadrilateral ABCD circumscribes a circle with centre O.

To Prove : $\angle AOB + \angle COD = 180°$ and $\angle BOC + \angle AOD = 180°$

Construction : Join OP, OQ., OR and OF

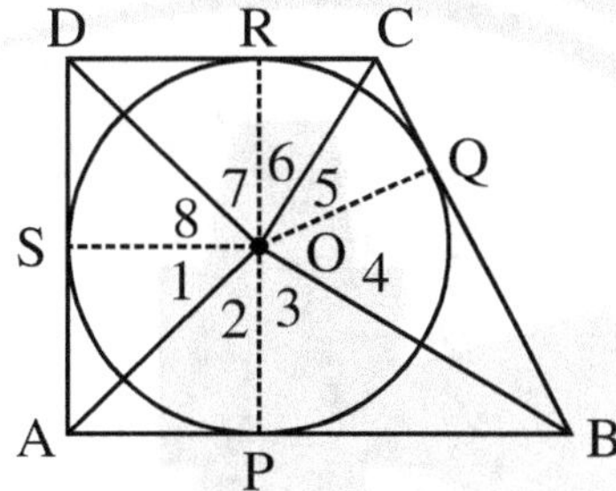

Proof : We know that the tangents drawn from an external point of a circle subtend equal angles at the centre.

$\therefore \quad \angle 1 = \angle 2, \angle 3 = \angle 4, \angle 5 = \angle 6$

and $\angle 7 = \angle 8$

And, $\angle 1 + \angle 2 + \angle 3 + \angle 4 + \angle 5 + \angle 6 + \angle 7 + \angle 8 = 360°$ $\hfill [\angle s \text{ at a point}]$

$\Rightarrow \quad 2(\angle 2 + \angle 3) + 2(\angle 6 + \angle 7) = 360°$ and $2(\angle 1 + \angle 8) + 2(\angle 4 + \angle 5) = 360°$

$\Rightarrow \quad \angle 2 + \angle 3 + \angle 6 + \angle 7 = 180°$ and $\angle 1 + \angle 8 + \angle 4 + \angle 5 = 180°$

$\Rightarrow \quad \angle AOB + \angle COD = 180°$ and $\angle AOD + \angle BOC = 180°$

Q. 13. **In the given figure, PQ. and RS are two parallel tangents to a circle with centre O and another tangent AB with point of contact C intersects PQ. at A and RS at B.**

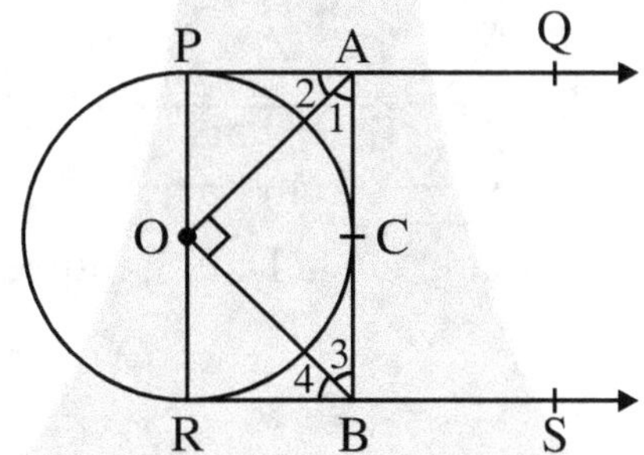

Prove that $\angle AOB = 90°$

Ans. **Given :** PQ. and RS are two parallel tangents to a circle with centre O and AB is a tangent to the circle at a point C, intersecting PQ. and RS at A and B respectively.

To prove : $\angle AOB = 90°$

Proof : Since PA and RB are tangents to the circle at P and R respectively and POR is a diameter of the circle, we have

$\quad \angle OPA = 90°$ and $\angle ORB = 90°$

$\Rightarrow$ $\angle OPA + \angle ORB = 180°$

$\Rightarrow$ $PA \parallel RB$

We know that the tangents to a circle from an external point are equally inclined to the line segment joining this point to the centre.

$\therefore$ $\angle 2 = \angle 1$ and $\angle 4 = \angle 3$

Now, PA $\parallel$ RB and AB is a transversal.

$\therefore$ $\angle PAB + \angle RBA = 180°$

$\Rightarrow$ $(\angle 1 + \angle 2) + (\angle 3 + \angle 4) = 180°$

$\Rightarrow$ $2\angle 1 + 2\angle 3 = 180°$ $[\therefore \angle 2 = \angle 1$ and $\angle 4$ and $\angle 3]$

$\Rightarrow$ $\angle 1 + \angle 3 = 90°$

From $\triangle AOB$, we have

$\angle AOB + \angle 1 + \angle 3 = 180°$ [Θ. sum of the $\angle$s of a triangle is $180°$]

$\Rightarrow$ $\angle AOB + 90° = 180°$

$\Rightarrow$ $\angle AOB = 90°$

Hence, $\angle AOB = 90°$

Q. 14. **ABC is a right triangle, right angled at B. A circle is inscribed in it. The lengths of the two sides containing the right angle are 6 cm and 8 cm. Find the radius of the incircle.**

Ans. Let the radius of the in circle be x cm.

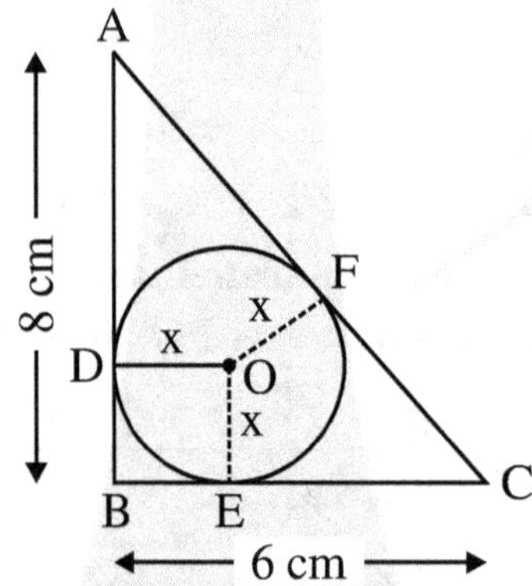

Let the in circle touch the side AB, BC and CA at D, E, F respectively. Let O be the centre of the circle.

Then, OD = OE = OF = x cm.

Also, AB = 8 cm and BC = 6 cm.

Since the tangents to a circle from an external point are equal, we have

AF = AD = (8 – x) cm, and CF = CE = (6 – x) cm.

$\therefore$ AC = AF + CF = (8 – x) cm + (6 – x) cm = (14 – 2x) cm.

Now, $AC^2 = AB^2 + BC^2$

$\Rightarrow$ $(14 - 2x)^2 = 8^2 + 6^2 = 100 = (10)^2$

$\Rightarrow$ $14 - 2x = \pm 10$ $\Rightarrow x = 2$ or $x = 12$

$\Rightarrow$ $x = 2$ [neglecting x = 12].

Hence, the radius of the in circle is 2 cm.

Q. 1. **A point P is 13 cm from the centre of the circle. The length of the tangent drawn from P to the circle is 12 cm. Find the radius of the circle.**

Ans. Since tangent to a circle is perpendicular to the radius through the point of contact.

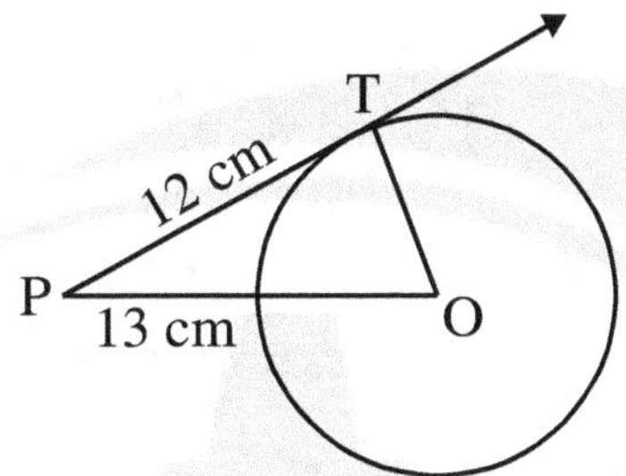

$\therefore \quad \angle OTP = 90°$

In right triangle OTP, we have

$$OP^2 = OT^2 + PT^2$$

$\Rightarrow \quad 13^2 = OT^2 + 12^2$

$\Rightarrow \quad OT^2 = 13^2 - 12^2$

$$= (13 - 12)(13 + 12) = 25$$

$\Rightarrow \quad OT = 5$

Hence, radius of the circle is 5 cm.

Q. 2. **Find the length of the tangent drawn from a point whose distance from the centre of a circle is 25 cm. Given that the radius of the circle is 7 cm.**

Ans. Let P be the given point, O be the centre of the circle and PT be the length of tangent from P. Then, OP = 25 cm and OT = 7 cm.

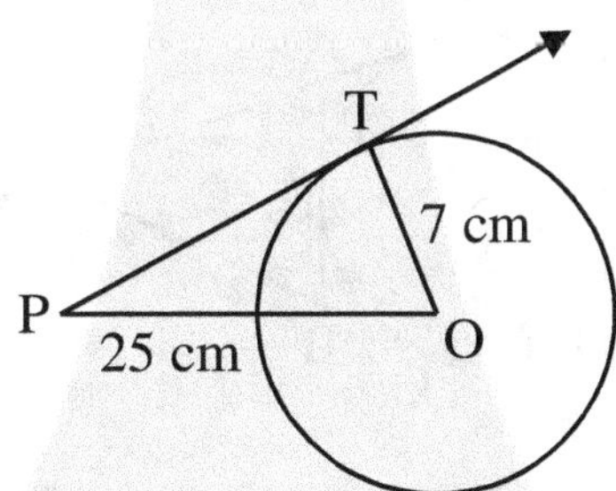

Since tangent to a circle is always perpendicular to the radius through the point of contact.

$\therefore \quad \angle OTP = 90°$

In right triangle OTP, we have

$$OP^2 = OT^2 + PT^2$$

$\Rightarrow \quad 25^2 = 7^2 + PT^2$

$\Rightarrow \quad PT^2 = 25^2 - 7^2$

$$= (25 - 7)(25 + 7)$$

$$= 576$$

$$\Rightarrow \quad PT = 24 \text{ cm}$$

Hence, length of tangent from $P = 24$ cm

Q. 3.   **In Fig., if AB = AC, prove that BE = EC**

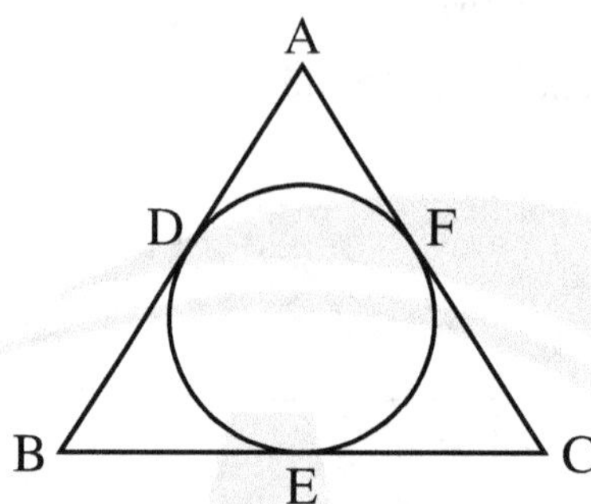

Ans.   Since tangents from an exterior point to a circle are equal in length

$\therefore$ $AD = AF$ [Tangents from A]

     $BD = BE$ [Tangents from B]

     $CE = CF$ [Tangents from C]

Now,

$$AB = AC$$

$\Rightarrow \quad AB - AD = AC - AD$ [Subtracting AD from both sides]

$\Rightarrow \quad AB - AD = AC - AF$ [Using (i)]

$\Rightarrow \quad BD = CF \quad \Rightarrow \quad BE = CF$ [Using (ii)]

$\Rightarrow \quad BE = CE$ [Using (iii)]

Q. 4.   **In fig. XP and XQ. are tangents from X to the circle with centre O, R is a point on the circle.**

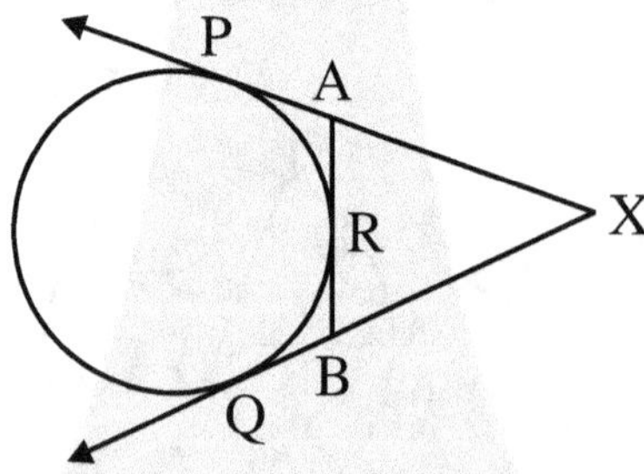

    Prove that, XA + AR = XB + BR.

Ans.   Since lengths of tangents from an exterior point to a circle are equal.

$\therefore$ $XP = XQ.$ .... (i) [From X]

     $AP = AR$ .... (ii) [From A]

     $BQ. = BR$ .... (iii) [From B]

Now, $XP = XQ.$

$\Rightarrow \quad XA + AP = XB + BQ.$

$\Rightarrow \quad XA + AR = XB + BR$ [Using equations (i) and (ii)]

Q. 5. **PA and PB are tangents from P to the circle with centre O. At point M, a tangent is drawn cutting PA at K and PB at N. Prove that KN = AK + BN.**

Ans. We know that the tangents drawn from an external point to a circle are equal in length.

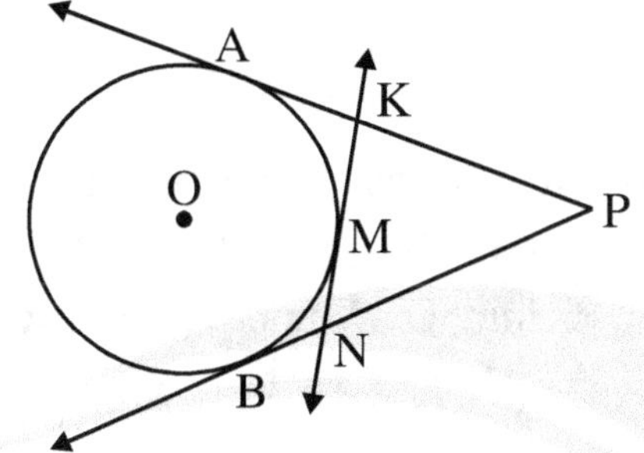

$$\therefore \quad PA = PB \qquad\qquad \text{.... (i) [From P]}$$
$$KA = KM \qquad\qquad \text{.... (ii) [From K]}$$
$$\text{and, } NB = NM \qquad\qquad \text{.... (iii) [From N]}$$

Adding equation (ii) and (iii), we get

$$KA + NB = KM + NM$$
$$\Rightarrow \quad AK + BN = KM + MN \Rightarrow AK + BN = KN$$

Q. 6. **ABCD is a quadrilateral such that $\angle D = 90°$. A circle (O, r) touches the sides AB, BC, CD and DA at P, Q., R and S respectively. If BC = 38 cm, CD = 25 cm and BP = 27 cm, find r.**

Ans. Since tangent to a circle is perpendicular to the radius through the point.

$$\therefore \quad \angle ORD = \angle OSD = 90°$$

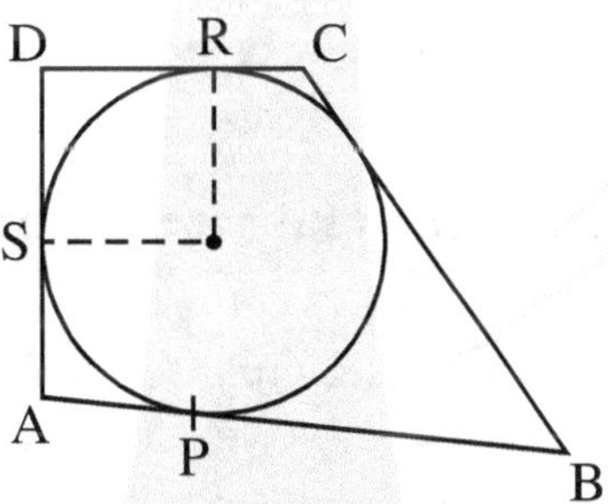

It is given that $\angle D = 90°$ Also, OR = OS. Therefore,

Therefore,

ORDS is a square.

Since tangents from an exterior point to a circle are equal in length.

$$\therefore \quad BP = BQ.$$
$$CQ. = CR \quad \text{and} \quad DR = DS$$

Now, BP = BQ.

$$\Rightarrow \quad BQ. = 27 \qquad\qquad [\odot\ BP = 27 \text{ cm (Given)}]$$
$$\Rightarrow \quad BC - CQ. = 27$$
$$\Rightarrow \quad 38 - CQ. = 27 \qquad\qquad [\odot\ BC = 38 \text{ cm}]$$
$$\Rightarrow \quad CQ. = 11 \text{ cm}$$
$$\Rightarrow \quad CR = 11 \text{ cm} \qquad\qquad [\odot\ CR = CQ.]$$
$$\Rightarrow \quad CD - DR = 11$$

$\Rightarrow \quad 25 - DR = 11$ [$\because$ CD = 25 cm]

$\Rightarrow \quad DR = 14$ cm

But, ORDS is a square.

Therefore, OR = DR = 14 cm

Hence, r = 14 cm

Q. 7. **Prove that the tangents at the extremities of any chord make equal angles with the chord.**

Ans. Let AB be a chord of a circle with centre O, and let AP and BP be the tangents at A and B respectively. Suppose the tangents meet at P. Join OP. Suppose OP meets AB at C. We have to prove that $\angle PAC = \angle PBC$. In triangles PCA and PCB, we have

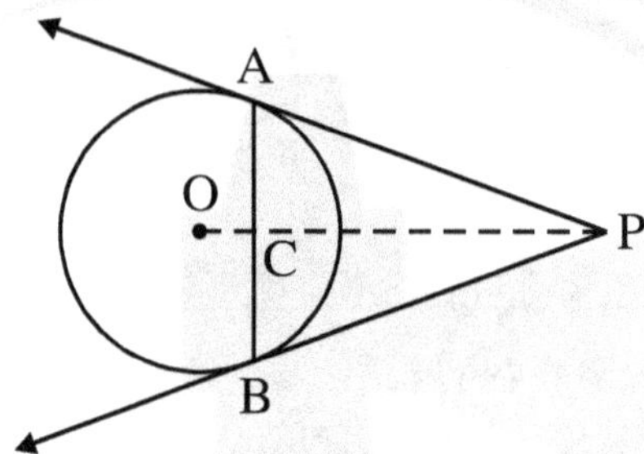

$PA = PB$ [$\because$ Tangents from an external point are equal]

$\angle APC = \angle BPC$ [$\because$ PA and PB are equally inclined to OP]

and, PC = PC [Common]

So, by SAS, criterion of congruence, we have

$\Delta PAC \cong \Delta PBC$

$\Rightarrow \quad \angle PAC = \angle PBC$

Q. 8. **In fig., O is the centre of the circle, PA and PB are tangent segments. Show that the quadrilateral AOBP is cyclic.**

Ans. Since tangent at a point to a circle is perpendicular to the radius through the point.

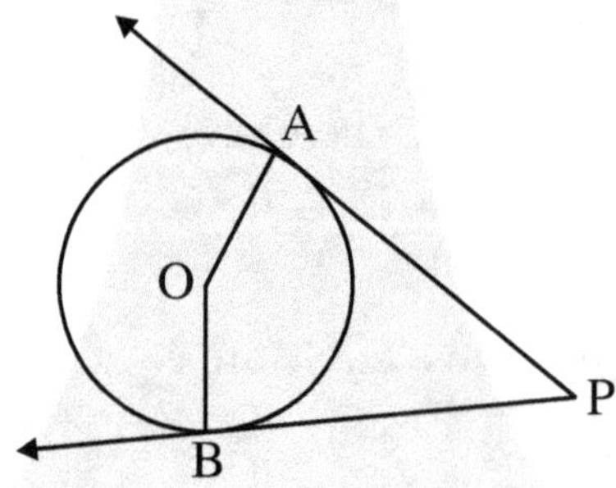

$\therefore \quad OA \perp AP$ and $OB \perp BP$

$\Rightarrow \quad \angle OAP = 90°$ and $\angle OBP = 90°$

$\Rightarrow \quad \angle OAP + \angle OBP = 90° + 90° = 180°$ ...(i)

In quadrilateral OAPB, we have

$\angle OAP + \angle APB + \angle AOB + \angle OBP = 360°$

$\Rightarrow \quad (\angle APB + \angle AOB) + (\angle OAP + \angle OBP) = 360°$

$\Rightarrow \quad \angle APB + \angle AOB + 180° = 360°$

$\angle APB + \angle AOB = 180°$ ...(ii)

 Innovative Mathematics X-10

From equations (i) and (ii), we can say that the quadrilateral AOBP is cyclic.

Q. 9. **In fig., circles C(O, r) and C(O′, r/2) touch internally at a point A and AB is a chord of the circle C (O, r) intersecting C(O′, r/2) at C, Prove that AC = CB.**

Ans. Join OA, OC and OB. Clearly, $\angle OCA$ is the angle in a semi-circle.

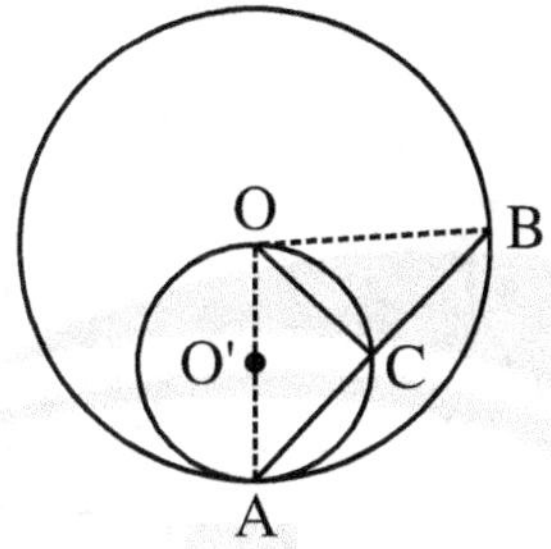

$\therefore \quad \angle OCA = 90°$

In right triangles OCA and OCB, we have

$$OA = OB = r$$

$$\angle OCA = \angle OCB = 90°$$

and OC = OC

So, by RHS criterion of congruence, we get

$$\triangle OCA \cong \triangle OCB$$

$$\Rightarrow \quad AC = CB$$

Q. 10. **In two concentric circles, prove that all chords of the outer circle which touch the inner circle are of equal length.**

Ans. Let AB and CD be two chords of the circle which touch the inner circle at M and N respectively.

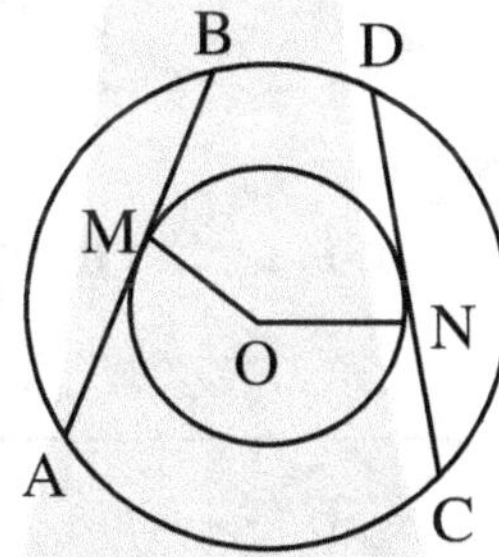

Then, we have to prove that

$$AB = CD$$

Since AB and CD are tangents to the smaller circle.

$\therefore \quad$ OM = ON = Radius of the smaller circle

Thus, AB and CD are two chords of the larger circle such that they are equidistant from the centre. Hence, AB = CD.

VERY SHORT ANSWER TYPE Q.UESTIONS

Q. 1. In fig., $\triangle$ABC is circumscribing a circle. Find the length of BC.

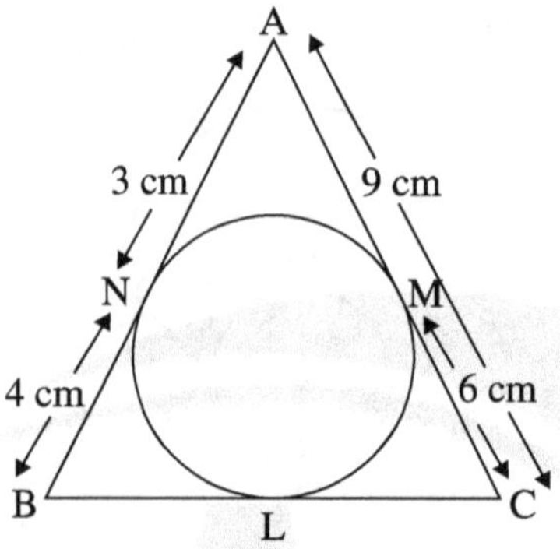

Sol.

Q. 2. The length of the tangent to a circle from a point P, which is 25 cm away from the centre, is 24 cm. What is the radius of the circle.

Sol.

Q. 3. In fig., ABCD is a cyclic quadrilateral. If $\angle$BAC = 50° and $\angle$DBC = 60°, then find $\angle$BCD.

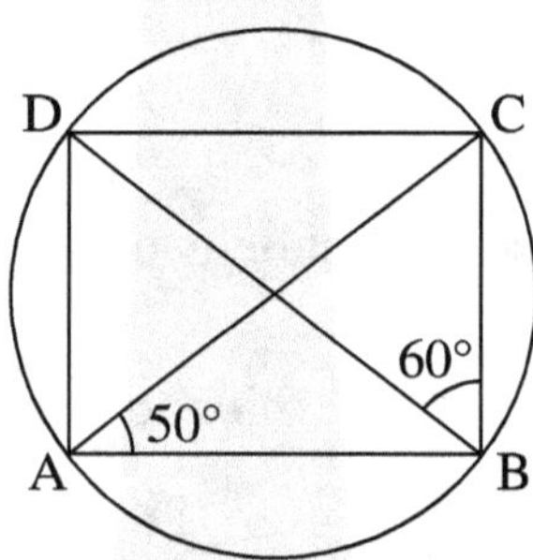

Sol.

Q. 4. In figure, O is the centre of a circle, PQ. is a chord and the tangent PR at P makes an angles of 50° with PQ.. Find $\angle$POQ..

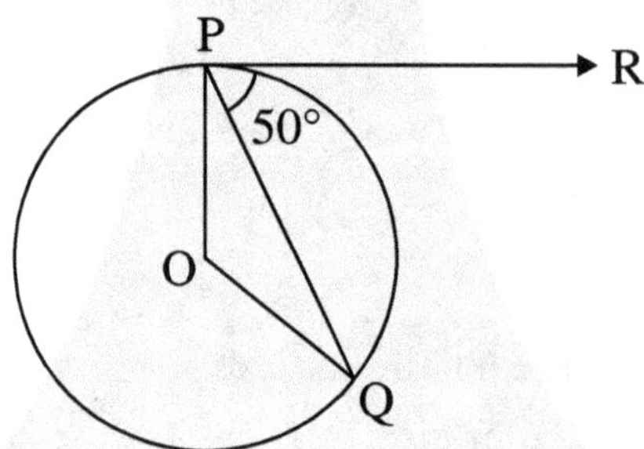

Sol.

Q. 5. If two tangents inclined at an angle 60° are drawn to a circle of radius 3 cm, then find the length of each tangent.

Sol.

Q. 6. If radii of two concentric circles are 4 cm and 5 cm, then find the length of the chord of that circle which is tangent to the other circles.

Sol.

Q. 7. In the given figure, PQ. is tangent to outer circle and PR is tangent to inner circle. If PQ. = 4 cm, OQ. = 3 cm and OR = 2 cm then find the length of PR.

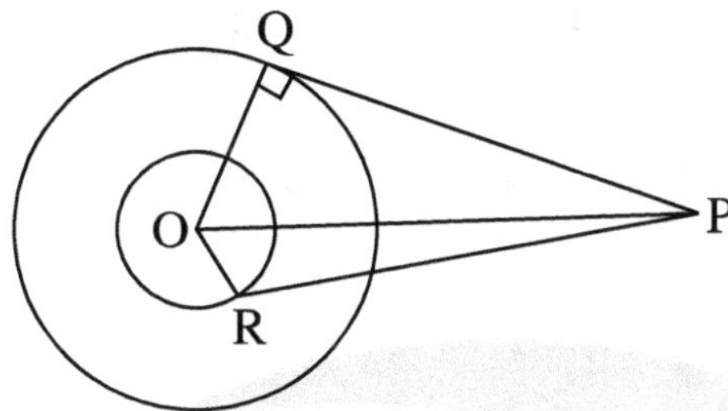

Sol.

Q. 8. In the given figure, O is the centre of the circle, PA and PB are tangents to the circle then find $\angle AQ.B$.   (CBSE 2016)

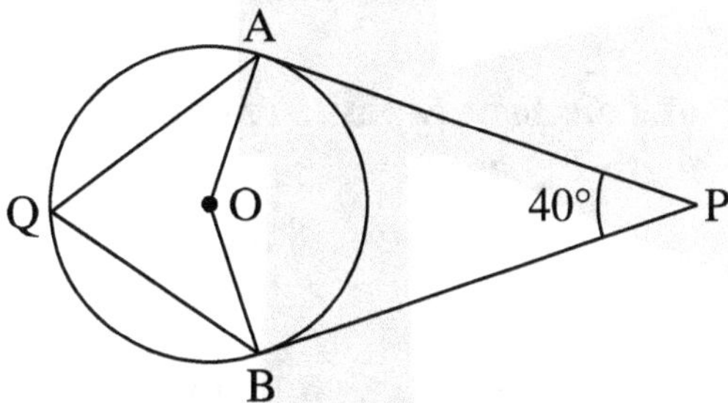

Sol.

Q. 9. In the given figure, If $\angle AOB = 125°$ then find $\angle COD$.

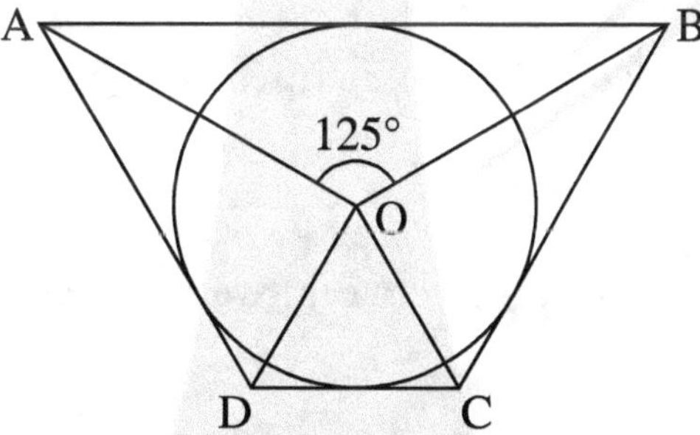

Sol.

Q. 10. If two tangent TP and TQ. are drawn from an external point T such that $\angle TQ.P = 60°$ then find $\angle OPQ..$

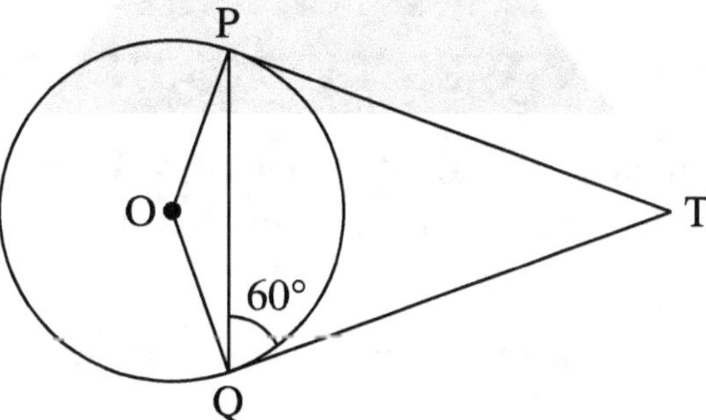

Sol.

Q. 11. How many tangents can a circle have?

Sol.

Q. 12. A tangent to a circle intersects it in _________ point.

Sol.

Q. 13. If PQ. is a tangent then find the value of $\angle POQ. + \angle Q.PO$.

Sol.

Q. 14. Choose the correct Answer.

A tangent PQ. at a point P of a circle of radius 5 cm meets a line through the centre O at a point Q. so that OQ. = 12 cm. Length PQ. is :

(a) 12 cm   (b) 13 cm   (c) 8.5 cm   (d) $\sqrt{119}$ cm

Sol.

Q. 15. A circle can have ______ parallel tangents at the most.

Sol.

Q. 16. The common point of a tangent to a circle and radius of the circle is called _______.

Sol.

Q. 17. Find the distance between two points of contact of two parallel tangents to a given circle of radius 9 cm.

Sol.

Q. 18. Find the radius of a circle, if distance between two parallel tangents be 10 cm.

Sol.

Q. 19. How many common tangents can be drawn to two circles touching internally.

Sol.

SHORT ANSWER TYPE Q.UESTIONS

Q. 20. If diameters of two concentric circles are d_1 and d_2 $(d_2 > d_1)$ and c is the length of chord of bigger circle which is tangent to the smaller circle. Show that $d_2^2 = c^2 + d_1^2$.

Sol.

Q. 21. The length of tangent to a circle of radius 2.5 cm from an external point P is 6 cm. Find the distance of P from the nearest point of the circle.

Sol.

Q. 22. TP and TQ. are the tangents from the external point T of a circle with centre O. If $\angle OPQ. = 30°$ then find the measure of $\angle TQ.P$.

Sol.

Q. 23. In the given fig. AP = 4 cm, BQ. = 6 cm and AC = 9 cm. Find the semi perimeter of $\triangle ABC$.

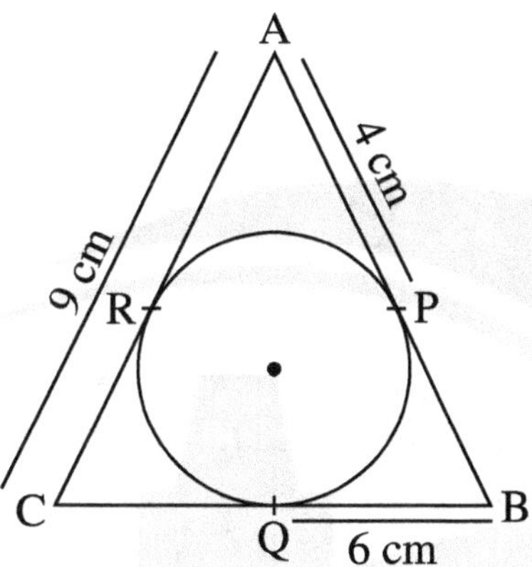

Sol.

Q. 24. A circle is drawn inside a right angled triangle whose sides are a, b, c where c is the hypotenuse, which touches all the sides of the triangle. Prove $r = \dfrac{a+b-c}{2}$ where r is the radius of the circle.

Sol.

Q. 25. Prove that in two concentric circles the chord of the larger circle which is tangent to the smaller circle is bisected at the point of contact.

Sol.

Q. 26. In the given Fig., AC is diameter of the circle with centre O and A is the point of contact, then find x.

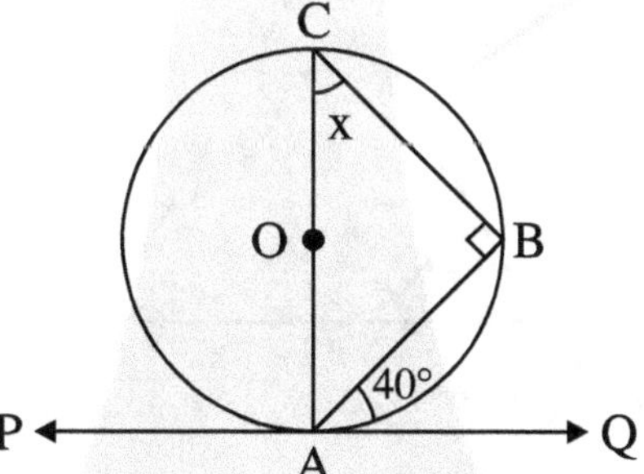

Sol.

Q. 27. In the given fig. KN, PA and PB are tangents to the circle. Prove that: KN = AK + BN.

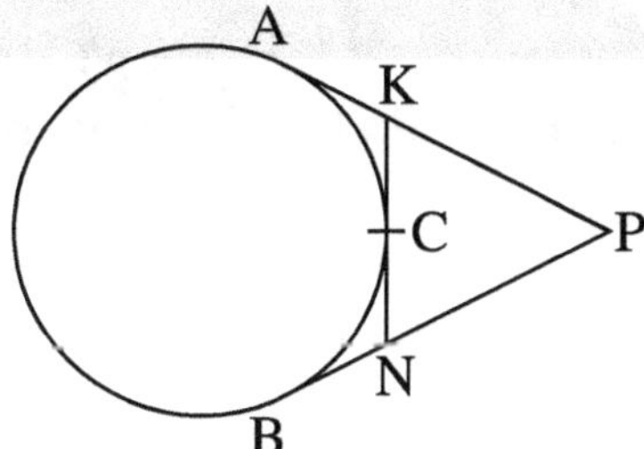

Sol.

Q. 28. In the given fig. PQ. is a chord of length 6 cm and the radius of the circle is 6 cm. TP and TQ. are two tangents drawn from an external point T. Find $\angle$PTQ..

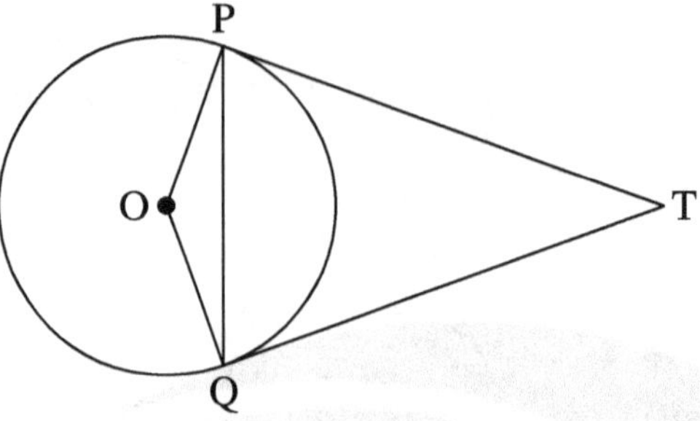

Sol.

Q. 29. In the given figure, ABC is a triangle in which $\angle$B = 90°, BC = 48 cm and AB = 14 cm. A circle is inscribed in the triangle, whose centre is O. Find the radius (r) of in the circle.

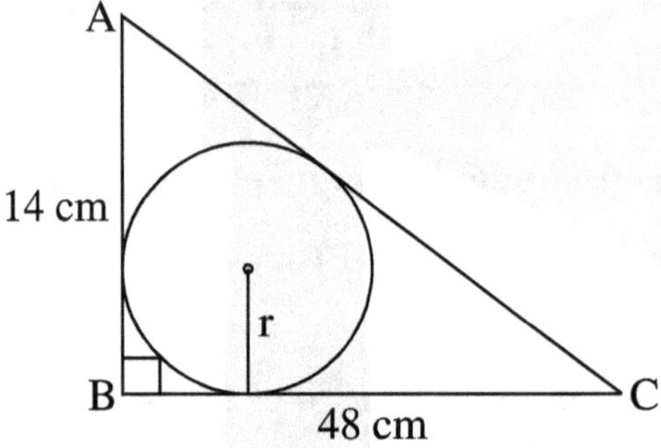

Sol.

Q. 30. If the inscribed circle of the $\triangle$ABC touches BC at D. Prove that AB – BD = AC – CD.

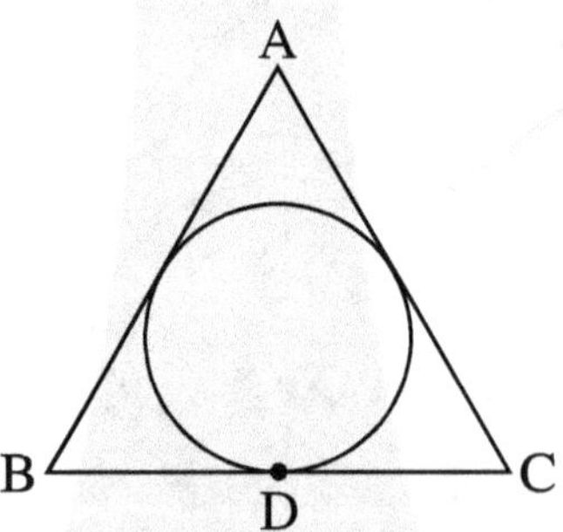

Sol.

Q. 31. From a point P which is at distance of 13 cm from the centre O of a circle of radius 5 cm, the pair of tangents PQ. and PR to the circle are drawn, then find the area of the quadrilateral PQ. OR.

Sol.

Q. 32. In the given figure, tangents AC and AB are drawn to a circle from a point A such. That $\angle$BAC = 30°, a chord BD is drawn parallel to the tangent AC, find $\angle$DBC.

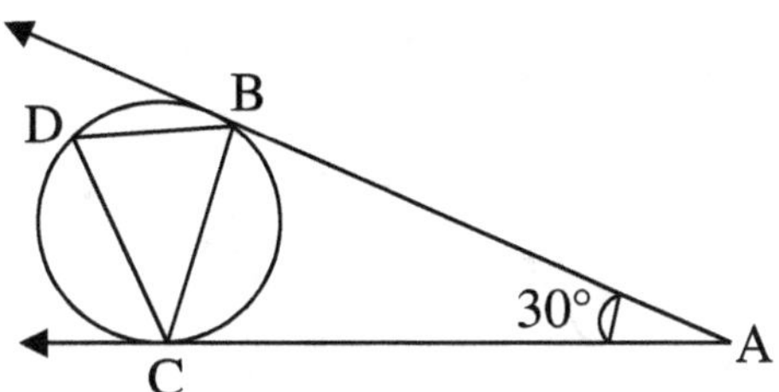

Sol.

Q. 33. Find the value of x°.

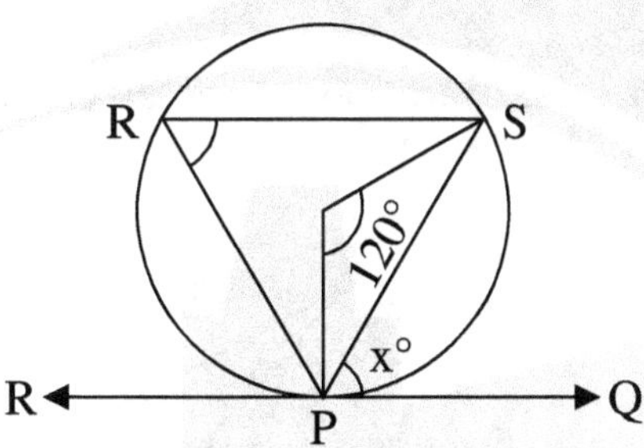

Sol.

Q. 34. PA and PB are tangents to the circle with centre at O. If ∠APB = 70°, then find ∠AQ.B.

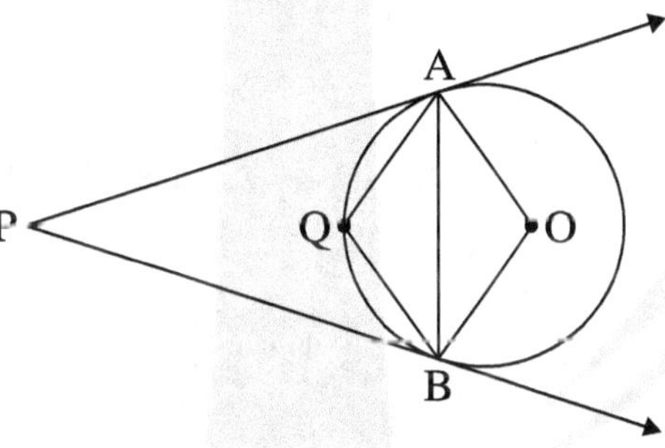

Sol.

Q. 35. In figure, CD is a tangent and AB is a drameter of the circle centrel at O. If ∠DCB = 30°, then find ∠ADC.

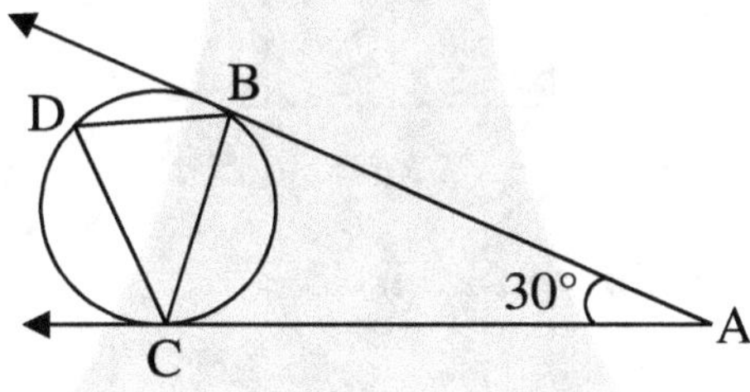

Sol.

LONG ANSWER TYPE Q.UESTIONS

Q. 36. In the given figure find AD, BE, CF where AB = 12 cm, BC = 8 cm and AC = 10 cm.

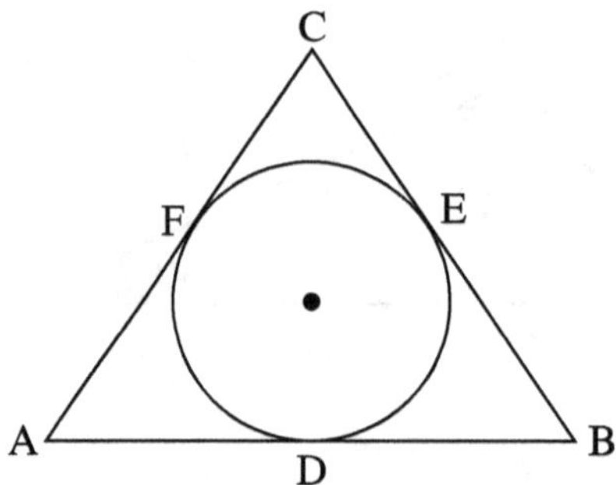

Sol.

Q. 37. In the given fig. OP is equal to the diameter of the circle with centre O. Prove that $\triangle ABP$ is an equilateral triangle.

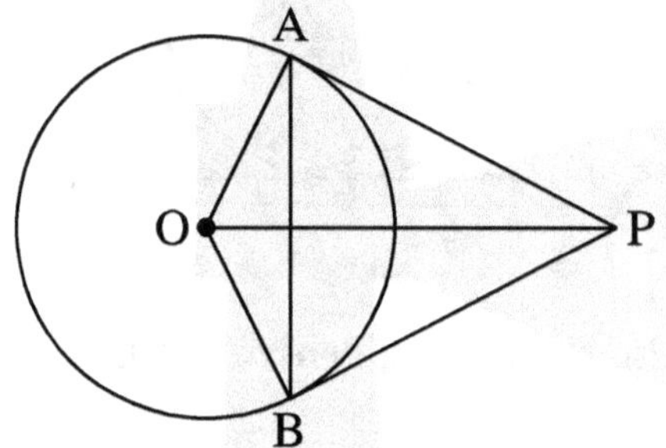

Sol.

Q. 38. In the given fig., find PC. If AB = 13 cm, BC = 7 cm and AD = 15 cm.

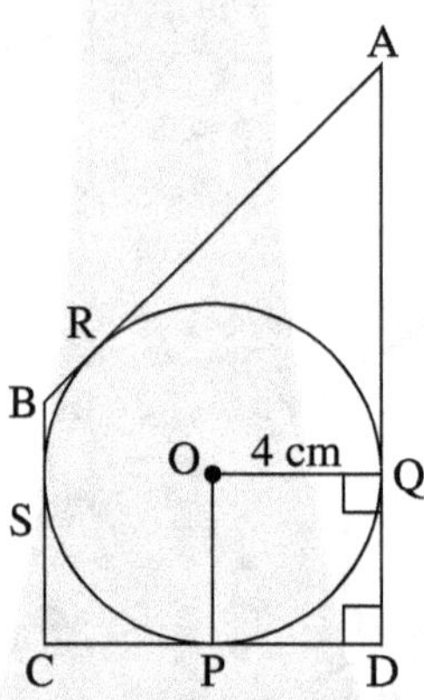

Sol.

Q. 39. In the given figure, find the radius of the circle.

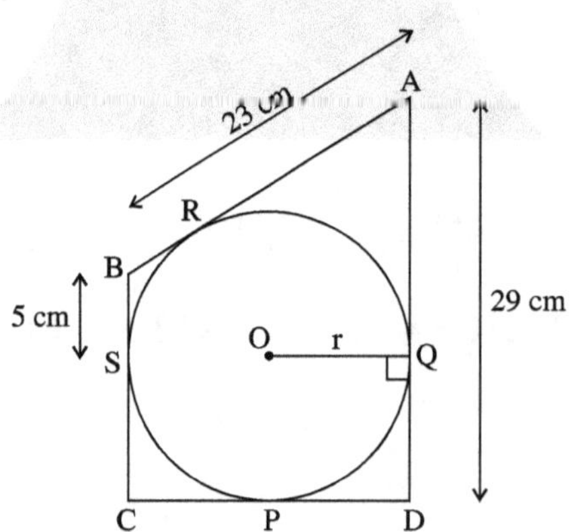

Innovative Mathematics X-10

Q. 40. In the given fig. PQ. is tangent and PB is diameter. Find the values of angles x and y.

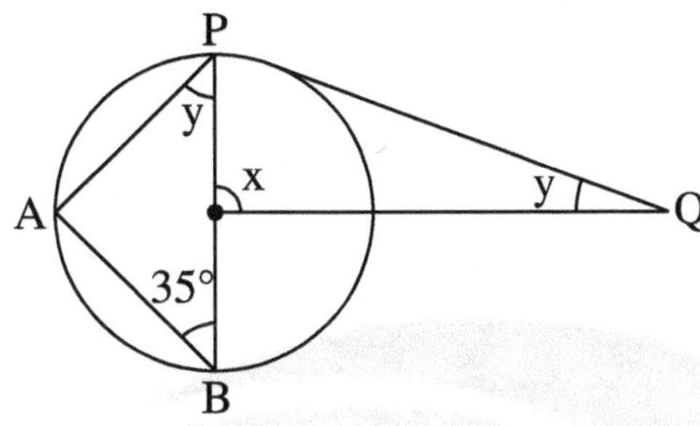

Sol.

Q. 41. In given figure, two circles touch each other at the point C. Prove that the common tangent to the circles at C, bisects the common tangent at P and Q..

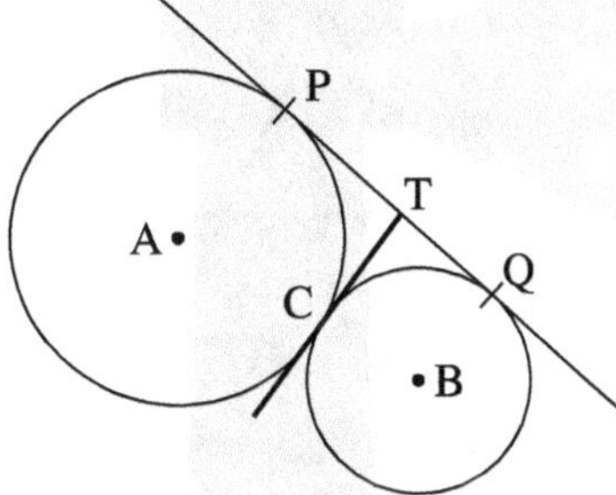

Sol.

Q. 42. In the given figure, a circle touches all the four sides of a quadrilateral ABCD. If AB = 6 cm, BC = 9 cm and CD = 8 cm, then find the length of AD.

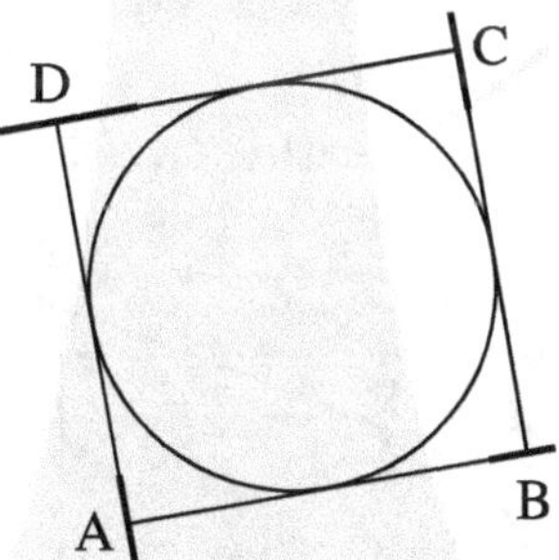

Sol.

Q. 43. In figure, PA is a tangent from an external point P to a circle with centre O, If ∠POB = 115°. Find ∠APO.

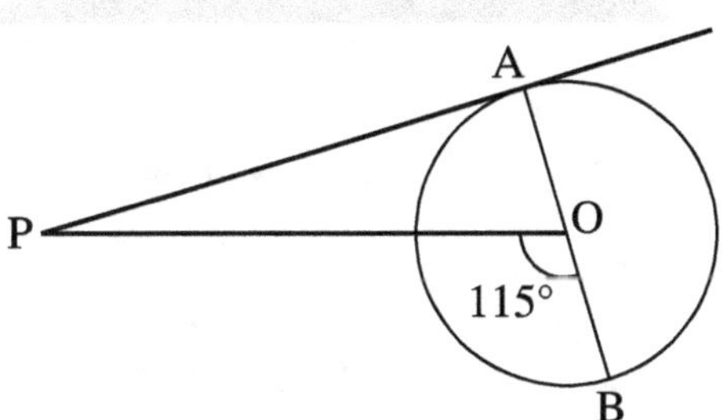

Sol.

Q. 44. In figure, XP and XQ. are tangents from X to the circle with centre O, R is a point on the circle and AB is tangent at R. Prove that:

$$XA + AR = XB + BR$$

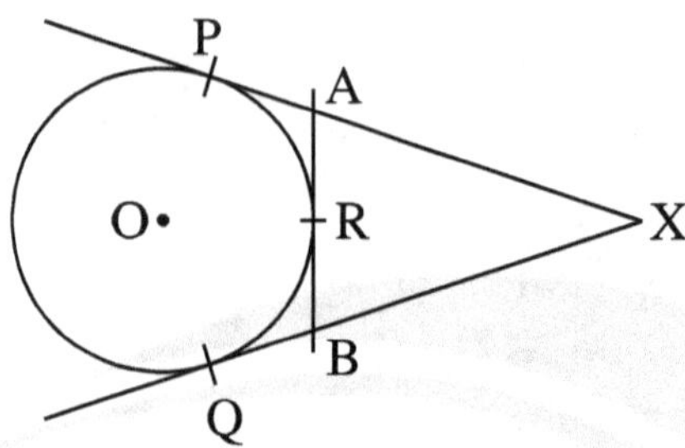

Sol.

Q. 45. In the given figure, find the perimeter of $\triangle ABC$, if AP = 12 cm.

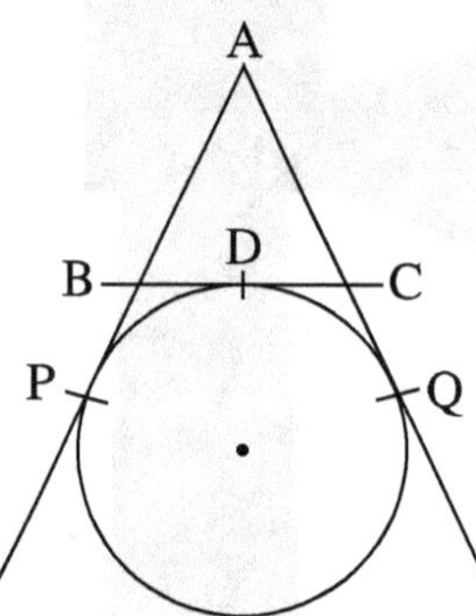

Sol.

Q. 46. From a point Q., the length of the tangent to a circle is 24 cm and the distance of Q. from the centre is 25 cm. Find the radius of the circle.

Sol.

Q. 47. In the given figure, if TP and TQ. are the two tangents to a circle with centre O so that $\angle POQ. = 110°$, then find $\angle PTQ.$.

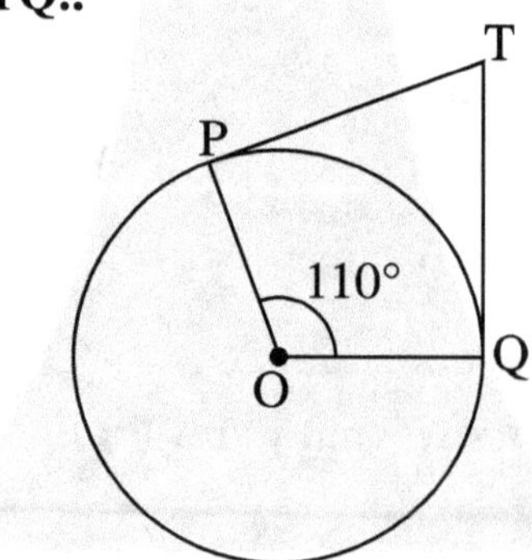

Sol.

Q. 48. If tangents PA and PB from a point P to a circle with centre O are inclined to each other at angle of 80°, then find $\angle POA$.

Sol.

Q. 49. Prove that the tangents drawn at the ends of a diameter of a circle are parallel.

Sol.

Q. 50. Prove that the perpendicular at the point of contact to the tangent to a circle passes through the centre.

Sol.

Q. 51. The length of a tangent from a point A at distance 5 cm from the centre of the circle is 4 cm. Find the radius of the circle.

Sol.

Q. 52. Two concentric circles are of radii 5 cm and 3 cm. Find the length of the chord of the larger circle which touches the smaller circle.

Sol.

Q. 53. In the given figure, a quadrilateral ABCD is drawn to circumscribe a circle. Prove that AB + CD = AD + BC.

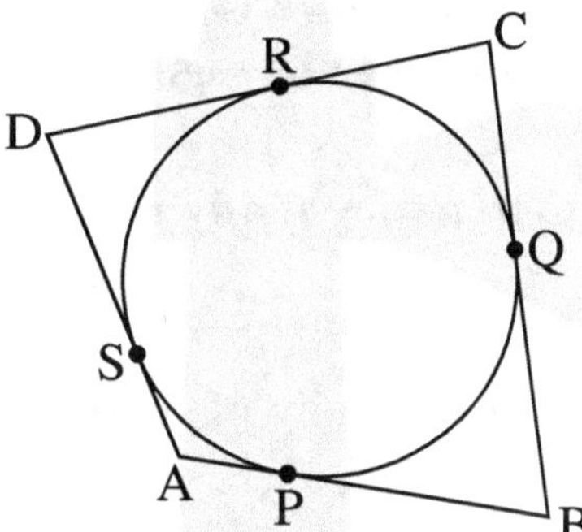

Sol.

Q. 54. In the given figure, XY and X′Y′ are two parallel tangents to a circle with centre O and another tangent AB with point of contact C intersecting XY and A and X′Y′ at B. Prove that $\angle AOB = 90°$.

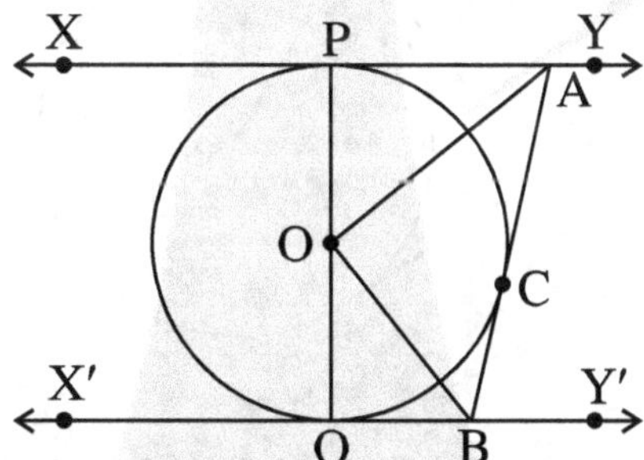

Sol.

Q. 55. Prove that the angle between the two tangents drawn from an external point to a circle is supplementary to the angle subtended by the line-segment joining the points of contact at the centre.

Sol.

Q. 56. Prove that the parallelogram circumscribing a circle is a rhombus.

Sol.

Q. 57. In the given figure, triangle ABC is drawn to circumscribe a circle of radius 4 cm such that the segments BD and DC into which BC is divided by the point of contact D are of lengths 8 cm and 6 cm respectively. Find the sides AB and AC.

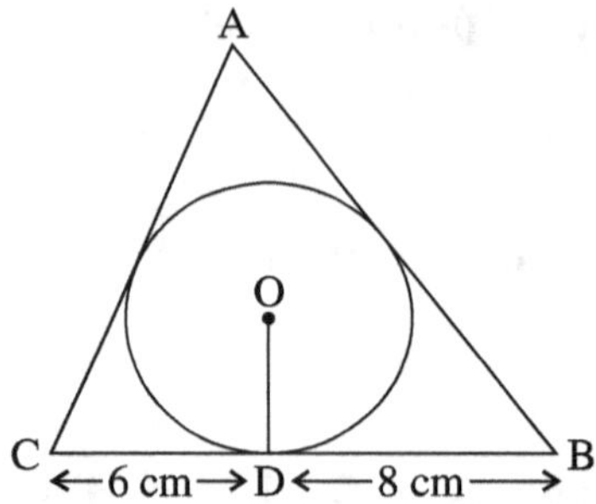

Sol.

Q. 58. Prove that opposite sides of a quadrilateral circumscribing a circle subtend supplementary angles at the centre of the circle.

Sol.

Q. 59. From a point P outside a circle with centre O, tangents PA and PB are drawn to the circle. Prove that OP is the right bisector of the line segment AB.

Sol.

Q. 60. Prove that the tangents at the extremities of any chord of a circle, make equal angles with the chord.

Sol.

Q. 61. Prove that the tangent drawn at the midpoint of an arc of a circle is parallel to the chord joining the end points of the arc.

Sol.

Q. 62. Two tangents PA and PB are drawn to a circle with centre O from an external point P. Prove that $\angle APB = 2\angle OAB$.

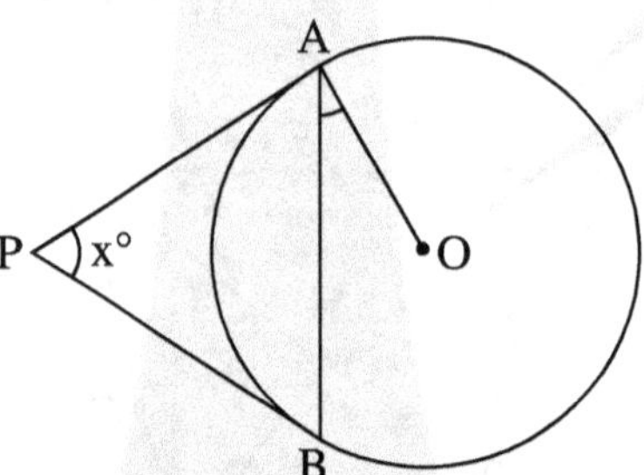

Sol.

Q. 63. In the given figure, a circle is inscribed in a triangle PQ.R. If PQ. = 10 cm, Q.R = 8 cm and PR = 12 cm, find the lengths of Q.M, RN and PL.

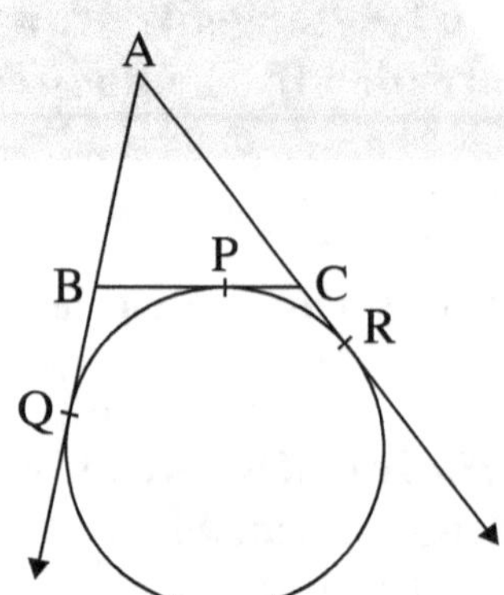

Innovative Mathematics X-10

Sol.

Q. 64. A circle is touching the side BC of 3ABC at P and touching AB and AC produced at Q. and R respectively. Prove that AQ. = $\dfrac{1}{2}$ (perimeter of 3ABC).

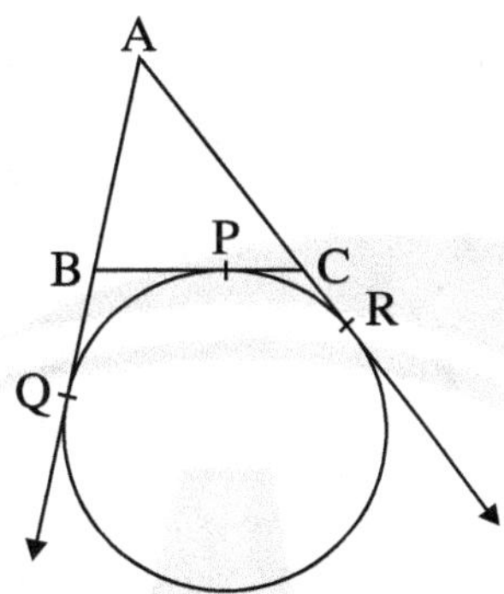

Sol.

1. 10 cm

2. 7 cm

3. 70°

4. 100°

5. $3\sqrt{3}$ cm

6. AB = 6 cm

7. OP = 5 cm, PR = $\sqrt{21}$ cm

8. $\angle$AQ.B = 70°

9. 55°

10. 30°

11. Infinite many

12. one

13. 90°

14. d($\sqrt{119}$ cm)

15. Two

16. Point of contact

17. 18 cm

18. 5 cm

19. 1 cm

20. $d_2^2 = c^2 + d_1^2$

21. Q.P = 4 cm

22. 60°

23. 15 cm

24. $r = \dfrac{a + b - c}{2}$

26. x = 40°

28. $\angle$PTQ. = 120°

29. r = 6 cm

31. 60 cm^2

32. 75°

33. x = 60°

34. $\angle$AQ.B = 125°

35. $\angle$ADC = 120°

38. PC = 5 cm

39. 11 cm

40. $\angle$x = 35°

42. AD = 5 cm

43. 25°

45. 24 cm

46.

47.

48.

49.

50.

51.

52.

53.

54.

55.

56.

57.

58.

59.

60.

61.

62.

63.

64.

PRACTICE TEST

SECTION-A

Q. 1. In the given figure find x, where ST is the tangent. (1)

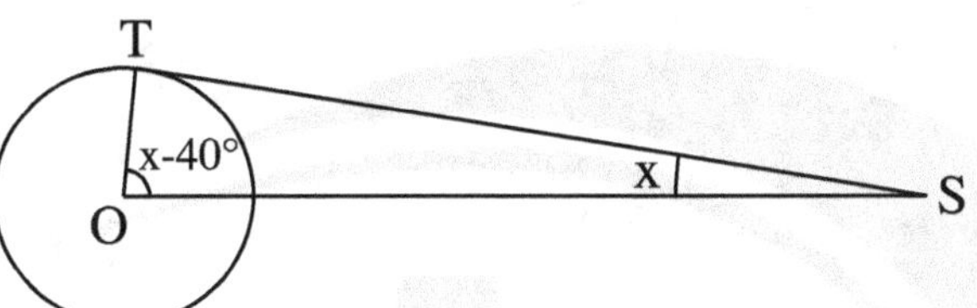

Sol. ( )

Q. 2. In the given figure if AC = 9 cm, find BD. (1)

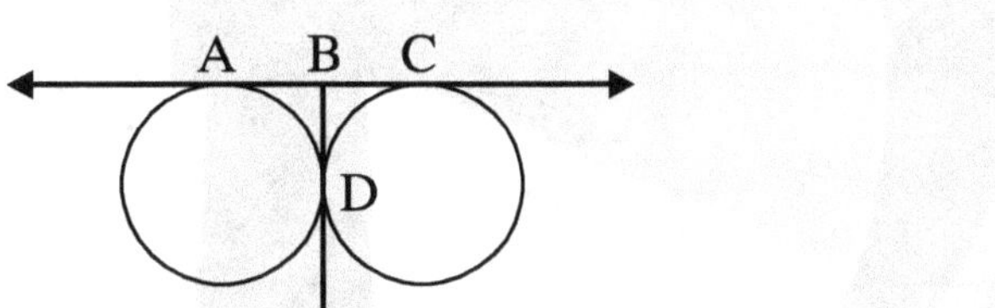

Sol. ( )

Q. 3. In the given figure, △ABC is circumscribing a circle, then find the length of BC. (1)

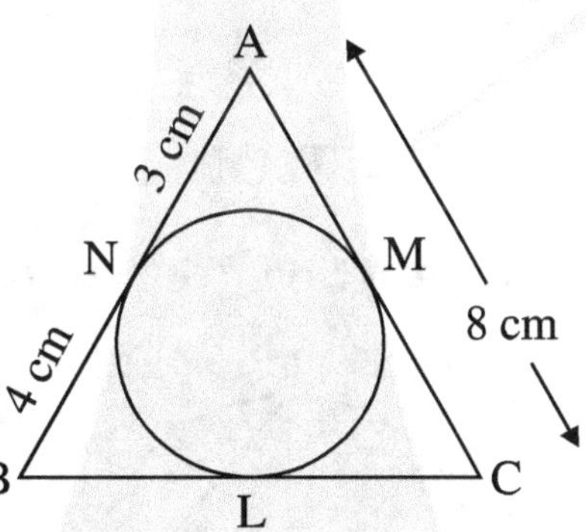

Sol. ( )

Q. 4. From the external point P, tangents PA and PB are drawn to a circle with centre O. If ∠PAB = 50°, then find ∠AOB. (1)

Sol. ( )

SECTION-B

Q. 5. If the angle between two tangents drawn from an external point P to a circle of radius a and centre O is 60° then find the length of OP. (2)

Sol. ( )

Q. 6. In the following figure find x. (2)

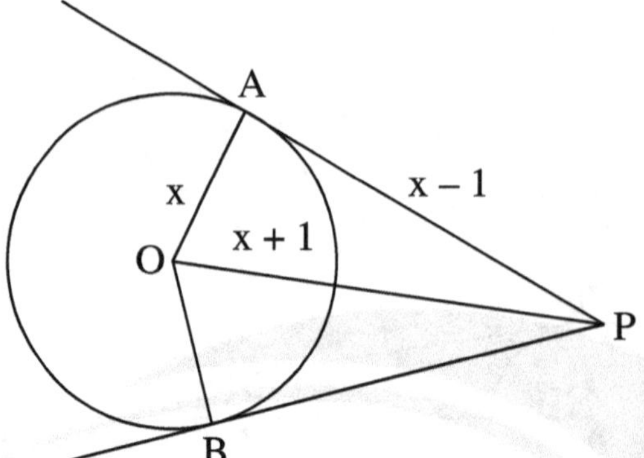

Sol. ( )

Q. 7. Two concentric circle with centre O are of radii 6 cm and 3 cm. From an external point P, tangents PA and PB are drawn to these circle as shown in the figure. If AP = 10 cm. Find BP (2)

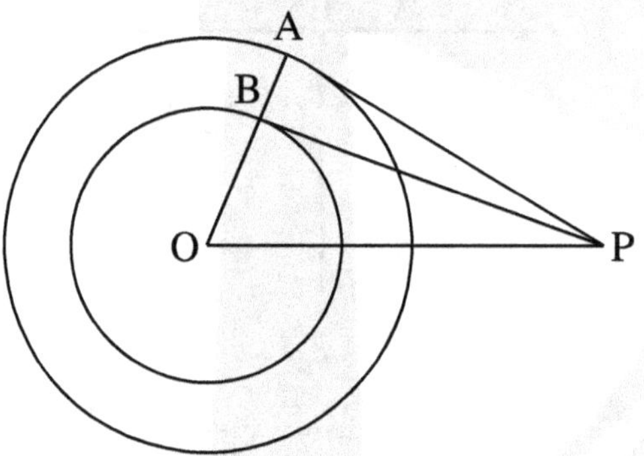

Sol. ( )

\

SECTION-C

Q. 8. In the given figure, AB is a tangent to a circle with centre O. Prove $\angle BPQ. = \angle PRQ..$

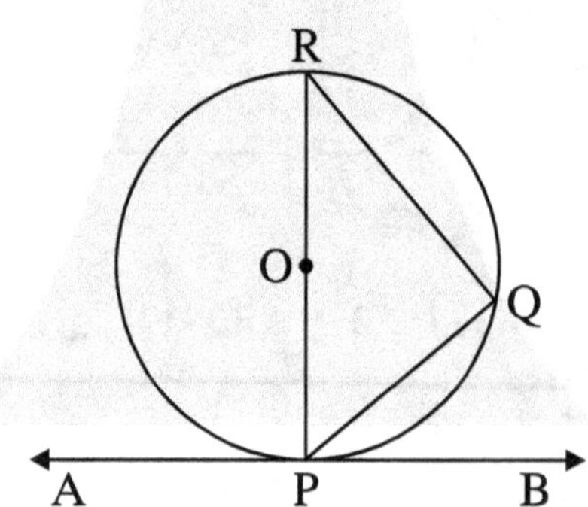

Sol. ( )

Q. 9. In the given figure $\triangle ABC$ is drawn to circumscribe a circle of radius 3 cm, such that the segment BD and DC into which BC is divided by the point of contact D are of length 6 cm and 8 cm respectively, find side AB if the ar($\triangle ABC$) = 63 cm² (3)

330 Innovative Mathematics X-10

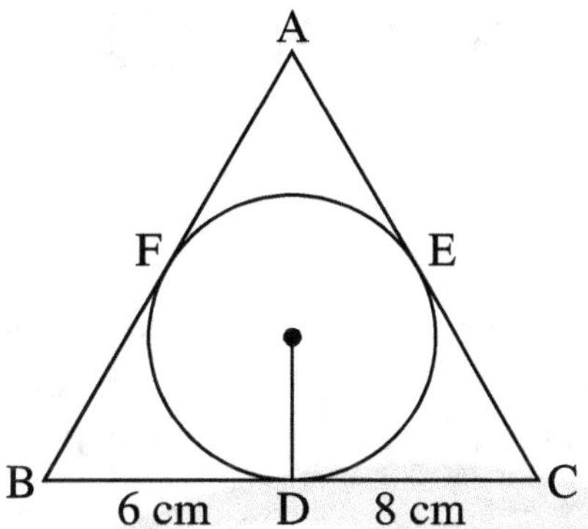

Sol. ( )

Q. 10. AB is a diameter of a circle with centre with centre O and AT is a tangent. If $\angle AOQ. = 58°$ find $\angle ATQ..$ **(4)**

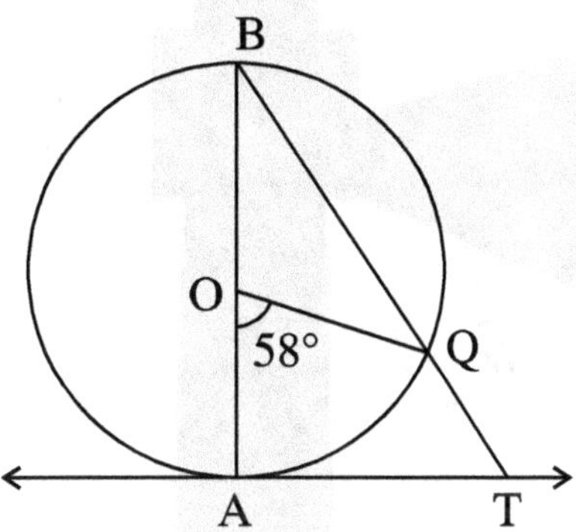

Sol. ( )

CASE STUDY

CASE STUDY 1.

A Ferris wheel (or a big wheel in the United Kingdom) is an amusement ride consisting of a rotating upright wheel with multiple passenger-carrying components (commonly referred to as passenger cars, cabins, tubs, capsules, gondolas, or pods) attached to the rim in such a way that as the wheel turns, they are kept upright, usually by gravity.

After taking a ride in Ferris wheel, Aarti came out from the crowd and was observing her friends who were enjoying the ride. She was curious about the different angles and measures that the wheel will form. She forms the figure as given below.

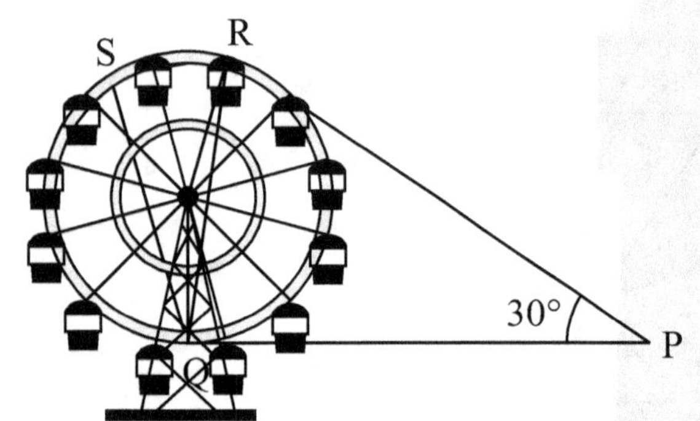 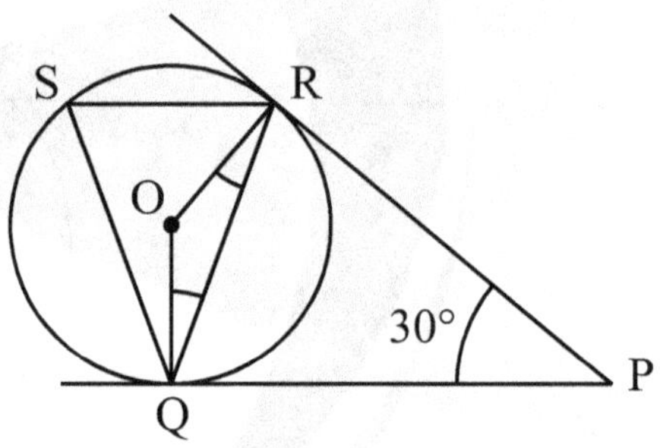

Q.. 1. **In the given figure find ∠ROQ.**

    (a) 60   (b) 100   (c) 150   (d) 90

Q.. 2. **Find ∠RQ.P**

    (a) 75   (b) 60   (c) 30   (d) 90

Q.. 3. **Find ∠RSQ.**

    (a) 60   (b) 75   (c) 100   (d) 30

Q.. 4. **Find ∠ORP**

    (a) 90   (b) 70   (c) 100   (d) 60

ANSWERS

1. (c) 150   **2.** (a) 75   **3.** (b) 75   **4.** (a) 90

CASE STUDY 2:

Varun has been selected by his School to design logo for Sports Day T-shirts for students and staff . The logo design is as given in the figure and he is working on the fonts and different colours according to the theme. In given figure, a circle with centre O is inscribed in a ΔABC, such that it touches the sides AB, BC and CA at points D, E and F respectively. The lengths of sides AB, BC and CA are 12 cm, 8 cm and 10 cm respectively.

Innovative Mathematics X-10

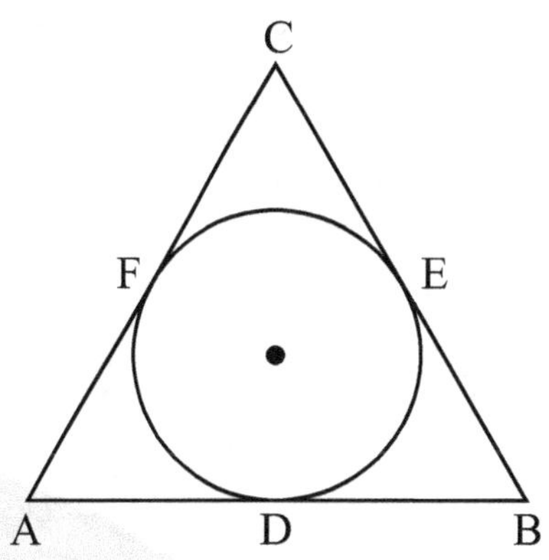

Q.. 1. **Find the length of AD**

    (a) 7   (b) 8   (c) 5   (d) 9

Q.. 2. **Find the Length of BE**

    (a) 8   (b) 5   (c) 2   (d) 9

Q.. 3. **Find the length of CF**

    (a) 9   (b) 5   (c) 2   (d) 3

Q.. 4. **If radius of the circle is 4 cm, Find the area of $\triangle OAB$**

    (a) 20   (b) 36   (c) 24   (d) 48

Q.. 5. **Find area of $\triangle ABC$**

    (a) 50   (b) 60   (c) 100   (d) 90

ANSWERS

1. (a) 7   **2.** (b) 5   **3.** (d) 3

4. (c) 24   **5.** (b) 60

CHAPTER-11

AREA RELATED TO CIRCLE

LEARNING OBJECTIVES

You have already learnt how to find the areas and perimeters of some plane figures like rectangles, squares, triangles and circles, etc. We begin this chapter with a review of the measurement of area and perimeter (circumference) of a circle, and learn further about the concept of a sector and a segment of a circle and their areas. Also, we shall learn about finding the areas of some combinations of plane figures involving circles or their parts.

12.1 Review of Perimeter and Area of a Circle

Circle:

Let us recall that a circle is the set of all those points in a plane each of which is at a constant distance from a fixed point in that plane.

The fixed point is called the centre and the constant distance is called the radius.

Perimeter of a Circle

Perimeter of a circle is called its circumference. It is the distance covered by travelling once around the circle.

We know that the ratio of circumference of a circle to its diameter is constant, and this constant ratio is denoted by π (a Greek letter, read as 'Pi').

$$\frac{\text{Circumference}}{\text{diameter}} = \pi$$

$\Rightarrow \qquad$ Circumference $= \pi \times d$, where d is the diameter of the circle.

or $\qquad$ Circumference $= \pi \times 2r = 2\pi r$, where r is the radius of the circle.

Area of a Circle:

Area of a circle is given by,

$$\text{Area} = \pi r^2, \text{ where r is the radius of the circle.}$$

We can easily verify this result by dividing a circle into a number of sectors and rearranging them in the shape of a figure which is nearly a rectangle.

The length of this rectangle is half of the circumference of the circle, i.e., πr ($\frac{1}{2} \times 2\pi r$) and its breadth is equal to the radius of the circle, i.e., r.

Hence, $\qquad$ its area $= \pi r \times r = \pi r^2$

(i)

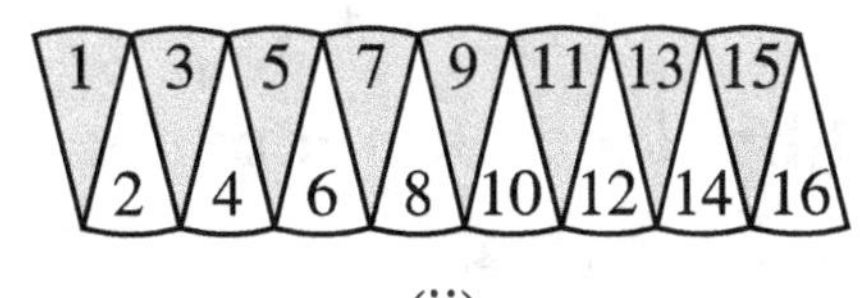

(ii)

$$\Rightarrow \qquad \text{Area of a circle} = \pi r^2$$

Perimeter and Area of a Semicircle:

If r is the radius of a circle,

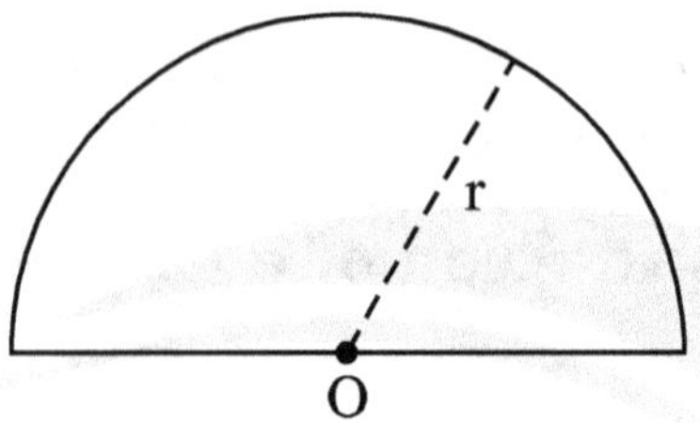

$$\text{Perimeter of the semicircle} = \frac{1}{2} \times 2\pi r + 2r$$

$$= \pi r + 2r$$

$$\text{Area of the semicircle} = \frac{1}{2}\pi r^2$$

Perimeter and Area of a Q.uadrant of a Circle:

If r is the radius of a circle,

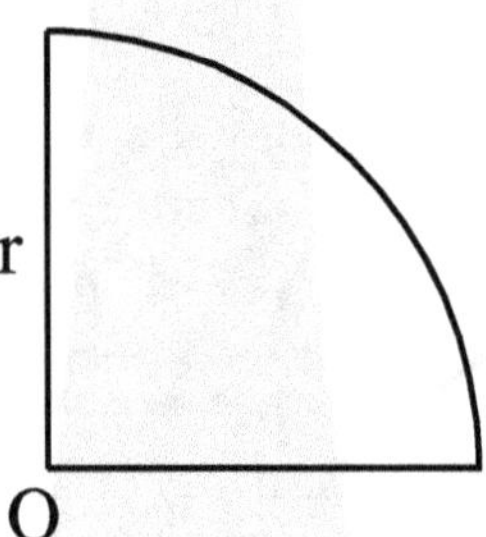

$$\text{Perimeter of the quadrant} = \frac{1}{4} \times 2\pi r + 2r$$

$$= \frac{\pi r}{2} + 2r$$

$$= r\left(\frac{\pi}{2} + 2\right)$$

$$\text{Area of the quadrant} = \frac{1}{4}\pi r^2$$

Area Enclosed by Two Concentric Circles or Area of a Circular Track:

If R and r are the radii of two concentric circles

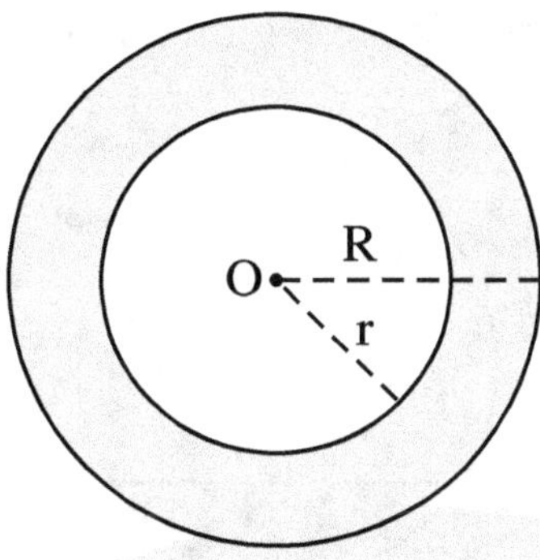

Area enclosed by them (area of circular track formed by them)

$$= \pi R^2 - \pi r^2$$
$$= \pi (R^2 - r^2)$$
$$= \pi (R + r)(R - r)$$

12.2 Areas Of Sector And Segment Of A Circle

Arc of the circle: Any part of a circle is called an arc of the circle. Two points A and B on a circle divide it into two arcs. In general, one arc is greater than other. The smaller arc is called minor arc and the greater arc is called the major arc. If the two arcs so formed are equal, then each of these arcs is called a semi-circle.

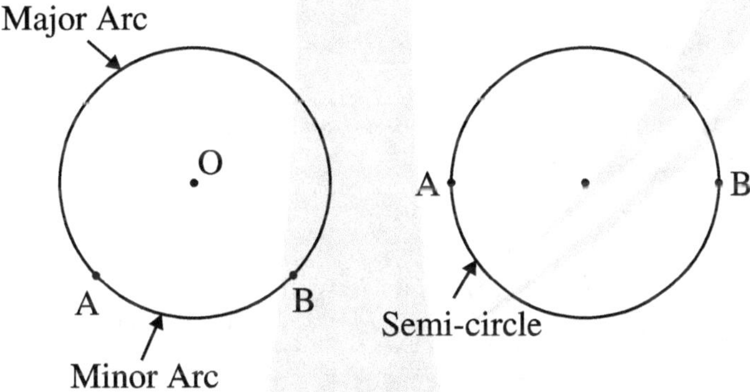

Sector of the Circle: The region bounded by an arc of a circle and its two bounding radii is called a sector of the circle.

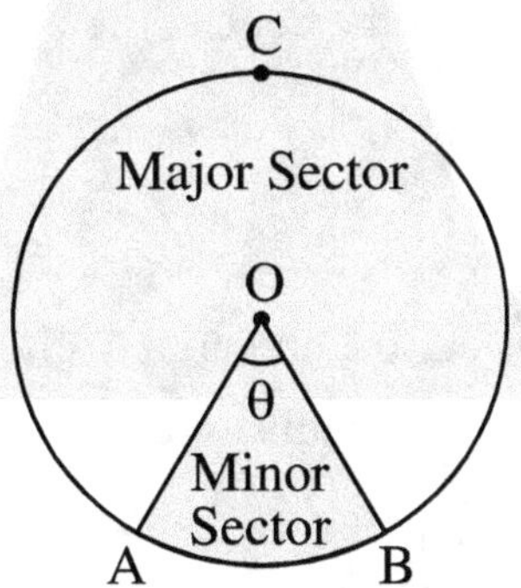

Let AB be an arc of a circle whose centre is O. Then region bounded by radii AO, BO and arc AB is called sector of the circle. The sector OAB is called minor sector if the bounding arc AB is minor. The minor arc AB of a minor sector AOB subtends an angle θ (Theta) in degrees at the centre ($\theta < 180°$). This angle is called central angle. The remaining part of the circle bounded by the major arc is called the major sector.

Segment of the Circle: A segment of a circle is the region bounded by an arc and a chord including the arc and the chord.

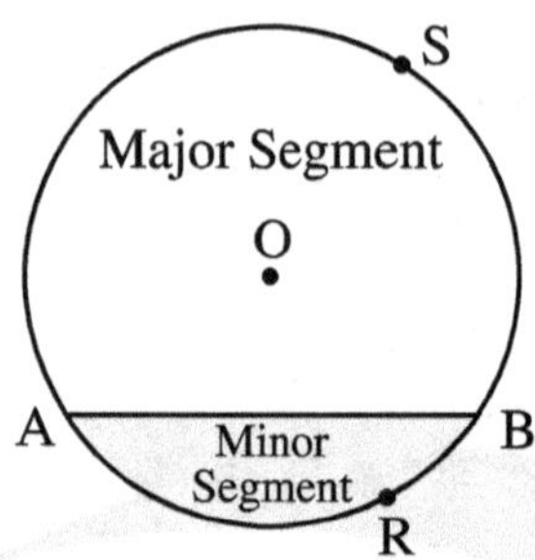

The segment containing the minor arc is called a minor segment and the remaining segment containing the major arc is called the major segment.

Length of an Arc and Area of a Sector of a Circle: Let r be the radius of a circle with centre θ. Let the arc AB of the circle subtend an angle θ at the centre, so that AOB is the corresponding sector.

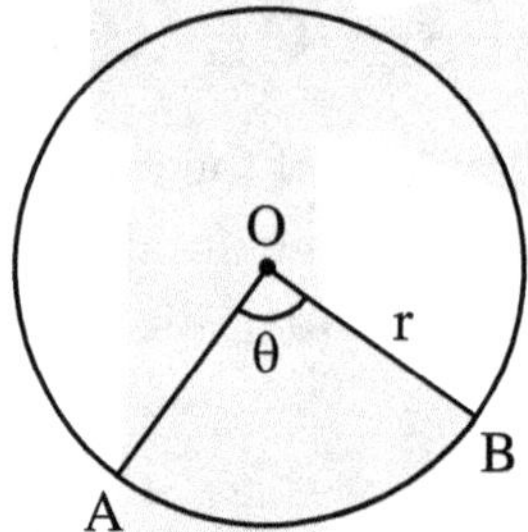

We know that as θ increases, the length of arc AB increases proportionally. When an arc subtends an angle 180° at the centre, its length = length of semi-circle

$$= \pi r$$

$\therefore$ When the arc subtends an angle θ at the centre, its length $= \dfrac{\pi r \theta}{180}$

If L is the length of the arc AB, we have the formula:

$$\boxed{L \;=\; \dfrac{\pi r \theta}{180}}$$

$$(1)$$

Similarly, when an arc subtends an angle 180° at the centre, the corresponding sector is a semi-circular region

of area $= \dfrac{\pi r^2}{2}$

$\therefore$ When the arc subtends an angle θ at the centre, the area of the corresponding sector $= \dfrac{\pi r^2}{2} \times \dfrac{\theta}{180} = \dfrac{\pi r^2 \theta}{360}$

If A is the area of the sector AOB, we have the formula:

$$\boxed{A \;=\; \dfrac{\pi r^2 \theta}{360}}$$

$$(2)$$

From (1) and (2), we observe:

$$\dfrac{A}{L} \;=\; \dfrac{\pi r^2 \theta}{360} \times \dfrac{180}{\pi r \theta} = \dfrac{r}{2}$$

Hence, we have:

$$A = \frac{Lr}{2} \tag{3}$$

Area of a Segment of a Circle: Let r be the radius of a circle with centre O. Let a chord AB make a minor segment ACB, i.e., the region bounded by the arc ACB and the chord AB. Let $\angle AOB = \theta$. Draw $BM \perp OA$.

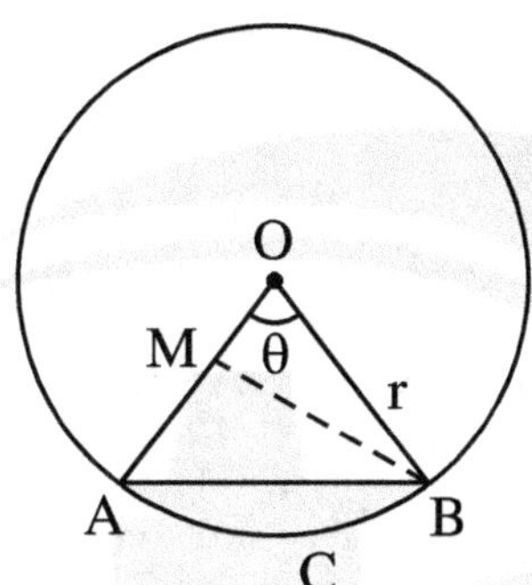

$$\text{Area of } \Delta OAB = \frac{1}{2}OA \times BM$$

$$= \frac{1}{2}OA \times OB \sin\theta \qquad \left[\because \text{ In } \Delta OMB, \frac{BM}{OB} = \sin\theta\right]$$

$$= \frac{1}{2}r \times r \sin\theta \qquad [\because \ OA = OB = r]$$

$$= \frac{1}{2}r^2 \sin\theta$$

$$\text{Area of segment ACB } = \text{ Area of sector OACB} - \text{Area of } \Delta OAB$$

$$= \frac{\pi r^2 \theta}{360} - \frac{1}{2}r^2 \sin\theta$$

Hence, we have the formula:

$$\boxed{\ \text{Area of minor segment } = \frac{\pi r^2 \theta}{360} - \frac{1}{2}r^2 \sin\theta\ }$$

Alternatively, we can find the area of segment ACB as follows:

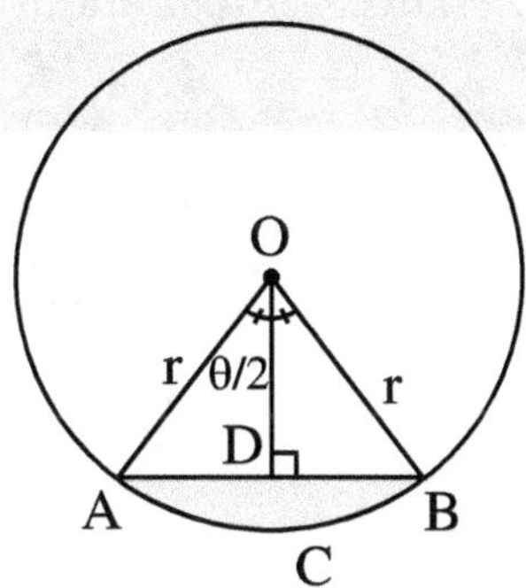

Draw the perpendicular OD from the centre O on the chord AB. Then OD bisects the angle θ, because ΔOAB is isosceles with $OA = OB = r$.

$$\therefore \quad \text{Area of } \Delta OAB \;=\; \frac{1}{2}AB \times OD = AD \times OD = r\sin\frac{\theta}{2} \times r\cos\frac{\theta}{2}$$

$$\left[\because \text{ In } \Delta OAD,\; \frac{AD}{OA} = \sin\frac{\theta}{2} \text{ and } \frac{OD}{OA} = \cos\frac{\theta}{2}\right]$$

$$= r^2 \sin\frac{\theta}{2}\cos\frac{\theta}{2}$$

$$\therefore \quad \text{Area of segment ACB} \;=\; \text{Area of sector OACB} - \text{Area of } \Delta OAB$$

$$= \frac{\pi r^2 \theta}{360} - r^2 \sin\frac{\theta}{2}\cos\frac{\theta}{2}$$

$$= r^2\left(\frac{\pi\theta}{360} - \sin\frac{\theta}{2}\cos\frac{\theta}{2}\right)$$

Hence, we have the formula:

$$\boxed{\;\text{Area of minor segment} \;=\; r^2\left(\frac{\pi\theta}{360} - \sin\frac{\theta}{2}\cos\frac{\theta}{2}\right)\;}$$

$$(5)$$

Also, since Area of minor segment + Area of major segment = Area of circle

We have, from (4) :

$$\boxed{\;\text{Area of major segment} \;=\; \pi r^2 - \left(\frac{\pi r^2 \theta}{360} - \frac{1}{2}r^2 \sin\theta\right)\;}$$

12.3 Areas of Combinations of Plane Figures

We know how to calculate the areas of different plane figures (triangles, squares, rectangles, circles, and sectors and segments of circles) separately. In our daily life we come across many interesting figures and designs which are a combination of two or more than two of such plane figures.

We can find out the areas of such figures by calculating the areas of its various components (parts) separately and then applying suitable mathematical operations. Illustrations given in the next section of this chapter, make the process further clear.

WARM-UP – LEVEL-I

Q. 1. **The area of a circle is 78.5 cm², find its circumference (Take $\pi = 3.14$).**

Ans. $\qquad$ Area $= \pi r^2 = 78.5$

$$\Rightarrow \qquad r^2 = \frac{78.5}{\pi} = \frac{78.5}{3.14} = 25$$

or $\qquad r = 5$ cm

$\therefore \qquad$ Circumference $= 2\pi r = 2 \times 3.14 \times 5 = 31.4$ cm

Q. 2. **The radii of two circles are 19 cm and 9 cm respectively. Find the radius of the circle which has circumference equal to the sum of the circumferences of the two circles.**

Ans. Circumference of the circle of radius 9 cm

$$= 2\pi \times 9 = 18\pi \text{ cm}$$

Circumference of the circle of radius 19 cm

$$= 2\pi \times 19 = 38\pi \text{ cm}$$

Sum of the circumferences of the two circles

$$= 18\pi + 38\pi$$
$$= 56\pi \text{ cm}$$

Let r be the radius of the circle which has circumference equal to the sum of the circumferences of the two circles.

$$\Rightarrow \qquad 2\pi r = 56\pi$$

$$\therefore \qquad r = 28 \text{ cm}$$

Q. 3. **The radii of two circles are 8 cm and 6 cm respectively. Find the radius of the circle having area equal to the sum of the areas of the two circles.**

Ans. Area of the circle of radius 8 cm $= \pi \times 8^2 = 64\pi$

Area of the circle of radius 6 cm $= \pi \times 6^2 = 36\pi$

Sum of the area of the two circles

$$= 64\pi + 36\pi$$
$$= 100\pi \text{ cm}^2$$

Let r be the radius of the circle which has area equal to the sum of the areas of two circles.

$$\Rightarrow \qquad \pi r^2 = 100\pi$$

or $\qquad r = \sqrt{100} = 10$ cm

Q. 4. **How many times will the wheel of a car rotate in a journey of 88 km if it is known that the diameter of the wheel is 56 cm?**

Ans. Given the diameter of the wheel $= 56$ cm

$$\therefore \qquad \text{The radius of the wheel} = \frac{1}{2} \times 56 = 28 \text{ cm}$$

$$\therefore \qquad \text{Circumference of the wheel} = 2\pi r$$

$$= \quad 2 \times \frac{22}{7} \times 28 = 176 \text{ cm}$$

$\therefore$ Distance covered by the wheel in one revolution = 176 cm

Since the distance covered = 88 km = 88 × 1000 × 100 cm

$\therefore$ the number of times the wheel will rotate $= \dfrac{88 \times 1000 \times 100}{176} = 50000$

Q. 5.   **The perimeter of a sheet of tin in the shape of a quadrant of a circle is 12.5 cm. Find its area.**

Ans.   Let r cm be the radius of the circle, then perimeter of a quadrant of the circle

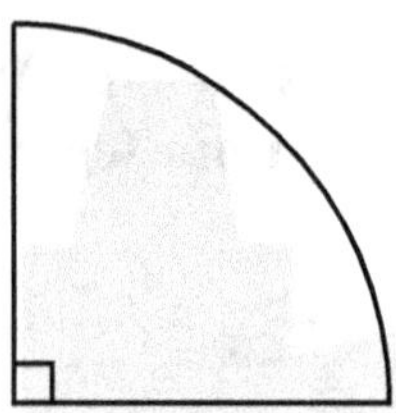

$$= \frac{1}{4} \times 2\pi r + 2r = \left(\frac{\pi}{2} + 2\right) r$$

According to given :

$$-\left(\frac{\pi}{2} + 2\right) r = 12.5$$

$$\Rightarrow \quad \left(\frac{1}{2} \times \frac{22}{7} + 2\right) r = \frac{25}{2}$$

$$\Rightarrow \quad \frac{25}{2} r = \frac{25}{2} \quad \Rightarrow \quad r = \frac{7}{2}$$

$\therefore$ Area of the quadrant $= \dfrac{1}{4}\pi r^2 = \dfrac{1}{4} \times \dfrac{22}{7} \times \left(\dfrac{7}{2}\right)^2$

$$= \frac{77}{8} \text{ cm}^2 = 9.625 \text{ cm}^2$$

Q. 6.   **The wheels of a car are of diameter 80 cm each. How many complete revolutions does each wheel make in 10 minutes when the car is travelling at a speed of 66 km per hour?**

Ans.   Diameter of a wheel = 80 cm

$\therefore$ Radius of the wheel $= \dfrac{1}{2} \times 80 = 40$ cm

Circumference of the wheel $= 2\pi r$

$$= 2 \times \frac{22}{7} \times 40$$

          Innovative Mathematics X-11

$$= \frac{1760}{7} \text{ cm}$$

Distance covered by the wheel is one revolution

$$= \frac{1760}{7} \text{ cm}$$

$$\text{Speed of the car} = 66 \text{ km/h}$$

$$= \frac{66 \times 1000 \times 100}{60} \text{ cm/min}$$

$$= 110000 \text{ cm/min}$$

Distance travelled by the car in 10 minutes

$$= 110000 \times 10 = 1100000 \text{ cm}$$

$\therefore$ Number of revolutions made by each wheel

$$= \frac{1100000}{\dfrac{1760}{7}} = \frac{1100000 \times 7}{1760}$$

$$= 4375$$

Q. 7. **The area and circumference of a circle are numerically equal. What is the radius of the circle?**

Ans. Let r be the radius of the circle,

According to given:

$$\pi r^2 = 2pr$$

$$\Rightarrow \qquad \pi r^2 - 2\pi r = 0$$

$$\Rightarrow \qquad \pi r(r - 2) = 0$$

$\therefore$ either $\qquad \pi r = 0 \quad \text{or} \quad (r - 2) = 0$

i.e., either $\qquad r = 0 \quad \text{or} \quad r = 2$

$\therefore \qquad r = 2 \text{ units}$ [neglecting $r = 0$]

Q. 8. **The area of a circular ring enclosed between two concentric circles is 286 cm². Find the radii of the two circles, given that their difference is 7 cm.**

Ans. Let the radii of the outer and the inner circles be R cm and r cm respectively.

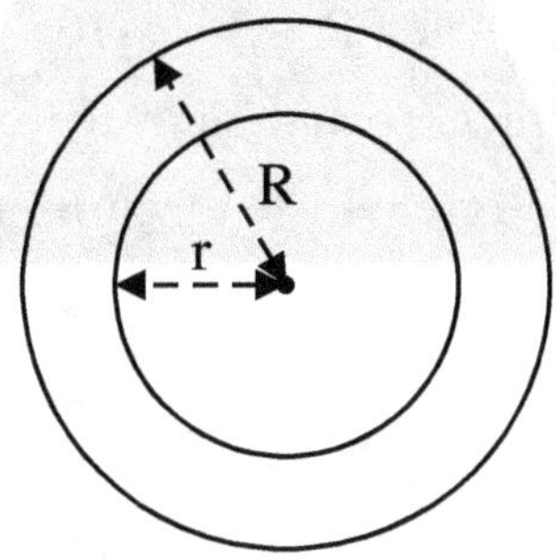

According to given :

$$R - r = 7 \qquad\qquad (1)$$

and $\qquad \pi(R^2 - r^2) = 286$

$$\Rightarrow \qquad \pi(R - r)(R + r) = 286$$

$$\Rightarrow \qquad \frac{22}{7} \times 7(R + r) = 286$$

$$\Rightarrow \qquad R + r = 13 \qquad\qquad\qquad (2)$$

Adding (1) from (2), we get:

$$2R = 20 \quad \Rightarrow \quad R = 10$$

Subtracting (1) and (2), we get:

$$2r = 6 \quad \Rightarrow \quad r = 3$$

$\therefore$ The radii of the two circles are 10 cm and 3 cm

Q. 9. **Two circles touch externally. The sum of their areas is 58π cm^2 and the distance between their centres is 10 cm. Find the radii of the two circles.**

Ans. Let R cm and r cm be the radii of two circles, then

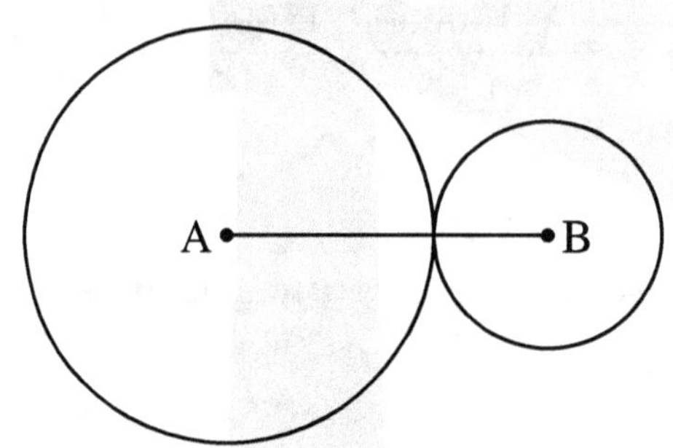

$$R + r = 10$$

$$\Rightarrow \qquad r = 10 - R \qquad\qquad\qquad (1)$$

Also, $\qquad \pi R^2 + \pi r^2 = 58\pi$

$$\Rightarrow \qquad R^2 + r^2 = 58$$

$$\Rightarrow \qquad R^2 + (10 - R)^2 = 58$$

$$\Rightarrow \qquad R^2 + 100 + R^2 - 20R - 58 = 0$$

$$\Rightarrow \qquad 2R^2 - 20R + 42 = 0$$

$$\Rightarrow \qquad R^2 - 10R + 21 = 0 \quad \Rightarrow \quad (R - 7)(R - 3) = 0$$

$$\Rightarrow \qquad R - 7 = 0 \quad \text{or} \quad R - 3 = 0 \Rightarrow R = 7 \text{ or } R = 3$$

When R = 7, then r = 10 − 7 = 3 and when R = 3, r = 10 − 3 = 7

Hence the radii of the two circles are 7 cm and 3 cm

Q. 10. **Find the area of a quadrant of a circle whose circumference is 22 cm.**

Ans. $\qquad$ Circumference $= 2\pi r$

$$\Rightarrow \qquad 2\pi r = 22$$

$$\therefore \qquad r = \frac{22}{2\pi} = \frac{22 \times 7}{44} = \frac{7}{2} \text{ cm}$$

$$\text{Area of the circle} = \pi r^2 = \frac{22}{7} \times \frac{7}{2} \times \frac{7}{2}$$

$$= 38.5 \text{ cm}^2$$

$\therefore$ Area of the quadrant of the circle

$$= \frac{38.5}{4} = 9.625 \text{ cm}^2$$

Q. 11. **Find the area of a ring shaped region enclosed between two concentric circles of radii 20 cm and 15 cm.**

Ans. Area of the ring shaped region (shaded)

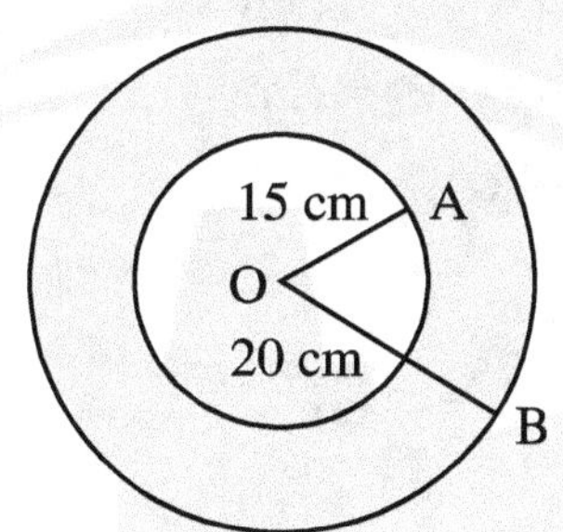

$$= \pi (R^2 - r^2)$$
$$= \pi (OB^2 - OA^2)$$
$$= \pi [(20)^2 - (15)^2]$$
$$= \frac{22}{7} [400 - 225]$$
$$= \frac{22 \times 175}{7}$$
$$= 550 \text{ cm}^2$$

Q. 12. **In the given Fig. 12.15, the area enclosed between the two concentric circles is 770 cm^2. Given that the radius of outer circle is 21 cm. Calculate the radius of the inner circle (Use $\pi = 22/7$).**

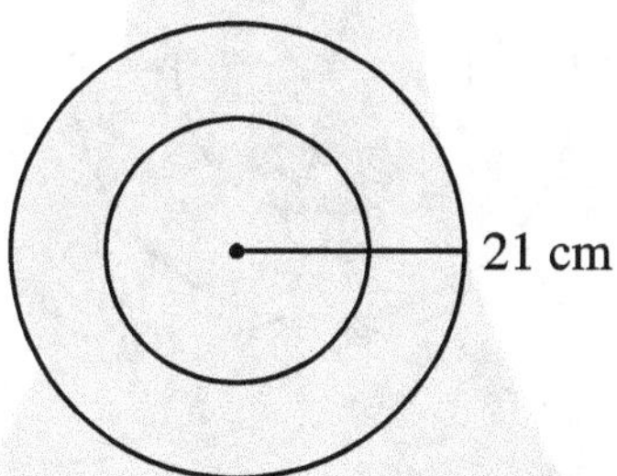

Ans. Area enclosed between two concentric circles $= \pi (R^2 - r^2) = 770 \text{ cm}^2$

$$\Rightarrow \quad \pi[(21)^2 - r^2] = 770$$

$$\Rightarrow \quad (21)^2 - r^2 = \frac{770 \times 7}{22} = 245$$

$$\Rightarrow \quad 441 - r^2 = 245$$

$$r^2 = 196$$

$\therefore$ r (radius of the inner circle) = 14 cm

Q. 1. A wire when bent into the form of a square encloses an area of 121 sq cm. If the wire is bent in the form of a circle, find the area of the circle (Use $\pi = 22/7$).

Ans.

$$\text{Area of square} = 121 \text{ cm}^2$$

$$\therefore \quad \text{Side of the square} = \sqrt{121} = 11 \text{ cm}$$

Hence, perimeter of the square $= 11 \times 4 = 44$ cm

When wire is bent in the form of a circle, perimeter of the square = circumference of the circle.

$\Rightarrow$ Circumference of the circle $= 44$ cm

$$\therefore \quad 2\pi r = 44$$

$$\text{or} \quad r = \frac{44 \times 7}{44} = 7 \text{ cm}$$

$$\text{Hence, area of the circle} = \pi r^2 = \frac{22}{7} \times 7 \times 7$$

$$= 154 \text{ cm}^2$$

Q. 2. A circular swimming pool is surrounded by a circular path which is 4 m wide. If the area of the path is $\dfrac{11}{25}$ th part of the area of the swimming pool, find the radius of the swimming pool.

Ans. Let radius of the pool = r

$$\therefore \quad \text{Radius of the outer circle} = r + 4$$

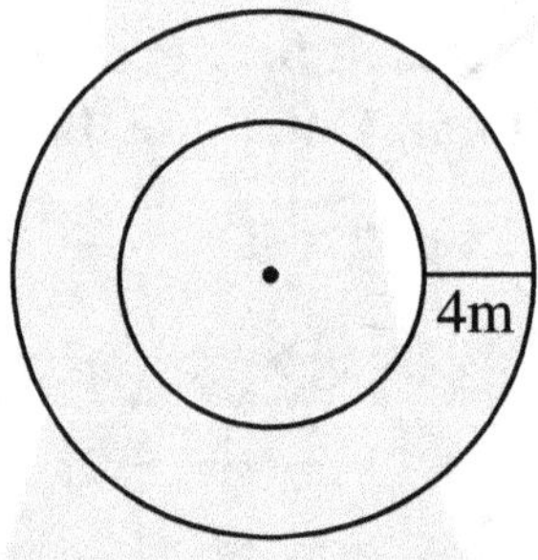

$$\text{Area of the pool} = \pi r^2$$

$$\text{Area of the path} = \pi(r + 4)^2 - \pi r^2$$

According to given:

$$\pi(r + 4)^2 - \pi r^2 = \frac{11}{25} \times \pi r^2$$

$$\Rightarrow \quad \pi r^2 + 16\pi + 8\pi r - \pi r^2 = \frac{11}{25} \pi r^2$$

$$\Rightarrow \quad \frac{11}{25}\pi r^2 - 8\pi r - 16\pi = 0$$

Innovative Mathematics X-11

$$\Rightarrow \qquad \frac{11}{25}r^2 - 8r - 16 = 0$$

$$\Rightarrow \qquad 11r^2 - 200r - 400 = 0$$
$$\Rightarrow \qquad 11r^2 - 220r + 20r - 400 = 0$$
$$\Rightarrow \qquad 11r\,(r - 20) + 20(r - 20) = 0$$
$$\Rightarrow \qquad (r - 20)\,(11r + 20) = 0$$

$$\therefore \quad r = 20 \ \text{or} \ r = \frac{-20}{11}$$

But r (radius) cannot be negative,

$$\therefore \qquad \text{Radius of the pool} = 20 \text{ m}$$

Q. 3. A race track is in the form of a ring whose inner circumference is 440 m and the outer circumference 506 m. Find the width of the track and also find the area of the track.

Ans. Let the inner radius = r

and outer radius = R

$$\therefore \qquad 2\pi r = 400$$

$$\Rightarrow \qquad r = \frac{440}{2 \times 22} \times 7 = 70 \text{ m}$$

and

$$2\pi R = 506 \Rightarrow R = \frac{506}{2 \times 22} \times 7 = 80.5 \text{ m}$$

$$\text{Width of the track} = R - r$$
$$= 80.5 - 70$$
$$= 10.5 \text{ m}$$
$$\text{Area of the track} = \pi R^2 - \pi r^2$$
$$= \pi\,(R + r)\,(R - r)$$
$$= \frac{22}{7} \times (80.5 + 70)\,(80.5 - 70)$$
$$= \frac{22}{7} \times 150.5 \times 10.5$$
$$= 4966.5 \text{ m}^2$$

Q. 4. PS is a diameter of a circle of radius 6 cm. Q. and R are points on the diameter such that PQ., Q.R and RS are equal. Semicircles are drawn with PQ. and Q.S as diameters, as shown in Fig. 12.17. Find the perimeter of the shaded region ($\pi = 3.14$)

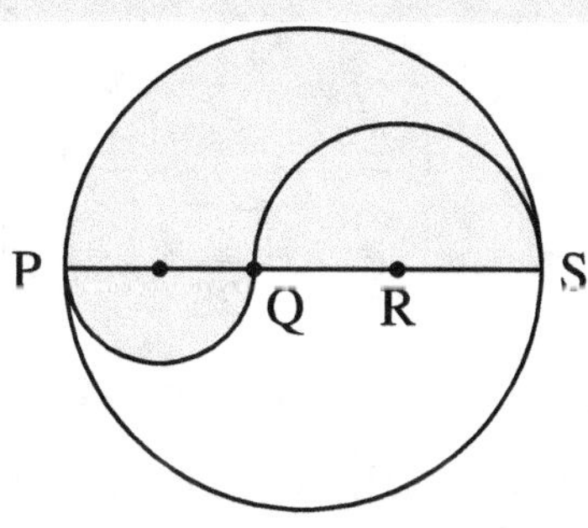

Also find the area of the shaded region.

Ans.

$$PS = 2 \times \text{radius} = (2 \times 6) = 12 \text{ cm}$$

Given $\qquad PQ. = Q.R = RS$

$\Rightarrow \qquad PQ. = \dfrac{1}{3} \text{ of } PS = \dfrac{1}{3} \times 12 = 4 \text{ cm}$

and $\qquad Q.S = 8 \text{ cm.}$

$\therefore$ The semicircles with PQ. and Q.S as diameters have radii 2 cm and 4 cm respectively.

$\therefore$ Perimeter of the shaded region

$$= \left(\frac{1}{2} \times 2\pi \times 6 + \frac{1}{2} \times 2\pi \times 4 + \frac{1}{2} \times 2\pi \times 2 \right)$$
$$= \pi(6 + 4 + 2) = 12\pi$$
$$= (12 \times 3.14) = 37.68 \text{ cm}$$
$$= \left(\frac{1}{2} \times \pi \times 6^2 - \frac{1}{2} \times \pi \times 4^2 + \frac{1}{2} \times \pi \times 2^2 \right)$$
$$= \frac{\pi}{2}(36 - 16 + 4) = 12\pi$$
$$= (12 \times 3.14) = 37.68 \text{ cm}^2$$

Q. 5. **There are two concentric circular tracks of radii 100 metres and 102 metres respectively. A runs on the inner track and goes once round the track in 1 minute 30 seconds; while B runs on the outer track in 1 minute 32 seconds. Who runs faster?**

Ans. Circumference of the inner track $= (2\pi \times 100) = 200\pi$ m

Circumference of the outer track $= (2\pi \times 102) = 204\pi$ m

A covers a distance equal to circumference of the inner track in 1 minute 30 seconds i.e., in $\dfrac{3}{2}$ minutes.

So the distance travelled by A in $\dfrac{3}{2}$ minutes $= 200\pi$ m

$\therefore$ The distance travelled by A in 1 minute $= \left(\dfrac{2}{3} \times 200\pi \right) = 133.33\pi$ m.

$\therefore$ Speed of A $= 133.33\pi$ m/min.

B covers a distance equal to circumference of the outer track in 1 minute 32 seconds i.e., in $\left(1 + \dfrac{32}{60} \right)$ min i.e., in $\dfrac{23}{15}$ min.

$\therefore$ The distance travelled by B in 1 minute $= \left(\dfrac{15}{23} \times 204\pi \right) = 133.04\pi$ m

$\therefore$ Speed of B $= 133.04\pi$ m/min.

Since speed of A is greater than speed of B, therefore, A runs faster.

Innovative Mathematics X-11

Q. 6. **Find the area of a sector of a circle with radius 6 cm if angle of the sector is 60°.**

Ans. Here, r = 6 cm and $\theta = 60°$

$$\text{Area of the sector} = \frac{\pi r^2 \theta}{360}$$

$$= \frac{22}{7} \times 6 \times 6 \times \frac{60}{360}$$

$$= \frac{132}{7} = 18.86 \text{ cm}^2$$

Q. 7. **The length of the minute hand of a clock is 14 cm. Find the area swept by the minute hand in 5 minutes.**

Ans. Area swept by the minute hand in 60 minutes $= \pi r^2$

$\therefore$ Area swept by the minute hand in 5 minutes $= \dfrac{\pi r^2 \times 5}{60}$

$$= \frac{22}{7} \times 14 \times 14 \times \frac{5}{60}$$

$$= \frac{154}{3} = 51.33 \text{ cm}^2$$

Q. 8. **A horse is tied to a peg at one corner of a square shaped grass-field of side 15 m by means of a 5 m long rope. Find**

(i) the area of that part of the field in which the horse can graze.

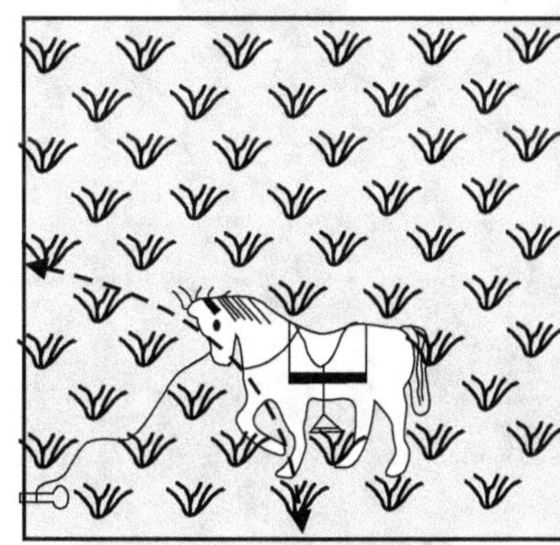

(ii) the increase in the grazing area if the rope were 10 m long instead of 5 m.

Ans. The area of the field in which the horse can graze is represented in the Fig. 12.19 by APQ., a quadrant of the circle with radius 5 m.

$\therefore$ Required area $= \dfrac{\pi r^2}{4} = \dfrac{22}{7} \times \dfrac{5 \times 5}{4} = 19.64 \text{ m}^2$

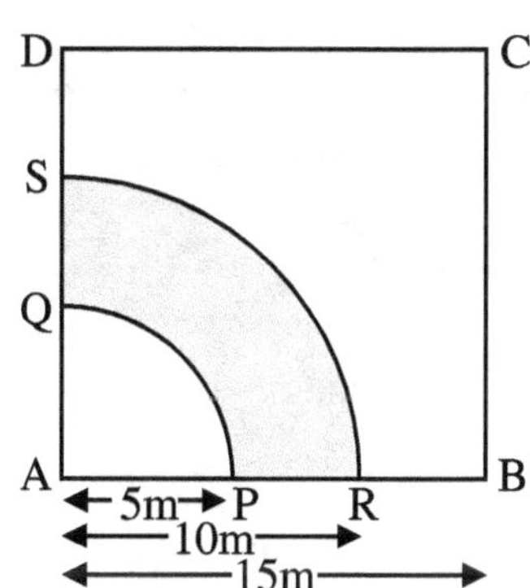

Increase in the area when the rope is 10 m long (shaped)

= Area of quadrant ARS – Area of quadrant APQ.

$$= \frac{\pi(10)^2}{4} - \frac{\pi(5)^2}{4} = \frac{75\pi}{4} = \frac{75}{4} \times \frac{22}{7}$$

$$= 58.93 \text{ m}^2$$

Q. 9. **In a circle of radius 21 cm, an arc subtends an angle of 60° at the centre. Find**

    (i) the length of the arc.

    (ii) area of the sector formed by the arc.

    (iii) the area of the segment formed by the corresponding chord.

Ans. Length of the arc $= \dfrac{\pi r \theta}{180} = \dfrac{22}{7} \times \dfrac{21 \times 60}{180} = 22$ cm

$$\text{Area of the sector} = \frac{Lr}{2} = \frac{22 \times 21}{2} = 231 \text{ cm}^2$$

$$\text{Area of the segment} = \text{Area of sector} - \text{Area of } \Delta OAB$$

$$= 231 - \frac{1}{2} r^2 \sin \theta$$

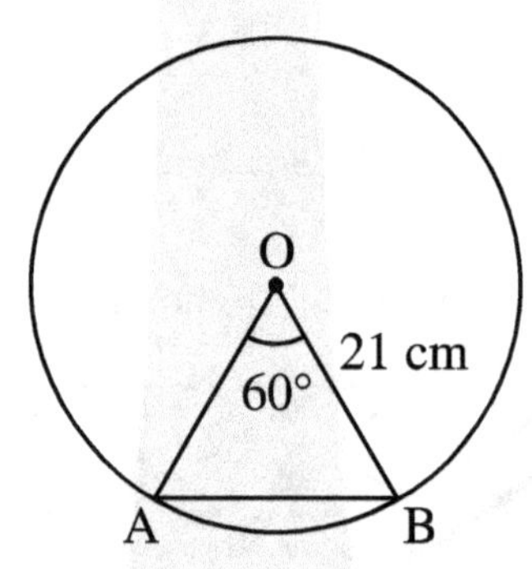

$$= 231 - \frac{1}{2} \times 21 \times 21 \times \sin 60°$$

$$= 231 - \frac{1}{2} \times 21 \times 21 \times \frac{\sqrt{3}}{2}$$

$$= 231 - 190.953$$

$$= 40.047 \text{ cm}^2$$

Q. 10. **Find the area of the sector of a circle if its radius is 21 cm and the length of the corresponding arc is 16.5 cm.**

Ans. We know that $\qquad A = \dfrac{Lr}{2}$

Here, $\qquad\qquad\qquad r = 21$ cm and $L = 16.5$ cm

$\therefore \qquad\qquad\qquad$ Area $= \dfrac{16.5 \times 21}{2} = 173.25 \text{ cm}^2$

Q. 11. Find the area of the segment of a circle if its radius is 14 cm and the area of the corresponding sector is 154 cm².

Ans. Area of the sector = 154 cm²

$$\Rightarrow \qquad \frac{\pi r^2 \theta}{360} = 154$$

$$\therefore \qquad \theta = \frac{154 \times 360}{\pi r^2} = \frac{154 \times 360}{14 \times 14} \times \frac{7}{22} = 90°$$

$\therefore$ Area of the segment

$$= \frac{\pi r^2 \theta}{360} - \frac{1}{2} r^2 \sin \theta$$

$$= 154 - \frac{1}{2} \times 14 \times 14 \times \sin 90°$$

$$= 154 - \frac{14 \times 14}{2} \times 1$$

$$= 154 - 98 = 56 \text{ cm}^2$$

Q. 12. A brooch is made with silver wire in the form of a circle with diameter 35 mm. The wire is also used in making 5 diameters which divide the circle into 10 equal sectors as shown in Fig. 12.26. Find:

(i) the total length of the silver wire required.

(ii) the area of each sector of the brooch.

Ans. Radius of the circle = $\dfrac{35}{2}$ mm

Circumference of the circle = $2\pi r$

$$= 2 \times \frac{22}{2} \times \frac{35}{2}$$

$$= 110 \text{ mm}$$

Area of the circle = πr^2

$$= \frac{22}{7} \times \frac{35}{2} \times \frac{35}{2}$$

$$= 962.5 \text{ mm}^2$$

Since the circle is divided into 10 equal sectors, the area of each sector

$$= \frac{\text{Area of the circle}}{10}$$

$$= \frac{962.5}{10} = 96.25 \text{ mm}^2$$

Total length of the wire required

$$= \text{Circumference of the circle} + \text{Length of 5 diameters}$$

$$= 110 + 5 \times 35 = 285 \text{ mm}.$$

Q. 13. **An umbrella has 8 ribs which are equally spaced. Assuming umbrella to be a flat circle of radius 45 cm, find the area between the two consecutive ribs of the umbrella.**

Ans. Area of the circle with radius 45 cm = $\pi(45)^2 = 2025\pi$ cm^2

Since the ribs divide the circle into 8 sectors of equal area,

$\therefore$ Area between the two consecutive ribs

$$= \text{Area of each sector of the circle}$$

$$= \frac{\text{Area of the circle}}{8}$$

$$= \frac{2025\pi}{8}$$

$$= \frac{2025}{8} \times \frac{22}{7} = 795.54 \text{ cm}^2$$

Q. 14. **A car has two wipers which do not overlap. Each wiper has a blade of length 25 cm sweeping through an angle of 115°. Find the total area cleaned at each sweep of the blades.**

Ans. Here, r = 25 cm and $\theta = 115°$

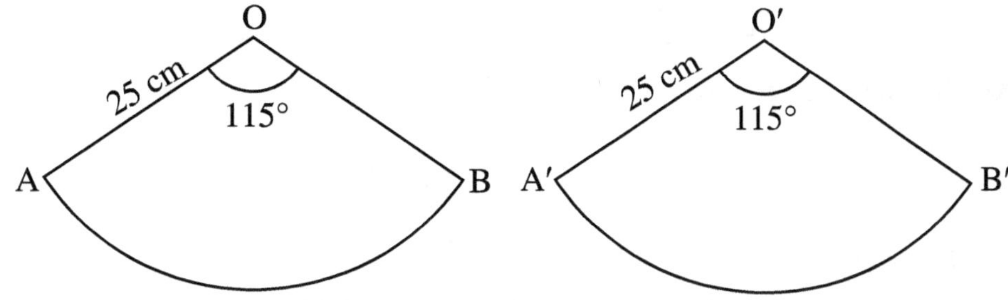

$\therefore \qquad$ Area of each sector $= \dfrac{\pi r^2 \theta}{360}$

$$= \dfrac{22}{7} \times 25 \times 25 \times \dfrac{115}{360}$$

$$= \dfrac{158125}{252} \text{ cm}^2$$

Total area cleaned by two wipers

$$= \dfrac{158125}{252} \times 2$$

$$= \dfrac{158125}{126}$$

$$= 1254.96 \text{ cm}^2$$

Q. 15. **A square of side 4 cm is inscribed in a circle. Find the area enclosed between the circle and the square.**

Ans. Let ABCD be the square of side 4 cm inscribed in a circle with centre O.

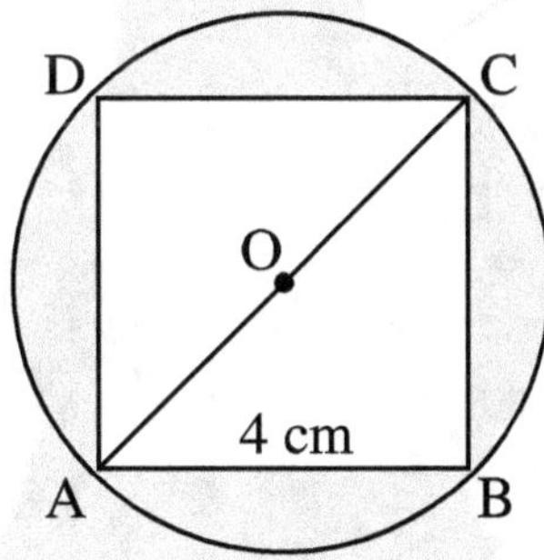

The diagonal AC of the square will be the diameter of the circle.

$$AC = \sqrt{AD^2 + DC^2}$$

$$= \sqrt{4^2 + 4^2} = \sqrt{32}$$

$$= 4\sqrt{2} \text{ cm}$$

$\therefore \qquad$ Radius of the circle $= \dfrac{1}{2}(4\sqrt{2})$

$$= 2\sqrt{2} \text{ cm}$$

$$\text{Area of the circle} = \pi r^2$$

$$= \frac{22}{7} \times 2\sqrt{2} \times 2\sqrt{2} = \frac{176}{7} \text{ cm}^2$$

$$\text{Area of the square} = 4 \times 4 = 16 \text{ cm}^2$$

$$\therefore \qquad \text{Required area} = \text{Area of the circle} - \text{Area of the square}$$

$$= \left(\frac{176}{7} - 16\right)$$

$$= \frac{64}{7} = 9\frac{1}{7} \text{ cm}^2$$

Q. 16. A square ABCD is inscribed in a circle of radius 10 units. Find the area of the circle, not included in the square (Use $\pi = 3.14$).

Ans. Required area (shaded) = Area of the circle – Area of the square.

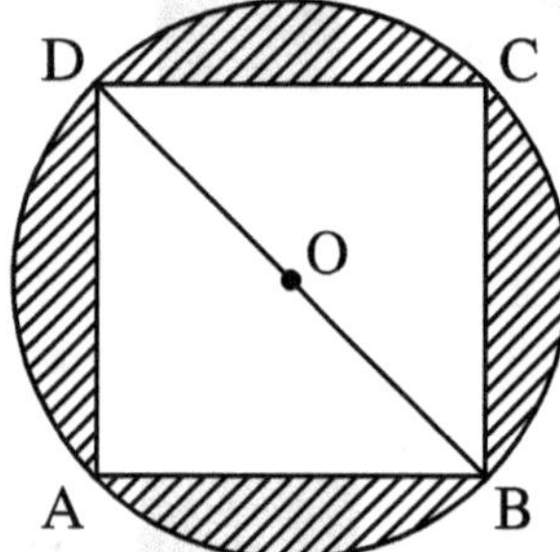

$$\text{Radius of the circle (OB)} = 10 \text{ units}$$

$$\therefore \qquad \text{Area of the circle} = \pi r^2 = 3.14 \times 100$$

$$= 314 \text{ sq units}$$

$$\text{Diameter of the circle (BD)} = 20 \text{ units}$$

$$\text{Let a side of the square} = a \text{ units}$$

By Pythagoras Theorem:

$$AB^2 + AD^2 = BD^2$$

$$\Rightarrow \qquad a^2 + a^2 = (20)^2$$

$$\Rightarrow \qquad 2a^2 = 400$$

$$\therefore \qquad a^2 = 200$$

$$\therefore \qquad \text{Area of the square} = a^2 = 200 \text{ sq units}$$

Hence, required area = $314 - 200 = 114$ sq units

Q. 17. In Fig. 12.31, OACB is a quadrant of a circle with centre O and radius 3.5 cm. If OD = 2 cm, find the area of the

(i) quadrant OACB,   (ii) shaded region.

Ans. Here, $r = 3.5$ cm, $\theta = 90°$

$\therefore$ Area of the quadrant OACB $= \dfrac{\pi r^2 \theta}{360}$

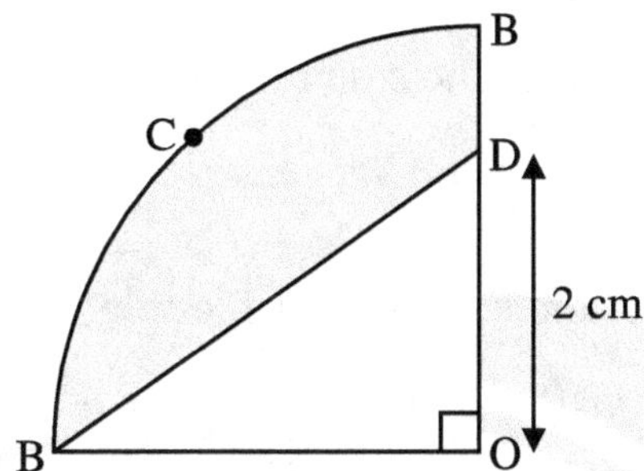

$$= \dfrac{22}{7} \times \dfrac{7}{2} \times \dfrac{7}{2} \times \dfrac{90}{360}$$

$$= 9.625 \text{ sq cm}$$

$$\text{Area of } \Delta OBD = \dfrac{1}{2} \times \text{Base} \times \text{Altitude}$$

$$= \dfrac{1}{2} \times 3.5 \times 2$$

$$= 3.5 \text{ sq cm}$$

$\therefore$ Area of shaded portion $=$ area of OACB $-$ area ΔOBD

$$= 9.625 - 3.5$$

$$= 6.125 \text{ sq cm}$$

Q. 18. **The area of an equilateral triangle is 17300 cm². With each vertex as centre, a circle is described with radius equal to half the length of the side of the triangle. Find the area of the triangle not included in the circles (Use $\pi = 3.14$ and $\sqrt{3} = 1.73$).**

Ans. Area of equilateral $\Delta ABC \left(\dfrac{\sqrt{3}}{4} a^2 \right) = 17300 \text{ cm}^2$ [where a is the side of the Δ]

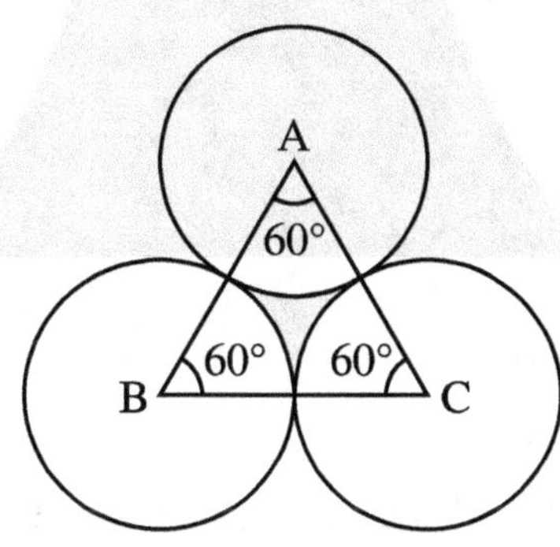

$$\dfrac{\sqrt{3}}{4} a^2 = 17300$$

$$a^2 = \frac{17300 \times 4}{\sqrt{3}} = \frac{17300 \times 4}{1.73} = 40000$$

$$\therefore \quad a = 200 \text{ cm}$$

Radius of each circle = Half of the side of Δ

$$= \frac{200}{2} = 100 \text{ cm}$$

$\therefore$ Area of sector (with central angle 60°) of each circle

$$= \frac{\pi r^2 \theta}{360} = \frac{3.14 \times 100 \times 100 \times 60}{360}$$

$$= 5233.33 \text{ cm}^2$$

Hence, required area (shaded) = $(17300 - 3 \times 5233.33)$

$$= 1600 \text{ cm}^2$$

Q. 19. **A paper is in the form of a rectangle ABCD, where AB = 22 cm and BC = 14 cm. A semicircular portion with BC as diameter is taken away. Find the area of remaining portion.**

Ans. Radius of the semi-circle (r) = $\dfrac{BC}{2} = \dfrac{14}{2} = 7$ cm

Area of the rectangle ABCD = AB × BC

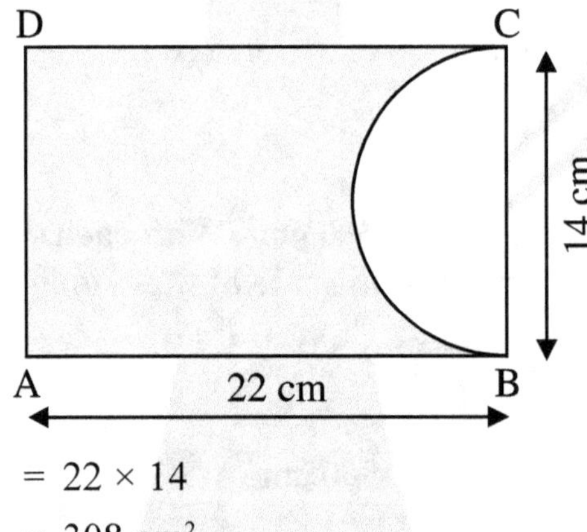

$$= 22 \times 14$$

$$= 308 \text{ cm}^2$$

Area of the semi-circle $= \dfrac{1}{2} \times \pi r^2$

$$= \dfrac{1}{2} \times \dfrac{22}{7} \times 7 \times 7$$

$$= 77 \text{ cm}^2$$

$\therefore$ Area of the remaining portion

$$= \text{Area of the rectangle} - \text{Area of the semicircle}$$

$$= 308 - 77$$

$$= 231 \text{ cm}^2$$

VERY SHORT ANSWER Q.UESTIONS

Q. 1. If the diameter of a semi circular protactor is 14 cm, then find its perimeter.

Sol.

Q. 2. If circumference and the area of a circle are numerically equal, find the diameter of the circle.

Sol.

Q. 3. Find the area of the circle 'inscribed' in a square of side a cm.

Sol.

Q. 4. Find the area of a sector of a circle whose radius is r and length of the arc is l.

Sol.

Q. 5. The radius of a wheel is 0.25 m. Find the numberof revolutions it will make to travel a distance of 11 kms.

Sol.

Q. 6. If the area of a circle is 616 cm^2, then what is its circumference?

Sol.

Q. 7. What is the area of the circle that can be inscribe in a square of side 6 cm?

Sol.

Q. 8. What is the diameter of a circle whose area is equal to the sum of the areas of two circles of radii 24 em and 7 cm?

Sol.

Q. 9. A wire can be bent in the form of a circle of radius 35 cm. If it is bent in the form of a square, then what will be its area?

Sol.

Q. 10. Whatis the angle subtended at the centre of a circle of radius 6 cm by an arc of length 3π cm?

Sol.

Q. 11. Write the formula for the area of a sector of angle θ (in degrees) of a circle of radius r.

Sol.

Q. 12. If the circumference of two circles are in the ratio 2 : 3, what is the ratio of their areas?

Sol.

Q. 13. If the difference between the circumference and radius of a circle is 37 cm, then find the circumference of the circle. (Use $\pi = 22/7$)

Sol.

Q. 14. If diameter of a circle is increased by 40%, find by how much percentage its area increases?

Sol.

Q. 15. The minute hand of a clock is 6 cm long. Find the area swept by it between 11:20 am and 11:55 am.

Sol.

Q. 16. The perimeter of a sector of a circle of radius 14 cm is 68 cm. Find the area of the sector.

(CBSE 2020)

Sol.

Q. 17. The circumference of a circle is 39.6 cm. Find its area. (Use $\pi = 22/7$) (CBSE 2020)

Sol.

Q. 18. The length of the minute hand of a clock is 14 cm. Find the area swept by the minute hand in one minute. (Use $\pi = 22/7$)

Sol.

Q. 19. Area of a sector having length of corresponding arc 'l' and radius 'r' is ________ .

Sol.

Q. 20. Circumference of a circle of radius s is ____________ .

Sol.

Q. 21. Area of a circle of radius is ________ .

Sol.

Q. 22. Length of an arc of a sector of a circle with radius r and angle θ is ________ .

Sol.

Q. 23. Area of a sector with radius r and angle with degrees measure θ is ________ .

Sol.

Q. 24. Area of segment of a circle = Area of the corresponding sector ________ .

Sol.

SHORT ANSWER TYPE QUESTIONS (1)

Q. 25. Find the area of a quadrant of a circle whose circumference is 22 cm. (Use $\pi = 22/7$)

Sol.

Q. 26. What is the angle subtended at the centre of a circle of radius 10 cm by an arc of length 51 cm?

Sol.

Q. 27. If a square is inscribed in a circle, what is the ratio of the area of the circle and the square?

Sol.

Q. 28. Find the area of a circle whose circumference is 44 cm. (CBSE 2020)

Sol.

Q. 29. If the perimeter of a circle is equal to that of square, then find the ratio of their areas.

Sol.

Q. 30. What is the ratio of the areas of a circle and an equilateral triangle whose diameter and a side are respectively equal?

Sol.

Q. 31. In fig., O is the centre of a circle. The area of sector OAPB is $\dfrac{5}{18}$ of the area of the circle. Find x.

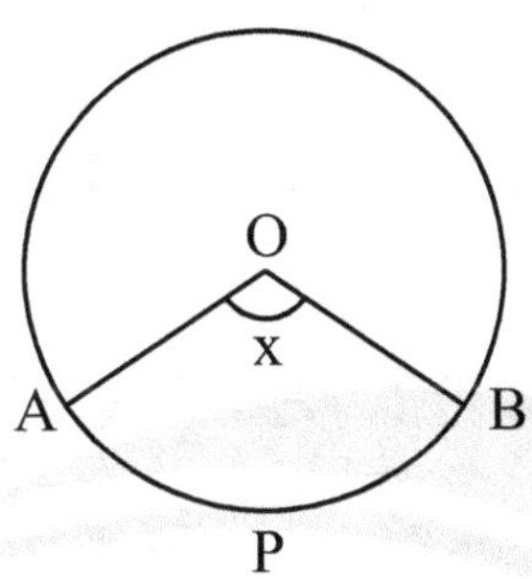

Sol.

Q. 32. Find the perimeter of the given fig, where AED is a semicircle and ABCD is a rectangle.

(CBSE 2015)

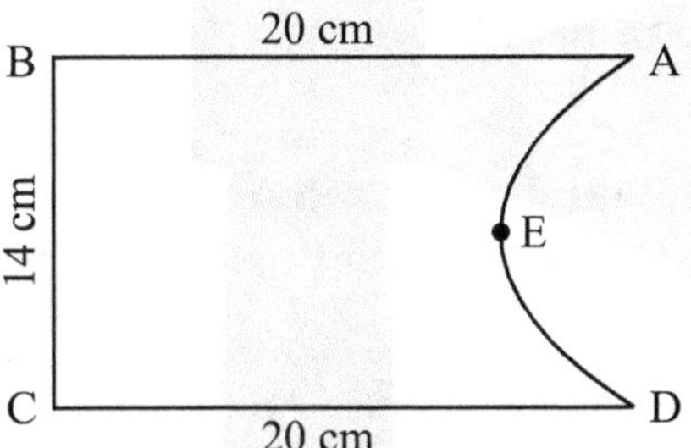

Sol.

Q. 33. In fig. OAPBO is a sector of a circle of radius 10.5 cm. Find the perimeter of the sector.

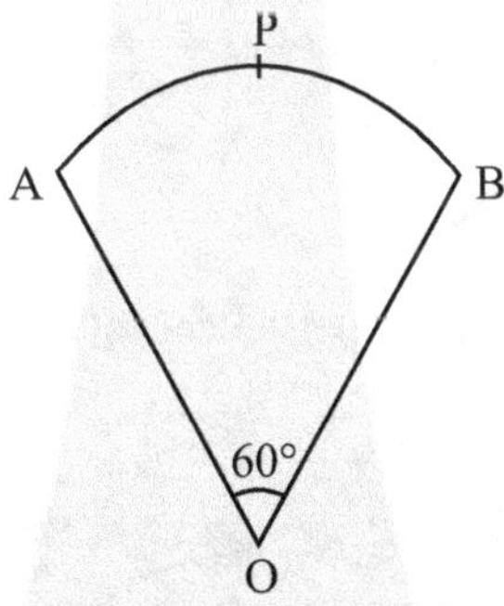

Sol.

Q. 34. A Japenese fan can be made by sliding open its 7 small sections, each of which is in the form of sector of a circle having central angle of 15°. If the radius of this fan is 24 cm, find the length of the lace that is required to cover its entire boundary. (Use $\pi = 22/7$)

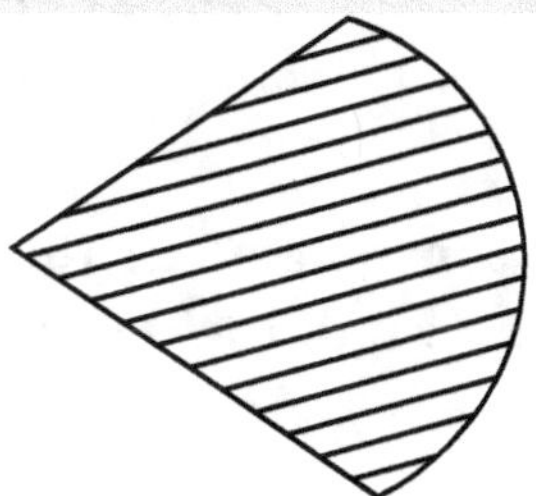

Sol.

Q. 35. The perimeter of a sector of circle of radius 6.3 cm is 25.8 cm. Find the area of the sector.

Sol.

Q. 36. Find the area of a circle in which a square of area 64 cm? is inscribed.

Sol.

Q. 37. Find the area of a circle which is inscribed in a square of area 64 cm^2.

Sol.

SHORT ANSWER TYPE QUESTIONS (2)

Q. 38. Area of a sector of a circle of radius 36 cm is 54 π cm? . Find the length of the corresponding arc of the sector.

Sol.

Q. 39. The length of the minute hand of a clock is 5 cm. Find the area swept by the minute hand during the time period 6:05 am to 6:40 am.

Sol.

Q. 40. Find the area of the segment bounded by a chord AB and the arc ACB of the circle with centre O having radius 7 cm and sector angle equal to 90°, as shown in the fig.

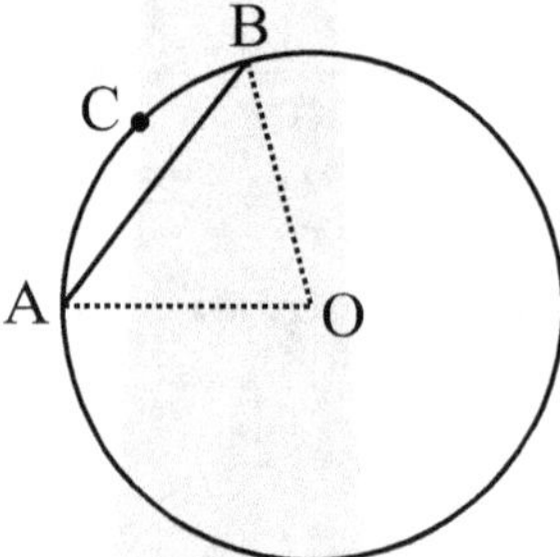

Sol.

Q. 41. In fig, OAPB is a sector of a circle of radius 3.5 cm with the centre at O and $\angle AOB = 120°$. Find the length of OAPBO.

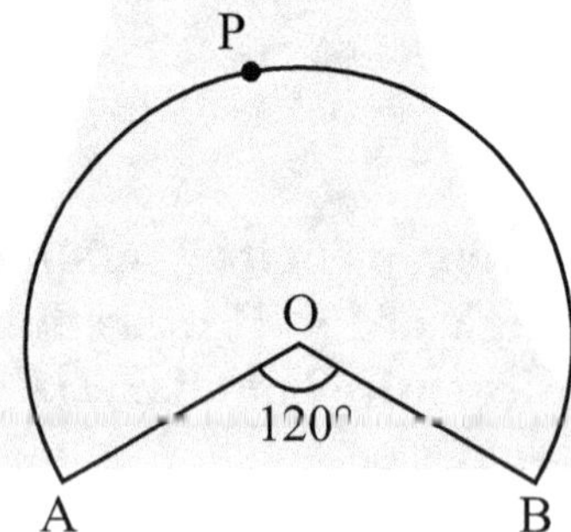

Sol.

Q. 42. Circular footpath of width 2 m is constructed at the rate of ₹ 20 per square meter, around a circular park of radius 1500 m. Find the total cost of construction of the foot path.

(Take $\pi = 13.14$)

Sol.

Innovative Mathematics X-11

Q. 43. A boy is cycling such that the wheels of the cycle are making 140 revolutions per minute. If the diameter of the wheel is 60 cm. Calculate the speed of cycle.

Sol.

Q. 44. In a circle with centre O and radius 4 cm, and of angle 30°. Find the area of minor sector and major sector AOB. (Use $\pi = 13.14$)

Sol.

Q. 45. Find the area of the largest triangle that can be inscribed in a semi circle of radius r unit.

(NCERT Exemplar)

Sol.

Q. 46. In a square park of side 8 m two goats are tied at opposite vertices with a rope of length 1.4 m and a cow is tied in the centre with a rope of length 2.1 m. Calculate the area of park which cannot be grazed by them.

Sol.

Q. 47. A sector of 100° cut off from a circle contains area 70.65 cm? Find the radius of the circle.

(Use $\pi = 3.14$)

Sol.

Q. 48. The hour and minute hand of a 12 hour clock are 3.5 cm and 7 cm long respectively. Find the sum of distance travelled by their tips in a day. (Use $\pi = 22/7$)

Sol.

Q. 49. A square water tank has its each side equal to 40 m. There are four semi circular grassy plots all around it. Find the cost of turfing the plot at Rs 1.25 per sq. m. (Use $\pi = 3.14$)

Sol.

Q. 50. Length of a chord of a circle of a radius of 4 cm is 4 cm. Find the area of the sector and segment formed by the chord.

Sol.

Q. 51. Find the area of the minor segment of a circle of radius 21 cm, when the angle of the corresponding sector is 120°.

Sol.

Q. 52. A piece of wire 11 cm long is bent into the form of an arc of a circle subtending an angle of 45° at its centre. Find the radius of the circle.

Sol.

Q. 53. The circumference of a circle exceeds the diameter by 16.8 em. Find the radius of the circle.

Sol.

Q. 54. A pendulum swings through an angle 0f45° and describes an arc of 22 cm in length. Find the length of the pendulum. (Use $\pi = 22/7$)

Sol.

Q. 55. Two circles touch externally. The sum of their area is 130π sq. cm and the distance between their centres is 14 cm. Find the radii of the circles.

Sol.

LONG ANSWER TYPE Q.UESTIONS

Q. 56. Two circles touch externally. The sum of their areas is 130π sq. cm and the distance between their centres is 14 cm. Find the radii of the circles.

Sol.

Q. 57. Find the number of revolutions made by a circular wheel of area 6.16 m² in rolling a distance of 572 m.

Sol.

Q. 58. Three horses are tied at the vertices of a triangular park of sides 35 m, 84 m and 91 m with the help of a rope of length 14 m cach. Calculate the ratio of the arca which can be grazed to the area which can't be grazed.

Sol.

Q. 59. Two circle touch each other internally, The sum of their area is 116m cm? and distance between their centres is 6 cm. Find the radii of the circles. **(CBSE = 2017)**

Sol.

Q. 60. You are required to create a model of a circular wall clock and paste the numbers from 1 to 12 on its dial. What is the angle made at the centre between 3 and 7? Find the area of this region, if the length of the minute hand is 21 cm. **(CBSE)**

Sol.

Q. 61. In the given fig, APB and CQD are semi circles of diameter 7 cm each, while ARC and BSD are semicircles of diameter 14 cm each. Find the perimeter of the shaded region. (Use $\pi = \dfrac{22}{7}$)

(Delhi, 2011)

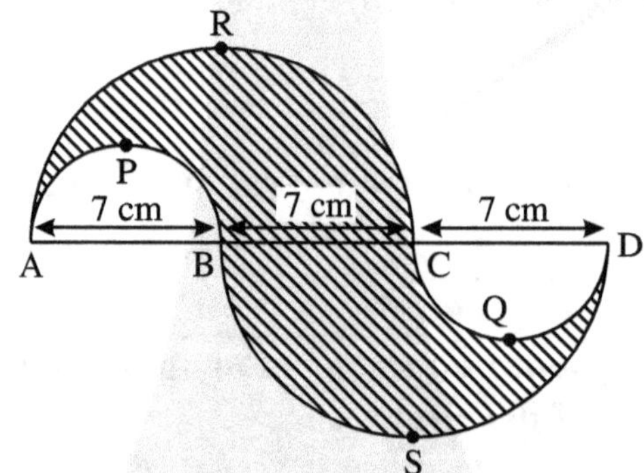

Sol.Q. 62. In fig. from a rectangular region ABCD with AB = 20 cm, a right triangle AED with AE = 9 cm and DE = 12 cm, is cut off. On the other end, taking BC as diameter, a semi circle is added on outside the region. Find the area of the shaded region.

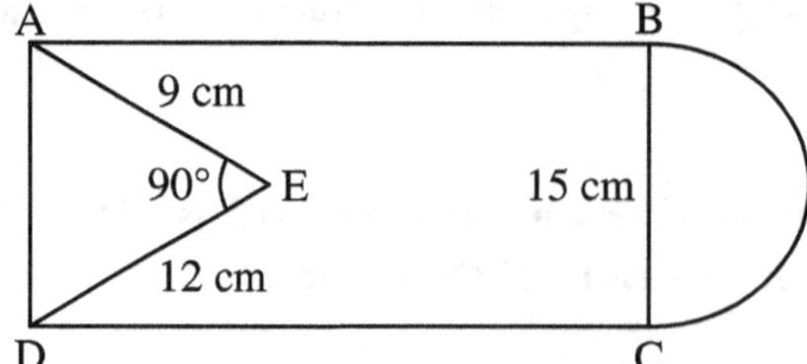

Sol.

Innovative Mathematics X-11

Q. 63. All the vertices of a rhombus lie on a circle. Find the area of the rhombus, if area of the circle is 2464 cm².

Sol.

Q. 64. With vertices A, B and C of a triangle ABC as centres, arcs are drawn with radius 6 cm each in fig. If AB = 20 cm, BC = 48 cm and CA = 52 cm, then find the area of the shaded region.

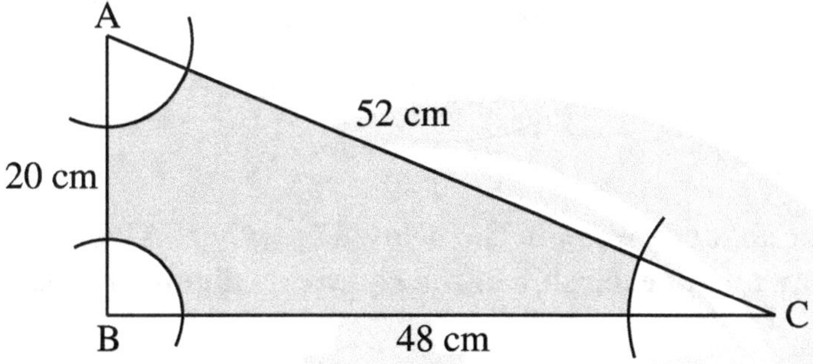

Sol.

Q. 65. ABCDEF is a regular hexagon. With vertices A, B, C, D, E and F as the centres, circles of same radius 'r' are drawn. Find the area of the shaped portion shown in the given figure.

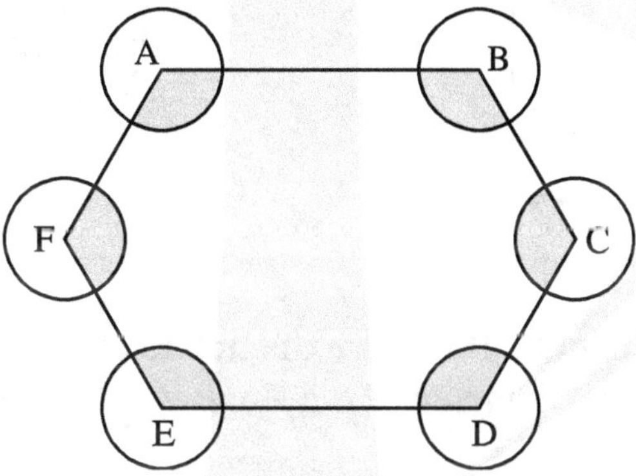

Sol.

Q. 66. ABCD is a diameter of a circle of radius 6 cm. The lengths AB, BC and CD are equal. Semicircles are drawn on AB and BD as diameter as shown in the fig. Find the perimeter and area of the shaded region.

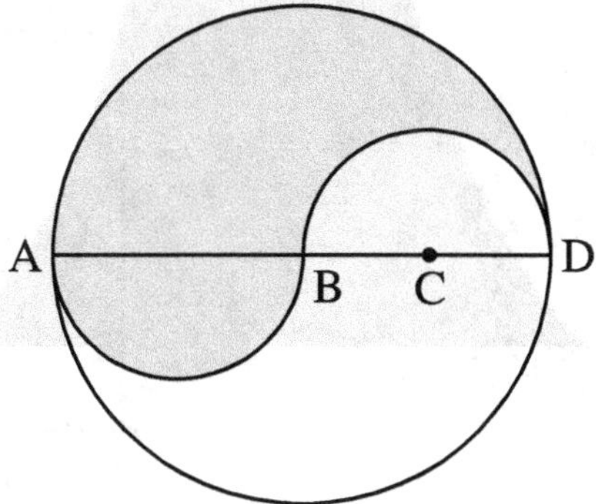

Sol.

Q. 67. In Fig, two circular flower beds have been shown on two sides of a square lawn ABCD of side 56 m. If the centre of each circular flower bed is the point of intersection O of the diagonals of the square lawn, find the sum of the areas of the lawn and the flower beds.

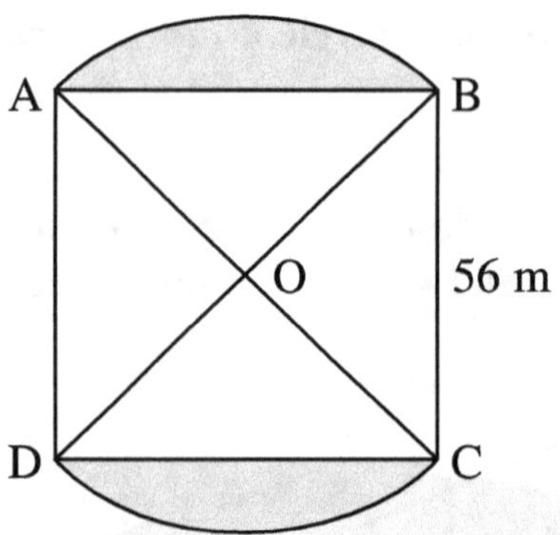

Sol.

Q. 68. Find the area of the shaded design in the below Fig., where ABCD is a square of side 10 cm and semicircles are drawn with each side of the square as diameter. (Use $\pi = 3.14$)

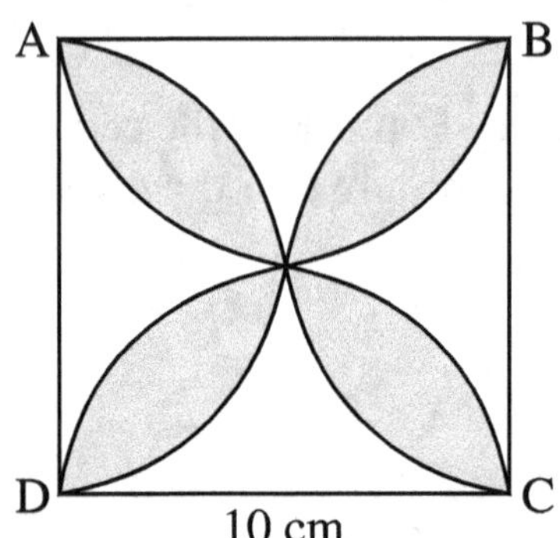

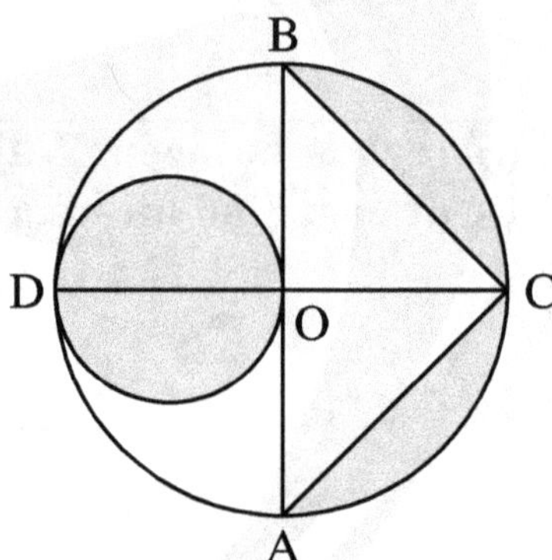

Sol.

Q. 69. In the above sided Fig., AB and CD are two diameters of a circle (with centre O) perpendicular to each other and OD is the diameter of the smaller circle. If OA = 7 cm, find the area of the shaded region.

Sol.

Q. 70. In a circular table cover of radius 32 cm, a design is formed leaving an equilateral triangle ABC in the middle as shown in Fig. Find the area of the design (shaded region).

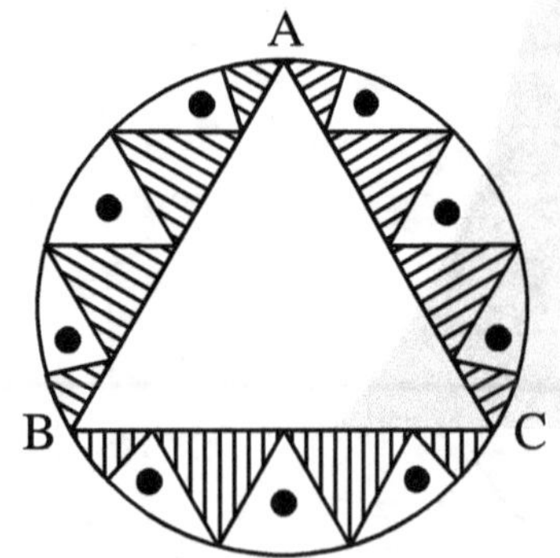

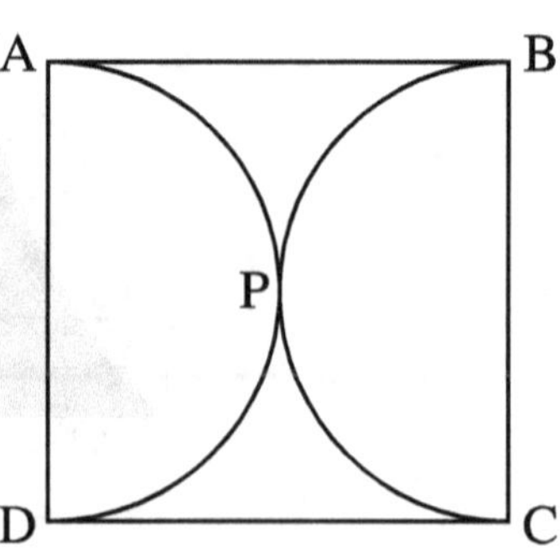

Sol.

Q. 71. In the given figure, $\triangle$ABC is right angled at A. Semicircles are drawn on AB, AC and BC as diameters. It is given that AB = 3 cm and AC = 4 cm. Find the area of the shaded region.

Innovative Mathematics X-11

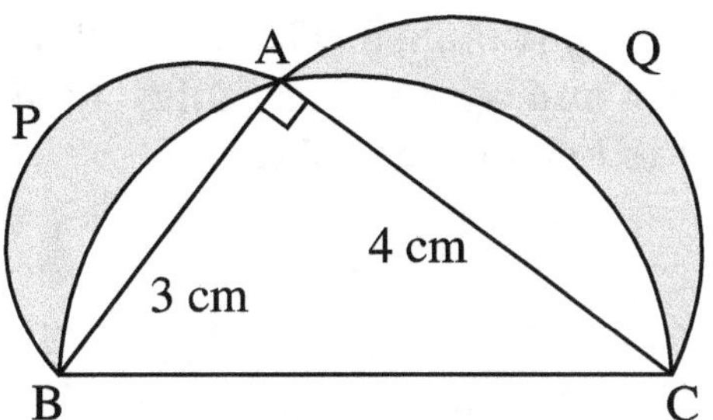

Sol.

Q. 72. Find the area of the shaded region in the below figure, if PQ = 24 cm, PR = 7 cm and O is the centre of the circle.

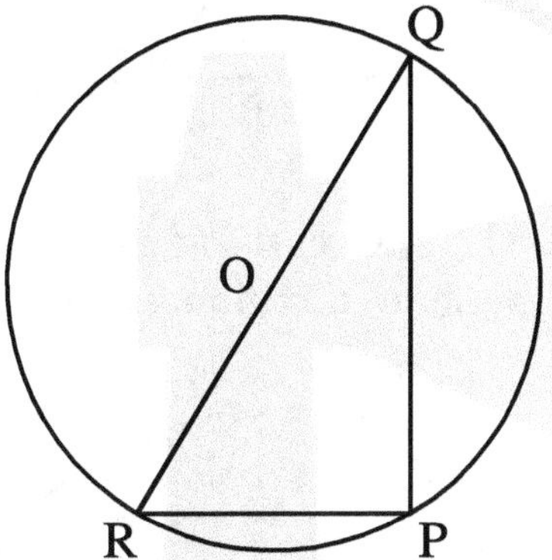

Sol.

Q. 73. Find the areas of the shaded region in the above right sided figure.

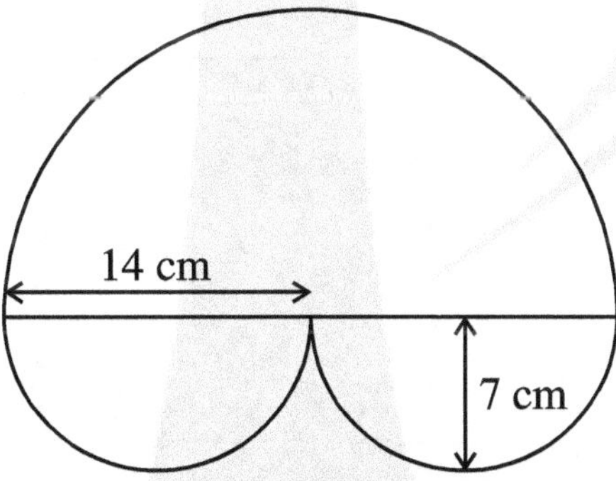

Sol.

Q. 74. Find the area of the shaded region given in below Figure

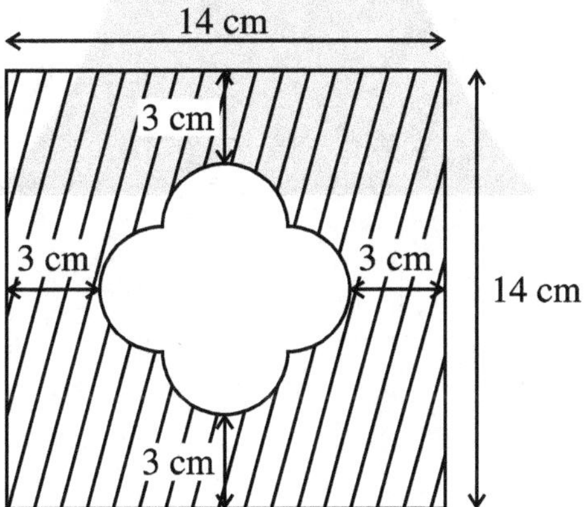

Sol.

Q. 75. In the below figure, ABC is a right angled triangle at B, AB = 28 cm and BC = 21 cm. With diameter a semicircle is drawn and with BC as radius a quarter circle is drawn. Find the area of the shaded region correct to two decimal places.

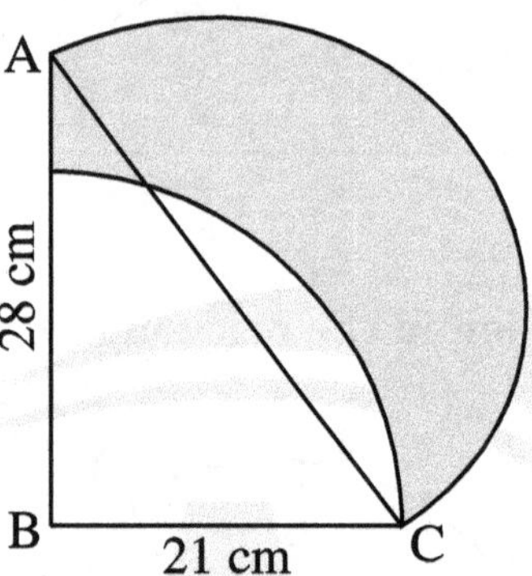

Sol.

Q. 76. In the above right-sided figure, O is the centre of a circular arc and AOB is a straight line. Find the perimeter and the area of the shaded region. (Use $\pi = 3.142$)

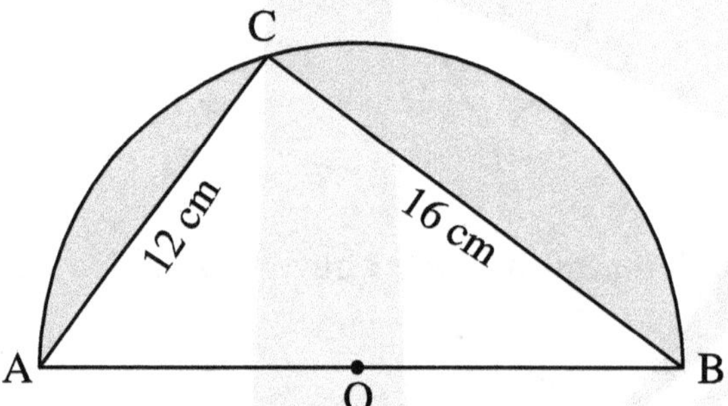

Sol.

Innovative Mathematics X-11

1. 36 cm

2. 4 units

3. $\pi r^2 = \pi\left(\dfrac{a^2}{4}\right)$

4. $\dfrac{lr}{2}$ unit^2

5. 7000

6. $d = 88$ cm, $r = 44$ cm

7. $r = 3$ cm, $\pi r^2 = a\pi$ cm^2

8. $R = 25$ and $d = 50$ cm

9. Side $= 55$ cm, Area $= 3025$ cm^2

10. $\theta = 90°$

11. $\dfrac{\theta}{300} \times \pi r^2$

12. $4 : 9$

13. 44 cm

14. 96%

15. 35 minutes

16. 280 cm^2

17. 124.74 cm^2

18. 10.27 cm^2

19. $A = \dfrac{1}{2}lr$

20. $2\pi r$

21. πs^2

22. $\dfrac{\theta}{360} \times 2\pi r$

23. $\dfrac{\theta}{300} \times \pi r^2$

24. Area of corresponding triangle

25. $r = \dfrac{7}{2}$, Area $= 9.625$ cm^2

26. $\theta = 90°$

27. $11 : 7$

28. 154 cm^2

29. $14 : 11$

30. $\sqrt{3} : \neq$

31. $\theta = 100°$

32. 76 cm

33. 32 cm

34. 92 cm

35. 13.2 cm, 41.58 cm^2

36. $r - 4\sqrt{2}$ cm, Area $= 32\pi$ cm^2

37. $r = 4$ cm, Area $= 16\pi$ cm^2

38. $l = 3\pi$ cm.

39. Area $= 45\dfrac{5}{6}$ cm^2, ($\theta = 210°$ in 35 minutes)

40. 14 cm^2

41. 21.67 cm

42. ₹ 377051.2

43. 15.84 km/h

44. 46.1 cm^2

45. r^2 sq.unit

46. 43.06 m^2

47. $r = 9$ cm

48. 1100 cm

49. ₹ 3140

50. $\left(\dfrac{8\pi}{3} - 4\sqrt{3}\right)$ cm^2

51. $l = \dfrac{\theta}{360} \times 2\pi r$

52. $r = 14$ cm

53. 3.92 cm

54. $r = 28$ cm

55. $r_2 = 11$ cm, $r_1 = 3$ cm

56. 65 revolutions

57. $22 : 83$

58. $R = 10$ cm, $r = 4$ cm

59. 462 cm^2

60.

61.

62.

63.

64.

65.

66.

67.

68.

69.

70.

71.

72.

73.

74.

75.

76.

Innovative Mathematics X-11

PRACTICE-TEST

SECTION-A

Q. 1. If the area of sector is $\dfrac{7}{18}$ of the area of the circle. Find the measure of central angle of the sector. (1)

Sol.

Q. 2. The diameter of a circle whose area is equal to the sum of the areas of the two circles of radii 24 cm and 7 cm is: (1)

    (a) 48 cm   (b) 31cm   (c) 25cm   (d) 17cm

Sol.

Q. 3. The area of sector whose perimeter is four times its radius of measure 7 units is __________ . (1)

Sol.

Q. 4. If the area of a sector of a circle bounded by an arc of length 5π cm is equal to 20π cm^2, then find the radius of the circle. (1)

Sol.

SECTION-B

Q. 5. The perimeter of a sector of circle of radius 5.7 cm is 27.2 cm. Find the area of the sector. (2)

Sol.

Q. 6. The minute hand of a clock is 12 cm long, Find the area of the face of the clock described by the minute hand between 6:10 pm and 6:45 pm. (2)

Sol.

Q. 7. Two circular pieces of equal radii and maximum area, touching each other are cut out from a rectangular cardboard of dimensions 16 cm × 8 cm. Find the area of the remaining cardboard. (2)

Sol.

SECTION-C

Q. 8. The length of a rope by which a cow is tied is increasd from 12m to 19m. How much more area can the cow graze now? (Use $\pi = 22/7$) (3)

Sol.

Q. 9. A chord of a circle of radius 14 cm subtends an angle of 60° at the centre. Find the area of the corresponding minor segment. (Use $\pi = 22/7$) (3)

Sol.

SECTION-D

Q. 10. Find the area of minor and major segments of a circle of radius 42 cm, if the length of the arc is 88 cm. (4)

Sol.

Innovative Mathematics X-11

CASE STUDY

CASE STUDY 3.

Pookalam is the flower bed or flower pattern designed during Onam in Kerala. It is similar as Rangoli in North India and Kolam in Tamil Nadu.

During the festival of Onam, your school is planning to conduct a Pookalam competition. Your friend who is a partner in competition, suggests two designs given below.

Observe these carefully.

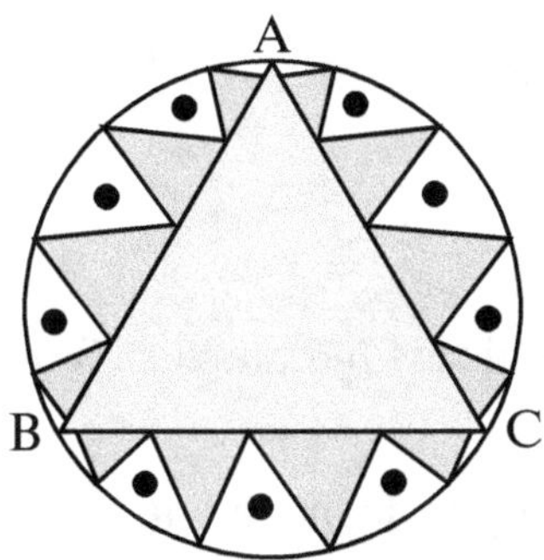
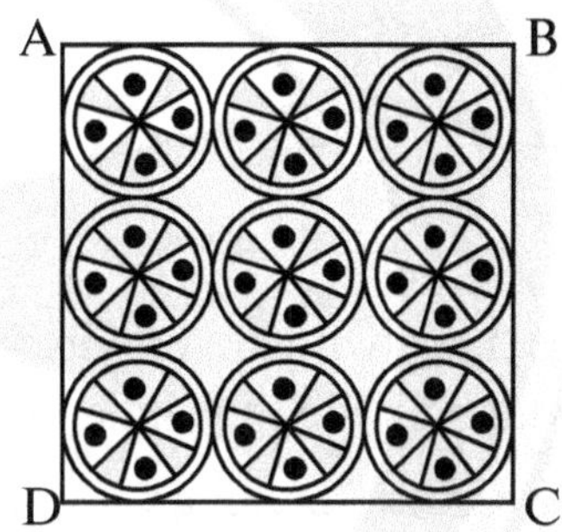

Design I: This design is made with a circle of radius 32 cm leaving equilateral triangle ABC in the middle as shown in the given figure.

Design II: This Pookalam is made with 9 circular design each of radius 7 cm.

Refer Design I:

Q.. 1.   The side of equilateral triangle is

    (a) $12\sqrt{3}$ cm   (b) $32\sqrt{3}$ cm   (c) 48 cm   (d) 64 cm

Q.. 2.   The altitude of the equilateral triangle is

    (a) 8 cm   (b) 12 cm   (c) 48 cm   (d) 52 cm

Refer Design II:

Q.. 3.   The area of square is

    (a) 1264 cm²   (b) 1764 cm²   (c) 1830 cm²   (d) 1944 cm²

Q.. 4.   Area of each circular design is

    (a) 124 cm²   (b) 132 cm²   (c) 144 cm²   (d) 154 cm²

Q.. 5.   Area of the remaining portion of the square ABCD is

    (a) 378 cm²   (b) 260 cm²   (c) 340 cm²   (d) 278 cm²

ANSWERS

1. (b) $32\sqrt{3}$ cm   **2.** (c) 48 cm   **3.** (b) 1764 cm²

4. (d) 154 cm²   **5.** (a) 378 cm²

A Brooch

CASE STUDY 4.

A brooch is a small piece of jewellery which has a pin at the back so it can be fastened on a dress, blouse or coat. Designs of some brooch are shown below. Observe them carefully.

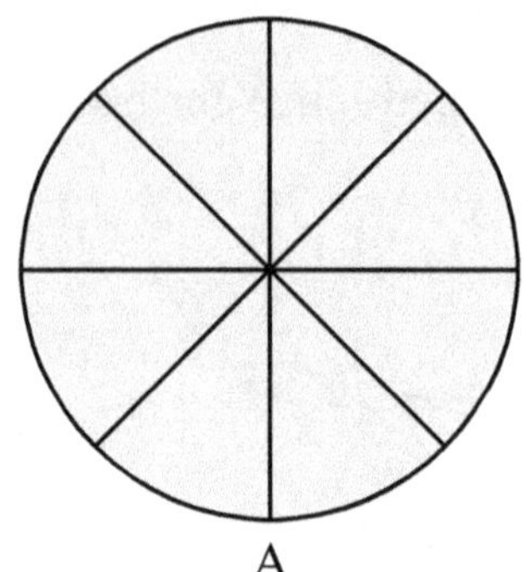

A

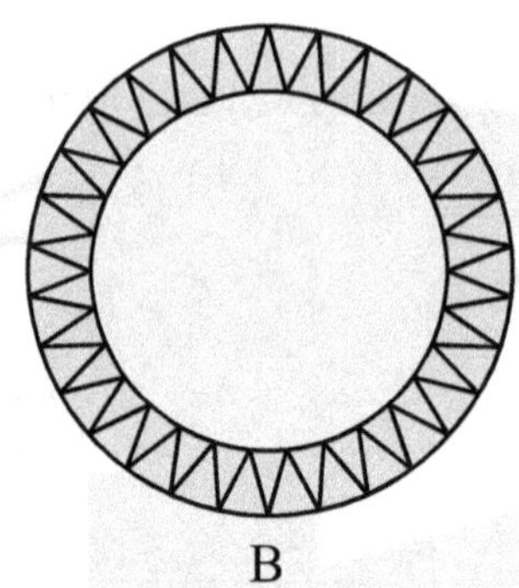

B

Design A: Brooch A is made with silver wire in the form of a circle with diameter 28 mm. The wire used for making 4 diameters which divide the circle into 8 equal parts.

Design B: Brooch b is made two colours_ Gold and silver. Outer part is made with Gold. The circumference of silver part is 44 mm and the gold part is 3 mm wide everywhere.

Refer to Design A

Q.. 1. **The total length of silver wire required is**

    (a) 180 mm   (b) 200 mm   (c) 250 mm   (d) 280 mm

Q.. 2. **The area of each sector of the brooch is**

    (a) 44 mm^2 (b) 52 mm^2 (c) 77 mm^2 (d) 68 mm^2

Refer to Design B

Q.. 3. **The circumference of outer part (golden) is**

    (a) 48.49 mm   (b) 82.2 mm   (c) 72.50 mm   (d) 62.86 mm

Q.. 4. **The difference of areas of golden and silver parts is**

    (a) 18   (b) 44   (c) 51   (d) 64

Q.. 5. **A boy is playing with brooch B. He makes revolution with it along its edge. How many complete revolutions must it take to cover 80 mm ?**

    (a) 2   (b) 3   (c) 4   (d) 5

ANSWERS

1. (b) 200 mm   **2.** (c) 77m m^2 **3.** (d) 62.86 mm

4. (c) 51   **5.** (c) 4

CHAPTER-12

SURFACE AREAS OF VOLUMES

LEARNING OBJECTIVES

FUNDAMENTALS

In your previous classes you have been introduced to different types of solids, such as cuboid, cube, right circular cylinder, right circular cone, sphere, etc. Also, you have learnt about the surface areas and volumes of such solids.

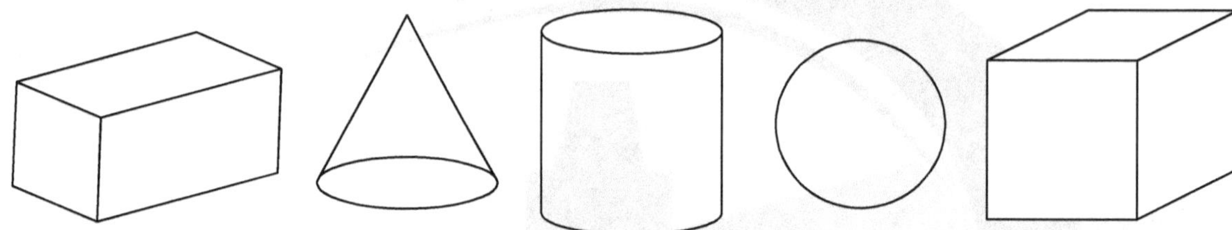

Here, in this chapter, we shall use our previous knowledge about these solids and learn further about solving problems:

(i) On finding volumes and surface areas of combinations of right circular cone, right circular cylinder, hemisphere, sphere, cube and cuboid.

(ii) On finding volume and surface area of frustum of a cone, and

(iii) Those involving converting one type of metallic solid into another.

Before we go ahead, let us recall our knowledge about shapes of such solids and the various formulae related to their surface areas and volumes.

CUBOID

A parallelopiped whose faces are rectangles and adjacent faces are perpendicular is called a Cubid or a rectangular parallelopiped.

If l, b and h denote resp~ctively the length, breadth and height of a cuboid, then :

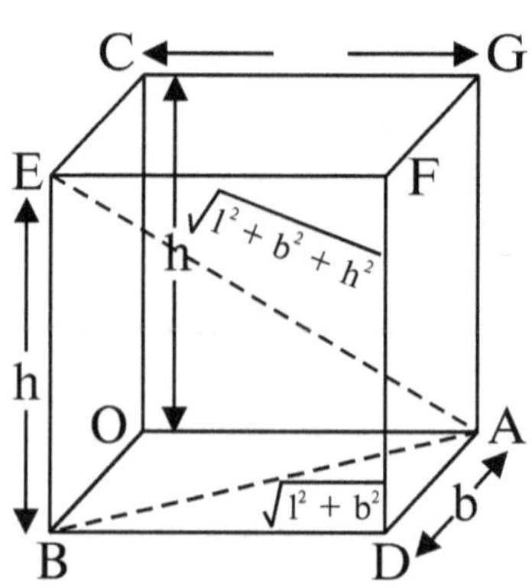

$$\text{Volume of cuboid} = l \times b \times h$$

$$\text{Total surface area of cuboid} = 2(lb + bh + hl)$$

Surface area of 4 walls of a room

$$= 2(bh + hl) = 2(l + b) \times h$$

Surface area of cuboid, in which top face is open

$$= lb + 2(bh + hl)$$

$$\text{Diagonals of faces of cuboid} = \sqrt{l^2 + b^2}, \sqrt{b^2 + h^2}, \sqrt{h^2 + l^2}$$

$$\text{Diagonal of cuboid} = \sqrt{l^2 + b^2 + h^2}$$

$$\text{Height of cuboid} = \frac{\text{Volume}}{\text{Base area}}$$

Area of base $= \dfrac{\text{Volume}}{\text{Height}}$

Remark: If l, b, h are extern.al dimensions of a closed cuboid of thickness 'a' then internal dimensions are:

$$l - 2a, \ b - 2a, \ h - 2a$$

If the cuboid is an open cuboid then Internal Dimensions are $l - 2a, \ b - 2a, \ h - a$.

CUBE

When all the edges of a cuboid are equal in length, it is called a Cube.

If the length of each edge of a cube is a, then:

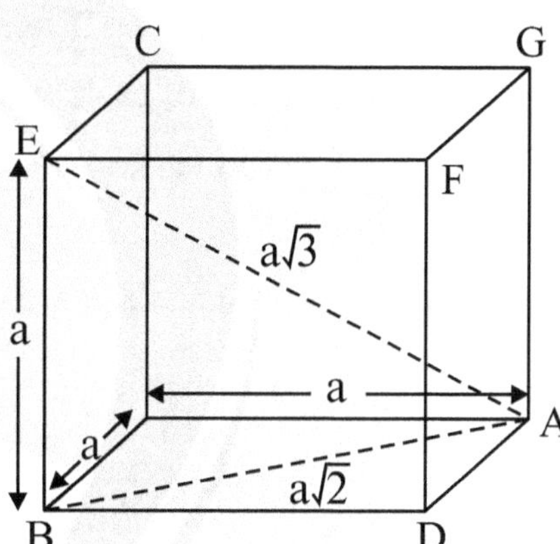

Volume of cube	$= (\text{edge})^3 = a^3$
Total surface area of cube	$= 2\,(aa + aa + aa)$
	$= 2\,(a^2 + a^2 + a^2)$
	$= 2\,(3a^2) = 6a^2 = 6\,(\text{edge})^2$
Digonal of face of the cube	$= \sqrt{2} \times a$
Diagonal of cube	$= \sqrt{3} \times a$
Edge of cube	$= (\text{Volume})^{1/3}$

RIGHT CIRCULAR CONE

If a right-angled triangle is revolved about one of the sides containing a right angle, the solid thus formed is called a Right Circular Cone.

In Fig. 13.4, V is the vertex of the right circular cone. Its base is a circle with centre O and radius OA. The length Q. is called the vertical height or height of the cone, and VA is known as the slant height of the cone.

If radius of base is r, height h and slant height l, then:

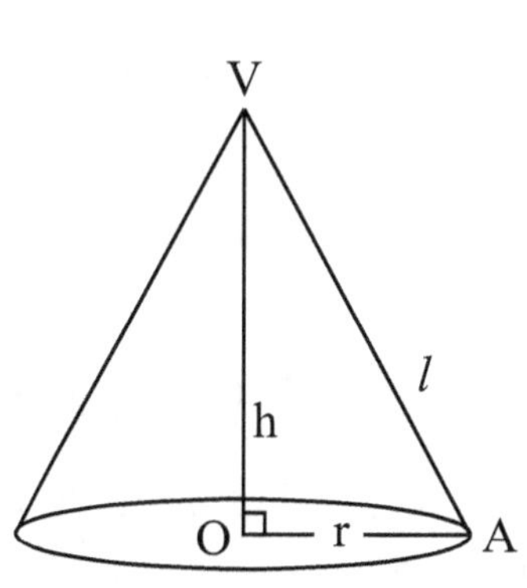

$$\text{Volume} = \frac{1}{3}\,(\text{Area of base}) \times \text{height}$$

$$= \frac{1}{3}(\pi r^2) \times h = \frac{1}{3}\pi r^2 h$$

Curved Surface Area $= \pi r l = \pi r \sqrt{r^2 + h^2}$

Total Surface Area $=$ Area of circular base + Curved surface

$$= \pi r^2 + prl$$

$$= \pi r\,(r + l).$$

RIGHT CIRCULAR CYLINDER

If a rectangle is revolved about one of its sides, the solid thus formed is called Right Circular Cylinder.

In the Fig. 13.5, the line AB is called the axis of the cylinder. BC is the radius and the distance between the ends i.e., the length AB is known as the height or t the length of the cylinder. I h

If radius of the base is r, and height h then :

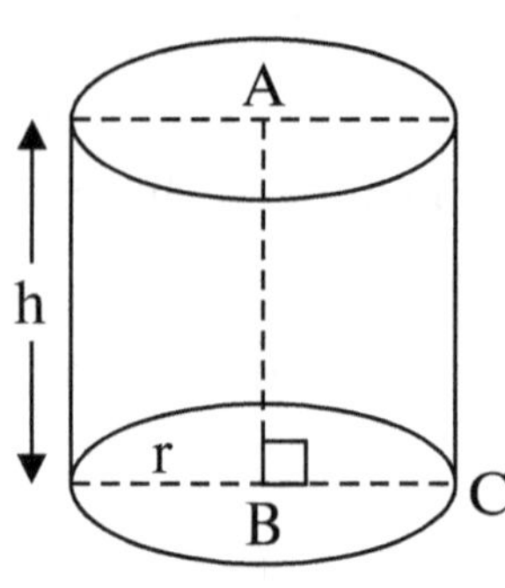

$$\text{Volume} = (\text{Area of base}) \times \text{height t}$$
$$= (\pi r^2) \times h$$
$$\text{Curved surface} = (\text{Perimeter of base}) \times \text{height}$$
$$= (2\pi r) \times h = 2\pi rh$$
$$\text{Total surface area} = \text{Area of circular ends} + \text{Curved Surface Area}$$
$$= 2\pi r^2 + 2\pi rh$$
$$= 2\pi r\ (r + h)$$

In case of Hollow Cylinders : If h is the height and external and internal radii are respectively R and r, then :

$$\text{Area of each end} = \pi(R^2 - r^2)$$
$$\text{Curved surface area} = \text{External surface} + \text{Internal surface}$$
$$= 2\pi Rh + 2\pi rh = 2\pi h\ (R + r)$$
$$\text{Total surface area} = \text{Curved surface area} + 2\ (\text{Area of base ring})$$
$$= 2\pi Rh + 2\pi rh + 2\ (\pi R^2 - \pi r^2)$$
$$= 2\pi\ (R + r)\ (h + R - r)$$
$$\text{Volume of the material} = \text{Exterior volume} - \text{Interior volume}$$
$$= \pi R^2 h - \pi r^2 h = \pi h\ (R^2 - r^2)$$

SPHERE

A sphere is the set of all points in space which are equidistant from a fixed point. The fixed point is called the centre of the sphere and the constant distance is called its radius.

Alternatively,

If a circle is revolved about a diameter, the solid thus generated is called a sphere.

If r is the radius of a sphere, then :

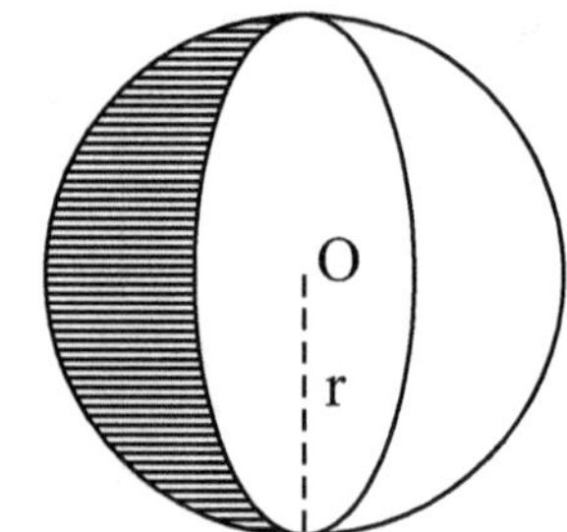

$$\text{Volume} = \frac{4}{3}\ \pi r^3$$

$$\text{Surface area} = 4\pi r^2$$

IN CASE OF HEMISPHERE

(A plane passing through the centre of a sphere divides the sphere into two equal parts. Each part is called a hemi-sphere)

If r is the radius of a hemi-sphere, then :

$$\text{Curved surface area} = 2\pi r^2$$

$$\text{Volume} = \frac{2}{3}\ \pi r^3$$

Total surface area of surface area of a solid hemi-sphere

$$= 2\pi r^2 + \pi r^2 = 3\pi r^2$$

IN CASE OF SPHERICAL SHELL

If R and r, are ·respectively the outer and inner radii of a spherical shell, then:

$$\text{Outer surface area} = 4\pi R^2$$

$$\text{Volume of material} = \frac{4}{3}\pi(R^3 - r^3)$$

SURFACE AREAS AND VOLUMES OF COMBINATION OF SOLIDS

In our daily life we come across various solids which are combinations of different basic solids (right circular cone, right circular cylinder, sphere, hemisphere, etc.). For a better understanding, a few examples are given below:

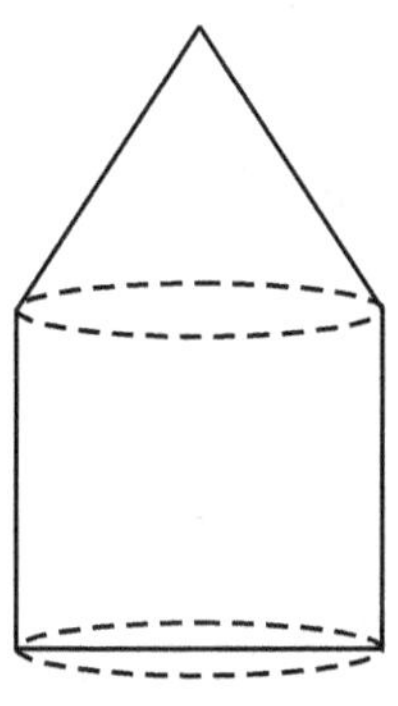
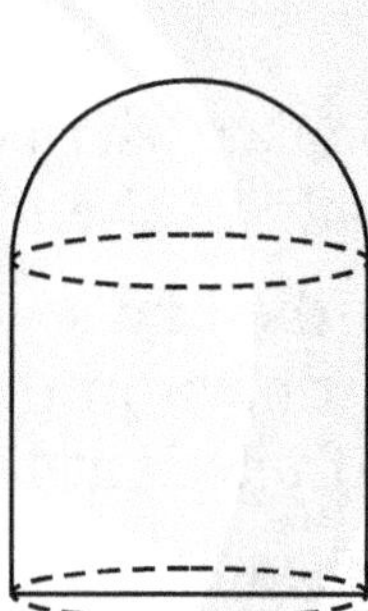

(i)   A circus tent consisting of a cylindrical base surmounted by a conical roof (Fig. 13.7).

(ii) A circus tent consisting of a cylindrical base surmounted by a hemispherical roof (Fig. 13.8).

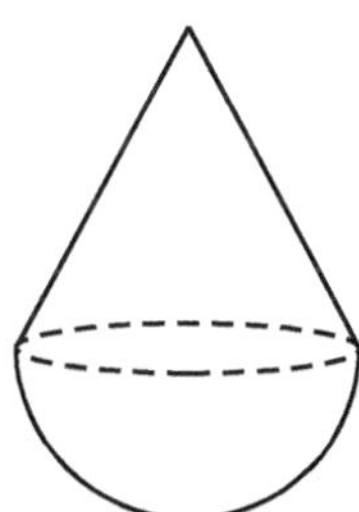
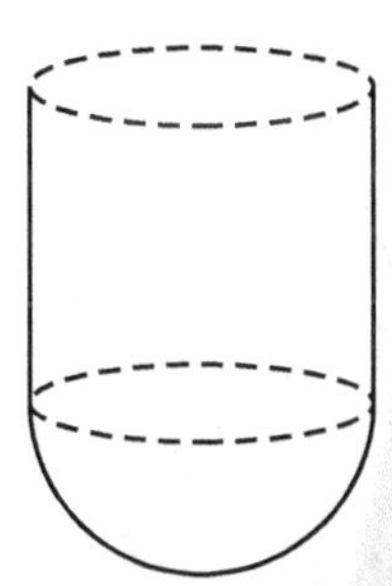
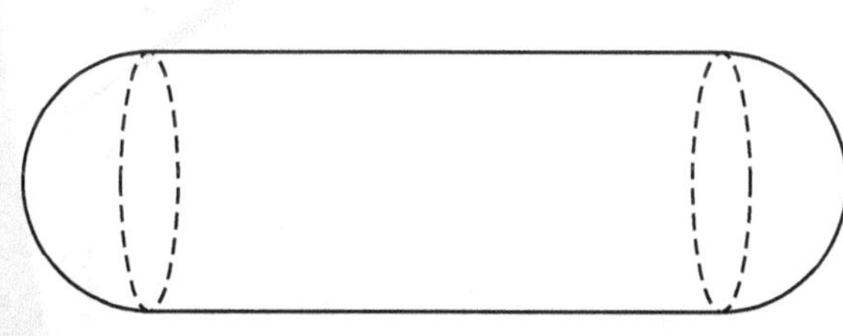

(v)   A capsule in the form of a cylinder with hemispherical ends (Fig. 13.11).

In order to determine the surface areas/volumes of such solids we calculate the surface areas/ volumes of the two or more parts separately and find the sum. Illustrations given in sect'i on 2 of the chapter, make the method further clear.

CONVERSION OF SOLIDS

At times we may convert a solid into another solid with different shape. For a better und er. standing, a few examples are given below :

(i)   A lead ball may be melted and converted into a number of smaller balls.

(ii) A silver sphere may be melted and recast into a wire (for making ornaments, etc.)

(iii) Making of metallic coins from a piece of metal in a different shape.

For solving problems involving conversion of one type of solid into another, we must keep in mind that with change of shape, only the surface area changes and no change of volume takes place. Hence :

Volume of the earlier solid = Volume of the new solid

Here also, illustrations given in Section 2, make the method further clear.

FRUSTUM OF A RIGHT CIRCULAR CONE

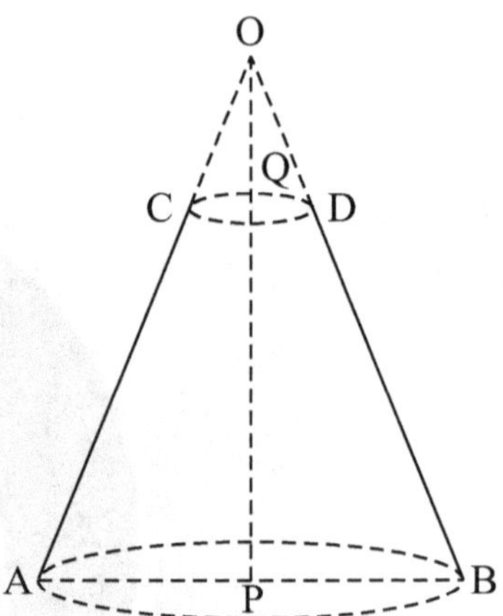

If a right circular cone is cut by a plane parallel to its base, and the portion containing the vertex is removed, the remaining part is called frustum of the cone. The shape of this part (frustum of the cone) is like that of a bucket or a glass-tumbler. Note that :

(i) A frustum of a right circular cone has two unequal circular ends.

(ii) The perpendicular dist1mce between the two circular ends is called the height of the frustum.

(iii) The length of the line segment joining the extremities of two parallel radii, drawn in the same direction, of the circular ends, is called the slant height of the frustum.

In Fig. 13.12, ABDC is the frustum cut out of the cone AOB; PQ.is the height of the frustum; and line-segments BD and AC represent the slant height.

VOLUME AND SURFACE AREA OF FRUSTUM

Let h be the height, l the slant height and R and r the radii of the two circular ends of the frustum (as shown in Fig. 13.13).

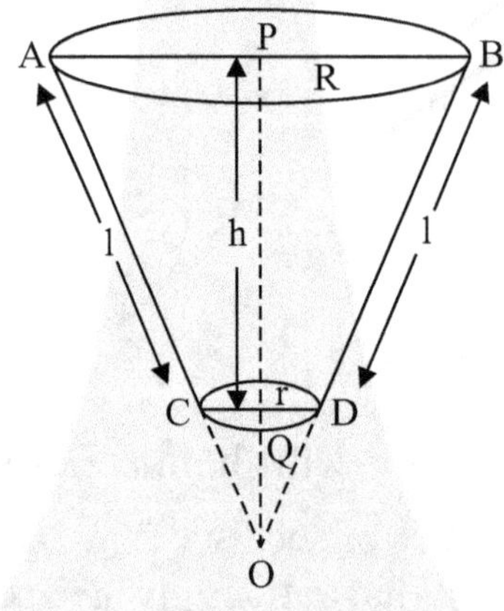

Its:

$$\text{Volume} = \pi h/3 \; (R^2 + Rr + r^2)$$

$$= \frac{h}{3}(A_1 + \sqrt{A_1 A_2} + A_2)$$

[where A_1, A_2 are the surface areas of the two circular ends]

$$\text{Curved Surface Area} = \pi (R + r)l$$

$$\text{Total Surface Area} = \pi (R + r)l + \pi R^2 + \pi r^2$$

$$\text{Slant height,} \quad l = \sqrt{h^2 + (R - r)^2}$$

Innovative Mathematics X-12

WARM UP – LEVEL-I

Q. 1. **Find the volume and whole surface area of a right circular cylinder whose height is 15 cm and radius of the base is 7 cm.**

Ans. Height of cylinder = h = 15 cm

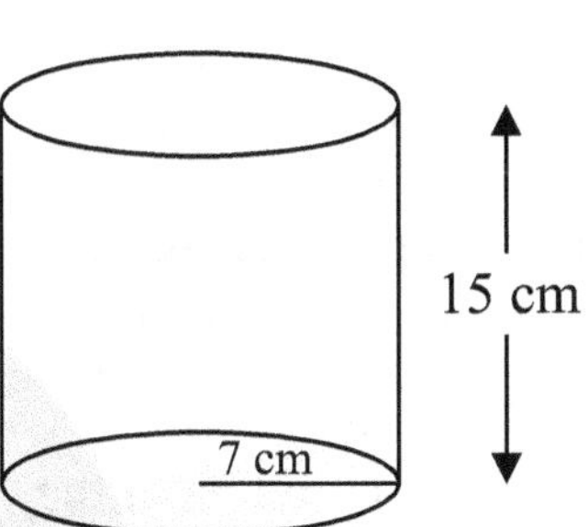

$$\text{Radius of the base} = r = 7 \text{ cm}$$

$$\text{Volume of cylinder} = \pi r^2 h$$

$$= \pi \times (7)^2 \times 15$$

$$= \frac{22}{7} \times 7 \times 7 \times 15$$

$$= 2310 \text{ cm}^3$$

$$\text{Whole surface area} = 2\pi rh + 2\pi r^2$$

$$= 2\left(\frac{22}{7} \times 7 \times 15\right) + \left(2 \times \frac{22}{7} \times 7 \times 7\right)$$

$$= 660 + 308 = 968 \text{ cm}^2$$

Q. 2. **The diameter and height of a conical tent are 24 m and 16 m. Find the canvas required to make it.**

Ans.

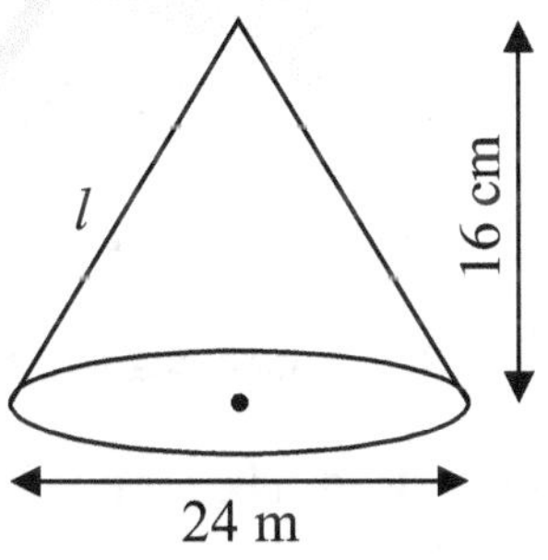

$$r = \text{radius of base} = \frac{24}{2} = 12 \text{ m}$$

$$l = \text{slant height of tent}$$

$$= \sqrt{(12)^2 + (16)^2} = 20 \text{ m}$$

Curved surface area of the tent = πrl

$$= \frac{22}{7} \times 12 \times 20 = 754.28 \text{ m}^2$$

$$\text{Hence, canvas required} = 754.28 \text{ m}^2$$

Q. 3. **The circumference of the base of a 9 m high wooden solid cone is 44 m. Find the volume of the cone.**

Ans. Circumference of the base of cone

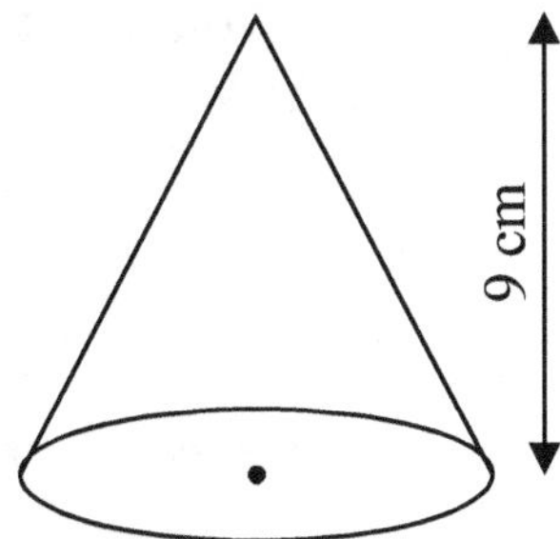

$$= 2\pi r$$

$$\therefore \qquad 2\pi r = 44 \text{ m}$$

$$r = \frac{44}{2\pi} = 44 \times \frac{7}{44} = 7\text{m}$$

$$\text{height (h)} = 9 \text{ m} \quad (\text{given})$$

$$\therefore \qquad \text{Volume of cone} = \frac{1}{3}\pi r^2 h$$

$$= \frac{1}{3} \times \frac{22}{7} \times 7 \times 7 \times 9 = 1386 \text{ m3}$$

Q. 4. **A cylindrical tank has a capacity of 6160 m³. Find its depth, if the diameter of the base is 28 m.**

Ans. Let height of the cylinder = h

$$\text{radius, } r = \frac{28}{2} = 14 \text{ m}$$

$$\text{Volume of the cylinder} = \pi r^2 h = 6160 \text{ m}^3$$

$$\therefore \quad \frac{22}{7} \times 14 \times 14 \times h = 6160$$

$$\therefore \quad h = \frac{6160 \times 7}{22 \times 14 \times 14} = 10 \text{ m}$$

Q. 5. **The volume of a cylinder is 69300 cm³ and its height is 50 cm. Find the curved surface area of the cylinder.**

Ans.
$$h = 50 \text{ cm}$$

$$\text{Let radius of the cylinder} = r \text{ cm}$$

$$\therefore \quad \text{Volume} = 69300 \text{ cm}^3$$

$$\Rightarrow \quad \pi r^2 h = 69300$$

$$\therefore \quad \frac{22}{7} \times r^2 \times 50 = 69300$$

$$r^2 = \frac{69300 \times 7}{22 \times 50} = 441 \text{ cm}^2$$

$$\therefore \quad r = 21 \text{ cm}$$

$\therefore$ Curved surface area of the cylinder

$$= 2\pi rh = 2 \times \frac{22}{7} \times 21 \times 50 = 6600 \text{ cm}^2$$

Q. 6. **A spherical tank is 21 cm in diameter. Find its surface area and volume.**

Ans. Radius of the sphere $= \dfrac{21}{2}$ cm

$$\text{Surface area of the sphere} = 4\pi r^2 = \frac{22}{7} \times \left(\frac{21}{2}\right) \times \left(\frac{21}{2}\right) = 1386 \text{ cm}^2$$

$$\text{Volume of the sphere} = \frac{4}{3}\pi r^3$$

$$= \frac{4}{3} \times \frac{22}{7} \times \left(\frac{21}{2}\right)^3 = 4851 \text{ cm}^3$$

Q. 7. **Find the surface area of a sphere whose volume is 4851 m³.**

Ans. For a sphere of radius r,

$$\text{Volume} = \frac{4}{3}\pi r^3$$

$$\frac{4}{3}\pi r^3 = 4851$$

$$r^3 = 4851 \times \frac{3}{4} \times \frac{7}{22} = \frac{441 \times 21}{8} = \left(\frac{21}{2}\right)^3$$

$$r = \frac{21}{2} \text{ m}$$

$$\text{Surface area of the sphere} = 4\pi r^2 = 4 \times \frac{22}{7} \times \frac{21}{2} \times \frac{21}{2} = 1386 \text{ m}^2$$

Q. 8. **The radius of the base and the height of a right circular cone are 7 cm and 24 cm respectively. Find the volume and total surface area of the cone.**

Ans. Given that,

$$\text{radius (r)} = 7 \text{ cm}$$

$$\text{height (h)} = 24 \text{ cm}$$

$$\therefore \quad \text{Slant height (l)} = \sqrt{r^2 + h^2} = \sqrt{49 + 576} = 25 \text{ cm}$$

$$\text{Volume} = \frac{1}{3} \pi r^2 h$$

$$= \frac{1}{3} \times \frac{22}{7} \times 7 \times 7 \times 24 = 1232 \text{ cm}^3$$

$$\text{Total surface area} = \pi r l + \pi r^2$$

$$= \pi r (l + r)$$

$$= \frac{22}{7} \times 7 (25 + 7)$$

$$= 22 \times 32 = 704 \text{ cm}^2$$

Q. 9. **The curved surface area of a right circular cone is 12320 cm². If the radius of its base is 56 cm, find its height.**

Ans. Let h be the height of the cone and l, its slant height.

$$\text{radius (r)} = 56 \text{ cm}$$

$$\text{Curved surface area} = \pi r l$$

$$\therefore \quad 12320 = \frac{22}{7} \times 56 \times l$$

$$\text{or} \quad l = \frac{12320}{22 \times 8} = 70 \text{ cm}$$

Also,

$$l^2 = r^2 + h^2$$

$$\therefore \quad 70 \times 70 = 56 \times 56 + h^2$$

$$\text{or} \quad h^2 = 4900 - 3136 = 1764$$

$$\therefore \quad h = 42 \text{ cm}$$

Q. 10. **If the surface area of a sphere is 616 cm², find its volume.**

Ans. Let r be the radius of the sphere

$$\text{Surface area} = 4\pi r^2$$

$$\Rightarrow \quad 616 = 4 \times \frac{22}{7} \times r^2$$

$$\therefore \qquad r^2 = \frac{616 \times 7}{88} = 49$$

or $\qquad\qquad r = 7 \text{ cm}$

$$\text{Volume} = \frac{4}{3}\pi r^3 = \frac{4}{3} \times \frac{22}{7} \times 7 \times 7 \times 7 = 1437.33 \text{ cm}^3$$

Q. 11. The internal and external diameters of a hollow hemispherical vessel are 42 cm and 45.5 cm, respectively. Find its capacity and also its outer curved surface area.

Ans. External radius (R) = $\dfrac{45.5}{2}$ cm

Internal radius (r) = $\dfrac{42}{2} = 21$ cm

$$\text{Capacity} = \frac{2}{3}\pi(R^3 - r^3)$$

$$= \frac{2}{3} \times \frac{22}{7} \times \left[\left(\frac{45.5}{2}\right)^3 - (21)^3\right]$$

$$= \frac{44}{21}\left[\left(\frac{91}{4}\right)^3 - (21)^3\right]$$

$$= 5266.48 \text{ cm}^3$$

$$\text{Outer curved surface area} = 2\pi R^2$$

$$= 2 \times \frac{22}{7} \times \left(\frac{45.5}{2}\right)^2$$

$$= 3253.25 \text{ cm}^2$$

Q. 12. The radius of a solid hemispherical toy is 3.5 cm. Find its total surface area.

Ans. $\qquad\qquad$ radius (r) = 3.5 cm

$$\text{Total surface area} = 2\pi r^2 + \pi r^2 = 3\pi r^2$$

$$= 3 \times \frac{22}{7} \times (3.5)^2$$

$$= 115.5 \text{ cm}^3$$

Q. 1. **2 cubes each of volume 64 cm³ are joined end to end. Find the surface area of the resulting cuboid.**

Ans. Volume of each of the cubes = a³ = 64 cm³

$$\therefore \qquad a = 4 \text{ cm}$$

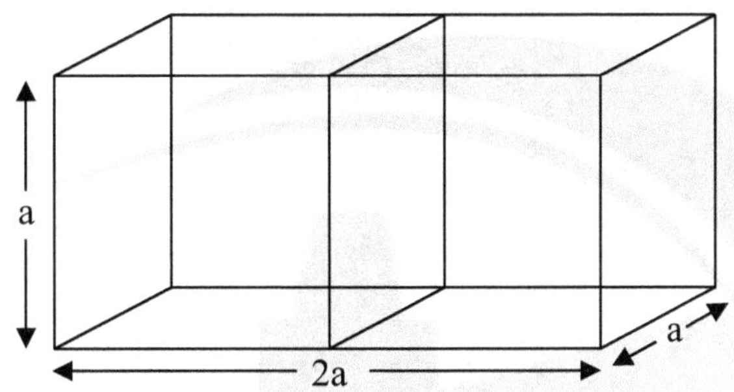

Length of the resulting cuboid = l = a + a = 4 + 4 = 8 cm

Breadth of the resulting cuboid = b = a = 4 cm

Height of the resulting cuboid = h = a = 4 cm

Surface area of the resulting cuboid

$$= 2(lb + bh + hl)$$
$$= 2(8 \times 4 + 4 \times 4 \times 4 \times 8)$$
$$= 2(32 + 16 + 32)$$
$$= 160 \text{ cm}^2$$

Q. 2. **A pen stand made of wood is in the shape of a cuboid with four conical depressions to hold pens. The dimensions of the cuboid are 15 cm by 10 cm by 3.5 cm. The radius of each of the depressions is 0.5 cm and the depth is 1.4 cm. Find the volume of wood in the entire stand (see Fig. 13.18).**

Ans. Volume of cuboid = lbh

$$= 15 \times 10 \times 3.5$$
$$= 525 \text{ cm}^3$$

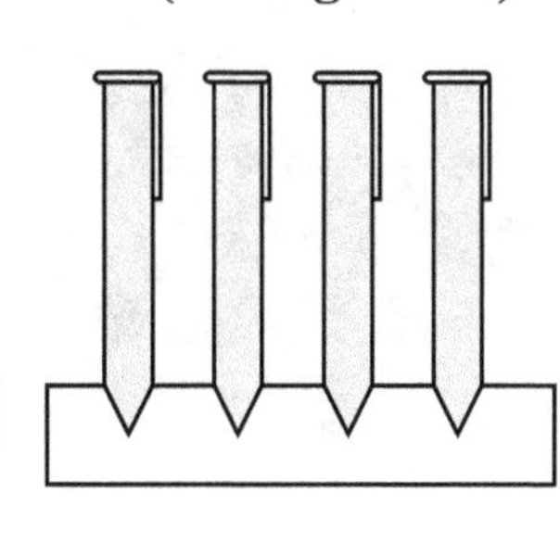

Radius of the conical depression = 0.5 cm

Depth of the conical depression = 1.4 cm

$$\therefore \quad \text{Volume of each conical depression} = \frac{1}{3}\pi r^2 h$$

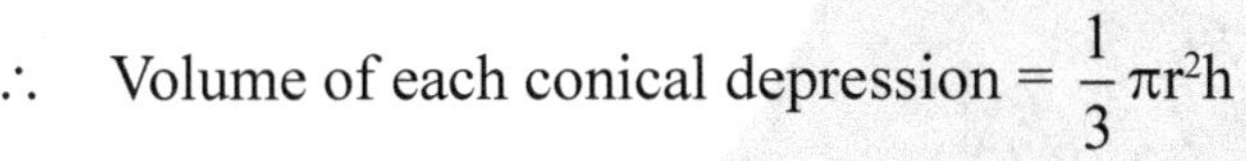

$$= \frac{1}{3} \times \frac{22}{7} \times 0.5 \times 0.5 \times 1.4$$
$$= 0.367 \text{ cm}^3$$

Volume of 4 conical depressions = 4 × 0.367

$$= 1.468 \text{ cm}^3$$

$$\therefore \qquad \text{Volume of wood} = 525 - 1.468$$
$$= 523.532 \text{ cm}^3$$

Q. 3. A cubical block of side 7 cm is surmounted by a hemisphere. What is the greatest diameter the hemisphere can have? Find the surface area of the solid.

Ans. Since the cubical block is surmounted by the hemisphere, and the side of the block is 7 cm, the greatest diameter that the hemisphere can have

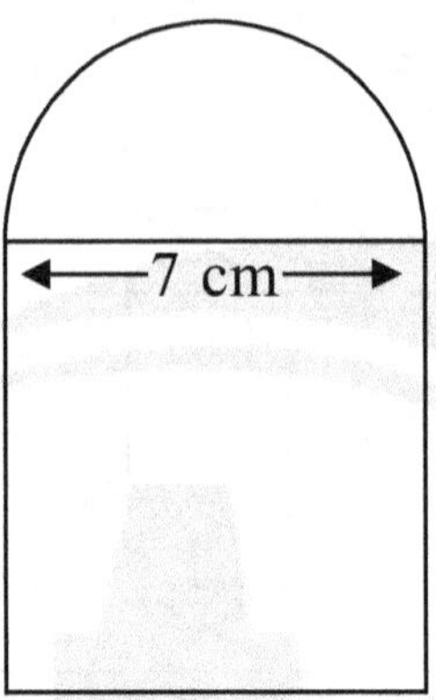

$$\text{Surface area of the solid} = 5a^2 + 2\pi r^2$$

$$\text{Here, } a = 7 \text{ cm and } r = 7/2 = 3.5 \text{ cm}$$

$$\therefore \quad \text{Required surface area} = 5 \times 7^2 + 2 \times \frac{22}{7} \times (3.5)^2$$

$$= 245 + 77$$

$$= 322 \text{ cm}^2$$

Q. 4. A wooden article was made by scooping out a hemisphere from each end of a solid cylinder. If the height of the cylinder is 10 cm, and its base is of radius 3.5 cm, find the total surface area of the article.

Ans. Curved surface area of the cylinder $= 2\pi rh$

$$= 2 \times \frac{22}{7} \times 3.5 \times 10$$

$$= 220 \text{ cm}^2$$

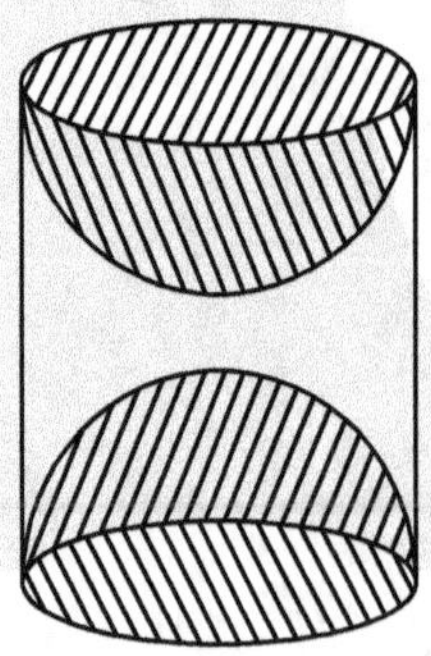

Curved surface area of each hemisphere $= 2\pi r^2$

$$= 2 \times \frac{22}{7} \times 3.5 \times 3.5$$

$$= 77 \text{ cm}^2$$

Total surface area of the wooden article

$$= \text{Curved surface area of the cylinder + curved surface area of 2 equal hemispheres}$$

$$= 220 + 2 \times 77$$

$$= 374 \text{ cm}^2$$

Q. 5. **Determine the ratio of the volume of a cube to that of a sphere which will exactly fit inside the cube.**

Ans. Let s be the edge of the cube. Now, the sphere which will exactly fit inside the cube will have diameter equal to the edge of the cube.

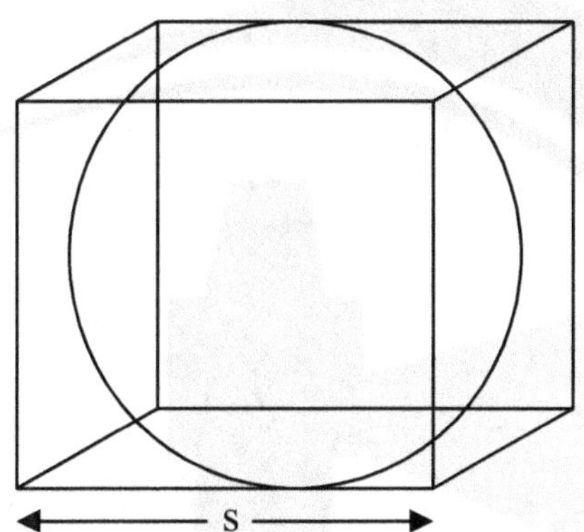

$$\therefore \quad \text{Radius of the sphere} = s/2$$

$$\text{Volume of the cube} = s^3$$

$$\text{and, Volume of the sphere} = \frac{4}{3}\pi(s/2)^3$$

$$\therefore \quad \text{Required ratio} = s^3 : \frac{4}{3}\pi\frac{s^3}{8}$$

$$= \frac{22}{7} \times \frac{441}{4} \times 24 = 8316 \text{ cm}^3$$

Q. 6. **A tent is in the form of a cylinder of diameter 20 m and height 2.5 m, surmounted by a cone of equal base and height 7.5 m. Find the capacity of the tent and the cost of the canvas at Rs 100 per square metre.**

Ans. Capacity of the tank = Volume of the cylinderical part + Volume of the conical part

$$= \pi(10)^2\,(2.5) + \frac{1}{3}\pi(10)^2\,(7.5)$$

$$= 100\pi\,(2.5 + 2.5) = 500\pi \text{ m}^3 = 1571.43 \text{ m}^3$$

Area of the canvas required = Curved surface area of the cylindrical part + Curved surface area of the conical part

$$= 2\pi\,(10)\,(2.5) + \pi(10)\left(\sqrt{10^2 + (7.5)^2}\right)$$

$$= 10\pi\,(5 + 12.5) = 10 \times \frac{22}{7} \times \frac{35}{2}$$

$$= 550 \text{ m}^2$$

Cost of canvas @ Rs 100 per m² = 550 × 100 = Rs. 55000

Q. 7. The interior of a building is in the form of a cylinder of base radius 12 m and height 3.5 m, surmounted by a cone of equal base and slant height 12.5 m. Find the internal curved surface area and the capacity of the building.

Ans. Internal curved surface area of the building

$$= 2\pi (12)(3.5) + \pi(12)(12.5)$$

$$= 12\pi (7 + 12.5)$$

$$= 12 \times \frac{22}{7} \times \frac{39}{2} = 735.43 \text{ m}^2$$

Capacity of the building $= [\pi \times 12 \times 12 \times 3.5] + \left[\frac{1}{3} \times \pi \times 12 \times 12 \times \sqrt{(12.5)^2 - (12)^2}\right]$

$$[h = \sqrt{l^2 - r^2}\,]$$

$$= 144\pi\left[3.5 + \frac{1}{3} \times 3.5\right]$$

$$= 144\pi\left(\frac{14}{3}\right) = 144 \times \frac{22}{7} \times \frac{14}{3} = 2112 \text{ m}^3$$

Q. 8. A boiler is in the form of a cylinder 2 m long with hemispherical ends each of 2 m diameter. Find the volume of the boiler.

Ans. Volume of the boiler = Volume of the cylindrical part + Volume of the two hemispherical ends

$$= \pi \times 1 \times 1 \times 2 + 2 \times \frac{2}{3}\,\pi \times 1 \times 1 \times 1$$

$$= 2\pi + \frac{4\pi}{3} = \frac{10\pi}{3} = \frac{10}{3} \times \frac{22}{7} = \frac{220}{21} = 10\frac{10}{21}\text{ m}^3$$

Q. 9. The dimensions of a rectangular field are 15 m × 12 m. A pit 8 m × 2.5 m × 2 m is dug in one corner of the field and the earth removed is evenly spread over the remaining area of the field. How much is the level of the field raised?

Ans. Length of the given field ABCD = 15 m

Breadth of the given field ABCD = 12 m

Area of the given field ABCD

$$= 15 \times 12 = 180 \text{ m}^2$$

Area of the pit CEFG $= 2.5 \times 8 = 20 \text{ m}^2$

Area of the remaining field

$$= 180 - 20 = 160 \text{ m}^2$$

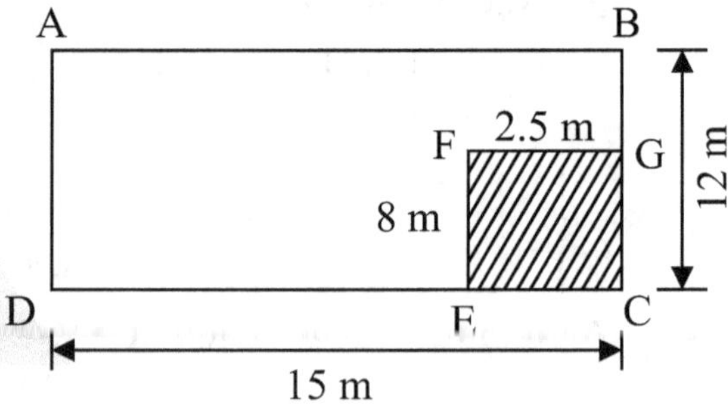

Total volume of earth removed

$$= 8 \times 2.5 \times 2 = 40 \text{ m}^3$$

Volume of earth removed = Volume of earth spread over the field

$$= (\text{area of the remaining field}) \times \text{height}$$

$$\therefore \qquad 40 = 160 \times \text{height}$$

Innovative Mathematics X-12

$$\therefore \qquad \text{height} \; = \; \frac{40}{160} = \frac{1}{4} \, \text{m} = 25 \, \text{cm}$$

Q. 10. **How many balls, each of radius 1 cm, can be made from a solid sphere of lead of radius 8 cm?**

Ans. Volume of solid sphere with radius 8 cm

$$= \frac{4}{3} \times \pi \times 8 \times 8 \times 8 \; \text{cm}^3$$

Volume of the ball with radius 1 cm

$$= \frac{4}{3} \times \pi \times 1 \times 1 \times 1$$

$$= \frac{4\pi}{3} \; \text{cm}^3$$

If x is the number of balls, we get

$$x \times \frac{4\pi}{3} \; = \; \frac{4\pi}{3} \times 8 \times 8 \times 8$$

$$x \; = \; 8 \times 8 \times 8 = 512$$

Q. 11. **A spherical shell of lead, whose external diameter is 18 cm, is melted and recast into a right circular cylinder, whose height is 8 cm and diameter 12 cm. Determine the internal diameter of the shell.**

Ans. External radius of the shell $(R) = \dfrac{18}{2} = 9$ cm

Let internal radius be r

$$\therefore \qquad \text{Volume of lead} \; = \; \frac{4}{3}\pi(R^3 - r^3)$$

$$= \frac{4\pi}{3}(9^3 - r^3)$$

$$= \frac{4\pi}{3}(729 - r^3) \; \text{cm}^3$$

Volume of right circular cylinder

$$= \pi r^2 h = \pi \cdot 6^2 \cdot 8$$

$$= \pi \cdot 36 \cdot 8 = 288\,\pi \; \text{cm}^3$$

Now, volume of lead (in the form of shell) = Volume of right circular cylinder

$$\therefore \qquad \frac{4\pi}{3}(729 - r^3) \; = \; 288\,\pi$$

$$(729 - r^3) \; = \; \frac{288 \times 3}{4} = 216$$

$$r^3 \; = \; 729 - 216 = 513$$

$$= \; 3 \times 3 \times 3 \times 19$$

$$\therefore \qquad r \; = \; 3\,(19)^{1/3}$$

Hence, internal diameter $= 2 \times 3\,(19)^{1/3} = 6\,(19)^{1/3}$ cm

Q. 12. The diameter of a metallic sphere is 6 cm. It is melted and drawn into a wire having diameter of the cross-section as 0.2 cm. Find the length of the wire.

Ans. Radius of the sphere $= \dfrac{6}{2} = 3$ cm

$$\therefore \quad \text{Volume of sphere} = \dfrac{4}{3}\pi \times 3 \times 3 \times 3 \text{ cm}^3 = 36\pi \text{ cm}^3$$

Radius of the cross-section of wire $= 0.1$ cm

Let h be the length of the wire

$$\text{Volume of the wire} = \pi r^2 h$$
$$= \pi \times (0.1)^2 \times h$$
$$= \dfrac{\pi h}{100} \text{ cm}^3$$

$$\text{Volume of the wire} = \text{Volume of the sphere}$$

$$\therefore \quad \dfrac{\pi h}{100} = 36\,\pi$$

$$h = 3600 \text{ cm} = 36 \text{ m}$$

Q. 13. A solid right circular cylinder of radius 8 cm and height 2 cm is melted and cast into a right circular cone of height 3 times that of cylinder. Find the curved surface area of the cone.

Ans. Radius of the cylinder $= 8$ cm

$$\text{Height of cylinder} = 2 \text{ cm}$$
$$\therefore \quad \text{Volume of cylinder} = \pi r^2 h$$
$$= \pi \times 8 \times 8 \times 2$$
$$= \dfrac{22}{7} \times 8 \times 8 \times 2 = \dfrac{2816}{7} \text{ cm}^3$$

$$\text{Height of circular cone} = 3 \times 2 = 6 \text{ cm}$$

$$\text{Volume of cone} = \dfrac{1}{3}\pi r^2 h$$

$$\text{Also, volume of cone} = \text{Volume of cylinder}$$

$$\dfrac{1}{3}\pi r^2 h = \dfrac{2816}{7}$$

$$\dfrac{1}{3} \times \dfrac{22}{7} \times r^2 \times 6 = \dfrac{2816}{7}$$

$$\therefore \quad r^2 = \dfrac{2816}{7} \times \dfrac{7 \times 3}{22 \times 6} = 64$$

$$\therefore \quad r = 8 \text{ cm}$$

$$\therefore \quad \text{Curved surface area of the cone} = \pi r l$$

$$\text{where } l = \text{slant height} = \sqrt{r^2 + h^2}$$

$$= \sqrt{64 + 36}$$

$$= \sqrt{100} = 10 \text{ cm}$$

Curved surface area of the cone $= \pi r l$

$$= \frac{22}{7} \times 8 \times 10 = 251\frac{3}{7} \text{ cm}^2$$

Q. 14. **A sphere of diameter 12.6 cm is melted and cast into a right circular cone of height 25.2 cm. Find the diameter of the base of the cone.**

Ans. Diameter of the sphere = 12.6 cm

$$\therefore \quad \text{Radius of the sphere} = \frac{12.6}{2} = 6.3 \text{ cm}$$

$$\therefore \quad \text{Volume of the sphere} = \frac{4}{3}\pi r^3$$

$$= \frac{4}{3} \times \frac{22}{7} \times (6.3)^3$$

$$= 1047.816 \text{ cm}^3$$

Since, volume of sphere = Volume of cone

$$\frac{1}{3}\pi r^2 h = 1047.816$$

$$\frac{1}{3} \times \frac{22}{7} \times (r)^2 \times 25.2 = 1047.816$$

$$(r)^2 = \frac{1047.816 \times 3 \times 7}{25.2 \times 22} = 39.69$$

$$r = 6.3$$

$$\text{diameter} = 2r = 2 \times 6.3 = 12.6 \text{ cm}$$

Q. 15. **A 20 m deep well with diameter 14 m is dug up and the earth from digging is spread evenly to form a platform 22 m × 14 m. Find the height of the platform.**

Ans. Radius of the well $= \dfrac{14}{2} = 7$ m

Volume of the earth dug = Volume of well

$$= \pi r^2 h = \pi (7)^2 \times 20$$

$$= \frac{22}{7} \times 7 \times 7 \times 20 = 3080 \text{ m}^3$$

Volme of platform = 22 × 14 × h

Also, volume of platform = Volume of well

$$22 \times 14 \times h = 3080$$

$$h = \frac{3080}{22 \times 14} = 10 \text{ m}$$

$\therefore$ Height of the platform $= 10$ m.

Q. 16. **If the diameter of the cross-section of a wire is decreased by 5%, how much percent will the length be increased so that the volume remains the same?**

Ans. Let r be the original radius and h the original length of the wire.

$$\therefore \qquad \text{Volume} = \pi r^2 h$$

Now radius (after 5% decrease) $= r \cdot \dfrac{19}{20} = \dfrac{19r}{20}$

Since volume remains same,

$$\therefore \qquad \text{New length} = \dfrac{V}{\pi \left(\dfrac{19r}{20}\right)^2}$$

$$= \dfrac{\pi r^2 h}{\pi} \times \dfrac{400}{361 r^2}$$

$$= h \times \dfrac{400}{361}$$

$$\therefore \qquad \text{increase in length} = h \times \dfrac{400}{361} - h$$

$$= h \left(\dfrac{39}{361}\right)$$

$$\text{Percentage increase} = h \times \dfrac{39}{361} \times \dfrac{100}{h}$$

$$= 10.8\%$$

Q. 17. **A cone is 8.4 cm high and the radius of its base is 2.1 cm. It is melted and recast into a sphere. Find the radius of the sphere.**

Ans. Volume of the cone $= \dfrac{1}{3} \times \pi \times (2.1)^2 \times 8.4 \text{ cm}^2$

If r is the radius of the sphere, its volume $= \dfrac{4}{3} \pi r^3 \text{ cm}^3$

Since volume of sphere $=$ Volume of cone

$$\therefore \qquad \dfrac{4}{3} \pi r^3 = \dfrac{\pi}{3} (2.1)^2 (8.4)$$

or $\qquad\qquad\qquad r = 2.1$ cm

Q. 18. **The radii of the internal and external surfaces of a metallic spherical shell are 3 cm and 5 cm, respectively. It is melted and recast into a solid right circular cylinder of height $10\dfrac{2}{3}$ cm. Find the diameter of the base of the cylinder.**

Ans. Volume of lead used to make the spherical shell $= \dfrac{4}{3} \pi [(5)^3 - (3)^3]$

$$= \frac{4}{3}\pi \times 98 = \frac{392\pi}{3} \text{ cm}^3$$

Let r be the radius of the right circular cylinder obtained by recasting the spherical shell.

$$\text{Height of the cylinder} = \frac{32}{3} \text{ cm}$$

$$\text{Its volume} = \pi r^2 \left(\frac{32}{3}\right) \text{ cm}^3$$

We get,
$$\pi r^2 \cdot \frac{32}{3} = \frac{392\pi}{3}$$

$$\Rightarrow \qquad r^2 = \frac{392}{32} = 12.25$$

or
$$r = 3.5 \text{ cm}$$

Hence, diameter of the base of the cylinder = 7 cm

Q. 19. **A spherical ball of lead 3 cm in diameter is melted and recast into three spherical balls. The diameter of two of these balls are 1 cm and 1.5 cm. Find the diameter of the third ball.**

Ans. Volume of the spherical ball with diameter 3 cm

$$= \frac{4}{3}\pi(1.5)^3 \text{ cm}^3$$

Let r be the radi of the third ball.

$\therefore$ Combined volume of the three balls obtained after recasting the original ball

$$= \frac{4}{3}\pi\left[\left(\frac{1}{2}\right)^3 + \left(\frac{1.5}{2}\right)^3 + r^3\right] \text{ cm}^3$$

We get,
$$\frac{4}{3}\pi\left[\left(\frac{1}{2}\right)^3 + \left(\frac{1.5}{2}\right)^3 + r^3\right] = \frac{4}{3}\pi(1.5)^3$$

$$\Rightarrow \qquad r^3 = \left(\frac{3}{2}\right)^3 - \left(\frac{1}{2}\right)^3 - \left(\frac{3}{4}\right)^3$$

$$= \frac{27}{8} - \frac{1}{8} - \frac{27}{64} = \frac{181}{64}$$

$$\therefore \qquad r = \left(\frac{181}{64}\right)^{1/3} = \frac{1}{4}(181)^{1/3} \text{ cm}$$

Hence, diameter of the third ball $= 2 \times \dfrac{1}{4}(181)^{1/3} = \dfrac{1}{2}(181)^{1/3}$ cm

Q. 20. **A hemispherical bowl of internal radius 9 cm is full of liquid. This liquid is to be filled into cylindrical shaped small bottles each of diameter 3 cm and height 4 cm. How many bottles are neccssary to empty the bowl?**

Ans. Volume of the liquid contained in the hemispherical bowl of interal radius

Innovative Mathematics X-12

$$9 \text{ cm} = \frac{2}{3}\pi(9)^3 = 486\pi \text{ cm}^3$$

Volume of a cylindrical shaped bottle with radius 3/2 cm and height 4 cm

$$= \pi(3/2)^2\,(4) = 9\pi \text{ cm}^3$$

$\therefore$ Number of bottles necessary to empty the bowl $= \dfrac{\text{Volume of liquid in the bowl}}{\text{Volume of one bottle}}$

$$= \frac{486\pi}{9\pi} = 54$$

Q. 21. **A conical flask is full of water. The flask has base-radius r and height h. The water is poured into a cylindrical flask of base-radius mr. Find the height of water in the cylindrical flask.**

Ans. Volume of water contained in the conical flask

$$= \frac{1}{3}\,\pi\,r^2\,h$$

Let H be the height of water in the cylindrical flask with base radius mr,

$\therefore$ Volume of water in cylindrical flask $= \pi(mr)^2.H$

We get, $\pi(mr)^2.H = \dfrac{1}{3}\,\pi r^2 h$

$$H = \frac{r^2 h}{3} \times \frac{1}{m^2 r^2} = \frac{h}{3m^2}$$

Q. 22. **Water is being pumped out through a circular pipe whose internal diameter is 7 cm. If the flow of water is 72 cm per second, how many litres of water are being pumped out in one hour?**

Ans. Internal radius of the circular pipe $= \dfrac{7}{2}$ cm

Since water is flowing at the rate of 72 cm/sec

$\therefore$ Volume of water being pumped out in one second $= \pi\left(\dfrac{7}{2}\right)^2.72 = 2772 \text{ cm}^3$

Volume of water being pumped out in one hour $= 2772 \times 3600$

$$= 9979200 \text{ cm}^3$$

$$= \frac{9979200}{100} \text{ litres}$$

$$= 9979.2 \text{ litres}$$

Q. 23. **Water flows out through a circular pipe, whose internal diameter is 2 cm, at the rate of 0.7 m per second into a cylindrical tank, the radius of whose base is 40 cm. By how much will be level of water rise in half an hour?**

Ans. Internal radius of the circular pipe $= 1$ cm

Since water is flowing at the rate of 0.7 m i.e., 70 cm per second.

$\therefore$ Volume of water that flows in one second

$$= \pi(1)^2 . 70 = 220 \text{ cm}^3$$

Volume of water that flows in half an hour

$$= 220 \times 1800 = 396000 \text{ cm}^3$$

Let water level rise by h cm, when water flows into the given cylindrical tank with base radius 40 cm.

$$\therefore \qquad \pi(40)^2\, h = 396000$$

$$\text{or} \qquad h = \frac{396000}{1600} \times \frac{7}{22} = 78.75 \text{ cm}$$

Q. 24. The rain water from a roof 22 m × 20 m drains into a cylindrical vessel having diameter of base 2 m and height 3.5 m. If the vessel is just full, find the rainfall in cm.

Ans. Let the rainfall be h (in metres)

According to given:

$$h \times 22 \times 20 = \pi\, (1)^2\, (3.5)$$

$$\therefore \qquad h = \frac{22}{7} \times \frac{7}{2} \times \frac{1}{22 \times 20} = \frac{1}{40} \text{ m} = 2.5 \text{ cm}$$

Q. 25. A rectangular tank 15 m long and 11 m broad is required to receive entire liquid contents from a full cylindrical tank of internal diameter 21 m and length 5 m. Find the least height of the tank that will serve the purpose.

Ans. Let the least height of the required tank be h m.

According to given:

$$15 \times 11 \times h = \pi \left(\frac{21}{2}\right)^2 (5)$$

$$\therefore \qquad h = \frac{22}{7} \times \frac{21}{2} \times \frac{21}{2} \times \frac{5}{15 \times 11} = \frac{21}{2} = 10.5 \text{ m}$$

Q. 26. A farmer connects a pipe of internal diameter 20 cm from a canal into a cylindrical tank in her field, which is 10 m in diameter and 2 m deep. If water flows through the pipe at the rate of 3 km/h, in how much time will the tank be filled?

Ans. Radius of the tank $= \dfrac{10}{2}$ m = 5 m

Depth of the tank = 2 m

Volume of the tank $= \dfrac{22}{7} \times 5^2 \times 2 = \dfrac{1100}{7}$ m^3

Internal radius of the cirtcular pipe

$$= \frac{20}{2} \text{ cm} = 10 \text{ cm} = \frac{1}{10} \text{ m}$$

Water is flowing through a circular pipe in the form of a right circular cylinder Length of this circular cylinder (in 1 hour)

$$= 3 \text{ km} = 3000 \text{ m}$$

Volume of water poured into the tank in one hour

$$= \pi r^2 h = \frac{22}{7} \times \frac{1}{10} \times \frac{1}{10} \times 3000 \text{ m}^3$$

$$= \frac{660}{7} \text{ m}^3$$

Time taken to fill the tank

$$= \frac{1100}{7} \sqrt{\frac{660}{7}}$$

$$= \frac{1100}{7} \times \frac{7}{660} \text{ hours} = \frac{5}{3} \text{ hours}$$

$$= 1 \text{ hour } 40 \text{ minutes}$$

Q. 27. **Water flows at the rate of 10 m per minute through a cylindrical pipe having its diameter as 5 mm. How much time will it take to fill a conical vewhose diameter of base is 40 cm and depth 24 cm?**

Ans. Volume of water that flows out of the cylindrical pipe in one minute

$$= \pi \left(\frac{2.5}{10} \right)^2 (10 \times 100) \text{ cm}^3$$

$$= \frac{22}{7} \times \frac{1}{4} \times \frac{1}{4} \times 1000 = \frac{1375}{7} \text{ cm}^3$$

Volume of the conical vessel $= \frac{1}{3} \pi (20)^{23} . 24$

$$= \frac{1}{3} \times \frac{22}{7} \times 400 \times 24$$

$$= \frac{70400}{7} \text{ cm}^3$$

Time required to fill the conical vessel

$$= \frac{\text{Volume of the vessel}}{\text{Volume of water flowing in one minute}}$$

$$= \frac{70400}{7} \times \frac{7}{1375}$$

$$= 512 \text{ minutes}$$

Innovative Mathematics X-12

VERY SHORT ANSWER TYPE Q.ESTIONS

Q. 1. **Match the following:**

Column I		Column II	
(a)	Surface area of a sphere	(i)	$2\pi rh$
(b)	Total surface area of a cone	(ii)	$\dfrac{1}{3}\pi r^2 h$
(c)	Volume of a cuboid	(iii)	$2\pi r(r + h)$
(d)	Volume of hemisphere	(iv)	$\dfrac{1}{3}\pi h(r^2 + R^2 + rR)$
(e)	Curved surface area of a cone	(v)	$\pi r\,(r + l)$
(f)	Total surface area of hemisphere	(vi)	$l \times b \times h$
(g)	Curved surface area of a cylinder	(vii)	$\dfrac{2}{3}\pi r^3$
(h)	Volume of a cone	(ix)	$3\,\pi r^2$
(j)	Volume of a frustum of a cone	(x)	$4\pi r^2$

Sol.

Q. 2. **Fill in the Blanks:**

(i)   The total surface area of cuboid of dimension $a \times a \times b$ is __________ .

(ii)   The volume of right circular cylinder of base radius r and height 2r is ____________ .

(iii)   The total surface area of a cylinder of base radius r and height h is ____________ .

(iv)   The curved surface area of a cone of base radius r and height h is ____________ .

(v)   If the height of a cone is equal to diameter of its base, the volume of cone is ______________ .

(vi)   The total surface area of a solid hemisphere of radius r is ____________ .

(vii)   The curved surface area of a hollow cylinder of outer radius R, inner radius r and height h is radius R, inner radius r and height h is __________ .

(viii)   If the radius of a sphere is doubled, its volume becomes __________ times the volume of original sphere.

(ix)   If the radius of a sphere is halved, its volume becomes __________ times the volume of original sphere.   **(NCERT Exemplar)**

Sol.

Q. 3. **Write 'True' or 'False' in the following:**

(i)   Two identical solid hemispheres of equal base radius r are stuck together along their base. The total surface area of the combination is $6\,\pi r^2$.

(ii)   A solid cylinder of radius r and height h is placed over other cylinder of same height and radius. The total surface area of the shape so formed is $(4\pi rh + 4\pi r^2)$.

(iii)   A solid cone of radius r and height h is placed over a solid cylinder having same base radius and height as that of a cone. The total surface area of the combined solid is $\pi r(\sqrt{r^2 + h^2} + 3r + 2h)$.

(iv)   A solid ball is exactly fitted inside the cubical box of side 'a'. The volume of the ball is $\frac{4}{3}\neq a^2$.

Sol.

Q. 4. **The total surface area of a solid hemisphere of radius r is**

(a) πr^2 (b) $2\pi r^2$ (c) $3\pi r^2$ (d) $4\pi r^2$

Sol.

Q. 5. **The volume and the surface area of a sphere are numerically equal, then the radius of sphere is**

(a) 0 units   (b) 1 unit   (c) 2 units   (d) 3 units

Sol.

Q. 6. **A cylinder, a cone and a hemisphere are of the same base and of the same height. The ratio of their volumes is**

(a) 1:2:3   (b) 2:1:3   (c) 3:1:2   (d) 3:2:1

Sol.

Q. 7. **A solid sphere of radius '7' is melted and recast into the shape of a solid cone of height '7'. Then the radius of the base of cone is**

(a) 2r   (b) r   (c) 4r   (d) 3r

Sol.

Q. 8. **Three solid spheres of diameters 6 cm, 8 cm and 10 cm are melted to form a single solid sphere. The diameter of the new sphere is**

(a) 6 cm   (b) 4.5 cm   (c) 3 cm   (d) 12 cm

Sol.

Q. 9. **A metallic spherical shell of internal and external diameters 4 cm and 8 cm, respectively is melted and recast into the form of a cone of base diameter 8 cm. The height of the cone is:**

(a) 12 cm   (b) 14 cm   (c) 15 cm   (d) 18 cm

Sol.

Q. 10. **Find total surface area of a solid hemi-sphere of radius 7 cm.**

Sol.

Q. 11. **Volume of two spheres is in the ratio 64 : 125. Find the ratio of their surface areas.**

Sol.

Q. 12. **A cylinder and a cone are of same base radius and of same height. Find the ratio of the volumes of cylinder to that of the cone.**

Sol.

Innovative Mathematics X-12

Q. 13. If the volume of a cube is 1331 cm^3, then find the length of its edge.

Sol.

Q. 14. Two cones have their heights in the ratio 1 : 3 and radii in the ratio 3 : 1. What is the ratio of their volumes? **(CBSE 2020)**

Sol.

Q. 15. The radius of a sphere is r cm. It is divided into two equal parts. Find the whole surface of the two parts. ($6\pi r^2$ cm^2 - Board 2012) **(HOTS)**

Sol.

Q. 16. 12 solid spheres of the same size are made by melting a solid metallic cylinder of base radius 1 cm and height $\dfrac{1}{3}$ of 48 cm. Find the radius of each sphere. (1cm–Board 2012)

Sol.

Q. 17. If the radius of the base of a right circular cylinder is halved, keeping the height same, find the ratio of the volume of the reduced cylinder to that of original cylinder. (1 : 4 - Board 2012) **(HOTS)**

Sol.

Q. 18. Three cubes of iron whose edges are 3 cm, 4 cm and 5 cm, respectively are melted and formed into a single cube. Find the edge of the new cube so formed. (6cm - Board 2013)

Sol.

Q. 19. Volumes of two spheres are in the ratio 64 : 27. Find the ratio of their surface areas.

(16 : 9 - Board 2013)

Sol.

SHORT ANSWER TYPE Q.ESTION (I)

Q. 20. How many cubes of side 2 cm can be cut from a cuboid measuring (16cm × 12cm × 10cm)?

Sol.

Q. 21. The circumference of the base of a conical tent is 44 cm. If the height of tent is 24 m, find the length of the canvas used in making the tent, if the width of the canvas is 2 m. (use $\pi = 22/7$)

(275 m - Board 2013)

Sol.

Q. 22. A spherical shell of lead whose external and internal diameters are 24 cm and 18 cm respectively is melted and recast into a right circular cylinder 37 cm high. Find the radius of the base of the cylinder. (6cm - Board 2013)

Sol.

Q. 23. A rectangular sheet of paper of dimensions 44 cm × 18 cm is rolled along its length and a cylinder is formed. Find the volume of the cylinder so formed. (use $\pi = 22/7$)

Sol.

Q. 24. Find the volume of the largest solid right circular cone that can be cut out of a solid cube of side 14 cm. (719 cm^3 - Board 2014)

Sol.

Q. 25. A solid right circular cylinder has a total surface of 462 sq. cm. Its curved surface area is one-third of its total surface area. Find the volume of the cylinder. (539 cm^3 - Board 2014)**(HOTS)**

Sol.

Q. 26. Find the height of largest right circular cone that can be cut out of a cube whose volume is 729 cm^3.

Sol.

Q. 27. Two identical cubes each of volume 216 cm^3 are joined together end to end. What is the surface area of the resulting cuboid?

Sol.

Q. 28. Two cones with same base radius 8 cm and height 15 cm are joined together along with their bases. Find the surface area of the shape so formed.   (NCERT examplar)

Sol.

Q. 29. The total surface area of a right circular cone is 90π cm? If the radius of the base of the cone is 5 cm, find the height of the cone.   (CBSE – 2011)

Sol.

Q. 30. The volume of a right circular cylinder with its height equal to the radius is $25\dfrac{1}{7}\text{cm}^3$. Find the height of the cylinder. (Use $\pi = 22/7$)   (CBSE 2020)

Sol.

Q. 31. Find the volume of the largest right circular cone that can be cut off from a cube of edge 4.2 cm.

Sol.

SHORT ANSWER TYPE Q.ESTION (II)

Q. 32. A sphere of maximum volume is cut out from a solid hemisphere of radius 6 cm. Find the volume of the cut out sphere.   (CBSE-2012)

Sol.

Q. 33. Find the depth of a cylindrical tank of radius 10.5 cm, if its capacity is equal to that of a rectangular tank of size 15 cm × 11 cm × 10.5 cm.

Sol.

Q. 34. Volume of two spheres are in the ratio 64:27, find the ratio of their surface areas.   (CBSE-2012)

Sol.

Q. 35. A petrol tank is a cylinder of base diameter 28 cm and length 24 cm filted with conical ends each of axis length 9 cm. Determine the capacity of the tank.

Sol.

Q. 36. A cylinder, a conc and a hemisphere have same base and same height. Find the ratio of their volumes.

Sol.

Q. 37. A solid is in the form of a cylinder with hemispherical ends. The total height of the solid is 20 cm and the diameter of the cylinder is 7 cm. Find the total volume of the solid.   (Use $\pi = 22/7$)

Sol.

Q. 38. The diameter of a roller 120 cm long is 64 cm. If it takes 500 complete revolutions to level a playground, determine the cost of levelling it at the rate of 30 paise per square meter.   (CBSE 2013)

Sol.

          Innovative Mathematics X-12

Q. 39. The sum of the radius of base and height of a solid right circuler cylinder is 37 cm. If the total surface area of the solid cylinder is 1628 square cm., find the volume of the cylinder. (Use $\pi = 22/7$) (CBSE-2016)

Sol.

Q. 40. A juice seller was serving his customers using glasses as shown in figure. The inner diameter of the cylindrical glass was 5 cm but bottom of the glass had a hemispherical raised portion which reduced the capacity of the glass. If the height of a glass was 10 cm, find the apparent and actual capacity of the glass. [Use $\pi = 3.14$] (NCERT, CBSE 2019, 2009)

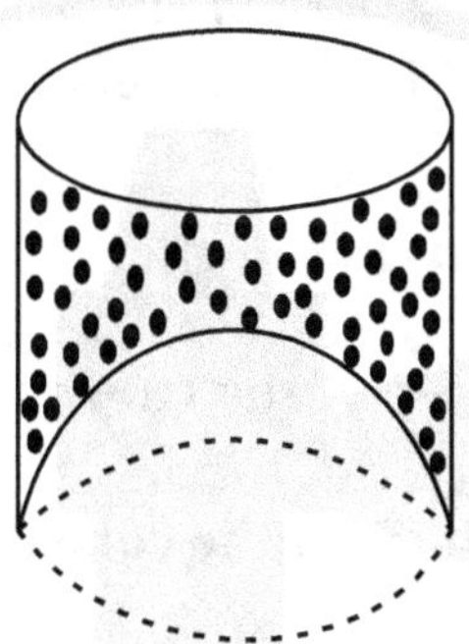

Sol.

Q. 41. The internal and external diameters of a hollow hemispherical vessel are 12 cm and 16 cm respectively. If the cost of painting 1 cm^2 of the surface area is ₹ 5.00, find the total cost of painting the vessel all over. (Use $\pi = 3.14$) (CBSE 2019)

Sol.

Q. 42. Suresh decided to donate canvas for 10 tents conical in shape with base diameter 14 m and height 24 m to a centre for handicapped person's welfare. If the cost of 2 m wide canvas is ₹ 40 per metre, find the amount by which Suresh helped the centre. (CBSE 2017)

Sol.

Q. 43. A cone of maximum size is curved out from a cube edge 14 cm. Find the surface area of remaining solid after the cone is curved out.

Sol.

LONG ANSWER TYPE Q.ESTIONS

Q. 44. A solid iron pole consists of a cylinder of height 220 cm and base diameter 24 cm, which is surmounted by another cylinder of height 60 cm and radius 8 cm. Find the mass of the pole, given that 1 cm^3 of iron has approximately 8 cm mass. (Use $\pi = 3.14$) (NCERT, CBSE 2019)

Sol.

Q. 45. Water is flowing through a cylindrical pipe, of internal diameter 2 cm, into a cylindrical tank of base radius 40 cm, at the rate of 0.4 m/s. Determine the rise in level of water in the tank in half an hour. (4.5 cm - Board 2013)

Sol.

Q. 46. A right cylindrical container of radius 6 cm and height 15 cm is full of icecream, which has to be distributed to 10 children in equal cones having hemispherical shape on the top. If the height of the conical portion is four times its base radius, find the radius of the ice-cream cone.

(CBSE 2019)

Sol.

Q. 47. Water is flowing at the rate of 15 km/hr through a cylindrical pipe of diameter 14 cm into a cuboidal pond which 50 m long and 44 m wide. In what time the level of water in pond rise by 21 cm? (2hrs. – Board 2012)

Sol.

Q. 48. A right angled triangle, whose sides are 3 cm, 4 cm, and 5 cm, is revolved about the longest side. Find the surface area of the figure (double cone) obtained. (52.8 cm^2 - Board 2012) **(HOTS)**

Sol.

Q. 49. A hollow cone is cut by a plane parallel to the base and the upper portion is removed. If the curved surface of the remainder is $\dfrac{8}{9}$ of the curved surface of the whole cone, find the ratio of the line segments in which the altitude of the cone is divided by the plane.

(1 : 2 - Board 2004) **(EXEMPLAR)**

Sol.

Q. 50. The height of a cone is 30 cm. A small cone is cut off at the top by a plane parallel to the base. If its volume be $\dfrac{1}{27}$ th of the volume of the given cone, at what height above the base is the section made? (20 cm - Board 2005) **(EXEMPLAR)**

Sol.

Q. 51. A cone is divided into two parts by drawing a plane through the midpoint of its axis, parallel to its base. Compare the volumes of the two parts. $(\dfrac{1}{7}$ – Board 2003) **(EXEMPLAR)**

Sol.

Q. 52. A wooden article as shown in the fig. was made from a cylinder by scooping out a hemisphere from one end and a cone from the other end. Find the total surface area of the remaining article, (NCERT, CBSE 2019)

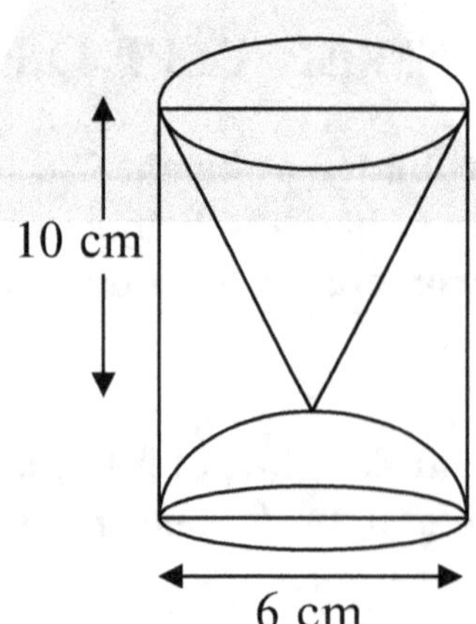

Sol.

Q. 53. The height of a solid cylinder is 15 cm and its diameter is 7 cm. Two equal conical holes of radius 3 cm and height 4 cm are cut off. Find the volume and surface area of the solid.

Sol.

Q. 54. If h, c and V respectively represent the height, curved surface area and volume of a cone, prove that $\qquad$ (CBSE 2015)

$$c^2 = \frac{3\pi Vh^3 + 9V^2}{h^2}$$

Sol.

Q. 55. A solid wooden toy is in the form of a hemi-sphere surmounted by a cone of same radius. The radius of hemi-sphere is 3.5 cm and the total wood used in the making of toy is $166\frac{5}{6}$ cm³. Find the height of the toy. Also, find the cost of painting the hemi-spherical part of the toy at the rate of ₹ 10 per cm? (Use $\pi = 22/7$) $\qquad$ (CBSE, 2015)

Sol.

Q. 56. In the given figure, from a cuboidal solid metalic block of dimensions 15 cm × 10 cm × 5 cm a cylindrical hole of diameter 7 cm is drilled out. Find the surface area of the remaining block. (Use $\pi = 22/7$) $\qquad$ (CBSE – 2015)

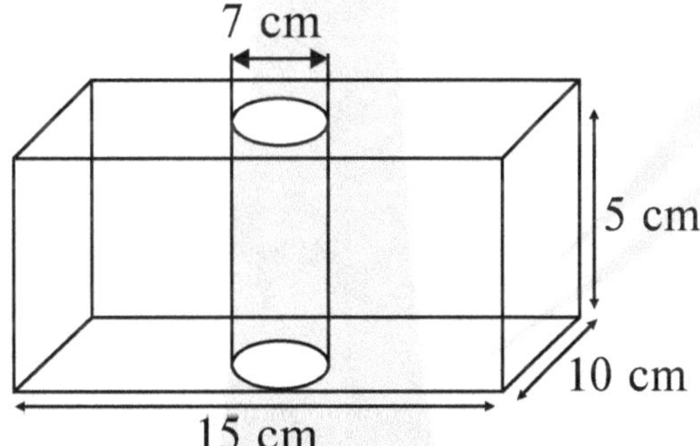

Sol.

Q. 57. A solid toy is the form of a right circular cylinder with a hemispherical shape at one end and a cone at the other end. Their diameter is 4.2 cm and the heights of the cylindrical and conical portions are 12 cm and 7 cm respectively. Find the volume of the toy.

Sol.

Q. 58. A tent is in the shape of a right circular cylinder upto a height of 3 m and conical above it. The total height of the tent is 13.5 m and radius of base is 14 m. Find the cost of cloth required to make the tent at the rate of ₹ 80 per m².

Sol.

Q. 59. The difference between outer and inner curved surface areas of a hollow right circular cylinder, 14 cm long is 88 cm². If the volume of the metal used in making the cylinder is 176 cm³. Find the outer and inner diameters of the cylinder. $\qquad$ (HOTS)

Sol.

Q. 60. A solid is in the shape of a cone surmounted on a hemisphere. The radius of each of them being 3.5 cm and the total height of the solid is 9.5 cm. Find the volume of the solid. **(CBSE 2020)**

Sol.

Q. 61. A hemispherical depression is cut out from one face of a cubical wooden block of edge 21 cm, such that the diameter of the hemisphere is equal to edge of the cube. Determine the volume of the remaining block. **(CBSE 2020)**

Sol.

1. (a) $4\pi r^2$ (b) $\pi r(l + r)$ (c) $l \times b \times h$ (d) $\frac{2}{3}\pi r^3$

   (e) πrl (f) $3\,\pi r^2$ (g) $2\,\pi rh$ (h) $\frac{1}{3}\pi r^2 h$

   (i) $2\pi r(r + h)$ (j) $\frac{1}{3}\pi h\,(r_1^2 + R^2 + Rr)$

2. (i) $2a^2 + 4ab$ (ii) $2\,\pi r^3$ (iii) $2\pi r\,(r + h)$ (iv) $\pi r\sqrt{r^2 + h^2}$

   (v) $\frac{2}{3}\pi r^3$ (vi) $3\pi r^2$ (vii) $2\pi h\,(R + r)$ (viii) 8

   (ix) $\frac{1}{8}$

3. (i) False   (ii) False   (iii) False   (iv) False

4. $3\pi r^2$ 5. 3 units   6. $3 : 1 : 2$ 7. $2r$

8. 12 cm   9. 14 cm   10. $462\ \text{cm}^2$ 11. $16 : 25$

12. $3 : 1$ 13. 11 cm   14. $3 : 1$ 15. 240

16. 9 cm   17. $360\ \text{cm}^2$ 18. $854.85\ \text{cm}^2$ 19. 12 cm

20. 2 cm   21. $19.4\ \text{cm}^3$ 22. $113.14\ \text{cm}^3$ 23. 5 cm

24. $16 : 9$ 25. $18480\ \text{cm}^2$ 26. $3 : 1 : 2$ 27. $680\frac{1}{6}\ \text{cm}^3$

28. ₹ 362.06   29. $4620\ \text{cm}^3$ 30. $163.54\ \text{cm}^3$ 31. ₹ 3579.60

32. ₹ 1,10,000   33. $1366.3\ \text{cm}^2$ 34. 892.2624 kg   35. $r = 3$ cm

36. $r = 3$ cm, $A = \pi(78 + 3\sqrt{58})\ \text{cm}^2$ 37. $444.7\ \text{cm}^2$ 39. $77\ \text{cm}^2$

40. $583\ \text{cm}^2$ 41. $218.064\ \text{cm}^3$ 42. ₹ 82720

43. $R = 2.5$ cm, $r = 1.5$ cm   44. $V = 166.83\ \text{cm}^3$ 45. $6835.5\ \text{cm}^3$ 46.

47.   48.   49.   50.

51.   52.   53.   54.

55.   56.   57.   58.

59.   60.   61.

SURFACE AREA AND VOLUMES

SECTION-A

Q.1. The total surface area ofa hemisphere of radius 2r is _______ (1)

Sol.

Q.2. The radius of the largest right circular cone that can be cut out from a cube of edge 4.2 cm is (1)

    (a) 4.2 cm   (b) 8.4 cm   (c) 2.1 cm   (d) 1.05 cm

Sol.

Q.3. The volume of a cube is $1l$. Find the length of the side of the cube. (1)

Sol.

Q.4. Volume of two cubes are in the ratio 27 : 125. The ratio of their surface areas is _______. (1)

Sol.

SECTION-B

Q.5. A cube and a sphere have equal total surface area. Find the ratio of the volume of sphere and cube. (2)

Sol.

Q.6. Two cubes, each of side 8 cm are joined end to end. Find the surface area of the resulting figure. (2)

Sol.

Q.7. The volume of a hemi-sphere is 2156 cm³. Find its curved surface area. (2)

Sol.

SECTION-C

Q.8. A circus tent is in the shape of a cylinder surmounted by a conical roof. If the common diameter is 56 m, the height of the cylindrical portion is 6 m and the height of the roof from the ground is 30 m, find the area of the canvas used for the tent. (3)

Sol.

Q.9. A metallic cylinder has radius 3 cm and height 5 cm. To reduce its weight, a conical hole of radius $\frac{3}{2}$ cm and depth $\frac{8}{9}$ cm is drilled in the cylinder. Calculate the ratio of the volume of metal left in the cylinder to the volume of metal taken out in conical shape. (3)

Sol.

SECTION-D

Q.10. A decorative block is made up by joining a cube and a hemisphere. The base of the block is a cube of side 6 cm and the hemisphere fixed on the top has a diameter of 4 cm. Find the cost of painting it at a price of ₹ 2.5 per cm². **(4)**

Sol.

CASE STUDY

CASE STUDY 1.

Adventure camps are the perfect place for the children to practice decision making for themselves without parents and teachers guiding their every move. Some students of a school reached for adventure at Sakleshpur. At the camp, the waiters served some students with a welcome drink in a cylindrical glass and some students in a hemispherical cup whose dimensions are shown below. After that they went for a jungle trek. The jungle trek was enjoyable but tiring. As dusk fell, it was time to take shelter. Each group of four students was given a canvas of area 551 m². Each group had to make a conical tent to accommodate all the four students. Assuming that all the stitching and wasting incurred while cutting, would amount to 1 m², the students put the tents. The radius of the tent is 7 m.

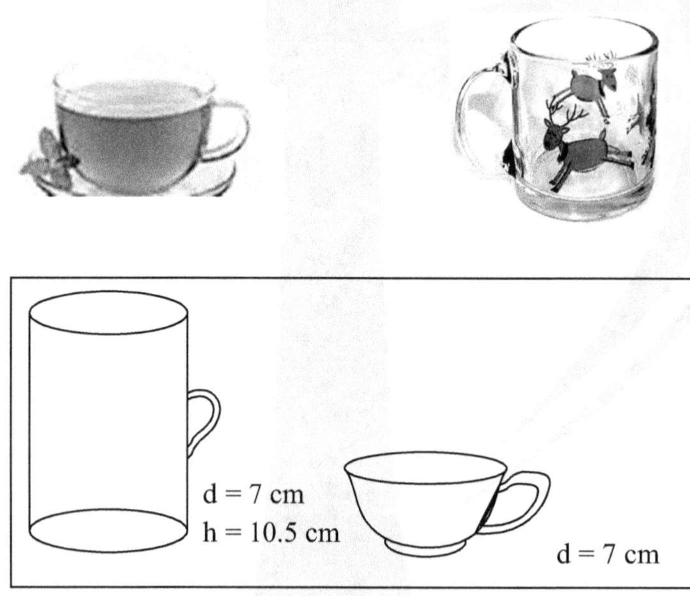

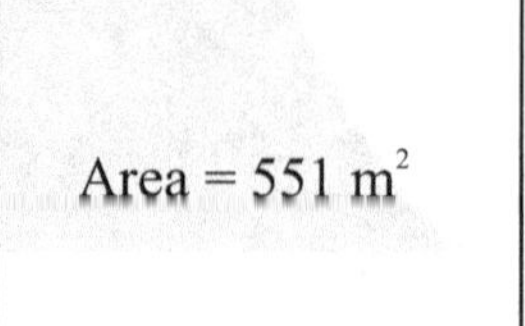

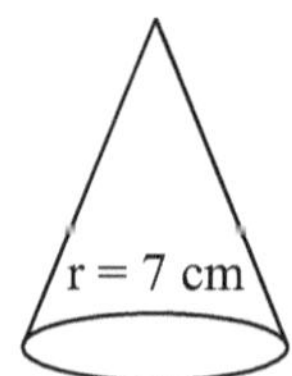

Q. 1. **The volume of cylindrical cup is**

    (a) 295.75 cm³   (b) 7415.5 cm³   (c) 384.88 cm³   (d) 404.25 cm³

Q. 2. **The volume of hemispherical cup is**

    (a) 179.67 cm³   (b) 89.83 cm³   (c) 172.25 cm³   (d) 210.60 cm³

Q. 3. **Which container had more juice and by how much?**

    (a) Hemispherical cup, 195 cm³   (b) Cylindrical glass, 207 cm³

    (c) Hemispherical cup, 280.85 cm³   (d) Cylindrical glass, 314.42 cm³

Q. 4. **The height of the conical tent prepared to accommodate four students is**

    (a) 18 m   (b) 10 m   (c) 24 m   (d) 14 m

Q. 5. **How much space on the ground is occupied by each student in the conical tent**

    (a) 54 m²   (b) 38.5 m²   (c) 86 m²   (d) 24 m²

ANSWERS

1. (d) 404.25 cm³   **2.** (b) 89.83 cm³   **3.** (d) Cylindrical glass, 314.42 cm³

4. (c) 24 m   **5.** (b) 38.5 m²

CASE STUDY 2.

The Great **Stupa** at **Sanchi** is one of the oldest stone structures in India, and an important monument of Indian Architecture. It was originally commissioned by the emperor Ashoka in the 3rd century BCE. Its nucleus was a simple hemispherical brick structure built over the relics of the Buddha. It is a perfect example of combination of solid figures. A big hemispherical dome with a cuboidal structure mounted on it. (Take π = 22/7)

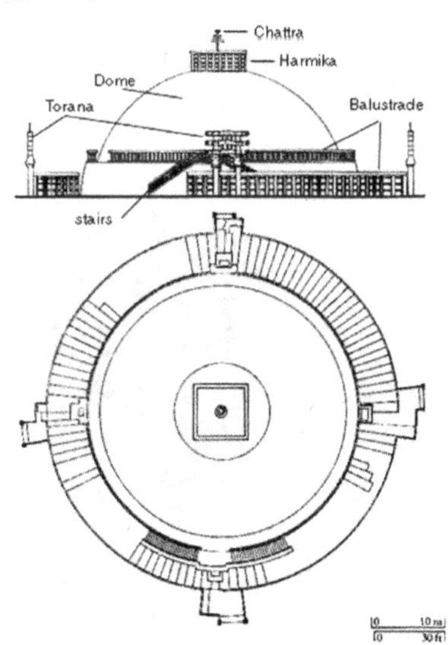

Q. 1. **Calculate the volume of the hemispherical dome if the height of the dome is 21 m –**

    (a) 19404 cu. m   (b) 2000 cu. m   (c) 15000 cu. m   (d) 19000 cu. m

Q. 2. **The formula to find the Volume of Sphere is-**

    (a) $\dfrac{2}{3}\neq r^3$ (b) $\dfrac{4}{3}\neq r^3$ (c) $4\pi r^2$ (d) $2\pi r^2$

Q. 3. The cloth require to cover the hemispherical dome if the radius of its base is 14 m is

(a) 1222 sq.m   (b) 1232 sq.m   (c) 1200 sq.m   (d) 1400 sq.m

Q. 4. The total surface area of the combined figure i.e. hemispherical dome with radius 14 m and cuboidal shaped top with dimensions 8 m × 6 m × 4 m is

(a) 1200 sq. m   (b) 1232 sq. m   (c) 1392 sq.m   (d) 1932 sq. m

Q. 5. The volume of the cuboidal shaped top is with dimensions mentioned in question 4

(a) 182.45 m^3 (b) 282.45 m^3 (c) 292 m^3 (d) 192 m^3

ANSWERS

1. (a) 19404 cu. m
2. (b) $\dfrac{4}{3}\pi r^3$
3. (b) 1232 sq.m
4. (c) 1392 sq.m
5. (d) 192 m^3

CASE STUDY 3.

On a Sunday, your Parents took you to a fair. You could see lot of toys displayed, and you wanted them to buy a RUBIK's cube and strawberry ice-cream for you. Observe the figures and answer the questions-:

Q. 1. The length of the diagonal if each edge measures 6 cm is

(a) $3\sqrt{3}$ (b) $3\sqrt{6}$ (c) $\sqrt{12}$ (d) $6\sqrt{3}$

Q. 2. Volume of the solid figure if the length of the edge is 7 cm is-

(a) 256 cm^3 (b) 196 cm^3 (c) 343 cm^3 (d) 434 cm^3

Q. 3. What is the curved surface area of hemisphere (ice cream) if the base radius is 7 cm?

(a) 309 cm^2 (b) 308 cm^2 (c) 803 cm^2 (d) 903 cm^2

Q. 4. Slant height of a cone if the radius is 7 cm and the height is 24 cm________

    (a) 26 cm   (b) 25 cm   (c) 52 cm   (d) 62 cm

Q. 5. The total surface area of cone with hemispherical ice cream is

    (a) 858 cm^2 (b) 885 cm^2 (c) 588 cm^2 (d) 855 cm^2

ANSWERS

1. (d) $6\sqrt{3}$ **2.** (c) 343 cm^3 **3.** (b) 308 cm^2

4. (b) 25 cm   **5.** (a) 858 cm^2

CHAPTER-13

STATISTICS

LEARNING OBJECTIVES

Statistics is a branch of Mathematics which deals with collection, presentation, analysis, interpretation of data and drawing of inferences/conclusions there form.

DATA: Facts or figures, which are numerical or otherwise collected with a definite purpose.

TYPES OF DATA:

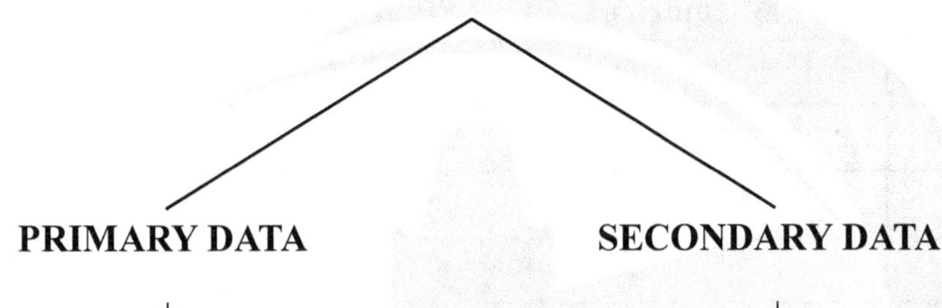

PRIMARY DATA

Data which an investigator collects for the first time for his own purpose.

SECONDARY DATA

Data which the investigator obtains from some other source, agency or office for his own purpose.

PRESENTATION:

Row or ungrouped data: The data obtained in original form and presented ungrouped without any re-arrangement or condensed form.

An array: The presentation of a data in ascending order or descending order of magnitude.

Grouped data: Rearrangement or condensed form of data into classes or groups.

Range of data: Difference between the highest and lowest values in the data.

Frequency: The number of times an observations occurs in data.

Class interval: Each group in which the observations/values of a data are condensed.

Class limits: Values by which each class interval is bounded. Value on the left is called lower limit and value on the right is called upper limit.

Class size: Difference between the upper limit and the lower limit.

Class mark: Mid value of class.

$$\text{interval} = \frac{\text{lower limit} + \text{upper limit}}{2}$$

Cumulative frequency of a class: Total of frequencies of a particular class and of all classes prior to that class.

GRAPHICAL REPRESENTATION OF DATA:-

Bar graph: A pictorial representation of data in which usually bars of uniform width are drawn with equal spacing between them on one axis and values of variable are shown on other axis.

Histogram: A pictorial representation like bar graph with no space between the bars, it is used for continuous grouped frequency distribution.

Q. 1. **Given below are the ages of 25 students of class IX in a school. Prepare a discrete frequency distribution.**

15, 16, 16, 14, 17, 17, 16, 15, 15, 16, 16, 17, 15, 16, 16, 14, 16, 15, 14, 15, 16, 16, 15, 14, 15.

Ans. Frequency distribution of ages of 25 students

Age	Tally marks	Frequency
14	IIII	4
15	ℕℍ III	8
16	ℕℍ ℕℍ	10
17	III	3
Total		25

Q. 2. **Form a discrete frequency distribution from the following scores:-**

Ans. 15, 18, 16, 20, 25, 24, 25, 20, 16, 15, 18, 18, 16, 24, 15, 20, 28, 30, 27, 16, 24, 25, 20, 18, 28, 27, 25, 24, 24, 18, 18, 25, 20, 16, 15, 20, 27, 28, 29, 16.

Frequency Distribution of Scores		
Variate	Tally marks	Frequency
15	IIII	4
16	ℕℍ I	6
18	ℕℍ I	6
20	ℕℍ I	6
24	ℕℍ	5
25	ℕℍ	5
27	III	3
28	III	3
29	I	1
30	I	1
Total		40

Q. 3. **The water tax bills (in rupees) of 30 houses in a locality are given below. Construct a grouped frequency distribution with class size of 10.**

30, 32, 45, 54, 74, 78, 108, 112, 66, 76, 88, 40, 14, 20, 15, 35, 44, 66, 75, 84, 95, 96, 102, 110, 88, 74, 112, 14, 34, 44.

Ans. Here the maximum and minimum values of the variate are 112 and 14 respectively.

$\therefore \quad$ Range $= 112 - 14 = 98$.

It is given that the class size is 10, and

$$\frac{\text{Range}}{\text{Class size}} = \frac{98}{10} = 9.8$$

So, we shoule have 10 classes each of size 10.

The minimum and maximum values of the variate are 14 and 112 respectively. So we have to make the classes in such a way that first class includes the minimum value and the last class includes the maximum value. If we take the first class as 14-24 it includes the minimum value 14. If the last class is taken as 104-114, then it includes the maximum value 112. Here, we form classes by exclusive method. In the class 14-24, 14 is included but 24 is excluded. Similarly, in other classes, the lower limit is included and the upper limit is excluded.

In the view of above discussion, we construct the frequency distribution table as follows:

Bill (in rupees)	Tally marks	Frequency
14-24	||||	4
24-34	||	2
34-44	|||	3
44-54	|||	3
54-64	|	1
64-74	||	2
74-84	⅂⅂⅂⅂⅂	5
84-94	|||	3
94-104	|||	3
104-114	||||	4
Total		30

Q. 4. **The marks obtained by 40 students of class IX in an examination are given below:**

18, 8, 12, 6, 8, 16, 12, 5, 23, 2, 16, 23, 2, 10, 20, 12, 9, 7, 6, 5, 3, 5, 13, 21, 13, 15, 20, 24, 1, 7, 21, 16, 13, 18, 23, 7, 3, 18, 17, 16.

Present the data in the form of a frequency distribution using the same class size, one such class being 15-20 (where 20 is not included)

Ans. The maximum and maximum marks in the given raw data are 0 and 24 respectively. It is given that 15-20 is one of the class intervals and the class size is same. So, the classes of equal size are

0-5, 5-10, 10-15, 15-20 and 20-25

Thus, the frequency distribution is as given under:

Frequency Distribution of Marks

Marks	Tally marks	Frequency
0-5	卌 I	4
24-34	卌 卌	2
34-44	卌 III	3
44-54	卌 III	3
54-64	卌 III	1
	Total	40

Q. 5. **The class marks of a distribution are:**

47, 52, 57, 62, 67, 72, 77, 82, 87, 92, 97, 102

Determine the class size, the class limits and the true class limits.

Ans. Here the class marks are uniformly spaced. So, the class size is the difference between any two consecutive class marks

$\therefore$ Class size = $52 - 47 = 5$

We know that, if a is the class mark of a class interval and h is its class size, then the lower and upper

limits of the class interval are $a - \dfrac{h}{2}$ and $a + \dfrac{h}{2}$ respectively.

$\therefore$ Lower limits of first class interval

$$= 47 - \frac{5}{2} = 44.5$$

And, upper limit of first class interval

$$= 47 + \frac{5}{2} = 49.5$$

So, first class interval is 44.5 – 49.5

Similarly, we obtain the other class limits as given under:

Class marks	Class limits
47	44.5 - 49.5
52	49.5 - 54.5
57	54.5 - 59.5
62	59.5 - 64.5
67	64.5 - 69.5
72	69.5 - 74.5
77	74.5 - 79.5
82	79.5 - 84.5
87	84.5 - 89.5
92	89.5 - 94.5
97	94.5 - 99.5
102	99.5 - 104.5

Since the classes are exclusive (continuous) so the true class limits are same as the class limits.

Q. 6. **The class marks of a distribution are 26, 31, 36, 41, 46, 51, 56, 61, 66, 71. Find the true class limits.**

Ans. Here the class marks are uniformly spaced. So, the class size is the difference between any two consecutive class marks.

$\therefore$ Class size $= 31 - 26 = 5$.

If a is the class mark of a class interval of size h, then the lower and upper limits of the class interval are $a - \dfrac{h}{2}$ and $a + \dfrac{h}{2}$ respectively.

Here $h = 5$

$\therefore$ Lower limit of first class interval

$$= 26 - \frac{5}{2} = 23.5$$

And, upper limit of first class interval

$$= 26 + \frac{5}{2} = 28.5$$

$\therefore$ First class interval is 23.5 – 28.5.

Thus, the class intervals are:

23.5 – 28.5, 28.5 – 33.5, 33.5 – 38.5, 38.5 – 43.5, 43.5 – 48.5, 48.5 – 53.5

Since the class are formed by exclusive method. Therefore, these limits are true class limits.

Q. 7. **Write down less than type cumulative frequency and greater than type cumulative frequency.**

Height (in cm)	Frequency
140 – 145	10
145 – 150	12
150 – 155	18
155 – 160	35
160 – 165	45
165 – 170	38
170 – 175	22
175 – 180	20

Ans. We have

Height (in cm)	140–145	145–150	150–155	155–160	160–165	165–170	170–175	175–180
Frequency	10	12	18	35	45	38	22	20
Height Less than type	145	150	155	160	165	170	175	180
Cumulative frequency	10	22	40	75	120	158	180	200
Height Greater than type	140	145	150	155	160	165	170	175
Cumulative frequency	200	190	178	160	125	80	42	20

Q. 8. **The distances (in km) covered by 24 cars in 2 hours are given below:**

125, 140, 128, 108, 96, 149, 136, 112, 84, 123, 130, 120, 103, 89, 65, 103, 145, 97, 102, 87, 67, 78, 98, 126

Represent them as a cumulative frequency table using 60 as the lower limit of the first group and all the classes having the class size of 15.

Ans. We have, Class size = 15

Maximum distance covered = 149 km.

Minimum distance covered = 65 km.

$\therefore$ Range = (149 – 65) km = 84 km.

So, number of classes = 6 $\left[\Theta\ \dfrac{84}{15} = 5.6\right]$

Thus, the class intervals are 60-75, 75-90, 90-105, 105-120, 120-135, 135-50.

The cumulative frequency distribution is as given below:

Class interval	Tally marks	Frequency	Cumulative frequency
60-75	||	2	2
75-90	||||	4	6
90-105	∖∖∖∖ |	6	12
105-120	||	2	14
120-135	∖∖∖∖ |	6	20
135-150	||||	4	24

Q. 9. **The following table gives the marks scored by 378 students in an entrance examination:**

Mark	No. of students
0-10	3
10-20	12
20-30	36
30-40	76
40-50	97
50-60	85
60-70	39
70-82	12
80-90	12
90-100	6

From this table form (i) the less than series, and (ii) the more than series.

Ans.   (i)   Less than cumulative frequency table

Marks obtained	Number of students (Cumulative frequency)
Less than 10	3
Less than 20	15
Less than 30	51
Less than 40	127
Less than 50	224
Less than 60	309
Less than 70	348
Less than 80	360
Less than 90	372
Less than 100	378

(ii)   More than cumulative frequency table

Marks obtained	Number of students (Cumulative frequency)
More than 0	378
More than 9	375
More than 19	363
More than 29	327
More than 39	257
More than 49	154
More than 59	69

More than 69	30
More than 79	18
More than 89	6

Q. 10. Find the unknown entires (a, b, c, d, e, f, g) from the following frequency distribution of heights of 50 students in a class:

Class intervals (Heights in cm)	Frequency	Cumulative frequency
150-155	12	a
155-160	b	25
160-165	10	c
165-170	d	43
170-175	e	48
175-180	2	f
Total	g	

Ans. Since the given frequency distribution is the frequency distribution of heights of 50 students. Therefore,

$$g = 50.$$

From the table, we have

$$a = 12, b + 12 = 25, 12 + b + 10 = c,$$
$$12 + b + 10 + d = 43,$$
$$12 + b + 10 + d + e = 48 \text{ and}$$
$$12 + b + 10 + d + e + 2 = f$$

Now, $b + 12 = 25$

$\Rightarrow \quad b = 13$

$\qquad 12 + b + 10 = c$

$\Rightarrow \quad 12 + 13 + 10 = c$ $\hfill [\because b = 13]$

$\Rightarrow \quad c = 35$

$\qquad 12 + b + 10 + d = 43$

$\Rightarrow \quad 12 + 13 + 10 + d = 43$ $\hfill [\because b = 13]$

$\Rightarrow \quad d = 8$

$\qquad 12 + b + 10 + d + e = 48$

$\Rightarrow \quad 12 + 13 + 10 + 8 + e = 48$ $\hfill [\because b = 13, d = 8]$

$\Rightarrow \quad e = 5$

and, $12 + b + 10 + d + e + 2 = f$

$\Rightarrow \quad 12 + 13 + 10 + 8 + 5 + 2 = f$

$\Rightarrow \quad f = 50.$

Hence, a = 12, b = 13, c = 35, d = 8

e = 5, f = 50 and g = 50.

Q. 11. **The marks out of 10 obtained by 32 students are:**

2, 4, 3, 1, 5, 4, 3, 8, 9, 7, 8, 5, 4, 3, 6, 7, 4, 7, 9, 8, 6, 4, 2, 1, 0, 0, 2, 6, 7, 8, 6, 1.

Array the data and form the frequency distribution

Ans. An array of the given data is prepared by arranging the scores in ascending order as follows:

0, 0,   1, 1, 1,   2, 2, 2,   3, 3, 3,   4, 4, 4, 4, 4   5, 5,   6,6,6,6,   7,7,7,7,   8,8,8,8, 9, 9.

Frequency distribution of the marks is shown below.

Marks	Tally marks	Frequency
0	||	2
1	|||	3
2	|||	3
3	|||	3
4	ⵑⵑ	5
5	||	2
6	||||	4
7	||||	4
8	||||	4
9	||	2

Q. 12. **Prepare a discrete frequency distribution from the data given below, showing the weights in kg of 30 students of class VI.**

39, 38, 42, 41, 39, 38, 39, 42, 41, 39, 38, 38, 41, 40, 41, 42, 41, 39, 40, 38, 42, 43, 45, 43, 39, 38, 41, 40, 42, 39.

Ans. The discrete frequency distribution table for the weight (in kg) of 30 students is shown below.

Weights (in kg)	Tally marks	Frequency
38	ⵑⵑⵑⵑⵑ |	6
39	ⵑⵑⵑⵑⵑ ||	7
40	|||	3
41	ⵑⵑⵑⵑⵑ |	6
42	ⵑⵑⵑⵑⵑ	5
43	||	2
45	|	1

Q. 13. **The class marks of a distribution are 82, 88, 94, 100, 106, 112 and 118. Determine the class size and the classes.**

Ans. The class size is the difference between two conseuctive class marks. ∴ Class size = 88 − 82 = 6. Now 82 is the class mark of the first class whose width is 6. ∴ Class limits of the first class are 82

$-\dfrac{6}{2}$ and $82 + \dfrac{6}{2}$ i.e. 79 and 85. Thus, the first class is 79-85. Similarly, the other classes are 85–91, 91–97, 97–103, 103–109, 109–115 and 115–121.

Q. 14. **The class marks of a distribution are 13, 17, 21, 25 and 29. Find the true class limits.**

Ans. The class marks are 13, 17, 21, 25 and 29.

The class marks are uniformly spaced.

Class size = difference between two consecutive class marks

$$= 17 - 13 = 4$$

Half of the class size $= \dfrac{4}{2} = 2$

To find the classes one has to subtract 2 from and add 2 to each of the class marks.

Hence, the classes are

$$11 - 15$$
$$15 - 19$$
$$19 - 23$$
$$23 - 27$$
$$27 - 31$$

Since the classes are exclusive, the true class limits are the same as the class limits. So the lower class limits as well as the true lower class limits are 11, 15, 19, 23 and 27. The upper class limits as well as the true upper class limits are 15, 19, 23, 27 and 31.

Q. 15. **Convert the given simple frequency series into a:**

(i) Less than cumulative frequency series.

(ii) More than cumulative frequency series.

Marks	No. of students
0-10	3
10-20	7
20-30	12
30-40	8
40-50	5

Ans. (i) Less than cumulative frequency series

Marks	No. of students
Less than 10	3
Less than 20	10 (= 3 + 7)
Less than 30	22 (= 3 + 7 + 12)
Less than 40	30 (= 3 + 7 + 12 + 8)
Less than 50	35 (= 3 + 7 + 12 + 8 + 5)

Innovative Mathematics X-13

(ii) More than cumulative frequency series

Marks	No. of students
More than 50	0
More than 40	5
More than 30	13
More than 20	25
More than 10	32
More than 0	35

Q. 16. Convert the following more than cumulative frequency series into simple frequency series.

Marks	No. of students
More than 0	40
More than 10	36
More than 20	29
More than 30	16
More than 40	5
More than 50	0

Ans. Simple frequency distribution table

Marks	Frequency
0-10	4 (= 40 − 36)
10-20	7 (= 36 − 29)
20-30	13 (= 29 − 16)
30-40	11 (= 16 − 5)
40-50	5 (= 5 − 0)

Q. 17. Drawn ogive for the following frequency distribution by less than method.

Marks	No. of Students
0–10	7
10–20	10
20–30	23
30–40	51
40–50	6
50–60	3

Ans. We first prepare the cumulative frequency distribution table by less than method as given below:

Marks:	0-10	10-20	20-30	30-40	40-50	50-60
No. of Students	7	10	23	51	6	2
Marks less than	10	20	30	40	50	60
Cumulative requency	7	17	40	91	97	100

Other than the given class intervals, we assume a class – 10-0 before the first class interval 0-10 with zero frequency.

Now, we mark the upper class limits (including the imagined class) along X-axis on a suitable scale and the cumulative frequencies along Y-axis a suitable scale. Thus, we plot the points

(0, 0), (10, 7), (20, 17), (30, 40), (40, 91), (50, 97), and (60, 100)

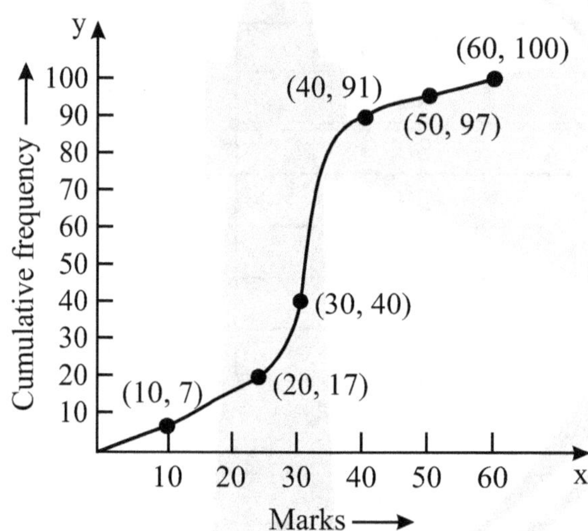

Now, we join the plotted points by a free hand curve to obtain the required ogive.

Q. 18. **Draw a cumulative frequency curve for the following frequency distribution by less than method**

Age (in years)	0 – 9	10 – 19	20 – 29	30 – 39	40 – 49	50 – 59	60 – 69
No. of person	5	15	20	23	17	11	9

Ans. The given frequency distribution is not continuous. So, we first make it continuous and prepare the cumulative frequency distribution as under:

Age (in years)	Frequency	Age less than	Cumulative frequency
– 0.5 – 9.5	5	9.5	5
9.5 - 19.5	15	19.5	20
19.5 - 29.5	20	29.5	40
29.5 - 39.5	23	39.5	63
39.5 - 49.5	17	49.5	80
49.5 - 59.5	11	59.5	91
59.5 - 69.5	9	69.5	100

Innovative Mathematics X-13

Now, we plot points (9.5, 5), (19.5, 20), (29.5, 40), (39.5, 63), (49.5, 80), (59.5, 91) and (69.5, 100) and join them by a free hand smooth curve to obtain the required ogive as shown in Fig.

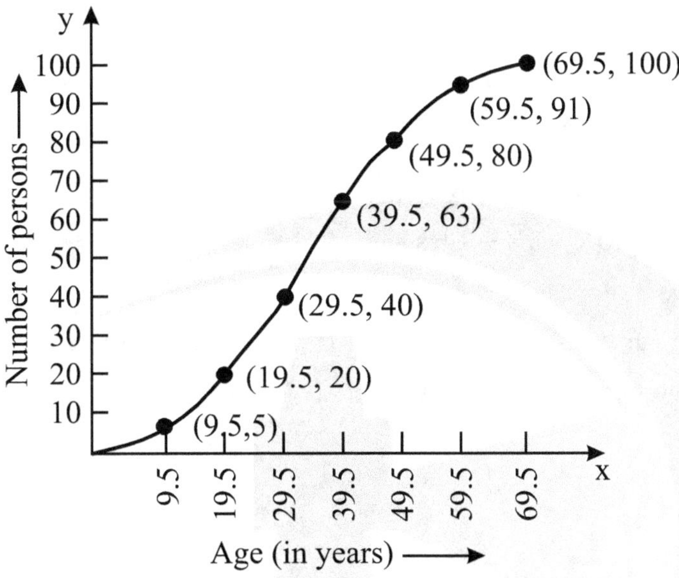

Q. 19. **The temperature of a patient, admitted in a hospital with typhoid fever, taken at different times of the day are given below. Draw the temperature-time graph to reprents the data:**

Time (in hours)	6:00	8:00	10:00	12:00	14:00	16:00	18:00
Temperature (in °F)	102	100	99	103	100	102	99

Ans. In order to draw the temperature-time graph, we represent time (in hours) on the x-axis and the temperature in °F on the y-axis. We first plot the ordered pairs (6, 102), (8, 100), (10, 99), (12, 103), (14, 100), (16, 102) and (18, 99) as points and then join them by line segments as shown in Fig.

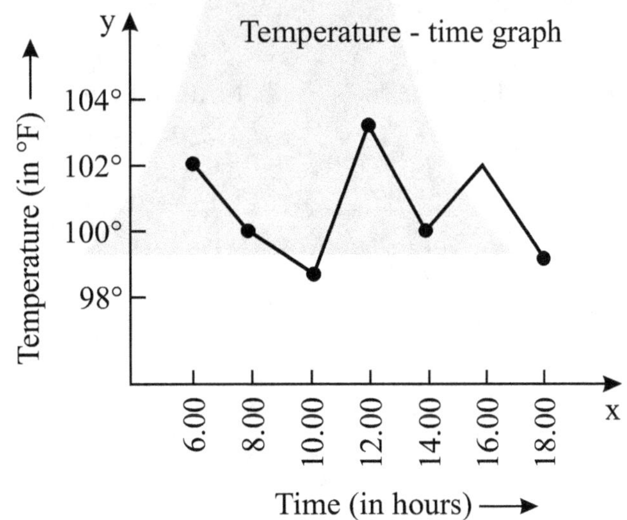

Q. 23. If the mean of n observations $ax_1, ax_2, ax_3,..., ax_n$ is a $\overline{X}$, show that
$$(ax_1 - a\overline{X}) + (ax_2 - a\overline{X}) +...+ (ax_n - a\overline{X}) = 0$$

Ans. We have

$$a\overline{X} = \frac{ax_1 + ax_2 +...+ ax_n}{n}$$

$$\Rightarrow \quad ax_1 + ax_2 +...+ ax_n = n(a\overline{X}) \qquad ....(i)$$

Now, $(ax_1 - a\overline{X}) + (ax_2 - a\overline{X}) +...+ (ax_n - a\overline{X})$

$= (ax_1 + ax_2 +...+ ax_n) - (a\overline{X} + a\overline{X} +...+ a\overline{X}) \qquad$ n-term

$= n(a\overline{X}) - n(a\overline{X}) = 0.$

Q. 24. The mean of n observations $x_1, x_2,...,x_n$ is X. If (a – b) is added to each of the observations, show that the mean of the new set of observations is $\overline{X}$ + (a – b)

Ans. We have,

$$\overline{X} = \frac{x_1 + x_2 +...+ x_n}{n} \qquad ....(i)$$

Let $\overline{X}'$, be the mean of $x_1 + (a - b), x_2 + (a - b), ...,x_n + (a - b)$. Then,

$$\overline{X} = \frac{\{x_1 + (a - b)\} + \{x_2 + (a - b)\} +...+ \{x_n + (a - b)\}}{n}$$

$$\Rightarrow \quad \overline{X}' = \frac{x_1 + x_2 +...+ x_n + n(a - b)}{n}$$

$$= \overline{X} + (a - b) \qquad \text{(using (i))}$$

Q. 25. Find the sum of the deviations of the variate values 3, 4, 6, 8, 14 from their mean.

Ans. Recall that the deviations of the values $x_1, x_2, x_3, ..., x_n$ about A are

$x_1 - A, x_2 - A, x_3 - A,..., x_n - A.$

Let $\overline{X}$ be the mean of the values 3, 4, 6, 8, 14. Then,

$$\overline{X} = \frac{3+4+6+8+14}{5} = \frac{35}{5} = 7$$

Now, sum of the deviations of the values 3, 4, 6, 8, 14 from their mean $\overline{X} = 7$ is given by

$= (3 - 7) + (4 - 7) + (6 - 7) + (8 - 7) + (14 - 7) = -4 - 3 - 1 + 1 + 7 = 0.$

<h1 style="text-align:center">WARM UP – LEVEL-II</h1>

Q. 1. The mean of 40 observations was 160. It was detected on rechecking that the value of 165 was wrongly copied as 125 for computation of mean. Find the correct mean.

Ans. $\odot$ Here, $n = 40$, $\overline{X} = 160$

So, $\overline{X} = \dfrac{1}{n}\left(\sum x_i\right) \Rightarrow 160 = \dfrac{1}{40}\left(\sum x_i\right)$

$\Rightarrow \quad \sum x_i = 160 \times 40 = 6400$

$\Rightarrow \quad$ Incorrect value of $\sum x_i = 6400$

Now,

$\sum x_i$

$= $ Incorrect value of $\sum x_i -$ Incorrect item $+$ Correct item

$\Rightarrow \quad$ Correct value of $\sum x_i = 6400 - 125 + 165 = 6440$

$\therefore \quad$ Correct mean

$= \dfrac{\text{Correct value of } \sum x_i}{n} = \dfrac{6440}{40} = 161.$

Q. 2. The mean of 10 numbers is 20. If 5 is subtracted from every number, what will be the new mean?

Ans. Let $x_1, x_2, \ldots, x_{10}$ be 10 numbers with their mean equal to 20. Then,

$$\overline{X} = \dfrac{1}{n}\left(\sum x_i\right)$$

$\Rightarrow \quad 20 = \dfrac{x_1 + x_2 + \ldots + x_{10}}{10}$

$\Rightarrow \quad x_1 + x_2 + \ldots + x_{10} = 200 \qquad \text{...(i)}$

New numbers are $x_1 - 5, x_2 - 5, \ldots, x_{10} - 5$.

Let X' be the mean of new numbers.

Then,

$$\overline{X}' = \dfrac{(x_1 - 5) + (x_2 - 5) + \ldots + (x_{10} - 5)}{10}$$

$$\overline{X}' = \dfrac{(x_1 + x_2 + \ldots + x_{10}) - 5 \times 10}{10} = \dfrac{200 - 50}{10} \qquad \text{[Using (i)]}$$

$$\overline{X}' = 15.$$

Q. 3. Neeta and her four friends secured 65, 78, 82, 94 and 71 marks in a test of mathematics. Find the average (arithmetic mean) of their marks.

Ans. Arithmetic mean or average

$$= \frac{65 + 78 + 82 + 94 + 71}{5} = \frac{390}{5} = 78$$

Hence, arithmetic mean = 78

Q. 4. The marks obtained by 10 students in physics out of 40 are 24, 27, 29, 34, 32, 19, 26, 35, 18, 21. Compute the mean of the marks.

Ans. Mean of the marks is given by

$$\bar{x} = \frac{24 + 27 + 29 + 34 + 32 + 19 + 26 + 35 + 18 + 21}{10}$$

$$= \frac{265}{10} = 26.50$$

Q. 5. If $\bar{x}$ denote the mean of $x_1, x_2, ..., x_n$, show that

$$\sum_{i=1}^{n} = (x_i - \bar{x})$$

Ans. $\bar{x} = \dfrac{x_1 + x_2 + ... + x_n}{n}$

$$= x_1 + x_2 + ... + x_n = n\bar{x} \qquad\qquad ...(i)$$
$$= \Sigma(x_1 - \bar{x}) = (x_1 - \bar{x}) + (x_2 - \bar{x}) + ... + (x_n - x_1)$$
$$= (x_1 + x_2 + ... + x_n) - n\bar{x} = n\bar{x} - n\bar{x}$$
$$= 0 \qquad\qquad\qquad\qquad\qquad \text{(from (i))}$$

Q. 6. Find the mean of the following distribution:

x :	4	6	9	10	15
f :	5	10	10	7	8

Ans. Calculation of Arithmetic Mean

x_i	f_i	$f_i x_i$
4	5	20
6	10	60
9	10	90
10	7	70
15	8	120
$N = \Sigma f_i = 40$		$\Sigma f_i x_i = 360$

$$\therefore \quad \text{Mean} = \bar{X} = \frac{\Sigma f_i x_i}{\Sigma f_i} = \frac{360}{40} = 9.$$

Q. 7. If the mean of the following data be 9.2, find the value of p.

x	4	6	7	p + 4	12	12
f	5	6	4	10	8	7

Ans. The table is rewritten as below:

x	f	$f_i x$
4	5	20
6	6	36
7	4	28
p + 4	10	10p + 40
12	8	96
14	7	98
Total	40	318 + 10p

Now, Mean $\bar{x} = \dfrac{\Sigma f.x}{\Sigma f} = \dfrac{318 + 10.p}{40}$

$\therefore \quad 9.2 = \dfrac{318 + 10.p}{40}$

$\Rightarrow \quad 318 + 10.p = 368 \Rightarrow 10p = 50 \Rightarrow p = 5$

Q. 8. The marks of 30 students are given below, find the mean marks.

Marks	Number of Students
10	4
11	3
12	8
13	6
14	7
15	2

Ans.

x	f	fx
10	4	40
11	3	33
12	8	96
13	6	78
14	7	98
15	2	30
	$\Sigma f = 30$	$\Sigma fx = 375$

Mean $= \dfrac{\Sigma fx}{\Sigma f} = \dfrac{375}{30} = 12.5$

Q. 9. Calculate the mean for the following distribution:

Variable	5	6	7	8	9
Frequency	4	8	14	11	3

Ans.

x	f	fx
5	4	20
6	8	48
7	14	98
8	11	88
9	3	27
Total	$N = \Sigma f = 40$	$\Sigma fx = 281$

$$\therefore \quad \text{Mean} = \frac{\Sigma fx}{\Sigma f} = \frac{281}{40} = 7.025$$

Q. 10. Find the mean of the following frequency distribution:

Class Interval	Frequency
10 – 30	90
30 – 50	20
50 – 70	30
70 – 90	20
90 – 110	40

Ans.

Class interval	f	Mid value (x)	fx
10 – 30	90	20	1800
30 – 50	20	40	800
50 – 70	30	60	1800
70 – 90	20	80	1600
90 – 110	40	100	4000
	$\Sigma f = 200$		$\Sigma fx = 10000$

$$\text{Mean } (\bar{x}) = \frac{\Sigma fx}{\Sigma f} = \frac{10000}{200} = 50$$

Q. 11. Find the mean of the following distribution by direct method.

Class interval	0 – 10	11 – 20	21 – 30	31 – 40	41 – 50
Frequency	3	4	2	5	6

Ans.

Class interval	Frequency f	Mid value x	fx
0 – 10	3	5.0	15.0
11 – 20	4	15.5	62.0
21 – 30	2	25.5	51.0
31 – 40	5	35.5	177.5
41 – 50	6	45.5	273.0
	$\Sigma f = 20$		$\Sigma fx = 578.5$

$$\therefore \quad \text{Mean} = \frac{\Sigma fx}{\Sigma f} = \frac{578.5}{20} = 28.9$$

Q. 12. For the following distribution, calculate mean using all the suitable methods.

Size of Item	1 – 4	4 – 9	9 – 16	16 – 27
Frequency	6	12	26	20

Ans.

Size of item	Mid value (x_i)	Frequency (f_i)	$f_i x_i$
1 – 4	2.5	6	15
4 – 9	6.5	12	78
9 – 16	12.5	26	32.5
16 – 27	21.5	20	430
		$\Sigma f_i = 64$	$\Sigma f_i x_i = 848$

$$\text{Mean} = \frac{\Sigma f_i x_i}{\Sigma f_i} = \frac{848}{64} = 13.25$$

Q. 13. Calculate the arithmetic mean of the following frequency distribution:

Class interval	50 – 60	60 – 70	70 – 80	80 – 90	90 – 100
Frequency	8	6	12	11	13

Ans. Let assumed mean = 75 i.e., a = 75

Class	frequency f_i	Mid value (x_i)	$d_i = x_i - 75$	$f_i d_i$
50 – 60	8	55	–20	–160
60 – 70	6	65	–10	–60
70 – 80	12	75	0	0
80 – 90	11	85	10	110
90 – 100	13	95	20	260
	$\Sigma f = 50$			$\Sigma f_i d_i = 150$

$a = 75$, $\Sigma f_i d_i = 150$, $\Sigma f_i = 50$

$$\text{Mean } (\overline{x}) = a + \frac{\Sigma f_i d_i}{\Sigma f_i} = 75 + \frac{150}{50} = 78$$

Q. 14. Thirty women were examined in a hospital by a doctor and the number of heart beats per minute were recorded and summarised as follows. Find the mean heart beats per minute for these women, choosing a suitable method.

Number of heart beats per minute	Frequency
65 – 68	2
68 – 71	4
71 – 24	3
74 – 77	8
77 – 80	7
80 – 83	4
83 – 86	2

Ans. Let assumed mean $a = 75.5$

No. of heart beats	No. of women f	Mid value (x)	$d = x - a$	fd
65 – 68	2	66.5	–9	–18
68 – 71	4	69.5	–6	–24
71 – 74	3	72.5	–3	–9
74 – 77	8	75.5	0	0
77 – 80	7	78.5	3	21
80 – 83	4	81.5	6	24
83 – 86	2	84.5	9	18
	$\Sigma f = 30$			$\Sigma fd = 12$

$$\text{Mean} = a + \frac{\Sigma fd}{\Sigma f} = 75.5 + \frac{12}{30} = 75.5 + 0.4 = 75.9$$

Q. 15. To find out the concentration of SO_2 in the air (in parts per million, i.e. ppm), the data was collected for 30 localities in a certain city and is presented below:

Concentration of SO_2 (in ppm)	Frequency
0.00 – 0.04	4
0.04 – 0.08	9
0.08 – 0.12	9
0.12 – 0.16	2
0.16 – 0.20	4
0.20 – 0.24	2

Find the mean concentration of SO_2 in the air.

Ans. Let the assumed mean a = 0.10.

Concentration of SO_2 (in ppm)	Frequency f	Mid value x_i	$u_i = \dfrac{x_i - 0.10}{0.04}$	$f_i u_i$
0.00 – 0.04	4	0.02	–2	–8
0.04 – 0.08	9	0.06	–1	–9
0.08 – 0.12	9	0.10	0	0
0.12 – 0.16	2	0.14	1	2
0.16 – 0.20	4	0.18	2	8
0.20 – 0.24	2	0.22	3	6
	$\Sigma f_i = 30$			$\Sigma f_i u_i = -1$

By step deviation method

$$\text{Mean} = a + \frac{\Sigma f_i u_i}{\Sigma f_i} \times h$$

$$= 0.10 + \frac{-1}{30} \times 0.04$$

$$= 0.10 - 0.0013$$

$$= 0.0987$$

$$= 0.099 \text{ ppm}$$

Q. 16. Find the mean marks from the following data by step deviation method

Marks	Number of students
Below 10	5
Below 20	9
Below 30	17
Below 40	29
Below 50	45
Below 60	60
Below 70	70
Below 80	78
Below 90	83
Below 100	85

Ans. Let assumed mean = 55 $\Rightarrow$ a = 55

Class interval	Frequency f_i	Mid value x_i	$u_i = \dfrac{x_i - 55}{10}$	$f_i u_i$
0 – 10	5	5	–5	–8
10 – 20	4	15	–4	–9
20 – 30	8	25	–3	0
30 – 40	12	35	–2	2
40 – 50	16	45	–1	8
50 – 60	15	55	0	0
60 – 70	10	65	1	10
70 – 80	8	75	2	16
80 – 90	5	85	3	15
90 – 100	2	95	4	8
	$\Sigma f_i = 85$			$\Sigma f_i u_i = -56$

Here, $a = 55$, $h = 10$,

$\Sigma f_i = 85$, $\Sigma f_i u_i = -56$

$$\text{Mean} (x) = a + \frac{\Sigma f_i u_i}{\Sigma f_i}$$

$$h = 55 + \frac{-56}{85} \times 10$$

$$= 55 - 6.59 = 48.41$$

Hence, mean mark = 48.41.

Q. 17. **The following distribution show the daily pocket allowance of children of a locality. The mean pocket allowance is Rs. 18.00. Find the missing frequency f.**

Class interval	11 – 13	13 – 15	15 – 17	17 – 19	19 – 21	21 – 23	23 – 25
Frequency	7	6	9	13	f	5	4

Ans. we have,

Class interval	Frequency	Mid value x	fx
11 – 13	7	12	84
13 – 15	6	14	84
15 – 17	9	16	144
17 – 19	13	18	234
19 – 21	f	20	20f
21 – 23	5	22	110
23 – 15	4	24	96
	$\Sigma f = 44+f$		$\Sigma fx = 752+20f$

$$\text{Mean } \overline{x} = \frac{\Sigma fx}{\Sigma f} \implies 18 = \frac{752 + 20f}{44 + f}$$

$\implies \quad 18\,(44 + f) = 752 + 20f$

$\implies \quad 752 + 20f = 792 + 18f$

$\implies \quad 2f = 40$

$\implies \quad f = 20$

Hence, the missing frequency is 20.

Q. 18. **The arithmetic mean of the following frequency distribution is 50. Find the value of p.**

Class interval	0 – 20	20 – 40	40 – 60	60 – 80	80 – 100
Frequency	17	P	32	24	19

Ans.

Class interval	Frequency (f)	Mid value (x)	fx
0 – 20	17	10	170
20 – 40	P	30	30 P
40 – 60	32	50	1600
60 – 80	24	70	1680
80 – 100	19	90	1710
	$\Sigma f = 92 + P$		$\Sigma fx = 5160 + 30P$

$$\text{Mean } \overline{x} = \frac{\Sigma fx}{\Sigma f} \implies 50 = \frac{5160 + 30P}{92 + P}$$

$\implies \quad 50\,(92 + P) = 5160 + 30\,P$

$\implies \quad 4600 + 50\,P = 5160 + 30\,P$

$\implies \quad 20\,P = 560 \qquad \implies P = 28$

Q. 19. **Find the median of the following data:**

25, 34, 31, 23, 22, 26, 35, 28, 20, 32

Ans. Arranging the data in ascending order, we get 20, 22, 23, 25, 26, 28, 31, 32, 34, 35

Here, the number of observation n = 10 (even).

$$\therefore \quad \text{Median} = \frac{\text{Value of } \left(\dfrac{10}{2}\right)^{th} \text{ observation } + \text{ Value of } \left(\dfrac{10}{2} + 1\right)^{th} \text{ observation}}{2}$$

$$\implies \quad \text{Median} = \frac{\text{Value of } 5^{th} \text{ observation } + \text{ value of } 6^{th} \text{ observation}}{2}$$

$$\therefore \quad \text{Median} = \frac{26 + 28}{2} = 27$$

Hence, median of the given data is 27.

Q. 20. Find the median of the following data : 19, 25, 59, 48, 35, 31, 30, 32, 51. If 25 is replaced by 52, what will be the new median.

Ans. Arranging the given data in ascending order, we have 19, 25, 30, 31, 32, 35, 48, 51, 59

Here, the number of observations n = 9 (odd)

Since the number of observations is odd. Therefore,

$$\text{Median} = \text{Value of} \left(\frac{9+1}{2} \right) \text{ the observations}$$

$\Rightarrow$ Median = value of 5th observations = 32.

Hence, Median = 32

If 25 is replaced by 52, then the new observations arranged in ascending order are:

19, 30, 31, 32, 35, 48, 51, 52, 59

$\therefore$ New median = Value of 5th observations = 35.

Q. 21. Find the median of the following data:

(i) 8, 10, 5, 7, 12, 15, 11

(ii) 12, 14, 10, 7, 15, 16

Ans. (i) 8, 10, 5, 7, 12, 15, 11

These numbers are arranged in an order

5, 7, 8, 10, 11, 12, 15

The number of observations = 7 (odd)

$$\Rightarrow \text{Median} = \frac{7+1}{2} = 4\text{th term}$$

$\Rightarrow$ Median = 10

(ii) 12, 14, 10, 7, 15, 16

These numbers are arranged in an order

7, 10, 12, 14, 15, 16

The number of observations = 6 (even)

The medians will be mean of $\dfrac{6}{2}$ = 3rd and 4th terms i.e., 12 and 14

$$\Rightarrow \text{The median} = \frac{12+14}{2} = 13$$

Q. 22. The following data have been arranged in desending orders of magnitude 75, 70, 68, x + 2, x – 2, 50, 45, 40

If the median of the data is 60, find the value of x.

Ans. The number of observations are 8, the median will be the average of 4th and 5th number

$$\Rightarrow \text{Median} = \frac{(x+2) + (x-2)}{2}$$

$$\Rightarrow 60 = \frac{2x}{2}$$

$$\Rightarrow \quad x = 60$$

Q. 23. **Find the median of 6, 8, 9, 10, 11, 12 and 13.**

Ans. Total number of terms = 7

The middle terms = $\dfrac{1}{2}(7 + 1) = $ 4th

Median = Value of the 4th term = 10.

Hence, the median of the given series is 10.

Q. 24. **Find the median of 21, 22, 23, 24, 25, 26, 27 and 28.**

Ans. Total number of terms = 8

Median

$$= \text{Value of } \frac{1}{2}\left[\frac{8}{2}\text{th term} + \left(\frac{8}{2} + 1\right)\text{th term}\right]$$

$$= \text{Value of } \frac{1}{2}\,[\text{4th term} + \text{5th term}]$$

$$= \frac{1}{2}\,[24 + 25] = \frac{49}{2} = 24.5$$

Q. 25. **Find the median of the following distribution:**

Wages (in Rs)	No. of labourers
200 – 300	3
300 – 400	5
400 – 500	20
500 – 600	10
600 – 700	6

Ans. We have,

Wages (in Rs)	No. of labours	Less than type cumulative frequency
200 – 300	3	3
300 – 400	5	8 = C
400 – 500	20 = f	28
500 – 600	10	38
600 – 700	6	44

Here, the median class is 400 – 500 as $\dfrac{44}{2}$ i.e. 22 belongs to the cumulative frequency of this class interval.

Lower limit of the median class = λ = 400 width of the class interval = h = 100

Cumulative frequency preceding median class frequency = C = 8

Frequency of Median class = f = 20

$$\text{Median} = \lambda + b\left(\dfrac{\dfrac{N}{2} - C}{f}\right) = 400 + 100\left(\dfrac{\dfrac{44}{2} - 8}{20}\right)$$

$$= 400 + 100\left(\dfrac{22 - 8}{20}\right) = 400 + 100\left(\dfrac{14}{20}\right)$$

$$= 400 + 70 = 470$$

Hence, the median of the given frequency distribution is 470.

Q. 26. **The following table shows the weekly drawn by number of workers in a factory:**

Weekly Wages (in Rs.)	0 – 100	100 – 200	200 – 300	300 – 400
No. of workers	40	39	34	30

Find the median income of the workers.

Ans.

Weekly Wages (in Rs.)	No. of workers	Less than type cumulative frequency
0 – 100	40	40
100 – 200	39	79 = C
200 – 300	34 = f	113
300 – 400	30	143
400 – 500	45	188

Since $\dfrac{188}{2} = 94$ belongs to the cumulative frequency of the median class interval (200 – 300), so 200 – 300 is the median class.

Lower limit of the median class interval = λ = 200.

Width of the class interval = h = 100

Total frequency = N = 188

Frequency of the median class = f = 34

Cumulative frequency preceding median class = C = 79

$$\text{Median} = \lambda + \left(\dfrac{\dfrac{N}{2} - C}{f}\right)h = 200 + \left(\dfrac{\dfrac{188}{2} - 79}{34}\right)100$$

$$= 200 + \left(\dfrac{94 - 79}{34}\right)100 = 200 + 44.117$$

$$= 244.117$$

Hence, the median of the given frequency distribution = 244.12.

Q. 27. **The following frequency distribution gives the monthly consumption of electricity of 68 consumers of a locality. Find the median and mode of the data and compare them.**

Monthly consumption	Number of consumers
65 – 85	4
85 – 105	5
105 – 175	13
125 – 145	20
145 – 165	14
165 – 185	8
185 – 205	4

Ans.

Monthly consumption	Number of consumers	Less than type cumulative frequency
65 – 85	4	4
85 – 105	5	9
105 – 125	13	22 = C
125 – 145	20 = f	42
145 – 165	14	56
165 – 185	8	64
183 – 205	4	68

Since $\dfrac{68}{2}$ belongs to the cumulative frequency (42) of the class interval 125 – 145, therefore 125 – 145 is the median class interval.

Lower limit of the median class interval = λ = 125.

Width of the class interval = h = 20

Total frequency = N = 68

Cumulative frequency preceding median class frequency = C = 22

Frequency of the median class = f = 20

$$\text{Median} = \lambda + \left(\frac{\frac{N}{2} - C}{f}\right) h = 125 + \left(\frac{\frac{68}{2} - 22}{20}\right) 20$$

$$= 125 + \frac{12 \times 20}{20} = 125 + 12 = 137$$

The frequency of class 125 – 145 is maximum i.e., 20, this is the modal class,

$x_k = 125$, $f_k = 20$, $f_{k-1} = 13$, $f_{k+1} = 14$, $h = 20$

$$\text{Mode} = x_k + \frac{f - f_{k-1}}{2f - f_{k-1} - f_{k+1}}$$

$$= 125 + \frac{20 - 13}{40 - 13 - 14} \times 20$$

$$= 125 + \frac{7}{40 - 27} \times 20 = 125 + \frac{7}{13} \times 20$$

$$= 125 + 10.77 = 135.77$$

Q. 28. **Compute the median from the marks obtained by the students of class X.**

Marks	Number of Students
40 – 49	5
50 – 59	10
60 – 69	20
70 – 79	30
80 – 89	20
90 – 99	15

Ans. First we will form the less than type cumulative frequency distribution and we make the distribution continuous by subtracting 0.5 from the lower limits and adding 0.5 to the upper limits.

Marks	Number of students	Less than type cumulative frequency
39.5 – 49.5	5	5
49.5 – 59.5	10	15
59.5 – 69.5	20	35 = C
69.5 – 79.5	30 = f	65
79.5 – 89.5	20	85
89.5 – 99.5	15	100

Since $\dfrac{100}{2}$ belongs to the cumulative frequency

(65) of the class interval 69.5 – 79.5, therefore 69.5 – 79.5 is the median class.

Lower limit of the median class = λ = 69.5.

Width of the class interval = h = 10

Total frequency = N = 100

Cumulative frequency preceding median class frequency = C = 35

Frequency of median class = f = 30

$$\text{Median} = \lambda + b\left(\dfrac{\dfrac{N}{2} - C}{f}\right) = 69.5 + 10\left(\dfrac{\dfrac{100}{2} - 35}{30}\right)$$

$$= 69.5 + 10\left(\dfrac{50 - 35}{30}\right) = 69.5 + \dfrac{10 \times 15}{30}$$

$$= 69.5 + 5 = 74.5$$

Hence, the median of given frequency distribution is 74.50.

Q. 29. **Recast the following cumulative table in the form of the ordinary frequency distribution and determine the median.**

No. of days absent	No. of students
less than 5	29
less than 10	224
less than 15	465
less than 20	582
less than 25	634
less than 30	644
less than 35	650
less than 40	653
less than 45	655

Ans.

No. of days	No. of students	No. of days absent	No. of students	Less than type cumulative frequency
less than 5	29	0 – 5	29	29
less than 10	224	5 – 10	195	224 = C
less than 15	465	10 – 15	241 = f	465
less than 20	582	15 – 20	117	582
less than 25	634	20 – 25	52	634
less than 30	644	25 – 30	10	644
less than 35	650	30 – 35	6	650
less than 40	653	35 – 40	3	653
less than 45	655	40 – 45	2	655

Since $\dfrac{655}{2}$ belongs to the cumulative frequency (465) of the class interval $10 - 15$, therefore $10 - 15$ is the median class.

Lower limit of the median class $= \lambda = 10$.

Width of the class interval $= h = 5$

Total frequency $= N = 655$

Cumulative frequency preceding median class frequency $= C = 224$

Frequency of median class $= f = 241$

$$\text{Median} = \lambda + h\left(\dfrac{\dfrac{N}{2} - C}{f}\right) = 10 + 5\left(\dfrac{\dfrac{655}{2} - 224}{241}\right)$$

$$= 10 + 5\left(\dfrac{327 \times 224}{241}\right) = 10 + \dfrac{5 \times 103.5}{241}$$

$$= 10 + 2.147 = 12.147$$

Hence, the median of given frequency distribution is 12.147.

VERY SHORT ANSWER TYPE (I) Q.UESTIONS

Q. 1.   **What is the mean of first 12 prime numbers?**

Sol.

Q. 2.   **The mean of 20 numbers is 18. If 2 is added to each number, what is the new mean?**

Sol.

Q. 3.   **The mean of 5 observations 3, 5, 7, x and 11 is 7, find the value of x.**

Sol.

Q. 4.   **What is the median of first 5 natural numbers?**

Sol.

Q. 5.   **What is the mode of the observations 5, 7, 8, 5, 7, 6, 9, 5, 10, 6?**

Sol.

Q. 6.   **The mean and mode of a data are 24 and 12 respectively. Find the median.**

Sol.

Q. 7.   **Write the class mark of the class 19.5 – 29.5.**

(i)   If the class intervals of a frequency distribution are $1 - 10, 11 - 20, 21 - 30, ..., 51 - 60$, then the size of each class is:

    (a)   9   (b)   10   (c)   11   (d)   5.5

(ii)   If the class intervals of a frequency distribution are $1 - 10, 11 - 20, 21 - 30 ....., 61 - 70$, Then the upper limit of $21 - 30$ is:

    (a)   21   (b)   30   (c)   30.5   (d)   20.5

(iii)   Consider the frequency distribution.

Class	0 – 5	6 – 11	12 – 17	18 – 23	24 – 29
Frequency	13	10	15	8	11

The upper limit of median class is:

    (a)   17   (b)   17.5   (c)   18   (d)   18.5

(iv)   Daily wages of a factory workers are recorded as:

Daily wages (in ₹)	121 – 126	127 – 132	133 – 138	139 – 144	145 – 150
No. of workers	5	27	20	18	12

The lower limit of Modal class is:

    (a)   ₹ 127   (b)   ₹ 126   (c)   ₹ 126.50   (d)   ₹ 133

(v)   For the following distribution

Class	0 – 5	5 – 10	10 – 15	15 – 20	20 – 25
Frequency	10	15	12	20	9

The sum of Lower limits of the median class and modal class is   (CBSE 2020)

(a) 15   (b) 25   (c) 30   (d) 35

(vi) The median and mode respectively of a frequency distribution are 26 and 29. Then, its mean is

(CBSE 2020)

(a) 27.5   (b) 24.5   (c) 28.4   (d) 25.8

Sol.

Q. 8. **Find the class-marks of the classes 10-25 and 35-55.** (CBSE 2020)

Sol.

Q. 9. **Fill in the blank**

(a) Mode = 3__________ – 2 __________

(b) Arithmetic mean of all factors of 20 is _________ .

(c) The cumulative frequency of a given class is obtained by adding the frequencies of all the classes __________ .

(d) The mode of a frequency distribution is determined graphically by _________ .

(e) If the mode is 8 and mean is also 8, then median will be _________ .

(f) The measure of central tendency which cannot be determined graphically is _________ .

(g) If the class marks of a continuous frequency distribution are 22, 30, 38, 46, 54, 62 then the class corresponding to class mark 46 is _________ .

(h) Construction of cumulative frequency distribution table is useful in determining _________ .

(i) The step deviation formula for finding mean is _________ .

(j) The formula to find median of grouped data is _________ .

(k) The formula to find mode of grouped data is _________ .

(l) The Range of the observations 255, 125, 130, 160, 185, 170, 103 is _________ .

(m) Class mark = f (_________ + _________) .

(n) The median of 1st ten prime numbers is _________ .

(o) The assumed mean method to find mean is _________ .

Sol.

SHORT ANSWER TYPE Q.UESTIONS (I)

Q. 10. **The mean of 11 observation is 50. If the mean of first Six observations is 49 and that of last six observation is 52, then find sixth observation.**

Sol.

Q. 11. **Find the mean of following distribution:**

x	12	16	20	24	28	32
f	5	7	8	5	3	2

Sol.

Q. 12. **Find the median of the following distribution:**

x	10	12	14	16	18	20
f	3	5	6	4	4	3

Sol.

Q. 13. Find the mode of the following frequency distribution:

Class	$0-5$	$5-10$	$10-15$	$15-20$	$20-25$	$25-30$
Frequency	2	7	18	10	8	5

Sol.

Q. 14. Convert the following deistribution in frequency distribution:

Marks	No. of students
Less than 20	0
Less than 30	4
Less than 40	16
Less than 50	30
Less than 60	46
Less than 70	66
Less than 80	82
Less than 90	92
Less than 100	100

Sol.

Q. 15. Write the following data into less than cummulative frequency distribution table:

Marks	$0-10$	$10-20$	$20-30$	$30-40$	$40-50$
No. of students	7	9	6	8	10

Sol.

Q. 16. Find mode of the following frequency distribution:

Class Interval	$25-30$	$30-35$	$35-40$	$40-45$	$45-50$	$50-55$
Frequency	25	34	50	42	38	14

Sol.

Q. 17. What is the median of the following data? (CBSE 2011)

x	10	20	30	40	50
f	2	3	2	3	1

Sol.

Q. 18. Mean of a frequency distribution ($\bar{x}$) is 45. If $\Sigma f_i = 20$ find $\Sigma f_i x_i$ (CBSE 2011)

Sol.

Q. 19. Find the mean of the following distribution: (CBSE 2020)

Class	$3-5$	$5-7$	$7-9$	$9-11$	$11-13$
Frequency	5	10	10	7	8

Sol.

Q. 20. Find the mode of the following data: (CBSE 2020)

Class	0 – 20	20 – 40	40 – 60	60 – 80	80 –100	100 – 120	120 –140
Frequency	6	8	10	12	6	5	3

Sol.

Q. 21. Compute the mode for the following frequency distribution: (CBSE 2020)

Size of items (in cm)	0 – 4	4 – 8	8 – 12	12 – 16	16 – 20	20 – 24	24 – 28
Frequency	5	7	9	17	12	10	6

Sol.

SHORT ANSWER TYPE Q.UESTIONS (II)

Q. 22. If the mean of the following distribution is 54, find the value of P.

Class	0–20	20–40	40–60	60–80	80–100
Frequency	7	P	10	9	13

Sol.

Q. 23. Find the median of the following frequency distribution:

C.I.	0–10	10–20	20–30	30–40	40–50	50–60
f	5	3	10	6	4	2

Sol.

Q. 24. The median of following frequency distribution is 24 years. Find the missing frequency x.

Age (In years)	0–10	10–20	20–30	30–40	40–50
No. of persons	5	25	x	18	7

Sol.

Q. 25. Find the median of the following data:

Marks	Below 10	Below 20	Below 30	Below 40	Below 50	Below 60
No. of student	0	12	20	28	33	40

Sol.

Q. 26. Find the mean weight of the following data:

Weight (In kg.)	30–35	35–40	40–45	45–50	50–55	55–60
No. of Students	2	4	10	15	6	3

Sol.

Q. 27. Find the mode of the following data:

Height (in cm)	Above 30	Above 40	Above 50	Above 60	Above 70	Above 80
No. of plants	34	30	27	19	8	2

Innovative Mathematics X-13

Sol.

Q. 28. The following table represent marks obtained by 100 students in a test:

Marks obtained	30–35	35–40	40–45	45–50	50–55	55–60	60–65
No. of students	14	16	28	23	18	8	3

Find mean marks of the students. (CBSE 2018-19)

Sol.

Q. 29. The following table represent pocket allowance of children of a colony. The mean pocket allowance is ₹ 18. Find the missing frequency. (CBSE – 2018)

Daily pocket allowance (in ₹)	11–13	13–15	15–17	17–19	19–21	21–23	23–25
No. of students	3	6	9	13	k	5	4

Sol.

Q. 30. Find mode of the following frequency distribution:

Class Interval	0–20	20–40	40–60	60–80	80–100
No. of Students	15	18	21	29	17

The mean of above distribution is 53. Use Empirical formula to find approximate value of median.

Sol.

LONG ANSWER TYPE Q.UESTIONS

Q. 31. The mean of the following data is 53, Find the values of f_1 and f_2.

C.I.	0–20	20–40	40–60	60–80	80–100	Total
f	15	f_1	21	f_2	17	100

Sol.

Q. 32. If the median of the distribution given below is 28.5, find the values of x and y.

C.I.	0–10	10–20	20–30	30–40	40–50	50–60	Total
f	5	8	x	15	y	5	60

Sol.

Q. 33. The median of the following distribution is 35, find the values of a and b.

C.I.	0–10	10–20	20–30	30–40	40–50	50–60	60–70	Total
f	10	20	a	40	b	25	15	170

Sol.

Q. 34. Find the mean, median and mode of the following data:

C.I.	11–15	16–20	21–25	26–30	31–35	36–40	41–45	46–50
f	2	3	6	7	14	12	4	6

Sol.

Q. 35. The rainfall recorded in a city for 60 days is given in the following table:

Raifall (in cm)	0–10	10–20	20–30	30–40	40–50	50–60
No. of Days	16	10	8	15	5	6

Calculate the median rainfall.

Sol.

Q. 36. Find the mean of the following distribution by step-deviation method:

Daily Expenditure (in ₹)	100–150	150–200	200–250	250–300	300–350
No. of Households	4	5	12	2	2

Sol.

Q. 37. The distribution given below show the marks of 100 students of a class:

Marks	0–5	5–10	10–15	15–20	20–25	25–30	30–35	35–40
No. of students	4	6	10	10	25	22	18	5

Find the median marks of the above distribution.

Sol.

Q. 38. The annual profit earned by 30 factories in an industrial area is given below:

Profit (₹ in lakh)	No. of Factories
More than or equal to 5	30
More than or equal to 10	28
More than or equal to 15	16
More than or equal to 20	14
More than or equal to 25	10
More than or equal to 30	7
More than or equal to 35	3
More than or equal to 40	0

Find the median of the above data.

Sol.

Q. 39. Find the mean and median of the following distribution: (CBSE 2018-19)

Class Interval	30 – 40	40 – 50	50 – 60	60 – 70	70 – 80	80 – 90	90 – 100
Frequency	7	5	8	10	6	6	8

Sol.

Q. 40. If mean of the given distirbution is 65.6 find the missing frequency. (CBSE 2017)

Class Interval	10 – 30	30 – 50	50 – 70	70 – 90	90 – 110	110 – 130	Total
Frequency	5	8	f_1	20	f_2	2	50

Sol.

Q. 41. The mode of the frequency distribution is 36. Find the missing frequency (f).   (CBSE 2020)

Class Interval	$0-10$	$10-20$	$20-30$	$30-40$	$40-50$	$50-60$	$60-70$
Frequency	8	10	f	16	12	6	7

Sol.

Q. 42. The mean of the following frequency distribution is 18. The frequency f in the class interval 19-21 is missing. Determine f.   (CBSE 2020)

Class Interval	$11-13$	$13-15$	$15-17$	$17-19$	$19-21$	$21-23$	$23-25$
Frequency	3	6	9	13	f	5	4

Sol.

Q. 43. The following table gives production yield per hectare of wheat of 100 farms of a village:

(CBSE 2020)

Production Yield	$40-45$	$45-50$	$50-55$	$55-60$	$60-55$	$65-70$
Frequency	4	6	16	20	30	24

Find the mode of the above data.

Sol.

Q. 44. Find the unknown entries a, b, c, d, e, f in the following distribution of heights of students in a class: (CBSE 2020)

Height (in cm)	150–155	155–160	160–165	165–170	170–175	175–180
Frequency	12	b	10	d	e	2
Cummulative Frequency	a	25	c	43	48	f

Find the mode of the above data.

Sol.

1. 16.4 approx.   2. 20   3. 9   4. 3

5. $x = 25$ 6. 5   7. Median $= 20$ 8. 24.5

9. (i) B (First make intervals continuous, Then find class size)

   (ii) C

   (iii) B

   (iv) C

   (v) B $\begin{bmatrix} \text{Modal class } 15 - 20 \\ \text{Median class } 10 - 15 \end{bmatrix}$

   (vi) B

10. 17.5 and 45

11. (a) 3 Median $-$ 2 mean   (b) 7   (c) Preceding the given classes

    (d) Histogram   (e) 8   (f) Mean

    (g) 42 $-$ 50 (as difference b/w 2 consecutive observation is 8

    $\therefore$ Subtract $\dfrac{8}{2}$ form 46 for Lower Limit and And $\dfrac{8}{2}$ to 46 for upper Limit)

    (h) Median   (i) $x = a + \dfrac{\Sigma f_i u_i}{\Sigma f_i} \times h$

12.   13.   14.   15.

16.   17.   18.   19.

20.   22.   23.   24.

25.   26.   27.   28.

29.   30.   31.   32.

33.   34.   35.   36.

37.   38.   39.   40.

41.   42.   43.   44.

PRACTICE TEST

SECTION-A (1 mark each)

Q. 1. Find the sum of lower limits of median class and modal class for the following distribution: (1)

Class:	0-5	5-10	10-15	15-20	20-25
Frequency:	10	15	12	20	9

Sol.

Q. 2. Find the upper limit of the median class of the following frequency distribution: (1)

Class:	0-5	6-11	12-17	18-23	24-29
Frequency:	13	10	15	8	11

Sol.

Q. 3. Find the mean of the numbers 1, 2, 3,...n. (1)

Sol.

Q. 4. Which measure of central tendency is obtained from the abscissa of the point of intersection of the less than type and the more than type cumulative frequency curves of a grouped data? (1)

Sol.

Q. 5. For the following distribution, find the modal class. (1)

Marks	Below 10	Below 20	Below 30	Below 40	Below 50	Below 60
No. of students	3	12	27	57	75	80

Sol.

SECTION-B (2 mark each)

Q. 6. The following is the distribution of weights (in kg) of 40 persons: (2)

Weight (kg)	40-45	45-50	50-55	55-60	60-65	65-70	70-75	75-80
No. of persons	4	4	13	5	6	5	2	1

Construct a cumulative frequency distribution (less than type) table for the above data.

Sol.

Q. 7. The frequency distribution table of agricultural holdings in a village is given below: (Exemplar)

Area of land (ha)	1-3	3-5	5-7	7-9	9-11	11-13
No. of families	20	45	80	55	40	12

Calculate the modal agricultural holdings of the village. (2)

Sol.

Q. 8. Construct a cumulative frequency distribution (more than type) of the following distribution: **(2)**

Class	12.5-17.5	17.5-22.5	22.5-27.5	27.5-32.5	32.5-37.5
Frequency	2	22	19	14	13

Sol.

SECTION-C (3 mark each)

Q. 9. The arithmetic mean of the following data is 14. Find the value of k. **(3)**

x_i	5	10	15	20	25
f_i	7	k	8	4	5

Sol.

Q. 10. Candidates of four schools appeared in mathematics test. The data were as follows: **(3)**

School	No. of candidates	Average score
A	60	75
B	Not available	55
C	48	80
D	40	50

If the average score of the candidates of all the four schools was 66, find the no. of candidates appeared from school B.

Sol.

Q. 11. The following table gives the no. of pages written by sarika for completing her own work for 30 days:

No. of pages written per day	16-18	19-21	22-24	25-27	28-30
No. of days	1	3	4	9	13

Find the average no. of pages written by her. **(3)**

Sol.

Q. 12. Find the mean, median and mode of the following frequency distribution. **(3)**

Class	0-10	10-20	20-30	30-40	40-50	50-60	60-70
Frequency	8	7	15	20	12	8	10

Sol.

SECTION-D (4 mark each)

Q. 13. The mean of the following frequency distribution is 57.6 and the sum of observation is 50. Find the missing frequency f_1 and f_2. **(4)**

Class	0-20	20-40	40-60	60-80	80-100	100-120
Frequency	7	f_1	12	f_2	8	5

Sol.

Q. 14. Compute the arithmetic mean for the following distribution. (4)

Marks obtained	No. of students
Below 10	5
Below 20	9
Below 30	17
Below 40	29
Below 50	45
Below 60	60
Below 70	70
Below 80	78
Below 90	83
Below 100	85

Sol.

Q. 15. Form a frequency distribution table and compute arithmetic mean for the following frequency distribution: (4)

Weight in kg	No. of persons
Above 80	0
Above 75	4
Above 70	11
Above 65	22
Above 60	38
Above 55	45
Above 50	48
Above 45	50

Sol.

Q. 16. The median of the following data is 50. Find the value of p and q, if the sum of all the frequencies is 90. (4)

Marks	20-30	30-40	40-50	50-60	60-70	70-80	80-90
No. of students	p	15	25	20	q	8	10

Sol.

Q. 17. Draw 'less than ogive' and 'more than ogive' for the following distribution and hence find its median. (4)

Class	20-30	30-40	40-50	50-60	60-70	70-80	80-90
Frequency	25	15	10	6	24	12	8

Sol.

 50 students enter for a school javelin throw competition. The distance (in metres) thrown are recorded below: (4)

Distance (m)	0-20	20-40	40-60	60-80	80-100
No. of students	6	11	17	12	4

(a) Construct a cumulative frequency table.

(b) Draw a cumulative frequency curve (less than type) and calculate the median distance thrown by using the curve.

(c) Calculate the median distance by using the formula for median.

(d) Are the median distance calculated in (b) and (c) same?

Sol.

ANSWERS

1. 25   **2.** 17.5   **3.** $\dfrac{n+1}{2}$ **5.** 30–40   **6.** 6.2 ha   **9.** 6

10. 52   **11.** 26   **12.** Mean:35.625;Median:35;Mode:33.85)

13. $(f_1 = 8; f_2 = 10)$ **14.** 48.41   **15.** 64.3 kg   **16.** $p = 5; q = 7$

17. Median $= 50$ **18.** 49 : 41

CASE STUDY

CASE STUDY 1.

COVID-19 Pandemic

The COVID-19 pandemic, also known as coronavirus pandemic, is an ongoing pandemic of coronavirus disease caused by the transmission of severe acute respiratory syndrome coronavirus 2 (SARS-CoV-2) among humans.

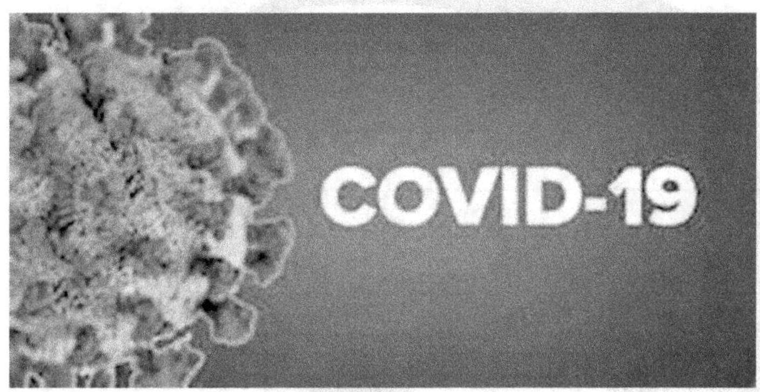

The following tables shows the age distribution of case admitted during a day in two different hospitals

Table 1

Age (in years)	5–15	15–25	25–35	35–45	45–55	55–65
No. of cases	6	11	21	23	14	5

Table 2

Age (in years)	5–15	15–25	25–35	35–45	45–55	55–65
No. of cases	8	16	10	42	24	12

Refer to table 1

Q.. 1. **The average age for which maximum cases occurred is**

(a) 32.24   (b) 34.36   (c) 36.82   (d) 42.24

Q.. 2. **The upper limit of modal class is**

(a) 15   (b) 25   (c) 35   (d) 45

Q.. 3. **The mean of the given data is**

(a) 26.2   (b) 32.4   (c) 33.5   (d) 35.4

Refer to table 2

Q.. 4. **The mode of the given data is**

(a) 41.4   (b) 48.2   (c) 55.3   (d) 64.6

Q.. 5. **The median of the given data is**

(a) 32.7   (b) 40.2   (c) 42.3   (d) 48.6

1. (c) 36.82   **2.** (d) 45   **3.** (d) 35.4

4. (a) 41.4   **5.** (b) 40.2

Electricity Energy Consumption

CASE STUDY 2.

Electricity energy consumption is the form of energy consumption that uses electric energy. Global electricity consumption continues to increase faster than world population, leading to an increase in the average amount of electricity consumed per person (per capita electricity consumption).

Tariff	: LT-Residental	Bill Number	: 384756
Type of Supply	: Single Phase	Connected Load	: 3 kw
Meter Reading Date	: 31-11-13	Meter Reading	: 65789
Previous Reading	: 31-10-13	Previous Meter Reading	: 65500
		Units Consumed	: 269

A survey is conducted for 56 families of a Colony A. The following tables gives the weekly consumption of electricity of these families.

Weekly consumption (in units)	0–10	10–20	20–30	30–40	40–50	50–60
No. of families	16	12	18	6	4	0

The similar survey is conducted for 80 families of Colony B and the data is recorded as below:

Weekly consumption (in units)	0–10	10–20	20–30	30–40	40–50	50–60
No. of families	0	5	10	20	40	5

Refer to data received from Colony A

Q.. 1. **The median weekly consumption is**

   (a) 12 units   (b) 16 units   (c) 20 units   (d) None of these

Q.. 2. **The mean weekly consumption is**

   (a) 19.64 units   (b) 22.5 units   (c) 26 units   (d) None of these

Q.. 3. **The modal class of the above data is I**

   (a) 0-10   (b) 10-20   (c) 20-30   (d) 30-40

Refer to data received from Colony B

Q.. 4.   **The modal weekly consumption is**

   (a)   38.2 units   (b)   43.6 units   (c)   26 units   (d)   32 units

Q.. 5.   **The mean weekly consumption is**

   (a)   15.65 units   (b)   32.8 units   (c)   38.75 units   (d)   48 units

ANSWERS

1. (c) 20 units   **2.** (a) 19.64 units   **3.** (c) 20-30 units

4. (b) 43.6 units   **5.** (c) 38.75 units

CHAPTER-14

PROBABILITY

LEARNING OBJECTIVES

FUNDAMENTALS

In class IX, you have been introduced to the 'Theory of Probability' as a branch of mathematics whcih deals with the measure of uncertainty. Also, you were told that there are different approaches to the study of probability, namely, empirical (or experimental) approach, theoretical (or classical) approach, and axiomatic approach. Not only that, you already have an idea about the empirical probability and the method to calculate the same. In this chapter, we introduce you to the theoretical approach to probability or we may say, 'theoretical probability'. Also, we shall solve some simple problems on probability, using this appraoch.

It may be noted here that empirical probability is based on the result of an actual experiment, while theoretical probability tries to predict what will happen without actully performing the experiment. It is observed that the empirical probability of an event approaches to its theoretical probability as the number of trials of an experiment is increased (or made very large).

Before we go ahead, let us recalJ some of the basic terms and results that we have learnt in the previous class alongwith a few new ones which are normally used in the study of probability.

15.1 Some Important Terms and Results Related to Probability

Experiment: An activity which ends in some well defined result is called an experiment.

Trial: Performing an experiment once is called a trial.

Outcome: The result of an experiment is known as out come. An experiment may result in one or several out comes. For example, when we toss a coin, we have two outcomes, a head and a tail. Similarly, when we throw a die, the possible outcomes are six (1, 2, 3, 4, 5 and 6).

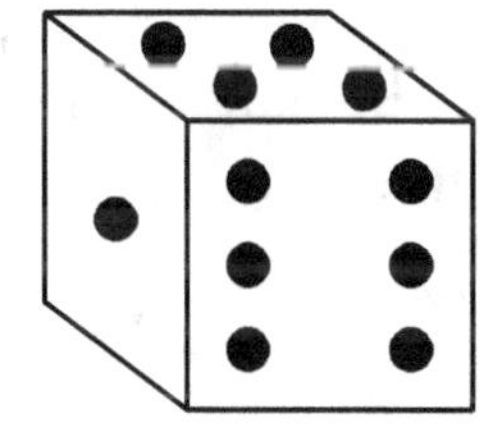

Note: A Die is a small cube used for playing indoor games. Its six faces are marked with numbers from 1 to 6, one number on one face. Many times dots are put in place of numbers. The plural of die is dice.

Event: The possible outcomes of an experiment are called events. Getting an even number on throwing a die is an event with outcomes 2, 4 and 6. We may say an Fig. 15.1 event for an experiment is the collection of some outcomes of the experiment.

Impossible Event: An event which cannot occur is called an impossible event. For example, When we throw a die, we cannot get a number greater than 6. Therefore the event, getting a number, say 7, on throwing a die, is an impossible event. The probability of an impossible event is zero (0).

Sure Event: An event which is certain to occur is called a sure event. For example, in case of throwmg a die, the event, getting a number less than 7, is a sure event. The probability of a sure event is one (1).

Equally Likely Events: Two events are said to be 'equally likely', if one of them cannot be expected in preference to the other. For example, when we throw a die once, each of the numbers (1, 2, 3, 4, 5 and 6) has the same possibility of showing up. So the equally likely outcomes of throwing a die are 1, 2, 3, 4, 5 and 6.

Elementary Event: An event having only one outcome of the experiment is called an elementary event.

Note that, the sum of the probabilities of nil the elementan; events of an experiment is always 1.

Complementry Events: If E denotes the happening of an event, and 'not E' (or E) its not happening, then 'E' and 'not E' are complementary events.

Note that, the sum of the probability of the happening of an event and the probability of its not happening is always 1.

i.e. $\qquad$ P(E) + P(not E) = 1

or $\qquad$ P(E) + P($\overline{\text{E}}$) = 1

or $\qquad$ P($\overline{\text{E}}$) = 1 – P(E)

15.2 Probability – A Theoretical Approach

The probability of a given event is an expression of the likelihood of its occurrence. It is given by a number whose value ranges from zero to one, zero (0) for an event which cannot.occur and one (1) for an event which is certain to occur. According to theoretical approach, this value (the value of probability) is measured as 'the ratio of the number of favourable' cases to the total number of equally likely cases'. If the probability of occurrence of an event E is denoted by P(E), we have:

$$P(E) \; = \; \frac{\text{Number of outcomes favourable to E}}{\text{Total number of possible outcomes of the experiment}}$$

For example, if a coin is tossed, there are two possible outcomes, a head or a tail, hence the probability of getting a head $= \dfrac{1}{2}$. Similarly, if a die is thrown, there are six equally likely or possible outcomes : 1, 2, 3, 4, 5, and 6.

Now, if want to calculate the probability of getting an even number on throwing a die, the favourable cases are three, i.e., 2, 4, and 6.

$\therefore \qquad$ P (an even number) $= \dfrac{3}{6} = \dfrac{1}{2}$

Remarks: The value of probability of an event cannot be negative or greater than 1.

15.3 About the Playing Cards

The following information related to playing cards may be useful in solving relevant problems:

*   A deck (or pack) of playmg cards consists of 52 cards.
*   Out of these 52 cards, 26 are red coloured and another 26 are black coloured.

- 26 red cards are further divided into 2 suits, i.e., Diamonds ($\blacklozenge$) and Hearts ($\heartsuit$), with 13 cards of each.

- Similarly, 26 black cards are also further divided into 2 suits, i.e., Clubs ($\clubsuit$) and Spades ($\spadesuit$), with 13 cards of each.

- We can also say that the 52 cards are divided into 4 suits (Diamonds, Hearts, Clubs and Spades) with 13 cards of each.

- The 13 cards in each suit are 1 (Ace), 2, 3, 4, 5, 6, 7, 8, 9, 10, Jack, Q.ueen and King.

- King, Q.ueen and Jack are called face cards. Since there are total 4 Kings, 4 Q.ueens and 4 Jack (one in each suit), therefore, in a pack of 52 cards, the total number of face cards is 12.

- The total number of non-face cards is $52 - 12 = 40$.

<h1 align="center">WARM-UP – LEVEL-I</h1>

Q. 1. **Find the probability of getting a number less than 5 in a single throw of a die.**

Ans. Favourable outcomes in the given case are four (1, 2, 3, and 4)

Total outcomes $= 6$

$\therefore$ Required probability $P(< 5)$

$$= \frac{\text{Number of favourable outcomes}}{\text{Total outcomes}}$$

$$= \frac{4}{6}$$

$$= \frac{2}{3}$$

Q. 2. **In a single throw of a die, find the probability of having**

(i)   an odd number   (ii)   a number greater than 4

Ans. In a single throw of a die, total outcomes are 6 (1, 2, 3, 4, 5, and 6)

(i) Favourable outcomes are three (1, 3, 5)

$\therefore$ required probability

$$= P(\text{an odd number})$$

$$= \frac{\text{Number of favourable outcomes}}{\text{Total outcomes}}$$

$$= \frac{3}{6} = \frac{1}{2}$$

(ii) Favourable outcomes are two (5, 6)

$\therefore \qquad P(> 4) = \frac{2}{6} = \frac{1}{3}$

Q. 3. **A die is thrown once. Find the probability of getting**

(i)   a prime number

(ii) a number lying between 2 and 6

(iii) an odd number.

Ans. In a single throw of a die, total outcomes are 6 (1, 2, 3, 4, 5 and 6)

(i)   Favourable outcomes for prime numbers are three (2, 3, 5)

$\therefore \qquad P(\text{a prime number}) = \frac{3}{6} = \frac{1}{2}$

(ii) Numbers lying between 2 and 6 are three (3, 4, 5)

$\therefore \qquad P(\text{a number lying between 2 and 6}) = \frac{3}{6} = \frac{1}{2}$

(iii) Favourable outcomes for odd numbers are three (1, 3, 5)

$$\therefore \qquad P(\text{an odd number}) = \frac{3}{6} = \frac{1}{2}$$

Q. 4. **A child has a die whose six faces show the letters as given below:**

Fig. 15.3

The die is thrown once. What is the probability of getting (i) A? (ii) D?

Ans. Total number of outcomes = 6

(i) Since there are two faces with letter A,

$$\therefore \qquad P(A) = \frac{2}{6} = \frac{1}{3}$$

(ii) $$P(D) = \frac{1}{6}$$

Q. 5. **Suppose you drop a die at random on the rectangular region shown in Fig. 15.4. What is the probability that it will land inside the circle with diameter 1m?**

Ans. Total area of the rectangular region $= 1 \times b$

$$= 3 \times 2 = 6 \text{ m}^2$$

Area of the circular region $= \pi r^2 = \pi \left(\frac{1}{2}\right)^2$

$$= \frac{\pi}{4} \text{ m}^2$$

P(the die lands inside the circle)

$$= \frac{\text{Area of the circular region}}{\text{Total area of the ractangular region}} = \frac{\frac{\pi}{4}}{6} = \frac{\pi}{24}$$

Q. 6. **A pair of dice is thrown once. Find the probability of getting a doublet.**

Ans. When a pair of dice is thrown once, there are 36 possible outcomes.

$$(1, 1), (1, 2), (1, 3), (1, 4), (1, 5), (1, 6)$$
$$(2, 1), (2, 2), (2, 3), (2, 4), (2, 5), (2, 6)$$
$$(3, 1), (3, 2), (3, 3), (3, 4), (3, 5), (3, 6)$$
$$(4, 1), (4, 2), (4, 3), (4, 4), (4, 5), (4, 6)$$
$$(5, 1), (5, 2), (5, 3), (5, 4), (5, 5), (5, 6)$$
$$(6, 1), (6, 2), (6, 3), (6, 4), (6, 5), (6, 6)$$

Out of these 36 possible outcomes,

doublets are: (1, 1), (2, 2), (3, 3), (4, 4), (5, 5), (6, 6)

$\therefore$ Number of favourable outcomes = 6

Hence, required probability $= \frac{6}{36} = \frac{1}{6}$

Q. 7. **A pair of dice is thrown once, find the probability of getting a total of 5 on two dice.**

Ans. Number of total outcomes = 36

Favourable outcomes are four : (1, 4), (2, 3), (3, 2), (4, 1)

$\therefore$ required probability $= \dfrac{4}{36} = \dfrac{1}{9}$

Q. 8. **A pair of dice is thrown once, find the probability of getting a number greater than 5 on each die.**

Ans. Total number of possible outcomes = 36

Favourable outcomes is only one i.e., (6, 6)

$\therefore$ required probability $= \dfrac{1}{36}$

Q. 9. **If two dice are thrown once, what is the probability that the sum of digits on the top faces of the dice is less than 6.**

Ans. Total number of possible outcomes = 36

Favourable outcomes are : (1, 1), (1, 2), (1, 3), (1, 4), (2, 1), (2, 2), (2, 3), (3, 1), (3, 2) and (4, 1)

$\therefore$ Number of favourable outcomes = 10

Hence, required probability $= \dfrac{10}{36} = \dfrac{5}{18}$

Q. 10. **A pair of dice is thrown once, find the probability of getting a total of 11.**

Ans. Number of total outcomes = 36

Favourable outcomes are two i.e., (5, 6) and (6, 5)

$\therefore$ Required probability $= \dfrac{2}{36} = \dfrac{1}{18}$

Q. 11. **A pair of dice is thrown once. Find the probability that neither a doublet nor a total of 7 will appear.**

Ans. Number of possible outcomes = 36

Outcomes when either a doublet or a total of 7 appears:

(1, 1), (2, 2), (3, 3), (4, 4), (5, 5), (6, 6), (1, 6), (2, 5), (3, 4), (4, 3), (5, 2) and (6, 1) = 12

$\therefore$ Number of favourable outcomes = 36 – 12 = 24

$\therefore$ required probability $= \dfrac{24}{36} = \dfrac{2}{3}$

Q. 12. **A die is thrown twice. What is the probability that**

(i) 5 will not come up either time? (ii) 5 will come up at least once?

Ans. Total number of possible outcomes = 36

(i) Out of these 36 outcomes, 5 comes up at least once in the following cases:

(1, 5), (2, 5), (3, 5), (4, 5), (5, 5), (6, 5), (5, 1), (5, 2), (5, 3), (5, 4), (5, 6)

Total number of these cases = 11

$\therefore$ Number of cases when 5 will not come up either time = 36 – 11 = 25.

$\therefore$ P(5 will not come up either time) $= \dfrac{25}{36}$

(ii) Number of cases when 5 comes up at least once = 11

∴   P(5 will not come up at least once) = $\dfrac{11}{36}$

Remarks: Throwing a die twice and throwing two dice simultaneously are treated as the same experiment.

Q. 13. **A coin is tossed twice. Find the probability of getting exactly one head.**

Ans. Possible outcomes are

(H, H), (H, T), (T, H) and (T, T)

For exactly one head, favourable outcomes are two i.e., (H, T) and (T, H)

∴   P (exactly one head) = $\dfrac{2}{4} = \dfrac{1}{2}$

Q. 14. **A coin is tossed twice. Find the probability of getting 'both tails'.**

Ans. Possible outcomes are four, i.e.,

(H, H), (H, T), (T, H) and (T, T)

Favourable outcome is only one i.e., (T, T)

∴   P (both tails) = $\dfrac{1}{4}$

Q. 15. **Three coins are tossed once. Find the probability of at least two heads.**

Ans. Possible outcomes are eight:

(H, H, H), (H, H, T), (H, T, H), (H, T, T)

(T, H, H), (T, H, T), (T, T, H) and (T, T, T)

Favourable outcomes for at least two heads are : (H, H, T), (H, T, H), (T, H, H) and (H, H, H)

∴   Number of favourable outcomes = 4

Hence P(at least two heads) = $\dfrac{4}{8} = \dfrac{1}{2}$

Q. 16. **A coin is tossed thrice. Find the probability of getting tail on the third throw.**

Ans. Possible outcomes are eight:

(H, H, H), (H, H, T), (H, T, H), (H, T, T)

(T, H, H), (T, H, T), (T, T, H) and (T, T, T)

Favourable outcomes for a tial on the third throw are (H, H, T), (H, T, T), (T, H, T) and (T, T, T)

Number of favourable outcomes = 4

∴   required probability = $\dfrac{4}{8} = \dfrac{1}{2}$

Q. 17. **Two coins are tossed simultaneously. Find the probability of getting**

(i)   two heads   (ii)   atleast one head   (iii)   no head

[**Hint :** Possible outcomes are HH, HT, TH, TT.]

Ans. When two coins are tossed simultaneously, the possible outcomes are HH, HT, TH, TT.

(i)   P (two heads) = 1/4

(ii)   P(atleast one head) $= 3/4$

(iii)   P(no head) $= 1/4$

Q. 18. **A game consists of tossing a one rupee coin 3 times and noting its outcome each time. Hanif wins if all the tosses give the same result i.e. three heads or three tails, and loses otherwise. Calculate the probability that Hanif will lose the game.**

Ans. When a coin is tossed three time, the possible outcomes are eight:

(H,H,H), (H,H,T), (H,T,H), (H,T,T), (T,H,H), (T,H,T), (T,T,H) and (T,T,T)

Number of cases when all the tosses give the same results are two, i.e., (H, H, H) and (T, T, T)

It is given that Hanif wins when all the tosses give the same results,

$\Rightarrow$ He loses in $8 - 2 = 6$ cases

$\therefore$ P(Hanif will lose the game) $= \dfrac{6}{8} = \dfrac{3}{4}$

Q. 19. **A bag contains 4 red balls, 6 green balls and 5 white balls. If one ball is drawn at random, find the probability that it is not green.**

Ans. Number of total outcomes $= 4 + 6 + 5 = 15$

Number of balls not green $= 15 - 6 = 9$

$\therefore$ No. of favourable outcomes $= 9$

$\therefore$ P (not green) $= \dfrac{9}{15} = \dfrac{3}{5}$

Q. 20. **A bag contains 3 red balls, 5 black balls and 4 white balls. A ball is drawn at random from the bag. What is the probability that the ball drawn is:**

(i)   white ?   (ii) red ?   (iii) black ?

Ans. Total number of possible outcomes $= 3 + 5 + 4 = 12$

(i) No. of favourable outcomes $= 4$

    $\therefore$ P(a white ball) $= 4/12 = 1/3$

(ii) No. of favourable outcomes $= 3$

    $\therefore$ P (a red ball) $= 3/12 = 1/14$

(iii) No. of favourable outcomes $= 5$

    $\therefore$ P (a black ball) $= 5/12$

Q. 1. **A box contains 5 red marbles, 8 white marbles and 4 green marbles. One marble is taken out of the box at random. What is the probability that the marble taken out will be**

(i) red ? (ii) white ? (iii) not green ?

Ans. Total number of possible outcomes

$$= 5 + 8 + 4 = 17$$

(i) Number of red marbles = 5

$$\therefore \quad P(\text{a red marble}) = \frac{5}{17}$$

(ii) Number of white marbles = 8

$$\therefore \quad P(\text{a red marble}) = \frac{8}{17}$$

(iii) Number of marbles which are not green = 5 + 8 = 13

$$\therefore \quad P(\text{a red marble}) = \frac{13}{17}$$

Q. 2. **A bag contains 5 red balls and some blue balls. If the probability of drawing a blue ball is double that of a red ball, find the number of blue balls in the bag.**

Ans. Let number of blue balls = x

$$\therefore \quad \text{Total number of balls} = 5 + x$$

$$P(\text{blue ball}) = \frac{x}{5 + x}$$

$$P(\text{red ball}) = \frac{5}{5 + x}$$

Given that P (blue) = 2. P (red)

$$\therefore \quad \frac{x}{5 + x} = 2.\frac{5}{5 + x}$$

On solving, x = 10

$\therefore$ Number of blue balls in the bag = 10

Q. 3. **A bag contains 12 balls out of which x are white.**

(i) If one ball is drawn at random, what is the probability that it will be a white ball?

(ii) If 6 more white balls are put in the bag, the probability of drawing a white ball will be double than that in (i). Find x.

Ans. (i) $$P(\text{White ball}) = \frac{x}{12}$$

(ii) When 6 more white balls are put in the bag.

Total number of balls = 12 + 6 = 18

Total number of white balls = x + 6

$$P(\text{white}) = \frac{x+6}{18}$$

According to given:

$$\frac{x+6}{18} = 2 \cdot \frac{x}{12}$$

$$= \frac{x}{6}$$

On solving : $x = 3$

Q. 4. **A jar contains 24 marbles, some are green and others are blue. If a marble is drawn at random from the jar, the probability that it is green is $\frac{2}{3}$. Find the number of blue balls in the jar.**

Ans. Total number of marbles = 24

$\Rightarrow$ Number of green marbles = 24 – x

$$P\ (\text{a green marble}) = \frac{24-x}{24}$$

Given that, P(a green marble) = $\frac{2}{3}$

$$\Rightarrow \qquad \frac{24-x}{24} = \frac{2}{3}$$

$$\Rightarrow \qquad 3(24-x) = 24 \times 2$$

$$\Rightarrow \qquad 72 - 3x = 48$$

$$\Rightarrow \qquad 3x = 24$$

$$\Rightarrow \qquad x = 8$$

$\therefore$ Number of blue marbles in the jar = 8

Q. 5. **A bag contains 30 balls numbered from 1 to 30. One ball is drawn at random. Find the probability that the number on the ball drawn is a multiple of 4 or 9.**

Ans. Number of total outcomes = 30

Favourable outcomes are balls with numbers 4, 8, 12, 16, 20, 24, 28, 9, 18 and 27

$\therefore$ Number of favourable outcomes = 10

Hence, required probability = $\frac{10}{30} = \frac{1}{3}$

Q. 6. **17 cards numbered 1, 2, 3, ..., 16, 17 are put in a box and mixed thoroughly. One person draws a card from the box. Find the probability that the number on the card is.**

(i) odd (ii) prime (iii) divisible by 3 (iv) divisible by 3 and 2 both

Ans. Total number of possible outcomes = 17

(i) Number of odd numbered cards (1, 3, 5, 7, 9, 11, 13, 15, 17) = 9

$\therefore$ required probability = 9/17

(ii) Number of prime numbered cards (2, 3, 5, 7, 11, 13, 17) = 7

$\therefore$ required probability = 7/17

(iii) Number of cards with numbers divisible by 3 (3, 6, 9, 12, 15) = 5

∴　required probability = 5/17

(iv) Number of cards with numbers divisible by 3 and 2 both (6, 12) = 2

∴　required probability = 2/17

Q. 7. **Cards marked with numbers 2 to 101 are placed in a box and mixed thoroughly. One card is drawn from this box. Find the probability that the number on the card is**

(i)　an even number

(ii)　a number less than 14.

(iii)　a number which is a perfect square.

(iv)　a prime number less than 20.

Ans.　Total number of cards with numbers 2 to 101 = 100

(i)　Even numbers from 2 to 101 = 50

∴　P (an even number) $= \dfrac{50}{100} = \dfrac{1}{2}$

(ii)　Numbers less than 14 (from 2 to 101) = 12

∴　P (a number less than 14) = 12/100 = 3/25

(iii)　Number from 2 to 101, that are perfect squares are 4, 9, 16, 25, 36, 49, 64, 81 and 100 (total = 9)

∴　P (a number which is a perfect square)

$$= 9/100$$

(iv)　Prime numbers less than 20 are 2, 3, 5, 7, 11, 13, 17, and 19 (total = 8)

∴　P (a prime number less than 20)

$$= 8/100 = 2/25$$

Q. 8. **A box contains 90 discs which are numbered from 1 to 90. If one disc is drawn at random from the box, find the probability that it bears (i) a two-digit number (ii) a perfect square number (iii) a number divisible by 5.**

Ans.　Total number of possible outcomes = 90

(i)　Two digit numbers from 1 to 90 are 10, 11, 12, ..., 90 (total = 81)

∴　P(a disc bearing a two digit number) $= \dfrac{81}{90} = \dfrac{9}{10}$

(ii)　Perfect square numbers from 1 to 90 are 1, 4, 9, 16, 25, 36, 49, 64, and 81 (total = 9)

∴　P(a disc bearing a perfect square number) $= \dfrac{9}{90} = \dfrac{1}{10}$

(iii)　Numbers divisible by 5 from 1 to 90 are 5, 10, 15, ..., 90 (total = 18)

∴　P(a disc bearing a number divisible by 5) $= \dfrac{18}{90} = \dfrac{1}{5}$

Q. 9. **A card is drawn at random from a well shuffled deck of 52 cards, what is the probability of drawing.**

(i)　a card of club　　　　　　(ii)　a king

Ans.　(i) Since there are 13 cards of club

$$\therefore \quad \text{favourable outcomes} = 13$$

$$\text{and,} \quad \text{total outcomes} = 52$$

$$P \text{ (a club)} = \frac{13}{52} = \frac{1}{4}$$

(ii) Since there are 4 kings in a deck

$$\therefore \quad \text{favourable outcomes} = 4$$

$$\text{and,} \quad \text{total outcomes} = 52$$

$$\therefore \quad P \text{ (a king)} = \frac{4}{52} = \frac{1}{13}$$

Q. 10. **One card is drawn from a well shuffled deck of 52 cards. Find the probability of getting a king of black suit.**

Ans. Out of total 4 kings in a deck of 52 cards, two are black (that of spade and club)

$$\therefore \quad \text{Number of favourable outcomes} = 2$$

$$\text{and,} \quad \text{total outcomes} = 52$$

$$P \text{ (a king of black suit)} = \frac{2}{52} = \frac{1}{26}$$

Q. 11. **One card is drawn from a well shuffled deck of 52 cards. Find the probability of drawing a jack, king or queen.**

Ans. In a deck of 52 cards, there are 4 jacks, 4 kings and 4 queens.

$$\therefore \quad \text{Favourable outcome are } 12$$

$$\text{and,} \quad \text{total outcomes} = 52$$

$$\text{Hence, required probability} = \frac{12}{52} = \frac{3}{13}$$

Q. 12. **Find the probability of getting a red coloured king from a well shuffled deck of 52 cards, when a card is drawn at random.**

Ans. Since there are 2 red coloured kings in a deck of 52 cards,

$$\therefore \quad \text{Number of favourable outcomes} = 2$$

$$\text{and,} \quad \text{total outcomes} = 52$$

$$\therefore \quad P \text{ (a red king)} = \frac{2}{52} = \frac{1}{26}$$

Q. 13. **A card is drawn from a well shuffled deck of 52 cards. Find the probability of getting a queen or a heart.**

Ans. In a deck of 52 cards, there are 4 queens and 13 hearts, but there is one card which is queen of heart (counted in both)

$$\therefore \quad \text{Number of favourable outcomes}$$

$$= 4 + 13 - 1 = 16$$

$$\text{and,} \quad \text{total outcomes} = 52$$

$$\therefore \quad \text{required probability} = \frac{16}{52} = \frac{4}{13}$$

Q. 14. **One card is drawn from a well shuffled deck of 52 cards. Find the probability of drawing:**

(i)   an ace. (ii) '2' of spades. (iii) '10' of a black suit.

Ans. Total number of possible outcomes = 52

(i)   Since there are total four aces

∴   Number of favourable outcomes = 4

$$P \text{ (an ace)} = \frac{4}{52} = \frac{1}{13}$$

(ii) Since there is only one '2' of spades

$$\therefore P \text{ ('2' of spades)} = \frac{1}{52}$$

(iii) Since there are two '10' of black suits

$$\therefore P \text{ (10 of a black suit)} = \frac{2}{52} = \frac{1}{26}$$

Q. 15. **All the kings and queens are removed from a pack of 52 cards and then well shuffled. One card is now drawn at random, find the probability of getting.**

(i)   a king (ii)   a heart

Ans. (i) Since all the kings are removed,

∴   Probability of getting a king = 0

(ii) Out of a total of 13 hearts, the king and queen of heart have been removed

∴ number of favourable outcomes = 13 – 2 = 11

and, number of total outcomes = 52 – 8 = 44

$$\therefore \quad \text{required probability} = \frac{11}{44} = \frac{1}{4}$$

Q. 16. **The king, queen and jack of clubs are removed from a deck of 52 playing cards and then well shuffled. One card is selected from the remaining cards. Find the probability of getting:**

(i)   a heart (ii)   a king

(iii) a club (iv)   the '10' hearts.

Ans. Total number of cards after removing the king, queen and jack of clubs = 49

(i) $$P \text{ (a heart)} = \frac{13}{49}$$

(ii) $$P \text{ (a king)} = \frac{3}{49}$$

(iii) $$P \text{ (a club)} = \frac{10}{49}$$

(iv) $$P \text{(the '10' of the hearts)} = \frac{1}{49}$$

Q. 17. **One card is drawn from a well shuffled deck of 52 cards. Find the probability of getting:**

(i)   a king of red colour (ii)   a face card (iii)   a red face card

(iv) the jack of hearts   (v) a spade   (vi)   the queen of diamonds

Ans.   Total number of possible outcomes = 52

(i)   There are two kings of red colour,

$$\therefore \quad P(\text{a king of red colour}) = \frac{2}{52} = \frac{1}{26}$$

(ii) There are 12 face cards (king, queen and jack of each suit),

$$\therefore \quad P(\text{a face card}) = \frac{12}{52} = \frac{3}{13}$$

(iii) There are 6 red face cards (king, queen and jack of diamonds and hearts),

$$\therefore \quad P(\text{a red face card}) = \frac{6}{52} = \frac{3}{26}$$

(iv) There is only one jack of hearts,

$$\therefore \quad P(\text{the jack of hearts}) = \frac{1}{52}$$

(v)   There are 13 cards of spade,

$$\therefore \quad P(\text{a spade}) = \frac{13}{52} = \frac{1}{4}$$

(vi) There is only one queen of diamonds

$$\therefore \quad P(\text{the queen of diamonds}) = \frac{1}{52}$$

Q. 18.   **Five cards — the ten, jack, queen, king and ace of diamonds, are well - shuffled with their face downwards. One card is then picked up at random.**

(i)   What is the probability that the card is a queen?

(ii) If the queen is drawn and put aside, what is the probability that the second card picked up is (a) an ace? (b) a queen?

Ans.   (i) Since the total number of cards is 5, the number of possible outcomes = 5

Out of these 5 cards, only one is queen,

$$\therefore \quad P(\text{a queen}) = \frac{1}{5}$$

(ii) After the queen has been drawn and put aside, the total number of cards = 5 − 1 = 4

$\therefore$ Number of possible outcomes = 4

These four cards include only one ace are no queen.

$$\therefore \quad P(\text{an ace}) = \frac{1}{4}$$

and $P(\text{a queen}) = 0$

Q. 19.   **It is known that a box of 600 electric bulbs contains 12 defective bulbs. One bulb is taken out at random from this box. What is the probability that it is a non-defective bulb?**

Ans.   Total outcomes = 600

Number of non-defective bulbs = 600 − 12 = 588

$\therefore$ P (a non-defective bulb)

$$= \frac{588}{600} = \frac{147}{150} = 0.98$$

Q. 20. 12 defective pens are accidentally mixed with 132 good ones. It is not possible to just look at a pen and tell whether or not it is defective. One pen is taken out at random from this lot. Determine the probability that the pen taken out is a good one.

Ans. Total number of pens = 12 + 132 = 144

$\therefore$ Number of possible outcomes = 144

Out of these 144 pens, 132 are good ones.

$\therefore$ $\qquad$ P(a good pen) $= \dfrac{132}{144} = \dfrac{11}{12}$

Q. 21. **(i)** A lot of 20 bulbs contain 4 defective ones. One bulb is drawn at random from the lot. What is the probability that this bulb is defective?

(ii) Suppose the bulb drawn in (i) is not defective and is not replaced. Now one bulb is drawn at random from the rest. What is the probability that this bulb is not defective?

Ans. (i) Total number of bulbs = 20

$\therefore$ Number of possible outcomes = 20

Since a total of 4 bulbs is defective in this lot,

$\therefore$ $\qquad$ P(a defective bulb) $= \dfrac{4}{20} = \dfrac{1}{5}$

(ii) According to given, the bulb drawn is not defective and it is not replaced.

$\therefore$ The lot now contains a total of 19 bulbs, out of which 4 are defective and 15 are good ones.

$\therefore$ P(a non-defective bulb) $= \dfrac{15}{19}$

Q. 22. 1000 tickets of a lottery were sold and there are 5 prizes on these tickets. If Saket has purchased one lottery ticket, what is the proabbility of winning a prize?

Ans. Total outcomes = 1000

Number of favourable outcomes = 5

$\therefore$ $\qquad$ Required probability $= \dfrac{5}{1000} = \dfrac{1}{200}$

$$= .005$$

Q. 23. A natural number is chosen at random from amongst the first 25. What is the probability that the number so chosen is divisible by 3 or 5?

Ans. Number of total outcomes = 25

Favourable outcomes so that the number chosen is divisible by 3 or 5 are 3, 6, 9, 12, 15, 18, 21, 24, 5, 10, 20, and 25.

$\therefore$ Number of favourable outcomes = 12

Hence, required probability $= \dfrac{12}{25}$

VERY SHORT ANSWER TYPE Q.UESTIONS

Q. 1. **Fill in the Blanks**

(a) The probability of an event is greater than or equal to .... and is less than or equal to ....

[NCERT]

(b) The probability of an impossible event is ....

(c) The probability of an event that is certain to happen is .... and such an event is called .... **[NCERT]**

(d) The sum of probabilities of all the elementary events of an experiment is .... **[NCERT]**

(e) Probability of an event E + probability of the event not E is equal to .... **[NCERT]**

(f) If probability of winning a game is $\dfrac{4}{9}$, then the probability of its losing is ....

(g) If coin is tossed twice, then the number of possible outcomes is ....

(h) If a die is thrown twice, then the number of possible outcomes is ....

Sol.

Q. 2. **State True/False**

(a) The probability of an event can be negative.

(b) The probability of an event is greater than 1.

Sol.

MULTIPLE CHOICE Q.UESTIONS

Q. 3. **Which of the following cannot be the probability of an event?** **[NCERT]**

(a) 0.7 (b) $\dfrac{2}{3}$ (c) -1.5 (d) 15%

Q. 4. **Which of the following can be the probability of an event?** **[NCERT Exemplar]**

(a) -0.04 (b) 1.004 (c) $\dfrac{18}{23}$ (d) $\dfrac{8}{7}$

Q. 5. **An event is very unlikely to happen, its probability is closest to** **[NCERT Exemplar]**

(a) 0.0001 (b) 0.001 (c) 0.01 (d) 0.1

Q. 4. **Out of one digit prime numbers, one number is selected at random. The probability of selecting an even number is:**

(a) $\dfrac{1}{2}$ (b) $\dfrac{1}{4}$ (c) $\dfrac{4}{9}$ (d) $\dfrac{2}{5}$

Q. 5. **When a die is thrown, the probability of getting an odd number less than 3 is:**

(a) $\dfrac{1}{6}$ (b) $\dfrac{1}{3}$ (c) $\dfrac{1}{2}$ (d) 0

Q. 6. **Rashmi has a die whose six faces show the letters as given below:**

| A | B | C | D | A | C |

If she throws the die once, then the probability of getting C is:

(a) $\dfrac{1}{3}$ (b) $\dfrac{1}{4}$ (c) $\dfrac{1}{5}$ (d) $\dfrac{1}{6}$

Q. 7. A card is drawn from a well shuffled pack of 52 playing cards. The event E is that the card drawn is not a face card. The number of outcomes favourable to the event E is:

(a) 51   (b) 40   (c) 36   (d) 12

Q. 8. If the probability of an even is 'p' then probability of its complementary event will be:

(a) $p - 1$ (b) p (c) $1 - p$ (d) $1 - \dfrac{1}{p}$

Q. 9. P(Winning) = x/12, P(Losing) = 1/3. Find x   [CBSE 2014]

(a) 6   (b) 8   (c) 7   (d) 9

Q. 10. The probability of a number selected at random from the numbers 1, 2, 3,.... 15 is a multiple of 4 is: [CBSE 2020]

(a) $\dfrac{4}{15}$ (b) $\dfrac{2}{15}$ (c) $\dfrac{1}{15}$ (d) $\dfrac{1}{5}$

Q. 11. The probability that a non-leap year selected at random will contains 53 Mondays is:

(a) $\dfrac{1}{7}$ (b) $\dfrac{2}{7}$ (c) $\dfrac{3}{7}$ (d) $\dfrac{5}{7}$

Q. 12. A bag contains 6 red and 5 blue balls. One ball is drawn at random. The probability that the ball is blue is:

(a) $\dfrac{2}{11}$ (b) $\dfrac{5}{6}$ (c) $\dfrac{5}{11}$ (d) $\dfrac{6}{11}$

Q. 13. One alphabet is chosen from the word MATHEMATICS. The probability of getting a vowel is:

(a) $\dfrac{6}{11}$ (b) $\dfrac{5}{11}$ (c) $\dfrac{3}{11}$ (d) $\dfrac{4}{11}$

Q. 14. Two coins are tossed simultaneously. The probability of getting at most one head is

(a) $\dfrac{1}{4}$ (b) $\dfrac{1}{2}$ (c) $\dfrac{2}{3}$ (d) $\dfrac{3}{4}$

Sol.

Q. 15. A card is drawn at random from a pack of 52 playing cards. Find the probability that the card drawn is neither an ace nor a king.

Sol.

Q. 16. Out of 250 bulbs in a box, 35 bulbs are defective. One bulb is taken out at random from the box. Find the probability that the drawn bulb is not defective.

Sol.

Q. 17. Non Occurance of any event is 3:4. What is the probability of Occurance of this event?

Sol.

Q. 18. If 29 is removed from (1, 4, 9, 16, 25, 29), then find the probability of getting a prime number.

Sol.

Q. 19. A card is drawn at random from a deck of playing cards. Find the probability of getting a face card.

Sol.

Q. 20. In 1000 lottery tickets, there are 5 prize winning tickets. Find the probability of winning a prize if a person buys one ticket.

Sol.

Q. 21. One card is drawn at random from a pack of cards. Find the probability that is is a black king. **(CBSE 2020)**

Sol.

Q. 22. A die is thrown once. Find the probability of getting a perfect square.

Sol.

Q. 23. Two dice are rolled simultaneously. Find the probability that the sum of the two numbers appearing on the top is more than and equal to 10.

Sol.

Q. 24. Find the probability of multiples of 7 in 1, 2, 3, ...,33, 34, 35.

Sol.

Q. 25. If a pair of dice is thrown once, then what is the probability of getting a sum of 8? **(CBSE 2020)**

Sol.

Q. 26. A letter of English alphabet is chosen at random. Determine the probability that chosen letter is a consonant. **(CBSE 2020)**

Sol.

Q. 27. If the probability of winning a game is 0.07, what is the probability of losing it? **(CBSE 2020)**

Sol.

SHORT ANSWER TYPE Q.UESTIONS

Q. 28. Two unbiased coins are tossed simultaneously. If the probability of getting no head is $\frac{a}{b}$ then find $(a + b)^2$? **[CBSE 2016]**

Sol.

Q. 29. Two different dice are rolled together. Find the probability

(a) of getting a doublet,

(b) of getting a sum of 10, of the numbers on the two dice. **[CBSE 2018]**

Sol.

Q. 30. A box contains 12 balls of which some are red in colour. If 6 more red balls are put in the box and a ball is drawn at random, the probability of drawing a red ball doubles than what it was before. Find the number of red balls in the box. **[CBSE 2018]**

Sol.

Q. 31. An integer is chosen random between 1 and 100. Find the probability that (i) it is divisible by 8, (ii) Not divisible by 8. **[CBSE 2018]**

Sol.

Q. 32. Three different coins are tossed together. Find the probability of getting (i) exactly two heads, (ii) at least two heads.

Sol.

Q. 33. Card from 11 to 30, are put in a box and mixed thoroughly. A card is then drawn from the box at random. Find the probability that the number on the drawn card is a prime number.

Sol.

Q. 34. A bag contains 5 red balls and some blue balls. If the probability of drawing a blue ball at random from the bag is three times that of a red ball, find the number of blue balls in the bag.

[CBSE 2020]

Sol.

Q. 35. Two different dice are thrown together, find the probability that the sum of the numbers appeared is less than 5. **[CBSE 2020]**

Sol.

Q. 36. Find the probability that 5 sundays occurs in the month of November of a randomly selected year. **[CBSE 2020]**

Sol.

Q. 37. In a family of three children. Find the probability of having at least two boys. **(CBSE 2020)**

Sol.

Q. 38. Two dice are thrown at the same time. Find the probability of getting different numbers on the two dice. **(CBSE 2020)**

Sol.

Q. 39. If a number x is chosen at random from the numbers $-3, -2, -1, 0, 1, 2, 3$. What is probability that $x^2 \leq 4$? **(CBSE 2020)**

Sol.

SHORT ANSWER TYPE Q.UESTIONS-II

Q. 40. A number x is selected at random from the numbers 1, 2, 3. Another number y is selected at random from the numbers 1, 4, 9. Find the probability that the product of x and y is less than 9.

Sol.

Q. 41. Two dice are thrown at the same time. Determine the probability that the difference of the numbers on the two dice is 2.

Sol.

Q. 42. An integer is chosen between 0 and 100. What is the probability that it is

(i)   divisible by 7?   (ii)   not divisible by 7?

Sol.

Q. 43. Two dice are rolled once. Find the probability of getting such numbers on the two dice.

    (a) whose product is 12.   (b) Sum of numbers on the two dice is atmost 5.

Sol.

Q. 44. Card with number 2 to 101 are placed in a box. A card is selected at random. Find the probability that the card has (i) an even number (ii) a square number.

Sol.

Q. 45. In a lottery, there are 10 prizes and 25 are empty. Find the probability of getting a prize. Also verify $P(E) + P(\overline{E}) = 1$ for this event.   **[CBSE 2020]**

Sol.

Q. 46. $P(\text{winning}) = \dfrac{x}{12}$, $P(\text{Losing}) = \dfrac{1}{3}$. Find x.

Sol.

LONG ANSWER TYPE Q.UESTIONS

Q. 47. Cards marked with numbers 3, 4, 5, ...,50 are placed in a box and mixed thoroughly. One card is drawn at random from the box, find the probability that the number on the drawn card is

    (i) divisible by 7   (ii) a two digit number.

Sol.

Q. 48. A bag contains 5 white balls, 7 red balls, 4 black balls and 2 blue balls. One ball is drawn at random from the bag. Find the probability that the balls drawn is

    (i) White or blue   (ii) red or black

    (iii) not white   (iv) neither white nor black

Sol.

Q. 49. The king, queen and jack of diamonds are removed from a pack of 52 playing cards and the pack is well shuffled. A card is drawn from the remaining cards. Find the probability of getting a card of

    (i) diamond   (ii) a jack

Sol.

Q. 50. The probability of a defective egg in a lot of 400 eggs is 0.035. Calculate the number of defective eggs in the lot. Also calculate the probability of taking out a non defective egg from the lot.

Sol.

Q. 51. Slips marked with numbers 3,3.5,7,7,7,9,9,9,11 are placed in a box at a game stall in a fair. A person wins if the mean of numbers are written on the slips. What is the probability of his losing the game?

Sol.

Q. 52. A box contains 90 discs which are numbered from 1 to 90. If one disc is drawn at random from the box, find the probability that it bears.

    (i) a two digit number   (ii) a perfect square number

    (iii) a number divisible by 5.

Sol.

Q. 53. A card is drawn at random from a well shuffled deck of playing cards. Find the probability that the card drawn is

    (i)   a card of spade or an ace   (ii) a red king

    (iii) neither a king nor a queen   (iv) either a king or a queen

Sol.

Q. 54. A card is drawn from a well shuffled deck of playing cards. Find the probability that the card drawn

    (i)   a face card   (ii) red colour face card

    (iii) black colour face card

Sol.

Q. 55. Ramesh got ₹ 24000 as Bonus. He donated ₹ 5000 to temple. He gave ₹ 12000 to his wife, ₹ 2000 to his servent and gave rest of the amount to his daughter. Calculate the probability of

    (i)   wife's share   (ii) Servant's Share   (iii) daughter's share.

Sol.

Q. 56. 240 students reside in a hostel. Out of which 50% go for the yoga classes early in the morning, 25% go for the Gym club and 15% of them go for the morning walk. Rest of the students have joined the laughing club. What is the probability of students who have joined laughing club?

Sol.

Q. 57. A box contains cards numbered from 11 to 123. A card is drawn at random from the box. Find the probability that the number on the drawn card is: **[CBSE 2018]**

    (i)   A square number   (ii) a multiple of 7.

Sol.

Q. 58. A die is thrown twice. Find the probability that:

    (i)   5 will come up at least once   (ii) 5 will not come up either time   **[CBSE 2019]**

Sol.

Q. 59. Cards marked 1, 3, 5 .... 49 are placed in a box and mixed thoroughly. One card is drawn from the box. Find the probability that the number on the card is: **[CBSE 2017]**

    (i)   divisible by 3   (ii) a composite number

    (iii) not a perfect square   (iv) multiple of 3 and 5

Sol.

Q. 60. 50 A child's game has 8 triangles of which 3 are blue and rest are red, and 10 squares of which 6 are blue and rest are red. One piece is lost at random. Find the probability that it is a   **[CBSE 2015]**

    (i)   triangle   (ii) square

    (iii) square of blue colour   (iv) triangle of red colour

Sol.

Q. 61. A box contain 24 balls of which x are red, 2x are white and 3x are blue. A ball is selected at random. What is the probability that it is

    (i)   not red?   (ii) White?

Sol.

1. (a) 0 and 1   (b) 0   (c) 1 and sure event   (d) 1

(e) 1   (f) $\dfrac{5}{9}$ (g) 4   (h) 36

2. (a) False   (b) False

3. (i) (c)   (ii) (c)   (iii) (a)   (iv) (b)

(v) (a)   (vi) (a)

4. (i) (c)   (ii) (b)   (iii) (d)   (iv) (a)

(v) (c)   (vi) (d)   (vii) (d)

5. $\dfrac{11}{13}$ **6.** $\dfrac{43}{50}$ **7.** $\dfrac{4}{7}$ **8.** 0

9. $\dfrac{3}{13}$ **10.** 0.005   **11.** $\dfrac{1}{26}$ **12.** $\dfrac{1}{3}$

13. $\dfrac{1}{6}$ **14.** $\dfrac{1}{7}$ **15.** $\dfrac{5}{36}$ **16.** $\dfrac{21}{26}$

17. 0.93   **18.** 25   **19.** $\dfrac{1}{12}$ **20.** 3

21. $\dfrac{6}{49}$ **22.** $\dfrac{1}{2}$ **23.** $\dfrac{3}{10}$ **24.** 15

25. $\dfrac{1}{6}$ **26.** $\dfrac{2}{7}$ **27.** $\dfrac{1}{2}$ **28.** $\dfrac{5}{6}$

29. $\dfrac{5}{7}$ **30.** $\dfrac{5}{9}$ **31.** $\dfrac{2}{9}$ **32.** $\dfrac{14}{101}$

33. (a) $\dfrac{1}{9}$ (b) $\dfrac{5}{18}$ **34.** (i) $\dfrac{1}{2}$ (ii) 0.09

35. 1   **36.** 8   **37.** (i) $\dfrac{7}{48}$ (ii) $\dfrac{41}{48}$

38. (i) $\dfrac{7}{18}$ (ii) $\dfrac{11}{18}$ (iii) $\dfrac{13}{18}$ (iv) $\dfrac{1}{2}$

39. $\dfrac{10}{49}, \dfrac{3}{49}$ **40.** 0.965   **41.** $\dfrac{7}{10}$

42. (i) $\dfrac{9}{10}$ (ii) $\dfrac{1}{10}$ (iv) $\dfrac{1}{5}$

43. (i) $\dfrac{4}{13}$ (ii) $\dfrac{1}{26}$ (iii) $\dfrac{11}{13}$ (iv) $\dfrac{2}{13}$

44. (i) $\dfrac{3}{13}$ (ii) $\dfrac{3}{26}$ (iii) $\dfrac{3}{26}$

45. (i) $\dfrac{1}{2}$ (ii) $\dfrac{1}{12}$ (iii) $\dfrac{5}{24}$ **46.** $\dfrac{1}{10}$

47. (i) $\dfrac{8}{113}$ (ii) $\dfrac{16}{113}$ **48.** (i) $\dfrac{11}{36}$ (ii) $\dfrac{25}{36}$

49. (i) $\dfrac{8}{25}$ (ii) $\dfrac{2}{5}$ (iii) $\dfrac{21}{25}$ (iv) $\dfrac{2}{25}$

50. (i) $\dfrac{4}{9}$ (ii) $\dfrac{5}{9}$ (iii) $\dfrac{1}{3}$ (iv) $\dfrac{5}{18}$

51. (i) $\dfrac{5}{6}$ (ii) $\dfrac{1}{3}$

52.   **53.**   **54.**   **55.**

56.   **57.**   **58.**   **59.**

60.   **61.**

SECTION-A

Q. 1. When a die is thrown once, the probability of getting an odd number less than 3 is: (1)

    (a) 1/6   (b) 1/3   (c) 1/2   (d) 0

Sol.

Q. 2. A bag contains 5 red, 8 green and 7 white balls. One ball is drawn at random from the bag, find the probability of getting neither green ball nor red ball. (1)

Sol.

Q. 3. One card is drawn at random from the well shuffled pack of 52 cards. Find the probability of getting a non face card. (1)

Sol.

Q. 4. Cards are marked with numbers 5, 6, 7, ... 50 are placed in the box and mixed thoroughly. One card is drawn at random from the box. What is the probability of getting a two digit number? (1)

Sol.

SECTION-B

Q. 5. A letter is chosen at random from 26 alphabets. Find the probability that the letter chosen is from the word 'ASSASSINATION'. (2)

Sol.

Q. 6. Out of 400 bulbs in a box, 15 bulbs are defective. One bulb is taken out at random from the box. Find the probability that the drawn bulb is not defective. (2)

Sol.

Q. 7. Find the probability of getting 53 Fridays or 53 Saturdays in a leap year. (2)

Sol.

SECTION-C

Q. 8. Daksh and Moksh are friends. What is the probability that both will have (i) different birthdays? (ii) the same birthday? [ignoring a leap year). (3)

Sol.

Q. 9. Two dice are thrown together. Find the probability that sum of two numbers will be a multiple of 4. (3)

Sol.

SECTION-D

Q. 10. Five cards- the ten, jack, queen, king and ace of diamonds, are removed from the well-shuffled 52 playing cards. One card is then picked up at random. Find the probability of getting:

    (a) neither a heart nor a king   (b) either a heart or a spade card

    (c) neither a red card nor a queen card   (d) a black card or an ace. (4)

Sol.

CASE STUDY 1.

On a weekend Rani was playing cards with her family .The deck has 52 cards. If her brother drew one card.

Q.. 1. **Find the probability of getting a king of red colour.**

   (a) $\dfrac{1}{26}$ (b) $\dfrac{1}{13}$ (c) $\dfrac{1}{52}$ (d) $\dfrac{1}{4}$

Q.. 2. **Find the probability of getting a face card.**

   (a) $\dfrac{1}{26}$ (b) $\dfrac{1}{13}$ (c) $\dfrac{2}{13}$ (d) $\dfrac{3}{13}$

Q.. 3. **Find the probability of getting a jack of hearts.**

   (a) $\dfrac{1}{26}$ (b) $\dfrac{1}{52}$ (c) $\dfrac{3}{52}$ (d) $\dfrac{3}{26}$

Q.. 4. **Find the probability of getting a red face card.**

   (a) $\dfrac{3}{13}$ (b) $\dfrac{1}{13}$ (c) $\dfrac{1}{52}$ (d) $\dfrac{1}{4}$

Q.. 5. **Find the probability of getting a spade.**

   (a) $\dfrac{1}{26}$ (b) $\dfrac{1}{13}$ (c) $\dfrac{1}{52}$ (d) $\dfrac{1}{4}$

ANSWERS

1. (a) $\dfrac{1}{26}$ **2.** (d) $\dfrac{3}{13}$ **3.** (b) $\dfrac{1}{26}$

4. (a) $\dfrac{3}{13}$ **5.** (d) $\dfrac{1}{4}$

CASE STUDY 2.

Rahul and Ravi planned to play Business (board game) in which they were supposed to use two dice.

Q.. 1. Ravi got first chance to roll the dice. What is the probability that he got the sum of the two numbers appearing on the top face of the dice is 8?

(a) $\dfrac{1}{26}$ (b) $\dfrac{5}{36}$ (c) $\dfrac{1}{18}$ (d) 0

Q.. 2. Rahul got next chance. What is the probability that he got the sum of the two numbers appearing on the top face of the dice is 13?

(a) 1 (b) $\dfrac{5}{36}$ (c) $\dfrac{1}{18}$ (d) 0

Q.. 3. Now it was Ravi's turn. He rolled the dice. What is the probability that he got the sum of the two numbers appearing on the top face of the dice is less than or equal to 12?

(a) 1 (b) $\dfrac{5}{36}$ (c) $\dfrac{1}{18}$ (d) 0

Q.. 4. Rahul got next chance. What is the probability that he got the sum of the two numbers appearing on the top face of the dice is equal to 7?

(a) $\dfrac{5}{9}$ (b) $\dfrac{5}{36}$ (c) $\dfrac{1}{6}$ (d) 0

Q.. 5. Now it was Ravi's turn. He rolled the dice. What is the probability that he got the sum of the two numbers appearing on the top face of the dice is greater than 8 ?

(a) 1 (b) $\dfrac{5}{36}$ (c) $\dfrac{1}{18}$ (d) $\dfrac{5}{18}$

ANSWERS

1. (b) $\dfrac{5}{36}$ **2.** (d) 0   **3.** (a) 1

4. (c) $\dfrac{1}{6}$ **5.** (d) $\dfrac{5}{18}$

Innovative Mathematics X-14